APPLICATIONS MANUAL

MARTINI

FUNDAMENTALS OF

Anatomy & Physiology

FIFTH EDITION

FREDERIC H. MARTINI, PH.D. ◦ KATHLEEN WELCH, M.D.

WITH

WILLIAM C. OBER, M.D. ◦ ART COORDINATOR AND ILLUSTRATOR

CLAIRE W. GARRISON, R.N. ◦ ILLUSTRATOR

RALPH T. HUTCHINGS ◦ BIOMEDICAL PHOTOGRAPHER

Upper Saddle River, NJ 07458

Library of Congress Cataloging-in-Publication Data

Martini, Frederic.
Applications manual for Fundamentals of anatomy & physiology / Frederic H. Martini, Kathleen Welch ; with William C. Ober, art coordinator and illustrator ; Claire W. Garrison, illustrator ; Ralph T. Hutchings, biomedical photographer.-- 5th ed.
p. ; cm.
Rev. ed. of: Fundamentals of anatomy & physiology. Applications manual. c1998.
Companion v. to: Fundamentals of anatomy physiology / Frederic H. Martini. 5th ed. 2001.
Includes index.
ISBN 0-13-31117-0
1. Human physiology. 2. Human anatomy. 3. Physiology, Pathological. I. Welch, Kathleen, M.D. II. Martini, Frederic. Fundamentals of anatomy & physiology. Applications manual. III. Martini, Frederic H. Fundamentals of anatomy & physiology. IV. Title.
[DNLM: 1. Anatomy--Examination Questions. 2. Anatomy--Handbooks. 3. Physiology--Examination Questions. 4. Physiology--Handbooks. QS 4 M3855f2001 Suppl.]
QP34.5 .M4615 2001 Suppl.
612--dc21

00-055084

Senior Acquisitions Editor: *Halee Dinsey*
Project Manager: *Don O'Neal*
Senior Development Editor: *Karen Karlin*
Production Editor: *David Munger, The Davidson Group*
Special Projects Manager: *Barbara A. Murray*
Supplements Production Editor: *Meaghan Forbes*
Assistant Vice President of Production & Manufacturing: *David W. Riccardi*
Editorial Director: *Paul F. Corey*
Editor in Chief for Biology: *Sheri Snavely*
Editors in Chief, Development: *Ray Mullaney, Carol Trueheart*
Senior Marketing Manager: *Martha McDonald*
Executive Marketing Manager for Biology: *Jennifer Welchans*
Editorial Assistant: *Susan Zeigler*
Marketing Assistant: *Andrew Gilfillan*
Manufacturing Manager: *Trudy Pisciotti*
Copy Editor: *Cathy McNamara*
Director, Creative Services: *Paul Belfanti*
Art Director: *Heather Scott*
Interior Design: *Anne Flanagan; David Munger, The Davidson Group*
Cover Designers: *Paul Gourhan, PM Workshop; John Christiana*
Art Manager: *Gus Vibal*
Art Editor: *Heather Hulett, The Davidson Group*
Illustrators: *William C. Ober, M.D.; Claire W. Garrison, R.N.*
Biomedical Photographer: *Ralph T. Hutchings*
Composition: *The Davidson Group*
Cover Photograph: *Jamey Hampton, Daniel Ezralow, Ashley Roland, and Sheila Lehner*
ISO Dance Company Photo © Lois Greenfield, 1991

Notice: Our knowledge in clinical sciences is constantly changing. The authors and the publisher of this volume have taken care that the information contained herein is accurate and compatible with the standards generally accepted at the time of publication. Nevertheless, it is difficult to ensure that all information given is entirely accurate for all circumstances. The authors and publisher disclaim any liability, loss, or damage incurred as a consequence, directly or indirectly, of the use and application of any of the contents of this volume.

Upper Saddle River, New Jersey 07458

Printed in the United States of America

10 9 8 7 6 5 4 3 2

ISBN 0-13-031117-0

Prentice-Hall International (UK) Limited, *London*
Prentice-Hall of Australia Pty. Limited, *Sydney*
Prentice-Hall Canada, Inc., *Toronto*
Prentice-Hall Hispanoamericana, S.A., *Mexico*
Prentice-Hall of India Private Limited, *New Delhi*
Prentice-Hall of Japan, Inc., *Tokyo*
Pearson Education Asia Pte. Ltd
Editora Prentice-Hall do Brasil, Ltda., *Rio de Janiero*

Contents

Preface

This *Applications Manual* is designed to complement *Fundamentals of Anatomy and Physiology* (FAP), Fifth Edition. Its primary goal is to help students apply the concepts introduced in FAP to their own lives. Although the main text covers all essential principles of anatomy and physiology, the *Applications Manual* introduces many clinical, diagnostic, and other topics of potential interest. Because most of these applied topics are discussed in the *Applications Manual* rather than in the textbook, they can be treated in greater depth here and provide greater reference value. The topics covered here are cross-referenced to the textbook, whereas the textbook uses the icon AM (in bright blue) and section titles to identify related topics in the *Applications Manual.*

In addition to the clinical topics and enrichment material, the *Applications Manual* also contains a Companion Atlas, which will help students bridge the gap between anatomical art and the real human body. The Companion Atlas includes (1) full-color Embryology Summaries, which integrate art and text to present the embryological and fetal development for each body system; (2) full-color photographs of surface anatomy, the skull and skeleton, prosected cadavers, and plastic models; and (3) black-and-white images produced by radiological procedures, including standard X-rays as well as as MRI and CT scans.

This new edition of the *Applications Manual* has been reorganized to improve its integration with the rest of the supplements package for FAP. Critical-Thinking Questions and Clinical Problems, features of the Fourth Edition of the *Applications Manual,* have been moved to the Companion Website, where students can review relevant material and test themselves online. Case Studies are now located on the *Martini Interactive* student CD-ROM (packaged with the textbook) and are structured to encourage critical thinking through the progressive disclosure of relevant clinical information. We feel that these self-testing materials are enhanced by the interactive environment of these media supplements.

Few instructors will cover all the material in the *Applications Manual.* Indeed, some instructors may choose not to cover all the boxed clinical discussions in the main text. Because courses differ in their emphases and students differ in their interests and backgrounds, the goal in designing the *Applications Manual* has been to provide maximum flexibility of use. The diversity of applied topics in the text material and boxes, the *Applications Manual,* and *The New York Times* Themes of the Times articles offers instructors a wide variety of ways in which to integrate the treatment of normal function, pathology, and other clinical or health-related topics. Boxed material in FAP and topics in the *Applications Manual* that are not covered in class can be assigned, recommended, used for reference, or left to the individual student. Experience indicates that each student will read those selections that deal with disorders that affect friends or family members, address topics of current interest and concern, or include information relevant to a chosen career path.

TO THE STUDENT

The topics covered in the *Applications Manual* will show you the relevance of the text material and will give you information that you can use in your daily life. When a family member becomes ill or a medical crisis develops on a TV show, this manual should help you make sense of the situation. The scope of coverage has been greatly influenced by student feedback. If you have suggestions or comments about this edition, please contact us by way of the Prentice Hall A & P Website or at the address on page viii.

Each discussion in this *Applications Manual* is cross-referenced to *Fundamentals of Anatomy and Physiology* (FAP) by page number. Corresponding chapter numbers in FAP are indicated here by thumb tabs that appear in the page margins.

This *Applications Manual* is organized into units that deal with a wide variety of applied topics:

- **An Introduction to Diagnostics** discusses the basic principles involved in the clinical diagnosis of disease states.
- **Applied Research Topics** considers principles of chemistry and cellular and molecular biology that are important to understanding, diagnosing, or treating homeostatic disorders.
- **The Body Systems: Clinical and Applied Topics** parallels FAP chapter by chapter and system by system. This section includes more-detailed discussions of many clinical topics that are introduced in the main text, along with discussions of diseases and diagnostic techniques that are not covered in the textbook.

This edition of the *Applications Manual* also contains a **Companion Atlas,** which includes images that supplement the artwork and photographs presented in the textbook. The Companion Atlas consists of the following:

- **Embryology Summaries,** which provide a unique combination of art and text to present the formation of significant organs, structures, and body systems during embryological and fetal development. These summary images are cross-referenced in the textbook as numbered AM *EMBRYOLOGY SUMMARIES,* in bright blue.

- A **Surface Anatomy and Cadaver Atlas,** which includes color photographs of living subjects, bones of the skull and skeleton, cadaver dissections, and plastic models. These color plates show the superficial and internal structures of all major body regions and organs. The images are cross-referenced in the textbook, within the relevant figure captions, as numbered AM *Plates,* in bright blue.
- A **Scanning Atlas,** which consists of images produced by various radiological techniques to show the interior of the human body from different perspectives. These scans will help you develop an understanding of three-dimensional relationships within the body. The images are cross-referenced in the textbook, within the relevant figure captions, as numbered AM *Scans,* in bright blue.

Your instructor may ask you to refer to the Embryology Summaries as you work through each chapter or may instead have you wait until you reach the FAP chapter on development and inheritance (Chapter 29). If you refer to the plates and scans when they are cited in the FAP figure captions, these images will help you understand the relationship between the human body and the diagrammatic art in the text.

ACKNOWLEDGMENTS

This was a complex project, and we thank everyone who helped us bring it to completion. Foremost on our thank-you list are the faculty and reviewers who offered suggestions that helped guide us through the revision process:

Leslie R. Carlson, *Iowa State University*
Susan J. Landesman, *Mt. Hood Community College*
Michael P. McKinley, *Glendale Community College*
Patricia Munn, *Longview Community College*
Michael Palladino, *Monmouth University*
Sandra Stewart, *Vincennes University*
Diane G. Tice, *SUNY Morrisville*
Bruce Wingerd, *San Diego State University*
Michael G. Wood, *Del Mar College*

Don O'Neal coordinated the assembly of all the components and still found time to manage the reviewing process. We acknowledge Dr. Eugene C. Wasson, III, the staff of Maui Radiology Consultants, and the Radiology Department of Maui Memorial Hospital for providing many of the MRI and CT scans used in the Scanning Atlas; Ralph Hutchings, who contributed most of the photographs for the Surface Anatomy and Cadaver Atlas; and Bill Ober, M.D., and Claire Garrison, R.N., who did the labeling and layout of those sections. Finally, we also express our thanks to Halee Dinsey, Senior Editor for Applied Biology, for supporting this project; to Karen Karlin, Senior Development Editor, for her attention to detail; to Dave Munger, Production Supervisor and Compositor, of The Davidson Group for coordinating the trafficking of materials between the textbook and the *Applications Manual* and for creating the design and page makeup; and to Barbara Murray, Special Projects Manager, Meaghan Forbes, Production Editor, and the rest of the Prentice Hall supplements production staff who worked on the assembly of this manual.

Frederic Martini, Ph.D. (martini@maui.net)
Kathleen Welch, M.D. (kwelch@maui.net)

A & P Website address:
www.prenhall.com/martini/fap5

Mailing address:
c/o ESM Editorial
Prentice Hall, Inc.
1 Lake Street
Upper Saddle River, NJ 07458

UNIT 1

An Introduction to Diagnostics

A **diagnosis** is a conclusion or decision based on a careful examination of relevant information. Each of us has made simple diagnoses in our everyday experiences. When the car won't start, the kitchen faucet leaks, or the checkbook doesn't balance, most of us will try to determine the nature of the problem. Sometimes the diagnosis is simple: The car battery is dead, the faucet is not completely turned off, or the amount of a check was recorded incorrectly. Once we make the diagnosis, we can take steps to remedy the situation.

Most of us use similar observational skills to diagnose simple medical conditions. For example, imagine that you awaken with a headache, feeling weak and miserable. Your face is flushed, your forehead is hot to the touch, and swallowing is painful. You know that these are the general symptoms of the flu, and you know also that your lab partner missed Tuesday's class because of the flu. You diagnose yourself as having the flu, and you open the medicine cabinet in search of appropriate medication.

The steps you took in arriving at the conclusion "I probably have the flu" were straightforward: (1) You made observations about your condition; (2) you compared your observations with available data; and (3) you determined the probable nature of the problem. **Clinical diagnosis,** or the identification of a disease, can be much more complicated, but these same steps are always required. In this section, we will examine the basic principles of diagnosis. The goal is not to train you to be a clinician but rather to demonstrate how these basic steps can be followed in a clinical setting.

Any diagnosis—of a disease or of a leaky kitchen faucet—requires an analysis that proceeds in a series of logical steps. Logical analysis, a process often called *critical thinking,* does not come naturally; it is too easy to become distracted or misled and then to make a hasty or incorrect decision. Critical thinking is a learned skill that follows rules designed to minimize the chances of error. Nowhere is critical thinking more important today than in the sciences, especially the medical sciences. In applying critical thinking to scientific investigation, we follow what is called the **scientific method,** a standardized means of organizing and evaluating information to reach valid conclusions.

The Scientific Method

Your course in anatomy and physiology should do more than simply teach you the names and functions of different body parts. It should provide you with a frame of reference that will enable you to understand new information, draw logical conclusions, and make intelligent decisions. A great deal of confusion and misinformation exists about just how medical science "works," and people make unwise and even dangerous decisions as a result. Nowhere is this more apparent than when a discussion drifts to health, nutrition, or cancer. Whether you are planning to work in a health-related profession or are just trying to make sound decisions about your own life, you must learn how to organize information, evaluate evidence, and draw logical conclusions.

FORMING A HYPOTHESIS

Science involves a lot more than just the collection of information. You could spend the rest of your life carefully observing the world around you, but such a task wouldn't reveal very much unless you could see some kind of pattern and come up with a **hypothesis,** an idea that explains your observations.

Hypotheses are ideas that may be correct or incorrect. To evaluate a hypothesis, you must have relevant data and a reliable method of data analysis. For example, you might propose the hypothesis that "radiation emitted by planet X confers immortality." Could anyone prove you wrong? Not likely, particularly if you didn't specify the location of the planet or the type of radiation. Would anyone believe you? If you were a "leading authority" on something (anything), a few people probably would.

That's not as ridiculous as it might seem. For almost 1500 years, "everyone knew" that inhaled air is transported from the lungs through blood vessels to the heart. They "knew" this because the famous Roman physician Galen had said so. But as we now know, Galen and all who agreed with him were quite wrong. Why were Galen's statements about the lungs believed? Because Galen was famous and much of what he said was correct, *all* his statements were accepted as true. To avoid making this kind of error, you must always remember to evaluate the hypothesis, not the individual who proposed it!

In the evaluation process, we must examine the hypothesis to see if it makes correct predictions about the real world. The steps in this process are diagrammed in Figure A-1•. A valid hypothesis has three characteristics: It is (1) testable, (2) unbiased, and (3) repeatable.

A testable hypothesis is one that can be studied by experimentation or data collection. Your assertion about planet X qualifies as a hypothesis, but it cannot be tested unless we find the planet and detect the radiation. An example of a testable hypothesis would be "left-handed airplane pilots have fewer crashes than

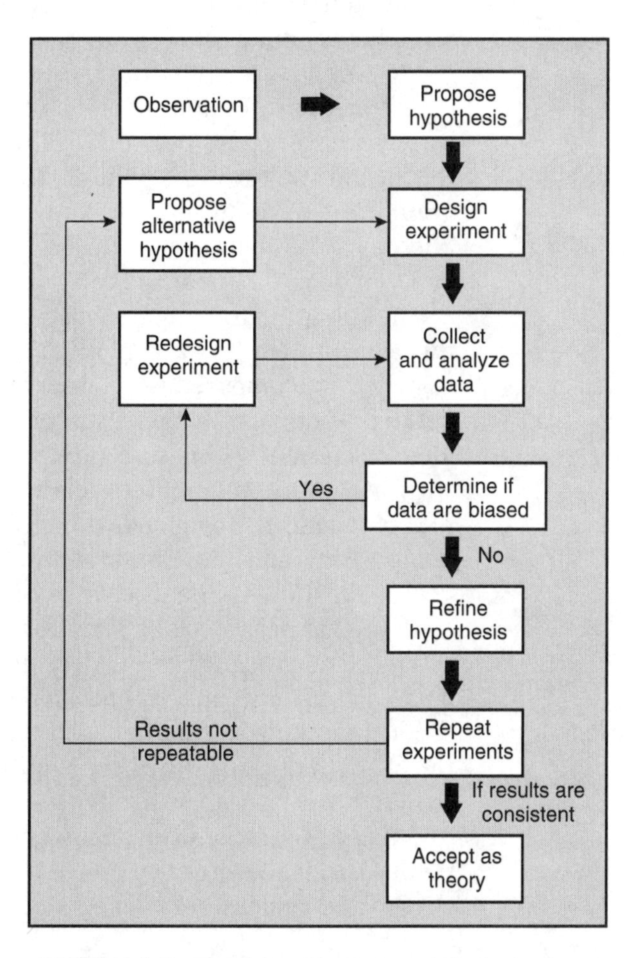

•FIGURE A-1 The Scientific Method. The basic sequence of steps involved in the development and acceptance of a scientific theory.

do right-handed pilots." That is testable because it makes a prediction about the world that can be checked—in this case by collecting and analyzing data.

AVOIDING BIAS

Suppose, then, that you collected information about all the plane crashes in the world and discovered that 80 percent of all crashed airplanes were flown by right-handed pilots. "Aha!" you might shout, "The hypothesis is correct!" The implications are obvious: Ban all right-handed airline pilots, eliminate four-fifths of all crashes, and sit back and wait for your prize from the Air Traffic Safety Association.

Unfortunately, you would be acting prematurely, because your data collection was biased. To test your hypothesis adequately, you need to know not only how many crashes involved right-handed or left-handed pilots but also how many right-handed and left-handed pilots were flying. If 90 percent of the pilots were right-handed, but they accounted for only 80 percent of the crashes, then left-handed pilots are the dangerous ones! Eliminating bias in this case is relatively easy, but health studies can have all kinds of complicating factors. Because 25 percent of us will probably develop cancer at some point in our lives, we will use cancer studies to exemplify the problems encountered.

The first example of bias in action concerns cancer statistics, which indicate that cancer rates in the United States and abroad vary by region. For example, although the estimated yearly cancer death rate in the United States was 206 per 100,000 population in 1999, the rate in Utah was only 114 per 100,000, whereas the rate in the District of Columbia was 268 per 100,000. It would be very easy to assume that this difference is the direct result of rural versus urban living. But these data alone should not convince you that moving from the District of Columbia to Utah would lower your risk of developing cancer. To draw that conclusion, you would have to be sure that the observed rates were the direct result of just a single factor, the difference in physical location. As you will find in later chapters, many factors can promote cancer development. To exclude all possibilities other than geography, you would have to be certain that the populations were alike in all other respects. Here are a few possible sources of variation that could affect that conclusion:

- *Different population profiles.* Cancer rates vary between males and females, among racial groups, and among age groups. Therefore, we need to know how the populations of Utah and the District of Columbia differ in each respect.
- *Different occupations.* Because chemicals used in the workplace are implicated in many cancers, we need to know how the populations of each region are employed and what occupational hazards they face.
- *Different mobilities.* Because the region in which a person dies may not be the region in which he or she developed cancer, we need to know whether people with cancer in Utah stay in the state or go elsewhere for critical care and whether people with cancer travel to the District of Columbia to seek treatment.
- *Different health care and habits.* Because cancer death rates reflect differences in patterns of health care, we need to know whether residents of Utah pay more attention to preventive health care and have more regular checkups, whether their medical facilities are better, and whether they devote a larger proportion of their annual income to health services than do residents in the District of Columbia. What is the proportion of smokers in both populations? (Cigarette smoke is the leading environmental carcinogen.)

You can probably think of additional factors, but the point is that avoiding experimental bias can be quite difficult!

A second example of the problem of bias comes from the collection of "miracle cures" that continue to appear and disappear at regular intervals. Pyramid power, coffee enemas, crystals, magnetic energy fields, and psychic healers come and go in the news. Wonder drugs are equally common, whether they are "secret formulas" or South American plant extracts discovered by colonists from other planets. The proponents of each new procedure or drug report glowing successes with patients who would otherwise have surely succumbed to the disease. And most of these remedies are said to have been suppressed or willfully ignored by the "medical establishment."

Even accepting that the claims aren't exaggerated, does the fact that 1, or 100, or even 1000 patients have been cured prove anything? No, because a list of successes doesn't mean much. To understand why, consider the questions you might pose to an instructor who announced on the first day of class that he or she had given 20 A's last semester. You would want to know how many students were in the class: only 20, or several hundred? You would also want to find out how the rest of the class performed—20 A's and 200 D's might be rather discouraging. You could see how the students were selected. If only students with A averages in other courses had enrolled, your opinion should change accordingly. Finally, you might check with the students and compare their grades with those given by other instructors who teach the same course.

With just a couple of modifications, the same questions could be asked about a potential cancer cure:

- How many patients were treated, how many were cured, and how many died?
- How were the patients selected? If selection depended on wealth, degree of illness, or previous exposure

to other therapeutic techniques, then the experimental procedure was biased from the start.

- How many might have recovered regardless of the treatment? Even "terminal" cancers sometimes simply disappear for no apparent reason. Such occurrences are rare, but they do happen. Thus, any treatment, however bizarre, will in some cases appear to work. If the frequency of recovery is lower than that among other patient groups, the treatment might actually be harmful despite the reported "cures."
- How do the foregoing statistics compare with those of traditional therapies when subjected to the same unbiased tests?

THE NEED FOR REPEATABILITY

Finally, let's examine the criterion of repeatability. It's not enough to develop a reasonable, testable hypothesis and collect unbiased data. Consider the hypothesis that every time a coin is tossed, it will come up heads. You could build a coin-tossing machine, turn it on, and find that in the first experiment of 10 tosses, the coin came up heads every time. Does this result prove the hypothesis?

No, despite the fact that it was an honest experiment and the data supported the hypothesis. The problem here is one of statistics, sample size, and luck. The odds that a coin will come up heads on any given toss are 50 percent, or 1 in 2—the same as the odds that it will come up tails. The odds that it will come up heads 10 times in a row are about 1 chance in 1×2^{10} (1 in 1024)—small but certainly not inconceivable. If that coin is tossed 50 times, however, the chance of getting 50 heads drops to 1 in 1×2^{50} (less than 1 chance in a thousand trillion)—a figure that most people would accept as vanishingly small. Proving that the hypothesis "a tossed coin always lands heads up" is false requires that the coin come up tails only once. So the truth could be revealed by running the experiment with more coin tosses or by letting other people set up identical experiments and toss their own coins.

For a hypothesis to be correct, anyone and everyone must get the same results when the experiment is performed. If the experiment isn't repeatable, you have to doubt the conclusions even when you have complete confidence in the abilities and integrity of the original investigator.

If a hypothesis satisfies all these criteria—it is testable, unbiased, and repeatable—it can be accepted as a scientific **theory.** The scientific use of this term differs from its use in general conversation. When people discuss "wild-eyed theories," they are usually referring to untested hypotheses. Hypotheses may be true or false, but by definition theories describe real phenomena and make accurate predictions about the world. Examples of scientific theories include the theory of gravity and the theory of evolution. The "fact" of gravity is not in question, and the theory of gravity accounts for the available data. But this does not mean that theories cannot change over time. Newton's original theory of gravity, though used successfully for more than two centuries, was profoundly modified and extended by Albert Einstein. Similarly, the theory of evolution has been greatly elaborated since it was first proposed by Charles Darwin in the middle of the nineteenth century. No one theory can tell the whole story, and all theories are continuously being modified and improved as we learn more about our universe.

[EXPLORE]

Additional resources, such as interactive tutorials, critical-thinking questions, clinical problems, and case studies, are available through the CD-ROM and Website for your textbook. To begin your exploration, launch the *Martini Interactive* CD-ROM or visit the Companion Website (http://www.prenhall.com/martini/fap5) and select Chapter 1.

The Diagnosis of Disease

Pathology is the study of disease, and *pathophysiology* is the study of functional changes caused by disease processes. Different diseases typically produce similar signs and symptoms. For example, a person whose lips are paler than normal and who complains of a lack of energy and breathlessness might have (1) respiratory problems that prevent normal oxygen transfer to the blood, as in *emphysema*; (2) cardiovascular problems that interfere with normal blood circulation to all parts of the body (heart failure); or (3) a reduced oxygen-carrying capacity of the blood (*anemia*). In such cases, doctors must ask questions and collect appropriate information to determine the source of the problem. A diagnosis is a decision about the nature of an illness. The diagnostic process is often a process of elimination, in which several potential causes are evaluated and the most likely one is selected. When uncertainties exist, the patient might need to undergo additional testing in order for the physician to reach a specific diagnosis.

If tests indicate that anemia is responsible for the symptoms described in our example, the specific type of anemia must then be determined before effective treatment can begin. After all, the treatment for anemia due to a dietary iron deficiency is very different from the treatment for anemia due to internal bleeding. You could not hope to identify the probable cause of the anemia unless you were already familiar with the physical and chemical structure of red blood cells and with their role in the transport of oxygen. This brings us to a key concept: *All diagnostic procedures assume an understanding of the normal structure and function of the human body.*

HOMEOSTASIS AND DISEASE FAP *p. 14*

The ability to maintain homeostasis depends on two interacting factors: (1) the status of the physiological systems involved and (2) the nature of the stress imposed. Homeostasis is a balancing act, and each of us is like a tightrope walker. Homeostatic systems must adapt to sudden or gradual changes in our environment, the arrival of pathogens, injuries, and many other factors, just as a tightrope walker must make allowances for gusts of wind, frayed segments of the rope, and thrown popcorn.

The ability to maintain homeostasis varies with age, general health, and genetic makeup. The geriatric patient or young infant with the flu is in much greater danger than an otherwise healthy young adult with the same viral infection. If homeostatic mechanisms cannot cope with a particular stress, physiological values will drift outside the normal range. This change can ultimately affect all other systems, with potentially fatal results. After all, a person unable to maintain balance will eventually fall off the tightrope.

Consider a person who is exercising heavily and has a heart rate of 180 beats per minute for several minutes. That would be acceptable in a young, healthy adult, but such a heart rate can be disastrous for an older person with cardiovascular and respiratory problems. If it is allowed to continue, cardiac muscle tissue will be damaged, leading to decreased pumping efficiency and a dangerous drop in blood pressure.

These changes represent a serious threat to homeostasis. Other systems will soon become involved. For example,the drop in blood pressure will suppress kidney function, and waste products will begin accumulating in the blood. The reduced blood flow in other tissues will result in a generalized *hypoxia*, or low tissue oxygen level. Cells throughout the body then begin to suffer from oxygen starvation. The person is now in serious trouble. Unless steps are taken to correct the situation, survival is threatened.

Disease is the failure to maintain homeostatic conditions. The disease process may initially affect a specific tissue, an organ, or an organ system, but it will ultimately lead to changes in the function or structure of cells throughout the body. A disease can often be overcome through appropriate, automatic adjustments in physiological systems. In a case of the flu, the disease develops because the immune system cannot defeat the flu virus before that virus has infected cells of the respiratory passageways. For most people, the physiological adjustments made in response to that viral invasion will lead to the elimination of the virus and the restoration of homeostasis. Some diseases cannot easily be overcome. In the case of the person with acute cardiovascular problems, some outside intervention must be provided to restore homeostasis and prevent fatal complications.

Diseases may result from the following:

- *Pathogens or parasites that invade the body.* Examples include the viruses that cause flu, mumps, or measles; the bacteria responsible for dysentery or pneumonia;

FAP Ch. 1

and pinworms, flukes, and tapeworms. The invasion process is called **infection**. Some parasites do not enter the body but instead attach themselves to the body surface. This process is called **infestation**.

- *Inherited genetic conditions that disrupt normal physiological mechanisms.* These conditions make normal homeostatic control difficult or impossible. Examples include the *lysosomal storage diseases*, *cystic fibrosis*, and *sickle cell anemia*.
- *The loss of normal regulatory control mechanisms.* For example, cancer involves the rapid, unregulated multiplication of abnormal cells. Many cancers have been linked to abnormalities in genes responsible for controlling the rates of cell division. A variety of other diseases, called *autoimmune disorders*, result when regulatory mechanisms of the immune system fail and healthy tissues are attacked.
- *Degenerative changes in vital physiological systems.* Many systems become less adaptable and less efficient as part of the aging process. For example, we experience significant reductions in bone mass, respiratory capacity, cardiac efficiency, and kidney filtration as we age. If the elderly are exposed to stresses that their weakened systems cannot tolerate, disease results.
- *Trauma, toxins, or other environmental hazards.* Accidents can damage organs, impairing their function. Toxins consumed in the diet or absorbed through the skin or lungs can disrupt normal metabolic activities.
- *Nutritional factors.* Diseases can result from diets inadequate in proteins, essential amino acids, essential fatty acids, vitamins, minerals, or water. *Kwashiorkor*, a protein deficiency disease, and *scurvy*, a disease caused by vitamin C deficiency, are two examples. Excessive consumption of high-calorie foods, fats, or fat-soluble vitamins can also cause disease.

SYMPTOMS AND SIGNS

An accurate diagnosis, or the identification of the disease, is accomplished through the observation and evaluation of symptoms and signs.

A **symptom** is the patient's perception of a change in normal body function. Examples of symptoms include nausea, malaise, and pain. Symptoms are difficult to measure, and a clinician must ask appropriate questions, such as the following:

"When did you first notice this symptom?"

"What does it feel like?"

"Does it come and go, or does it always feel the same?"

"Does anything make it feel better or worse?"

The answers provide information about the location, duration, sensations, recurrence, and triggering mechanisms of the symptoms important to the patient.

Pain, an important symptom of many illnesses, is often an indication of tissue injury. The flow chart in Figure A-2• indicates the types of pain and introduces important related terminology. Pain sensations and pathways are detailed in Chapter 17 of the text, and we shall consider the control of pain in related sections of the *Applications Manual.*

A **sign** is a physical manifestation of the disease. Unlike symptoms, signs can be measured and observed through sight, hearing, or touch. The yellow color of the skin caused by liver dysfunction and a detectable breast lump are signs of disease. A sign that results from a change in the structure of tissue or cells is called a **lesion**. We shall consider lesions of the skin in detail in a later section dealing with the integumentary system.

THE STEPS IN DIAGNOSIS FAP *p. 23*

A person experiencing serious symptoms usually seeks professional help and thereby becomes a patient. The clinician, whether a nurse, physician, or emergency medical technician, must determine the need for medical care on the basis of observation and assessment of the patient's symptoms and signs. This is the process of diagnosis: the identification of a pathological process by its characteristic symptoms and signs.

Diagnosis is a lot like assembling a jigsaw puzzle. The more pieces (clues) available, the more complete the picture will be. The process of diagnosis is one of deduction and follows an orderly sequence of steps:

1. *Obtain the patient's medical history.* The medical history is a concise summary of current and past medical disorders, general factors that may affect the functioning of body systems, and the health of the patient's family. This information provides a framework for considering the individual's current problem. The examiner gains information about the patient's concerns by asking specific questions and using good listening skills. Physical assessment begins here, and this is the time for unspoken questions such as, "Is this person moving, speaking, and thinking normally?" The answers will later be integrated with the results of more-precise observations. Other components of the medical history may include the following:
   - *Chief complaint.* The patient is asked to specify the primary problem that requires attention. This is recorded as the *chief complaint.* An example would be the entry "Patient complains of pain in the right lower quadrant."
   - *History of current illness.* Which areas of the body are affected? What kind of functional problems have developed? When did the patient first notice the symptoms? The duration of the disease process is an important factor. For example, an infection may have been present for months, only gradually increasing in severity. This would be called a *chronic infection.* A disease process may have been underway for some time before the person recognizes that a problem exists. Over the initial period, the individual ex-

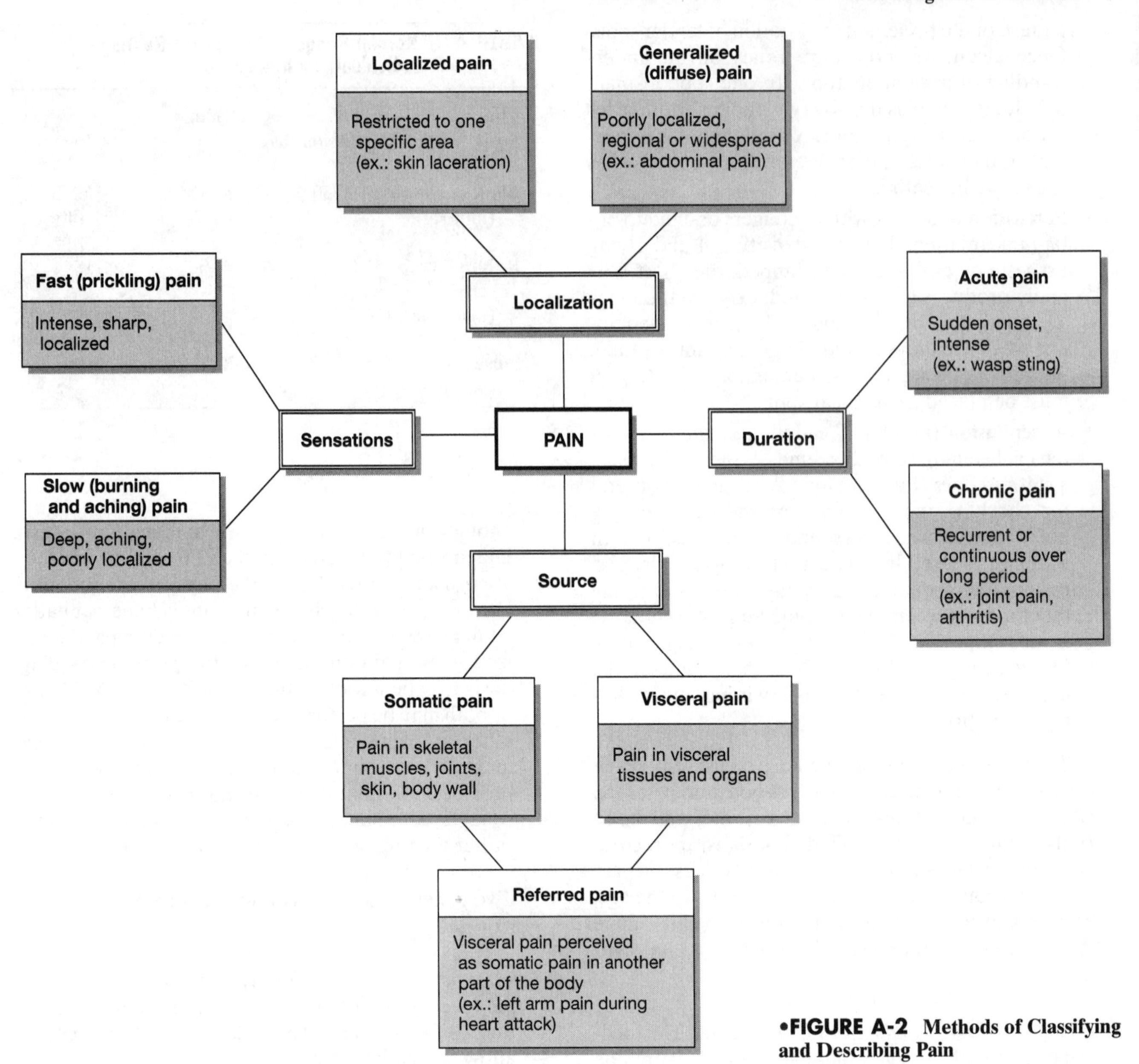

•**FIGURE A-2 Methods of Classifying and Describing Pain**

periences *subclinical symptoms,* so mild that they are usually ignored. Chronic infections commonly have different causes and treatments than do *acute infections*, which produce sudden, intense symptoms.

- *Review of systems.* The patient is asked questions that focus on the general status of each body system. This process may detect related problems or causative factors. For example, a chief complaint of headache pain may be *related* to visual problems (stars, spots, blurs, or blanks seen in the field of vision) or *caused* by visual problems (eyeglasses poorly fitted or the wrong prescription).

2. *Perform a physical examination.* The physical examination is a basic but vital part of the diagnostic process. The common techniques used in physical examination are *inspection* (vision), *palpation* (touch), *percussion* (tapping and listening), and *auscultation* (listening):

- **Inspection** is careful observation. A general inspection involves examining body proportions, posture, and patterns of movements. Local inspection is the examination of sites or regions of suspected disease. Of the four components of the physical exam, inspection is often the most important because it provides the largest amount of useful information. Many diagnostic conclusions can be made on the basis of inspection alone; most skin conditions, for example, are identified in this way. A number of endocrine problems and inherited metabolic disorders can produce changes in body proportions. Many neurologic disorders affect speech and movement in distinctive ways.
- **Palpation** is the clinician's use of hands and fingers to feel the patient's body. This procedure provides information about skin texture and temperature, the presence and texture of abnormal tissue masses, the

pattern of the pulse, and the location of tender spots. Once again, the procedure relies on an understanding of normal anatomy. In one spot, a small, soft, lumpy mass is a salivary gland; in another location it could be a tumor. A tender spot is important in diagnosis only if the observer knows what organs lie beneath it.

- **Percussion** is tapping with the fingers or hand to obtain information about the densities of underlying tissues. For example, when tapped, the chest normally produces a hollow sound, because the lungs are filled with air. That sound changes in pneumonia, when the lungs contain large amounts of fluid. To get the clearest chest percussions, the fingers must be placed in the right spots.
- **Auscultation** (aws-kul-TĀ-shun; *auscultare*, to listen) is listening to body sounds, typically by using a stethoscope. This technique is particularly useful for checking the condition of the lungs during breathing. The wheezing sound heard in people with asthma is caused by a constriction of the airways; pneumonia produces a gurgling sound, indicating that fluid has accumulated in the lungs. Auscultation is also important in diagnosing heart conditions. Many cardiac problems affect the sound of the heartbeat or produce abnormal swirling sounds during blood flow.

Every examination also includes measurements of certain vital body functions, such as body temperature, weight, blood pressure, respiratory rate, and heart (pulse) rate. The results, called **vital signs**, are recorded on the patient's chart. Each of these values can vary over a normal range that differs according to the age, gender, and general condition of the individual. Table A-1 indicates the representative ranges for vital signs in infants, children, and adults.

3. *If necessary, perform diagnostic procedures.* The medical history and physical examination may not provide enough information to permit a precise diagnosis. Diagnostic procedures can then be used to focus on abnormalities revealed by the physical examination. For example, if the chief complaint is knee pain after a fall and the examination reveals swelling and localized, acute pain on palpation, the **preliminary diagnosis** may be a torn cartilage. An X-ray or MRI scan or both may be performed to ensure that there are no other problems, such as broken bones or torn ligaments. With the information the diagnostic procedure provides, the **final diagnosis** can be made with reasonable confidence. Diagnostic procedures extend, rather than replace, the physical examination.

TABLE A-1 Normal Range of Values for Resting Individuals by Age Group

Vital Sign	Infant (3 months)	Child (10 years)	Adult
Blood pressure (mm Hg)	90/50	90–125/60	95/60 to 140/90
Respiratory rate (per minute)	30–50	18–30	8–18
Pulse rate (per minute)	70–170	70–110	50–95

Two general categories of diagnostic procedures are performed:

1. *Tests performed on the individual, generally in a hospital facility.* Information about representative tests of this type is summarized in Table A-2. These procedures allow the clinician to:

TABLE A-2 Representative Diagnostic Tests, Their Principles, and Uses

Procedure	Principle	Examples of Uses
Endoscopy	Insertion of fiber-optic tubing into a body opening or through a small incision (laparoscopy and arthroscopy); permits visualization of a body cavity or organ interior; allows direct visualization of internal structures and detection of abnormalities of surrounding soft tissue (FAP, *p. 804*)	*Bronchoscopy*: bronchi and lungs *Laparoscopy*: abdominopelvic organs *Cystoscopy*: urinary bladder *Esophagoscopy*: esophagus *Gastroscopy*: stomach *Colonoscopy*: colon *Arthroscopy*: joint cavity
Standard X-rays	A beam of X-rays passes through the body and then strikes a photographic plate; radiodense tissues block X-ray penetration, leaving unexposed (white) areas on the film negative. (FAP, *p. 23*)	Limb bones: to detect fracture, tumor, growth patterns Chest: to detect tumors, pneumonia, atelectasis, tuberculosis Skull: to detect fractures, sinusitis, metastatic tumors

TABLE A-2 Representative Diagnostic Tests, Their Principles, and Uses *(continued)*

Procedure	Principle	Examples of Uses
Contrast X-rays	X-rays taken after infusion or ingestion of radiodense solutions (FAP, *p. 23,* and Scans 10–17)	Barium swallow (upper GI): series of X-rays after the ingestion of barium, to detect abnormalities of esophagus, stomach, and duodenum Barium enema: series of X-rays after barium enema is given to detect abnormalities of colon IV pyelography: series of X-rays after intravenous injection of radiopaque dye filtered by kidneys; reveals abnormalities of kidneys, ureters, and bladder; allows assessment of renal function Mammogram: X-rays of each breast taken at different angles for early detection of breast cancer and other masses, such as cysts
Digital subtraction angiography	Produces strikingly clear images of blood vessel distribution by computer analysis of images taken before and after dye infusion (FAP, *pp. 25, 667*)	Analysis of blood flow to the heart, kidneys, and brain to detect blockages and restricted circulation
Computerized tomography (CT or CAT)	Produces cross-sectional images of body area viewed; together, all sections can produce a three-dimensional image for detailed examination. (FAP, *p. 24,* and Scan 9)	CT scans of the head, abdominal region (liver, pancreas, kidney), chest, and spine, to assess organ size and position, to determine progression of a disease, and to detect abnormal masses
Spiral CT scans	Produce three-dimensional images by computer reconstruction of CT data (FAP, *p. 25*)	Primarily a research tool but of clinical use at large regional hospitals and universities
Nuclear scans	Radioisotope ingested or injected into the body becomes concentrated in the organ to be viewed; gamma radiation camera records image on film. Area should appear uniformly shaded; dark or light areas suggest hyperactivity or hypoactivity of the organ. (p. 16)	Bone scan: to detect tumors, infections, and degenerative diseases Scans of the brain, heart, thyroid, liver, spleen, and kidney, to assess organ function and the extent of any disease
Radioactive iodine uptake test (RAI)	Radioactive iodine is given orally; thyroid scans are taken multiple times to determine thyroid percentage uptake of radioiodine	Aids in the determination of a hyperthyroid or hypothyroid condition and in detection of thyroid nodules
Positron emission tomography (PET)	Radioisotopes are given by injection or inhalation; gamma detectors absorb energy and transmit information to computers to generate cross-sectional images. (p. 16)	Used to measure metabolic activity of heart and brain and to analyze blood flow through organs; primarily a research tool
Magnetic resonance imaging (MRI)	A magnetic field is produced to align hydrogen protons and then exposed to radio waves that cause the aligned atoms to absorb energy. The energy is later emitted and captured to produce an image. (FAP, *p. 24* and Scans 1–8)	Gives excellent contrast of normal and abnormal tissue; tumor extent; demyelination and other brain and spinal cord abnormalities; obstructions or aneurysms in arteries; ligaments and cartilages at joints
Ultrasonography	A transducer contacting the skin or other body surface sends out sound waves and then picks up the echoes. (FAP, *p. 24*)	Used in obstetrics, to detect ectopic pregnancy, determine fetus size; check fetal rate of growth; upper abdominal ultrasound detects gallstones, visceral abnormalities, and measures kidneys
Echocardiography	Ultrasonography of the heart (p. 123)	Used to assess the structure and function of the heart and heart valves
Electrocardiography (ECG)	Graphed record of the electrical activity of the heart, using electrodes on the skin surface (FAP, *p. 673*)	Useful in detection of arrhythmias, such as premature ventricular contractions (PVCs) and fibrillation, and to assess damage after a heart attack
Electroencephalography (EEG)	Graphed record of electrical activity in the brain through the use of electrodes on the surface of the scalp (FAP, *p. 493*)	Analysis of brain wave frequency and amplitude aids in the diagnosis of tumors and seizure disorders

FAP Ch. 1

TABLE A-2 Representative Diagnostic Tests, Their Principles, and Uses *(continued)*

Procedure	Principle	Examples of Uses
Electromyography (EMG)	Graphed record of electrical activity resulting from skeletal muscle contraction, using electrodes inserted into the muscles	Determination of neural or muscular origin of muscle disorder; aids in the diagnosis of muscular dystrophy and peripheral neuropathies
Pulmonary function tests	Measurement of lung volumes and capacities by a spirometer or other device (FAP, *p. 822*)	Aids in the differentiation between obstructive and restrictive lung diseases; used to test for asthma
Pap smears	Removal of cells for laboratory analysis (FAP, *p. 115*)	Detects precancerous cells or infections; most often used to assess mucosal cells of cervix
Stress testing	Monitoring of blood pressure, pulse rate, and ECG during exercise; may include intravenous injection of radioisotopes	Aids in the determination of the extent of coronary artery disease, which may not be apparent while the individual is at rest
Skin tests	Injection of a substance under the skin, or placement of a substance on the skin surface, to determine the response of the immune system (p. 143)	Tuberculin test: injection of tuberculin protein under skin Allergen test: injection of allergen or application of a patch containing allergen

- Visualize internal structures (endoscopy; X-rays; scanning procedures such as CT, MRI, and DSA; ultrasonography; mammography)
- Monitor physiological processes (EEG, ECG, PET, RAI, pulmonary function tests)
- Assess the patient's homeostatic responses (stress testing, skin tests)

2. *Tests performed in a clinical laboratory on tissue samples, body fluids, or other materials collected from the patient.* Table A-3 includes details about a representative sample of such tests.

Many of the diagnostic procedures and disorders noted in these tables will be unfamiliar to you now. The main purpose here is to give you an overview; you can refer to these tables as needed throughout the course.

THE PURPOSE OF DIAGNOSIS

Several hundred years ago, a physician would arrive at a final diagnosis and consider the job virtually done. Once the diagnosis was made, the patient and family would know what to expect. In effect, the physician was more of an oracle than a healer. Wounds could be closed and limbs amputated, but few effective treatment options were available. Therapy consisted largely of some combination of bleeding (typically performed by barbers rather than by surgeons), dietary changes, and herbal medicines. Strong laxatives might have helped in cases of intestinal parasites, but the combination of bleeding and laxatives was potentially dangerous because it reduced both blood volume and blood pressure.

An incredible variety of treatment options is available today. An accurate diagnosis is vital, because it determines which treatment options will be selected. A modern physician with a new patient follows the *SOAP* protocol:

S is for *subjective.* The clinician obtains subjective information from the patient and completes the medical history.

O is for *objective.* The clinician performs the physical examination and obtains objective information about the physical condition of the patient. This may include the use of diagnostic procedures.

A is for *assessment.* The clinician arrives at a diagnosis and, if necessary, reviews the literature on the condition. A preliminary conclusion as to the **prognosis** (probable outcome) is made.

P is for *plan.* A treatment plan is designed. This can be very simple (take two aspirin) or very complex (radiation, chemotherapy, or surgery). If the treatment is complex, one or more treatment options are usually reviewed with the patient and, in many cases, the patient's family. Treatment begins only after informed decisions are made.

As you may have noticed, these are precisely the steps you followed in diagnosing the flu at the very beginning of this section: subjective (you felt ill), objective (flushed face, fever), assessment (flu-like symptoms), and plan (take medicine). The SOAP protocol is both simple to remember and remarkably effective.

The primary goal of an introductory anatomy and physiology course is to provide you with the foundation for other, more specialized courses. In the unit of

TABLE A-3 Laboratory Tests Performed on Samples Taken from the Body

Blood tests: Serum, plasma, or whole blood samples can be evaluated. Either venous or arterial blood is taken, depending on the blood constituent or chemical being monitored.

Laboratory Test	Significance	Notes
Complete blood count (CBC): RBC count Hemoglobin (Hb, Hgb) Hematocrit (Hct)	Data from this test series inform the practitioner about a change in the number of red blood cells; changes may indicate the presence of disease, hemorrhaging, malnutrition, or other problems. A CBC is generally performed during a normal physical exam to give the practitioner more information about the patient's general health. Changes may indicate blood loss, infections, or other problems.	For more information see Table A-27c, p. 110.
RBC indices: Mean corpuscular hemoglobin (MCH) Mean corpuscular hemoglobin concentration (MCHC)	Provide information about the status of hemoglobin production and red blood cell maturation	See Table A-27c, p. 110.
WBC count: Differential WBC count	The white blood cell count reflects the state of the body's immune system and the ability to fight infection. Increased white blood cell count could indicate the presence of infection.	See Table A-27c, p. 110.
Hemostasis tests: Platelet count Bleeding time Factors assay Plasma fibrinogen Plasma prothrombin time (PT) Plasminogen	A decreased number of platelets could result in uncontrolled bleeding. Other constituents, such as fibrinogen, clotting factors, and prothrombin, also contribute to the clotting process, and these can be assessed separately.	See Table A-27c, p. 110.
Electrolytes: Sodium Potassium Chloride Bicarbonate	Sodium, potassium, and chloride levels are electrolytes that function in nerve transmission, skeletal muscle contraction, and cardiac rhythm. Abnormal levels of bicarbonate indicate problems with acid–base balance.	See Tables A-27b, A-37, A-39, pp. 109, 160, 175.
Iron	Decreased levels cause iron deficiency anemia; increased levels may cause liver and heart damage.	
Arterial blood gases and pH: pH P_{CO_2} P_{O_2}	Respiratory acidosis and alkalosis can be monitored with these values. Decreased oxygen levels occur in respiratory and cardiovascular system dysfunction.	See Table A-34, p. 148.
Hemoglobin electrophoresis: Hemoglobin A Hemoglobin F Hemoglobin S	Electrophoresis separates the types of hemoglobin for quantitative measurement. Abnormal types of hemoglobin occur in sickle cell anemias and thalassemias.	See Table A-27c, p. 110.
ABO and Rh typing	Blood typing is critical for correct matching of blood types prior to transfusion. Rh typing during pregnancy is important to determine risk of fetal–maternal Rh incompatibility.	See Table A-42, p. 196.
Cholesterol	Elevated cholesterol levels reflect the potential for atherosclerosis and coronary artery disease.	See Table A-27b, p. 109.
Lipoproteins: LDL HDL	Electrophoresis is used to separate the LDL fraction of total cholesterol to determine the HDL and LDL levels, risk factors for coronary artery disease.	See Table A-27b, p. 109.

TABLE A-3 Laboratory Tests Performed on Samples Taken from the Body *(continued)*

Laboratory Test	Significance	Notes
Enzymes:		
Creatine phosphokinase (CPK or CK) Isoenzymes (CPK-MM, CPK-MB, CPK-BB)	Abnormal enzyme levels in the blood are generally due to cellular damage. CPK-MM is useful in the diagnosis of muscle disease; CPK-MB is used in the diagnosis of heart attacks.	See Table A-27b, p. 109.
Aspartate aminotransferase (AST)	AST levels are important to assess after a heart attack or liver damage.	
Lactate dehydrogenase (LDH)	Different isoenzymes of LDH can be useful in the detection of heart damage, liver problems, and pulmonary dysfunction.	
Rheumatoid factor	Measures presence of antibodies characteristic of rheumatoid arthritis and (less often) systemic lupus erythematosus and other autoimmune diseases	See Table A-16, p. 56.
Hormones	Varies with age and gender; abnormally increased or decreased levels reflect endocrine system disorders.	See Table A-26, p. 100.
Blood urea nitrogen (BUN)	Assesses kidney function	See Table A-39, p. 176.
Immunoglobin electrophoresis (IgA, IgG, IgD, IgE, IgM)	Monitors infections and allergic response	See Table A-29, p. 131.
Alcohol	Determines level of intoxication or detects metabolic problems or poisoning	
Human chorionic gonadotropin (hCG)	Detects pregnancy	See Table A-42, p. 196.
Phenylalanine	Detects phenylketonuria (PKU), a genetic disorder of amino acid metabolism	
Alpha fetoprotein	Identifies probability of fetal defects or presence of twins; elevated levels with liver tumors	
Glucose tolerance test	Detects hyperglycemia (diabetes mellitus)	See Table A-26, p. 101.
Blood culture	Presence of bacterial pathogens occurs in septicemia, pneumonia, and other infectious disorders.	See Tables A-27c, A-29, A-37, pp. 110, 131, 160.

Urine tests: A single urine sample may be tested, or urine may be collected over a period of time and tested (usually from 2 to 24 hours). A routine urinalysis aids in the detection of kidney dysfunction as well as metabolic imbalances and other disorders. Presence of abnormal cellular constituents in urine indicates urinary system disorder, including infection, inflammation, or the existence of a tumor.

Laboratory Test	Significance	Notes
Creatine clearance	Abnormal values indicate reduced renal function.	
Urine electrolytes: Sodium, Potassium	Abnormal levels reflect fluid or electrolyte imbalances and the effects of hormones on renal function.	See Table A-39, p. 176.
Uric acid	Increased levels occur with gout and some kidney disorders.	See Table A-39, p. 176.
Human chorionic gonadotropin (hCG)	Detects pregnancy	See Table A-42, p. 196.
Urine culture	Detects pathogens present in urinary tract infections	

TABLE A-3 Laboratory Tests Performed on Samples Taken from the Body *(continued)*

OTher laboratory testS: Additional laboratory tests can be used to monitor other body fluids, excretory products, or tissues. Here are several examples.

Laboratory Test	Significance	Notes
Cerebrospinal fluid (CSF)	Checks CSF pressure, color, sugar and protein content, and detects the presence of antibodies, pathogens, or blood cells	See Table A-19, p. 73.
Stool sample	Culturing and microscopic examination of sample to identify pathogenic bacteria or parasites	See Table A-37, p. 161.
Semen analysis	Useful in diagnosis of male infertility or in assessing success of vasectomy	See Table A-41, p. 186.
Tissue biopsy	The removal of tissue for microscopic examination	See Table A-17, p. 64.

this manual that deals with body systems, you will be introduced to clinical conditions that demonstrate the relationships between normal and pathological anatomy and physiology. The sections on diagnostic procedures and the Case Study exercises on your *Martini Interactive* CD-ROM demonstrate how clinicians extract and organize information to reach a reasonable tentative diagnosis. The goal is to acquaint you with the mechanics of the process. This knowledge will not enable you to make accurate clinical diagnoses; situations in the real world are much more complicated and variable than the examples provided here. Making an accurate clinical diagnosis is generally a complex process that demands a far greater level of experience and training than this course can provide.

For similar reasons, we will not discuss detailed treatment plans; the treatment of serious diseases requires specialized training and competence in advanced biochemistry, pharmacology, microbiology, pathology, and other clinical disciplines. However, many of the discussions in later sections include information about the use of specific drugs and other therapeutic procedures in the treatment of disease. These are representative examples intended to show potential treatment strategies rather than to endorse specific protocols and therapies.

[EXPLORE]

Additional resources, such as interactive tutorials, critical-thinking questions, clinical problems, and case studies, are available through the CD-ROM and Website for your textbook. To begin your exploration, launch the *Martini Interactive* CD-ROM or visit the Companion Website (http://www.prenhall.com/martini/fap5) and select Chapter 1.

UNIT 2

Applied Research Topics

Chapters 2 through 4 of your text provide background on basic concepts in chemistry, cell biology, and histology. This corresponding unit in the *Applications Manual* builds on that foundation by considering the relevance of those basic concepts to modern medicine.

We begin with topics in basic chemistry. We shall, among other things, consider both the use of radioactive materials to diagnose and treat disease and the factors that turn enzymes “on” or “off.”

In the next section, we look at the relationships between cell biology and disease. Throughout this course, as well as in daily life, you will hear about a variety of infectious diseases. As yet, however, you may not have a clear idea about the diversity or nature of the microorganisms that cause those diseases. This section begins with a broad overview of the major groups of disease-causing organisms. We then consider how normal and abnormal cells and tissues are examined. Finally, we discuss how abnormalities in a single cellular organelle can affect your overall health.

We end this unit with a discussion that highlights the importance of molecular biology. We begin with an overview of the now-prominent field of genetic engineering and learn how gene therapy can be used to synthesize drugs as well as treat disease. We then look at the origins, progression, and treatment of cancer, a deadly but common disease that begins when normal genetic control mechanisms break down.

The Application of Principles in Chemistry

Cells, tissues, organs, and organ systems are composed of chemicals. The survival of cells, tissues, organs, and systems depends on the control of chemical reactions, both within individual cells and in the extracellular fluids of the body. It is therefore not surprising that you cannot understand physiological principles without a familiarity with basic chemistry. As our understanding of physiological mechanisms has improved, physicians have become more adept at using chemical tests to diagnose and monitor disease. Physicians have also developed ways of manipulating intracellular and extracellular chemical reactions to help restore homeostasis. In this section, we will consider the practical application of some basic chemical principles introduced in Chapter 2 of the text.

THE MEDICAL IMPORTANCE OF RADIOISOTOPES FAP *p. 31*

A **radioisotope,** or radioactive isotope, is an isotope whose nucleus is unstable—that is, the nucleus spontaneously decays, or emits subatomic particles, in measurable amounts. Many recent technological advances in medicine have involved the use of radioisotopes for the diagnosis and treatment of disease. We will focus on two examples:

1. *The use of radioactive tracers to create diagnostic images.* Radioisotopes can be attached to organic or inorganic molecules and injected into the body. Within the body, these labeled molecules emit radiation energy that can be used to create images. These images provide information about tissue structure, tumors, blocked or weakened blood vessels, and other abnormalities in the body.
2. *The use of radioactive compounds to destroy abnormal cells and tissues.* If a suitable radiation source can be accurately delivered to a target site, the radioactivity can be used to treat many diseases.

RADIOISOTOPES AND CLINICAL TESTING

Alpha particles are subatomic particles that consist of a helium nucleus: two protons and two neutrons. These particles generally are emitted by the nuclei of large radioactive atoms, such as uranium. **Beta particles** are electrons, more typically released by radioisotopes of lighter atoms. **Gamma rays** are very-high-energy electromagnetic waves comparable to the X-rays used in clinical diagnosis.

The **half-life** of any radioisotope is the time required for half of a given amount of the isotope to decay. The half-lives of radioisotopes range from fractions of a second to billions of years.

Gamma rays, beta particles, and alpha particles—like X-rays—can damage or destroy living tissues. The danger posed by radiation exposure varies with the nature of the emission and the duration of exposure. But radiation also has a variety of beneficial uses in medical research and clinical diagnosis. Weakly radioactive isotopes with short half-lives are used to check the structural and functional states of an organ—something that would otherwise be impossible without surgery.

Radioisotopes can be incorporated into specific compounds that the body normally possesses. These compounds, called **tracers,** are said to be *labeled:* When introduced into the body, labeled compounds can be tracked by the radiation they release. After a labeled compound is swallowed, its uptake, distribution, and excretion can be determined by monitoring the radioactivity of samples taken from the digestive tract, body fluids, and waste products, respectively. For example, compounds labeled with radioisotopes of cobalt are used to monitor the intestinal absorption of vitamin B_{12}. Normally, cobalt-58, a radioisotope with a half-life of 71 days, is used.

Radioisotopes can also be injected into the blood or other body fluids to provide information about circulatory anatomy and the anatomy and function of specific target organs. In **nuclear imaging,** the radiation emitted by injected radioisotopes creates an image on a special photographic plate. Such a procedure is used to identify regions where particular radioactive materials are concentrated or to check the blood flow through vital organs. Radioisotopes can produce pictures of specific organs, such as the liver, spleen, thyroid, or bone, where different labeled compounds are preferentially removed from the bloodstream.

The thyroid gland sits below the larynx (voicebox) on the anterior portion of the neck (Figure A-3a•). A normal thyroid gland absorbs iodine, which is then used to produce thyroid hormones. As a result, the thyroid gland will actively absorb and concentrate

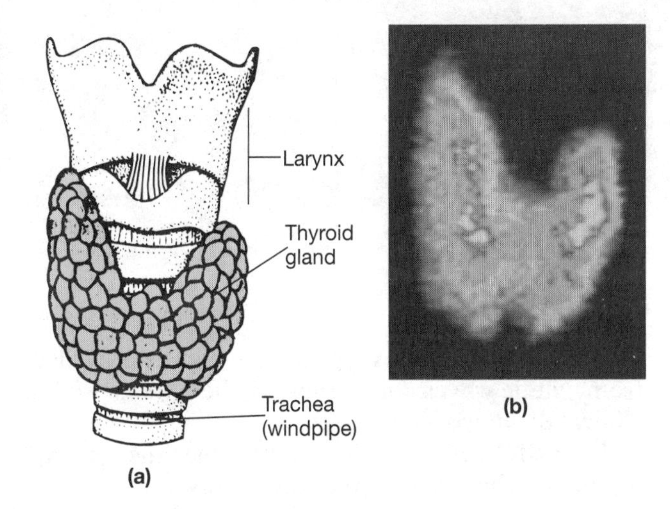

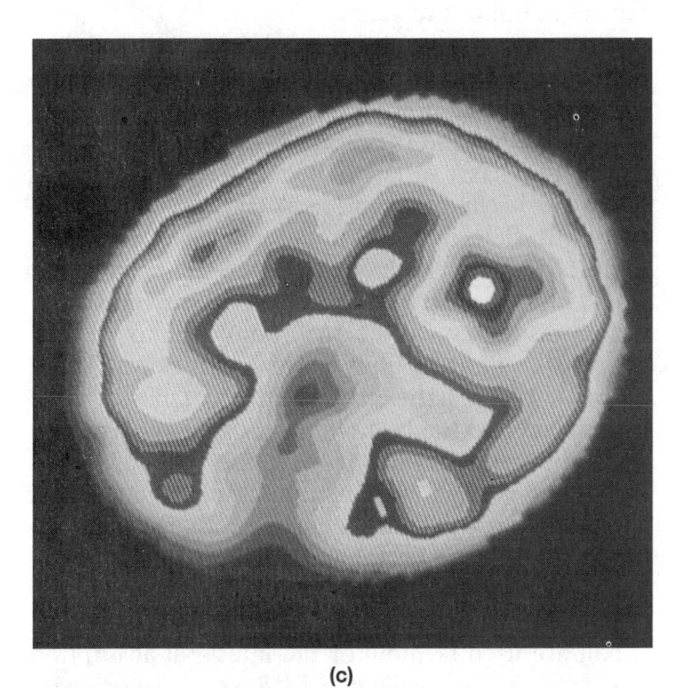

•**FIGURE A-3 Imaging Techniques.** **(a)** The position and contours of the normal thyroid gland as seen in dissection. **(b)** After it has been labeled with radioactive iodine, the thyroid can be examined by special imaging techniques. In this computer-enhanced image, different intensities indicate different concentrations of the radioactive tracer. **(c)** A PET scan of the brain in lateral view. The light areas indicate regions of increased metabolic activity.

radioactive iodine. The **thyroid scan** in Figure A-3b• was taken following the injection of iodine-131, a radioisotope with an 8-day half-life. This procedure, called a *thyroid radioactive iodine uptake measurement,* or **RAIU,** can provide information about (1) the size and shape of the thyroid gland and (2) the amount of iodine absorption. Comparing the rate of iodine uptake with the level of circulating hormones allows us to evaluate the functional state of the gland.

Radioactive iodine is an obvious choice for imaging the thyroid gland. For most other tissues and organs, a radioactive label must be attached to another compound. *Technetium* (^{99}mTc), a versatile tracer, is the primary radioisotope used in nuclear imaging today. The isotope is artificially produced and has a half-life of 6 hours. This brief half-life significantly reduces the patient's exposure to radiation. Technetium is used in more than 80 percent of all scanning procedures. The nature of the technetium-labeled compound varies with the identity of the target organ. Technetium scans are performed to examine the thyroid gland, spleen, liver, kidneys, digestive tract, bone, and a variety of other organs.

PET (**p**ositron **e**mission **t**omography) scans utilize the same principles as standard radioisotope scans, but the analyses are performed by computer. The scans are much more sensitive, and the computers can reconstruct sections through the body and provide extremely precise localization. Among other things, this procedure can analyze blood flow through organs and assess the metabolic activity in specific portions of an organ as complex as the human brain.

Figure A-3c• is a PET scan of the brain showing activity at a single moment in time. The scan is dynamic, however, and changing patterns of activity can be followed in real time. PET scans can be used to analyze normal brain function as well as to diagnose brain disorders. To date, the technique has served primarily as a research tool. Because the equipment is expensive and bulky, it is available only in large, regional medical centers or universities. The research advantages of PET scans have been challenged by the advent of real-time CT analysis *(cine-CT)* and the realization that MRI can be used to monitor small changes in blood flow and tissue activity without the use of radioactive tracers.

RADIOPHARMACEUTICALS

Nuclear medicine involving injected radioisotopes has been far more successful in producing useful images than in treating specific disorders. The problem is that the doses of radiation must be relatively large to destroy abnormal or cancerous tissues, and it is very difficult to control the distribution of these radioisotopes in the body with sufficient precision. As a result, radiation exposure can damage normal as well as abnormal tissues. It is also difficult to control the radiation dosage administered to the target tissues. Underexposure can have very little effect on the abnormal cells, whereas overexposure can destroy adjacent normal tissues.

Radioactive drugs, or *radiopharmaceuticals,* can be effective only if they are delivered precisely and selectively. One success story has been the treatment of *hyperthyroidism,* or thyroid oversecretion. As we noted earlier, the thyroid gland selectively concentrates iodine. Large doses of radioactive iodine (^{131}I) can be administered to treat hyperthyroidism. The radiation

released destroys the abnormal thyroid tissue and stops the excessive production of thyroid hormones. (Following this treatment, most individuals eventually become *hypothyroid*—deficient in thyroid hormone—but this condition can be treated by taking thyroid hormones in tablet form.) This is now the preferred treatment method for most adult hyperthyroid patients.

A relatively new application of nuclear medicine involves the attachment of a radioactive isotope to a *monoclonal antibody (MoAb)*. Antibodies are proteins produced in the body to provide a selective defense against foreign proteins, toxins, or pathogens. Monoclonal antibodies are produced by immune cells cultured under laboratory conditions. The antibodies these cells produce are then labeled with radioactive materials. If injected into the body, the antibodies will bind to their targets and expose the surrounding tissues to radiation. MoAbs specific for certain types of tumor cells have already been approved by the Food and Drug Administration (FDA). The amount of radiation emitted is low, however, and the procedure is used to produce diagnostic images rather than to treat the disease. Experiments continue, with the eventual goal of using radiolabeled MoAbs to destroy tumor cells.

SOLUTIONS AND CONCENTRATIONS

FAP pp. 42, 954

Physiologists and clinicians pay particular attention to ionic distributions across membranes and to the electrolyte composition of body fluids. Standard values for physiological tests are provided throughout the text and are summarized there in Appendix VI. Data must be analyzed from several perspectives, and physiological values can be reported in several ways. One method is to report the concentration of atoms, ions, or molecules in terms of weight per unit volume of solution. Although grams per liter (g/l) can be used, values are most often expressed in grams (g), milligrams (mg), or micrograms (μg) per 100 ml. Because 100 ml = 0.1 liter = 1 deciliter (dl), the abbreviations most often used in this text are **g/dl** *(grams per deciliter)* and **mg/dl** *(milligrams per deciliter)*.

Osmotic concentration, or osmolarity, depends on the total number of individual atoms, ions, and molecules in solution, without regard to molecular weight, electrical charge, or molecular identity. As a result, if fluid balance and osmolarity are being monitored, concentrations are usually reported in *moles per liter* (**mol/l** or **M**) or *millimoles per liter* (**mmol/l** or **mM**) rather than in g/dl or mg/dl. To convert g/dl to mol/l, multiply by 10 and divide by the atomic weight of the element. For example, a sample of plasma (blood with the cells removed) contains sodium ions at a concentration of roughly 0.32 g/dl (320 mg/dl). We convert this value to mmol/l as follows:

$$\frac{\text{g/dl} \times 10}{\text{atomic weight}} = \frac{0.32 \times 10}{22.99}$$
$$= 0.14 \text{ mol/l } (= 140 \text{ mmol/l})$$

Moles or millimoles per liter can also be used to indicate the concentration of molecules in solution. We can perform the same conversion by substituting molecular weight for atomic weight in the above equation. The total solute concentration of a solution can be determined by adding the concentrations of individual solutes, expressed in moles per liter or millimoles per liter. The resulting value is reported in **milliosmoles per liter (mOsm/l).** The use of mOsm rather than mmol indicates that multiple solutes are present, each contributing to the total osmolarity.

Because electrolyte concentrations have profound effects on cells, it is often important to know how many positive and negative charges the ions or molecules in a biological solution bear, not just how many ions or molecules are present. For example, a single calcium ion (Ca^{2+}) has twice the electrical charge of a single sodium ion (Na^{+}), although the two are identical in terms of their effects on osmolarity. One **equivalent (Eq)** is a mole of positive or negative charges. Physiological concentrations are often reported in **milliequivalents per liter (mEq/l).** You should become familiar with both methods of expression. Fortunately, the conversion from millimoles to milliequivalents is relatively easy to perform. For **monovalent ions,** those with a +1 or –1 charge, millimole and milliequivalent values are identical, so no calculation is needed. For **divalent ions,** with a +2 or –2 charge, the number of charges (mEq) is twice the number of ions (mmol). For an ion with a +3 or –3 charge, the number of milliequivalents is three times the number of millimoles. To convert mEq to mmol, simply divide by the ionic valence (number of charges).

Table A-4 compares the methods of reporting the concentration of major electrolytes in plasma in terms of weight, moles, and equivalents; the notation "nr"

TABLE A-4 A Comparison of Methods for Reporting Concentrations of Solutes in Blood*

Solute	mg/dl	mmol/l	mEq/l	SI Units
Electrolytes				
Sodium (Na^{+})	320	140	140	140 mmol/l
Potassium (K^{+})	16.4	4.2	4.2	4.2 mmol/l
Calcium (Ca^{2+})	9.5	2.4	4.8	2.4 mmol/l
Chloride (Cl^{-})	354	100	100	100 mmol/l
Metabolites				
Glucose	90	5	nr	5 mmol/l
Lipids, total	600	nr	nr	0.6 g/l
Proteins, total	7 g/dl	nr	nr	70 g/l

*nr = not reported in these units

indicates that concentrations are not reported in those terms. The tables included in Appendix VI of the text provide data in terms currently accepted for clinical laboratory reports.

Physiologists and clinicians surely would benefit from the use of standardized reporting procedures. It can be very frustrating to consult three references and find that the first reports electrolyte concentrations in mg/dl, the second in mmol/l, and the third in mEq/l. In 1984, the American Medical Association House of Delegates endorsed a plan to standardize clinical test results through the use of **SI (Système Internationale)** units, with a target date of July 1, 1987, for the switchover. Unfortunately, there was no mechanism for enforcing compliance, and the standardization attempt ultimately failed. As of 1997, all scientific and medical journals around the world report data in SI units, but most U.S. clinical laboratories and journals continue to use their traditional reporting methods.

The major problem is that the relationships to values currently in use are difficult to remember. Electrolyte concentrations, now most often given in mEq/l, are reported in mmol/l in the SI. Thus the values for sodium and potassium concentrations remain unchanged, but the normal values for calcium and magnesium are reduced by 50 percent. The situation becomes more confusing in terms of metabolite concentrations. Cholesterol and glucose concentrations are now most often reported in mg/dl, but the SI units are mmol/l. However, total lipid concentrations, also currently reported as mg/dl, and total protein concentrations, now given as g/dl, are reported in terms of g/l under the SI. For these units to be useful in a clinical setting, physicians must not only remember the definition of each SI unit but must also convert and relearn the normal ranges. As a result, it appears unlikely that the conversion to SI units will be completed in the immediate future.

THE PHARMACEUTICAL USE OF ISOMERS

FAP p. 45

A chemical compound is a combination of atoms bonded together in a particular arrangement. The chemical formula specifies the *number* and *types* of atoms that combine to form the compound. The *arrangement* of the atoms, which determines the specific shape of each molecule, is shown by the molecule's structural formula.

Isomers are chemical compounds that have the same chemical formula but different structural formulas. Isomers called *stereoisomers* are mirror images of each other. Stereoisomers are analogous to the left hand and right hand of the human body. The two hands contain the same palm bones and finger bones (metacarpals and phalanges), but each hand is a mirror image of the other. A glove designed for the left hand will not fit the right hand, and vice versa. Chemical compounds are also said to be left-handed or right-handed, depending on their structural configuration. For example, glucose has a left-handed (*levo-*) isomer and a right-handed (*dextro-*) isomer. Just as the right hand cannot fit into a left glove, receptors and enzymes in our cells cannot bind the levo-isomer of glucose. Our cells are therefore unable to metabolize levo-glucose as an energy source.

This pattern is common: Our cells and tissues will typically respond to only one structural form—either levo or dextro—but not to both. This feature can pose a problem for pharmaceutical chemists, because many of the chemical reactions used to synthesize a drug produce a mixture of levo- and dextro-isomers. In some cases, the inactive isomer is simply ignored; in others, the inactive isomer is removed. For example, the antibiotic *chloramphenicol* contains both levo- and dextro-isomers, but only the levo form is effective in killing bacterial pathogens. And only the levo form (the active form) of *ephedrine,* a drug that dilates the bronchioles of the lungs, is contained in the popular tablet *Primatene*®, which is sometimes used to treat *asthma* attacks. Finally, birth control pills containing the steroid *Levonorgestrel*®, the levo form, are effective at half the dosage of pills containing a mixture of levo- and dextro-isomers.

In some cases, both forms of an isomer are biologically active but have strikingly different effects. The drug *thalidomide* was given to pregnant women in the 1960s to alleviate symptoms of morning sickness. The sedative effect of one isomer was well documented, but the medication sold contained both forms. Unfortunately, the other isomer caused tragic abnormalities in fetal limb development. (We discuss the mechanisms that underlie thalidomide's effects on fetal development on p. 192.)

TOPICS IN METABOLISM

Metabolism is the sum of all the biochemical reactions that proceed in the body. Hundreds of thousands of reactions occur in each cell. At any moment, biochemical pathways may be producing phospholipids for the cell membrane or peptide hormones for secretion while breaking down carbohydrates to generate ATP. In the following sections, we will consider three aspects of metabolism:

1. Many disease processes are the result of a faulty biochemical pathway. For example, an enzyme may be missing or inactive, or the necessary enzymatic substrates may be unavailable. *Phenylketonuria, galactosemia, albinism,* and *hypercholesterolemia* are metabolic disorders that we will consider.
2. Enzymes control metabolic processes in our cells. The mechanisms that control enzymatic reactions are

therefore important, and problems with enzymatic regulation can cause severe metabolic disorders.

3. Diet and nutrition have an obvious impact on metabolic operations in the body. A substantial research effort is under way to manipulate metabolic operations by dietary changes. The control of cholesterol in the diet is only one of several pertinent examples.

ARTIFICIAL SWEETENERS FAP p. 46

Some people cannot tolerate sugar for medical reasons; others avoid it to comply with recent dietary guidelines that call for reduced sugar consumption or to lose weight. Thus, many people today use artificial sweeteners in their foods and beverages.

Artificial sweeteners are organic molecules that can stimulate taste buds and provide a sweet taste to foods without adding substantial amounts of calories to the diet. These molecules have a much greater effect on the taste receptors than do natural sweeteners, such as fructose or sucrose, so they can be used in minute quantities. For example, *saccharin* is about 300 times as sweet as sucrose. The popularity of this sweetener has declined since it was reported that saccharin can promote bladder cancer in rats. The risk is very small, however, and saccharin continues to be used. Several other artificial sweeteners, including *aspartame (NutraSweet®), sucralose,* and *acesulfame potassium* (*Ace-K,* or *Surette®*), are currently available. The market success of an artificial sweetener ultimately depends on its taste and its chemical properties. Stability in high temperatures (as in baking) and resistance to breakdown in an acidic pH (as in carbonated drinks) are important properties for any artificial sweetener.

Molecules of artificial sweeteners do not resemble those of natural sugars. Saccharin, acesulfame potassium, and sucralose cannot be broken down by the body, and they have no nutritional value. Aspartame consists of a pair of amino acids. Amino acids are the building blocks of proteins (as we will discuss later in this chapter), and they can be broken down in the body to provide energy. However, because aspartame is 200 times as sweet as sucrose, very small quantities are needed, so this artificial sweetener adds few calories to a meal. Because it does not produce the bitter aftertaste sometimes attributed to saccharin, aspartame is used in many diet drinks and low-calorie desserts.

Two recent entries into the market for artificial sweeteners, *thaumatin-1* and *monellin,* are proteins extracted from African berries. Thaumatin, roughly 100,000 times as sweet as sucrose, has been approved by the FDA for use in chewing gums.

FAT SUBSTITUTES FAP p. 47

The average diet in the United States contains more fat than do the diets of people in many other parts of the world. Diets high in fat have been linked to heart disease as well as to certain forms of cancer. Recent recommendations suggest that lowering the percentage of calories we derive from fat would benefit our health. This suggestion has led to an increased interest in the development of fat substitutes.

Fat substitutes provide the texture, taste, and cooking properties of natural fats. Two fat substitutes, *Simplesse®* and *Olestra®*, are in widespread use. Simplesse is made from proteins of egg white and skim milk or whey. The heated proteins are treated to form small spherical masses that have the taste and texture of fats. Simplesse can be used in place of fats in any application other than baking; it is used in low-calorie "ice creams" under the trade name *Simple Pleasures®*. These fat substitutes can be broken down in the body, but they provide less energy than do natural fats. For example, ice cream made with Simplesse has half the calories of ice cream that contains natural fats.

Olestra is made by chemically combining sucrose and fatty acids. The resulting compounds cannot be used by the body and so contribute no calories. Olestra has been approved as an ingredient in margarines, baked goods, and other snack foods despite some concerns. One of the problems is that Olestra droplets within the digestive tract collect lipid-soluble materials, including fat-soluble vitamins, and prevent their absorption. In addition, if eaten in large quantities, Olestra can cause diarrhea. These side effects pose a serious threat—a combination of fluid loss and vitamin deficiency. To prevent vitamin deficiencies among consumers, manufacturers of snack foods prepared with Olestra now fortify them with fat-soluble vitamins.

Olestra is not the only approved fat substitute derived from carbohydrates. *Oatrim®*, a fat substitute derived from soluble fiber and complex carbohydrates, is now used in muffins, cookies, fat-free cheeses, and lean hot dogs and luncheon meats.

The use of fat substitutes may also pose secondary metabolic problems. Fat-soluble vitamins (A, D, E, and K) are normally absorbed by the intestinal tract in company with dietary lipids. A drastic reduction in the lipid content of the diet can therefore lead to deficiencies of these vitamins unless vitamin supplements are taken. Several new drugs are available as alternatives to fat substitutes. These drugs, such as *Xenical®* and *Lipitor®*, block the action of the pancreatic enzymes responsible for fat digestion and prevent the absorption of dietary fats.

A CLOSER LOOK: THE CONTROL OF ENZYME ACTIVITY FAP p. 55

Four major factors determine the rate at which an enzymatic reaction occurs: (1) *substrate and product concentrations,* (2) *enzyme concentration,* (3) *competitive inhibition,* and (4) *enzyme activation states* (Figure A-4•). We will examine each of these factors individually.

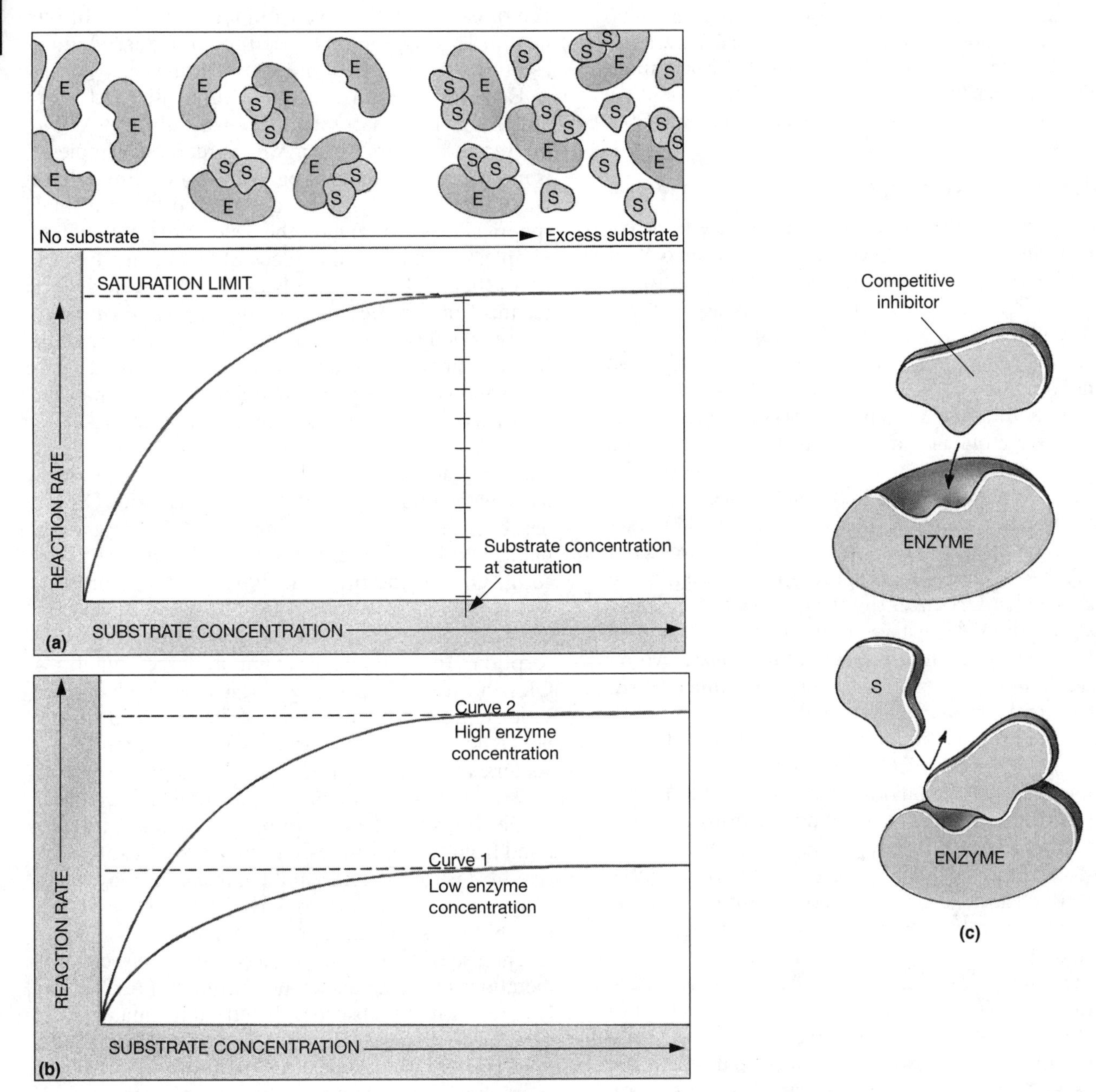

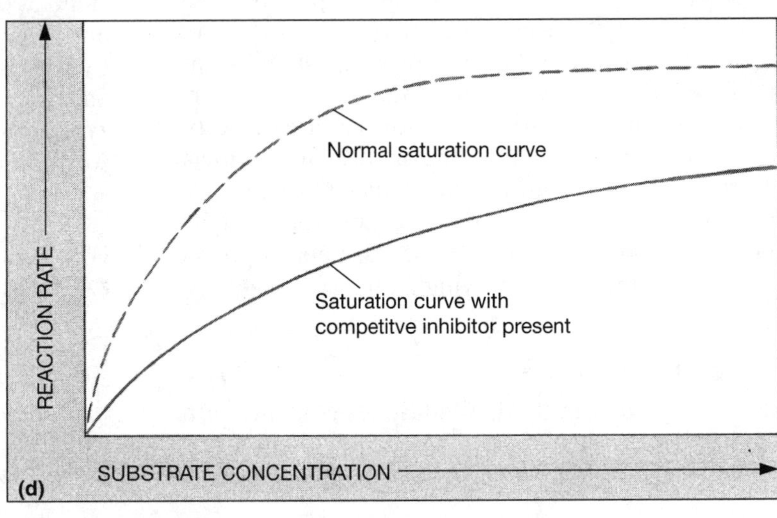

•**FIGURE A-4 Factors That Affect Enzymatic Reaction Rates.** **(a)** A saturation curve, showing the effects of changing substrate concentrations. Here E = enzyme and S = substrate. **(b)** The effect of changing the concentration of enzymes on a reaction rate. Curve 2 has a higher enzyme concentration and therefore a faster initial reaction rate than curve 1 has. **(c)** A competitive inhibitor blocks the enzyme's active site, so the substrate cannot bind. **(d)** The effects of competitive inhibitors on reaction rates.

SUBSTRATE AND PRODUCT CONCENTRATIONS

Figure A-4a• plots the effect of substrate concentration on the reaction rate. When many enzyme molecules are available but no substrate is available, no reaction can occur. At low substrate concentrations, substrate availability is the factor that limits the reaction rate. The higher the substrate concentration, the faster the reaction proceeds. This acceleration does not continue indefinitely; as product concentration rises, the chances increase that a product molecule, instead of substrate molecules, will contact the active site. When all the available active sites are bound either to substrate or to product molecules, the enzyme system is **saturated.** Any further increase in substrate concentration will have no effect on the reaction rate. The graph in Figure A-4a• is a typical enzyme **saturation curve.**

ENZYME CONCENTRATION

The concentration of enzymes has a direct effect on the rate of the reaction (Figure A-4b•). The higher the enzyme concentration, the faster the initial reaction rate. Varying rates of enzyme synthesis or destruction may change enzyme concentrations in a cell. Changing the cytoplasmic concentration of enzymes is a slow process. During that time, turning enzymes "on" or "off" regulates the concentration of functional enzymes.

COMPETITIVE INHIBITION

Enzyme specificity results from the shape and charge characteristics of the active site. That specificity is not perfect. Molecules that closely resemble the normal substrate can bind to the active site and interfere with substrate binding. This process is called **competitive inhibition,** because substrate molecules must compete for a space on the active site. The higher the concentration of competitive inhibitors, the lower the rate of reaction. Figure A-4c,d• indicates the effects of competitive inhibition on reaction rates.

ENZYME ACTIVATION STATES

An activated enzyme will catalyze a particular reaction; an inactivated enzyme will not. In an inactivated enzyme, the active site cannot interact with substrate molecules. During activation, the shape of the active site changes. Four types of factors may activate or inactivate a specific enzyme:

1. *Physical factors.* Environmental factors such as high or low temperature or a change in pH can alter the enzyme's tertiary or quaternary structure. Variations outside normal limits will temporarily or permanently inactivate enzymes. For example, body temperature regulation is essential, because temperature directly affects enzyme activity. At high body temperatures (over 40°C), enzymes begin to denature, becoming permanently nonfunctional. Enzymes are equally sensitive to pH changes. Each enzyme works best at an optimal combination of temperature and pH. For instance, *pepsin,* an enzyme that breaks down proteins in the stomach contents, works best at a pH of 2.0 (strongly acidic). The small intestine contains *trypsin,* another enzyme that attacks proteins, which works only in an alkaline environment, with an optimum pH of 9.5.
2. *The presence or absence of cofactors.* **Cofactors** are ions or molecules that must attach to the active site before substrates can bind (Figure A-5•). Examples of cofactors include mineral ions, such as calcium (Ca^{2+}) and magnesium (Mg^{2+}), and several vitamins. A **holoenzyme** is an enzyme activated by an appropriate cofactor. An enzyme without its cofactor is an inactive **apoenzyme** (ap-ō-EN-zīm).
3. *Allosteric effects.* An **allosteric** (*allos*, other + *stereo*, solid) **effect** is a change in the shape of the active site caused by an interaction between an enzyme and some other molecule. The molecule involved is called a

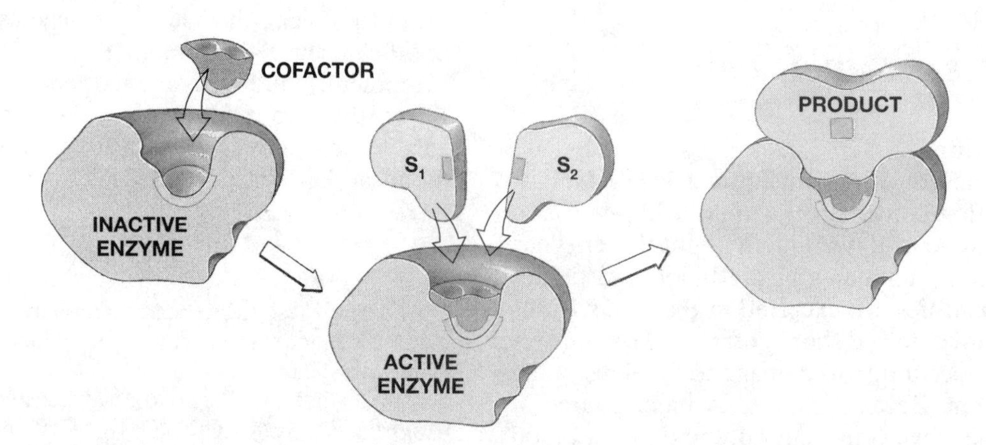

•**FIGURE A-5 Cofactors and Enzyme Activity.** In some cases, a cofactor must bind to the active site before the enzyme can bind substrate molecules and function normally.

modulator. Activators are modulators that turn an enzyme "on." This process is **allosteric activation. Inhibitors** turn an enzyme "off"; they perform **allosteric inhibition.** (This is how drugs such as Xenical inhibit fat digestion.) Complex proteins or small molecules may have powerful allosteric effects. Many hormones, such as *adrenaline,* activate enzymes in cell membranes throughout the body. The hormonal effects vary with the nature of the activated enzyme.

4. *Phosphorylation.* **Phosphorylation** is the enzymatic attachment of a phosphate group to a molecule. Some enzymes are activated or inactivated by phosphorylation. For example, the phosphorylation of a muscle enzyme results in glycogen breakdown and glucose release.

Chemicals that affect enzyme activity can have powerful effects on cells throughout the body. For instance, several deadly compounds, such as hydrogen cyanide (HCN) and hydrogen sulfide (H_2S), kill cells by inhibiting mitochondrial enzymes involved in ATP production. Many inhibitors with less drastic effects are in clinical use. For example, *warfarin* slows blood clotting by inhibiting liver enzymes responsible for the synthesis of clotting factors. Several of the beneficial effects of *aspirin*—notably, the reduction of inflammation—are related to the inhibition of enzymes involved with prostaglandin synthesis. Many important antibiotics, such as penicillin, kill bacteria by inhibiting enzymes that are essential to bacteria but are absent from our cells.

METABOLIC ANOMALIES

If enzymes are nonfunctional or are missing, metabolic disorders known as *metabolic anomalies* result. The effects are variable, depending on the enzyme involved, but in severe cases growth and development are impaired and vital tissues are damaged or destroyed. Additional information about many of these conditions is given elsewhere in the *Applications Manual* and in the text.

ANOMALIES IN AMINO ACID METABOLISM

Phenylketonuria

Persons diagnosed with **phenylketonuria (PKU)** lack the enzyme that converts the amino acid *phenylalanine* to the amino acid *tyrosine.* Without this enzyme, phenylalanine accumulates in the blood and tissues and large quantities are excreted in the urine. If this condition is not detected shortly after birth, mental retardation can occur due to damage to the developing nervous system. Because milk is a major source of phenylalanine, newborns generally undergo a blood test for PKU 48 hours after nursing begins. Abnormally high levels of phenylalanine in the bloodstream may indicate PKU. Once a diagnosis of PKU is made, the diet is controlled to avoid foods containing high levels of phenylalanine.

Albinism

Albinism is a genetic disorder that results in a lack of pigment production in the skin. The cause is a defective enzyme involved in the metabolism of the amino acid tyrosine. Because this enzyme is abnormal, the protein pigment *melanin* cannot be synthesized. The skin is white, and the hair and eyes are also affected. Among its other functions, melanin helps protect the skin from the effects of ultraviolet (UV) radiation. When outdoors, individuals with albinism must be careful to avoid skin damage from the UV radiation in sunlight.

ANOMALIES IN LIPID METABOLISM: HYPERCHOLESTEROLEMIA

Familial hypercholesterolemia is a genetic disorder resulting in a reduced ability to remove cholesterol from the bloodstream. As circulating levels rise, cholesterol accumulates around tendons, creating yellow deposits called *xanthomas* beneath the skin. The worst aspect of this disorder is the deposition of cholesterol in the walls of blood vessels. This condition, a form of *atherosclerosis*, can restrict the flow of blood through vital organs such as the heart and brain. Atherosclerosis can develop in individuals with normal cholesterol metabolism, but clinical symptoms do not ordinarily appear until age 40 or older. Individuals with hypercholesterolemia may develop acute coronary artery disease or even suffer a heart attack at or before 20 years of age.

ANOMALIES IN CARBOHYDRATE METABOLISM: GALACTOSEMIA

Milk contains the monosaccharide *galactose,* which can be converted to glucose within cells. The genetic disorder **galactosemia** is caused by the absence of the enzyme that catalyzes this reaction. Affected individuals have elevated levels of galactose in the blood and urine. Chronically high levels of galactose during childhood can cause abnormalities in nervous system development, jaundice, liver enlargement, and cataracts. Preventive treatment involves the early detection of galactosemia and a restriction of dietary intake of this monosaccharide.

[EXPLORE]

Additional resources, such as interactive tutorials, critical-thinking questions, clinical problems, and case studies, are available through the CD-ROM and Website for your textbook. To begin your exploration, launch the *Martini Interactive* CD-ROM or visit the Companion Website (http://www.prenhall.com/martini/fap5) and select Chapter 2.

Topics in Cellular Biology

Cells are the smallest living units in the body, but they are not the only forms of life. Discussions throughout the text and the *Applications Manual* assume that you are already familiar with the basic properties of cells and with the characteristics of potential **pathogens** (disease-causing organisms). This section begins with a review of the important distinctions among various types of pathogens and then proceeds to a discussion of other topics at the cellular level of organization.

A CLOSER LOOK: THE NATURE OF PATHOGENS

FAP *pp. 65, 752, 790*

Chapter 3 of the text presents the structure of a "typical" cell. The cellular organization depicted in Figure 3-2 (FAP *p. 66*) and described in that chapter is that of a *eukaryotic* cell (ū-kar-ē-OT-ik; *eu*, true + *karyon*, nucleus). The defining characteristic of eukaryotic cells is the presence of a nucleus. All eukaryotic cells have similar membranes, organelles, and methods of cell division. All multicellular animals, plants, and fungi (plus some single-celled organisms) are composed of eukaryotic cells.

The eukaryotic plan of organization is not the only one in the living world, however. Some organisms do not consist of eukaryotic cells. These organisms are of great interest to us, because they include most of the pathogens that can cause human diseases.

BACTERIA

Prokaryotic cells do not have nuclei or other membranous organelles. They do not have a cytoskeleton, and typically their cell membranes are surrounded by a semirigid cell wall made of carbohydrate and protein.

Bacteria are probably the most familiar prokaryotic cells. They are generally less than 2 µm in diameter. Many bacteria are quite harmless, and many more—including some that live within our bodies—are beneficial to us in a variety of ways. Other bacteria are dangerous pathogens that, given the opportunity, will destroy body tissues. These bacteria are dangerous because they absorb nutrients and release enzymes that damage cells and tissues. A few pathogenic bacteria also release toxic chemicals. Bacterial infections are responsible for many serious diseases, as indicated in Table A-5. We consider these and other bacterial infections in various chapters of the text and in other sections of the *Applications Manual*.

TABLE A-5 Examples of Bacterial Diseases and Primary Organ Systems Affected

Organism	Disease	Affected Organ System
Bacilli		
Bacillus anthracis	Anthrax	Integumentary system
Mycobacterium tuberculosis	Tuberculosis	Respiratory system
Corynebacterium diphtheriae	Diphtheria	Respiratory system
Cocci		
Staphylococcus aureus	Various skin infections	Integumentary system
Streptococcus pyogenes	Pharyngitis (strep throat)	Respiratory system
Neisseria gonorrhoeae	Gonorrhea	Reproductive system
Vibrios		
Vibrio cholerae	Cholera	Digestive system
Spirochetes		
Treponema pallidum	Syphilis	Reproductive, nervous systems
Borrelia burgdorferi	Lyme disease	Skeletal system (joints)
Rickettsias		
Rickettsia prowazekii	Epidemic typhus fever	Cardiovascular, integumentary systems
Coxiella burnetii	Q fever	Respiratory system
Chlamydias		
Chlamydia trachomatis	Trachoma (eye infections)	Integumentary system
	Lymphogranuloma venereum (LGV)	Reproductive system

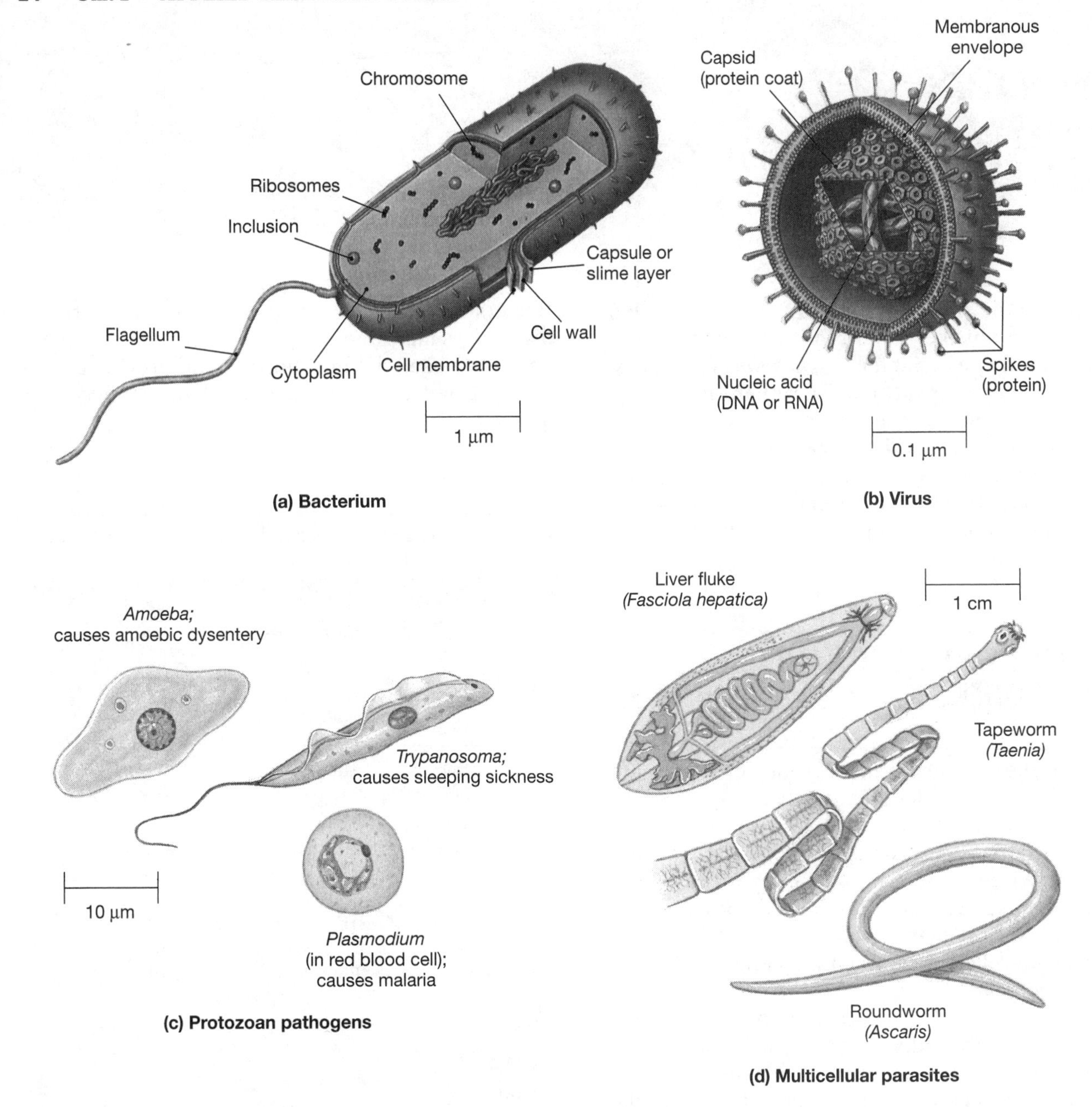

•**FIGURE A-6 Representative Pathogens.** **(a)** A bacterium, with prokaryotic characteristics indicated. Compare with Figure 3-2 (FAP *p. 66*), which shows a representative eukaryotic cell. **(b)** A typical virus. Each virus has an inner chamber containing nucleic acid, surrounded by a protein capsid or an inner capsid and an outer membranous envelope. The herpes viruses are enveloped DNA viruses; they cause chickenpox, shingles, and herpes. **(c)** Protozoan pathogens. Protozoa are eukaryotic single-celled organisms, common in soil and water. **(d)** Multicellular parasites. Several groups of organisms are human pathogens, and many have complex life cycles.

Figure A-6a• shows the structure of a representative bacterium. Figure A-7• shows the three basic shapes of bacteria: round, rodlike, and spiral. A round bacterium is called a **coccus** (KOK-us; plural, *cocci,* KOK-sē). A rodlike bacterium is a **bacillus** (ba-SIL-us; plural, *bacilli,* ba-SIL-ē). Shapes of spiral bacteria vary, and so do their names. A **vibrio** (VIB-rē-ō) is comma-shaped; a **spirillum** (spi-RIL-um; plural, *spirilla*) is rigid and wavy; and a **spirochete** (SPĪ-rō-kēt) is shaped like a corkscrew.

Some cocci and bacilli form groupings of cells. The Latin names used to describe these groupings are also used to identify specific bacteria. For instance, pairs of cocci are called *diplococci* (*diplo-*, double). *Streptococci* and *streptobacilli* form twisted chains of cells (*strepto-*, twisted), and *staphylococci* look like a bunch of grapes (*staphylo-*, grapelike).

VIRUSES

Another type of pathogen conforms neither to the prokaryotic nor to the eukaryotic organizational plan. These tiny pathogens, called **viruses,** are not cellular. In fact, when free in the environment, they do not show any of the characteristics of living organisms. They are classified as **infectious agents**—factors that

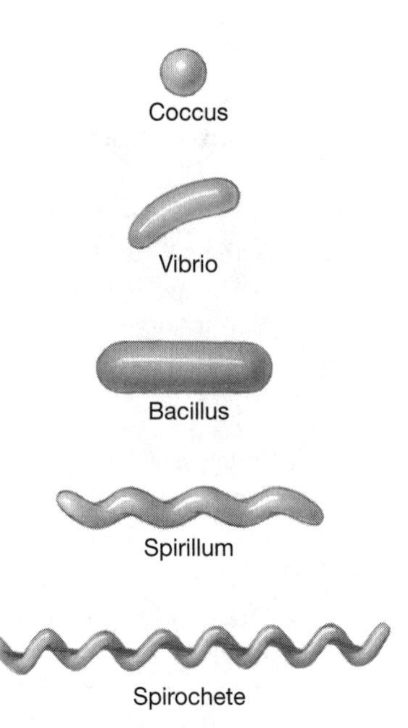

•**FIGURE A-7** **Common Bacterial Shapes**

cause infection—because they can enter cells (either prokaryotic or eukaryotic) and replicate themselves.

Viruses consist of a core of nucleic acid (DNA or RNA) surrounded by a protein coat called a *capsid.* (Some varieties have an *envelope,* a membranous outer covering, as well.) The structure of a representative virus is shown in Figures A-6b and A-8•. Important viral diseases include influenza (flu), yellow fever, some leukemias, AIDS, hepatitis, polio, measles, mumps, rabies, and the common cold (Table A-6).

To enter a cell, a virus must first attach to the cell membrane. Attachment occurs at one of the normal membrane proteins. Once the virus has penetrated the cell membrane, the viral nucleic acid takes over the cell's metabolic machinery. In the case of a DNA virus (Figure A-9a•), the viral DNA enters the cell nucleus, where transcription begins. The mRNA produced then enters the cytoplasm and is used in translation, in which the cell's ribosomes begin synthesizing viral proteins. The viral DNA replicates in the nucleus, "stealing" the cell's nucleotides. The replicated viral DNA and the new viral proteins then form new viruses that pass out of the cell through the cell membrane.

In an RNA virus, the situation is somewhat more complicated (Figure A-9b•). In the simplest RNA viruses, the viral RNA entering the cell functions as an mRNA strand that carries the information needed to direct the cell's ribosomes to synthesize viral proteins. These proteins include enzymes essential to the duplication of viral RNA. When the cell is packed with new viruses, the cell membrane ruptures and the RNA viruses are released into the interstitial fluid.

In *retroviruses,* a group that includes HIV (the virus responsible for AIDS), the replication process is even more complex. These viruses carry an enzyme called *reverse transcriptase,* which directs "reverse transcription"—the assembly of DNA according to the nucleotide sequence of an RNA strand. The DNA created in this way is then inserted into the infected

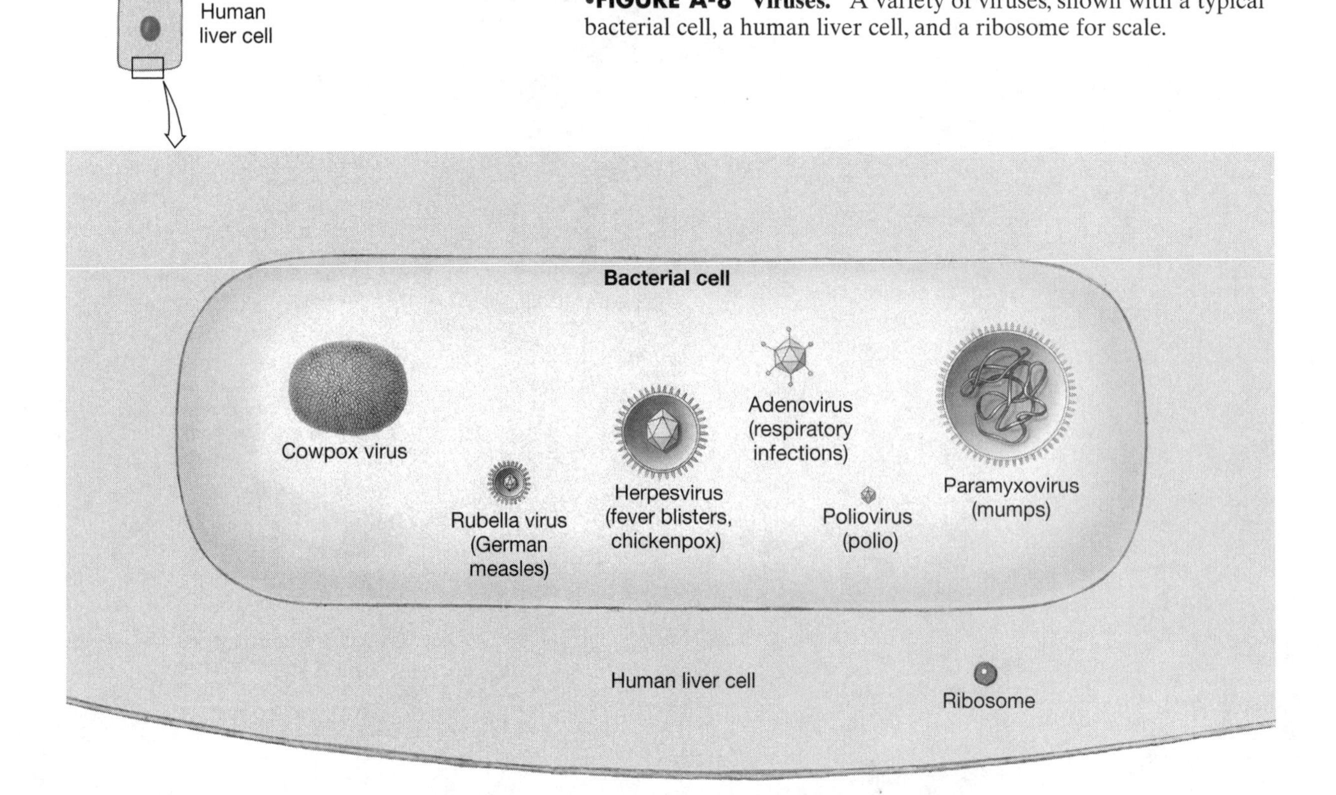

•**FIGURE A-8** **Viruses.** A variety of viruses, shown with a typical bacterial cell, a human liver cell, and a ribosome for scale.

TABLE A-6 Examples of Viral Diseases and Primary Organ Systems Affected

Nucleic Acid	Virus	Disease	Affected Organ System
RNA	Influenza A, B, C	Flu	Respiratory, reproductive systems
	Paromyxovirus	Mumps	Digestive system
	Hepatitis A, C, D, E, G	Infectious hepatitis	Digestive system (liver)
	Rhinovirus	Common cold	Respiratory system
	Human immunodeficiency virus (HIV)	AIDS	Lymphatic system
DNA	Herpesvirus		
	Herpes simplex 1	Cold sore/fever blister	Integumentary system
	Herpes simplex 2	Genital herpes	Reproductive system
	Varicella-zoster	Chickenpox	Integumentary system
	Varicella-zoster	Shingles	Nervous system
	Hepatitis B	Hepatitis	Digestive system (liver)
	Epstein–Barr	Mononucleosis	Respiratory system

cell's chromosomes. The viral genes are then activated, and the cell begins producing RNA by normal transcription. The RNA produced includes viral RNA, mRNA carrying the information for the synthesis of reverse transcriptase, and mRNA controlling the synthesis of viral proteins. These components then combine within the cytoplasm, which gradually becomes filled with viruses. The new RNA viruses are then shed at the cell surface. Two new anti-influenza medicines, *Relenza*® and *Tamiflu*®, inhibit a key enzyme involved in virus assembly and release by infected cells.

Even if the host cell is not destroyed by these events, normal cell function is usually disrupted. In effect, the metabolic activity of the cell is diverted to create viral components rather than performing tasks needed for cell maintenance and survival. Some virus-

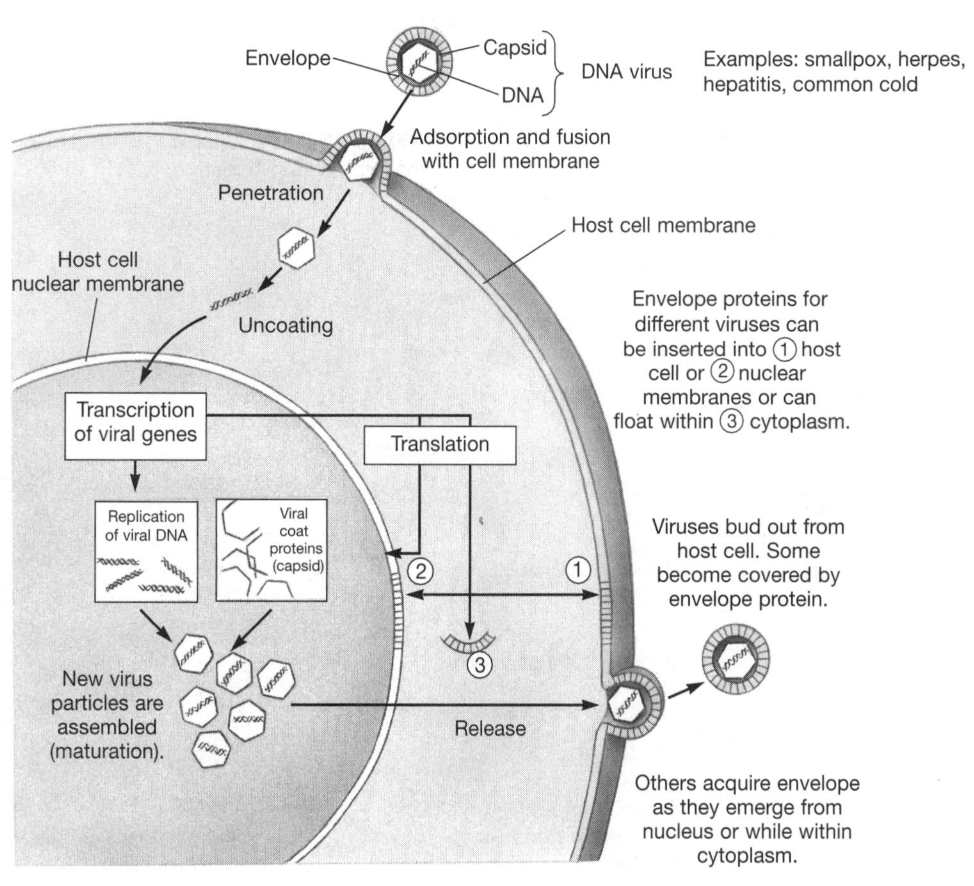

•**FIGURE A-9 Viral Replication.** **(a)** The replication of a DNA virus.

es, such as *herpes simplex*, can lie dormant within infected cells for long periods of time before initiating this process of replication.

Viruses are now becoming important as benefactors as well as adversaries. In *genetic engineering* procedures (p. 34), viruses whose nucleic acid structure has been intentionally altered can be used to transfer copies of normal human genes into the cells of individuals with inherited enzymatic disorders. This was the method used to insert the gene for the enzyme missing in persons with adenosine deaminase deficiency (p. 35). Attempts are now planned to treat cystic fibrosis in the same way. *Cystic fibrosis (CF)* is a debilitating genetic defect whose most obvious—and potentially deadly—symptoms involve the respiratory system. The underlying problem is an abnormal gene that carries instructions for a chloride ion channel that occurs in cell membranes throughout the body. Researchers have recently treated CF in laboratory animals by inserting the normal gene into a virus that infects cells lining the respiratory passageways.

PRIONS

Prions (PRĒ-onz) are controversial proteins that are either infectious agents or the result of infection by an as-yet unidentified virus. (Current evidence suggests that the proteins are themselves infectious.) Normal neurons produce a very similar protein that exists as individual molecules. These molecules are located in the cell membrane and are in contact with the extracellular fluid. When exposed to a prion, this protein changes shape; the individual molecules interact to form large insoluble complexes that are released into the extracellular fluid. These complexes are called *amyloid plaques.* Either the presence of the amyloid plaques or the absence of the normal protein then disrupts neural function.

The first prion disease described was *kuru,* a deadly disease affecting members of a cannibalistic society in New Guinea. The prions were passed from person to person when uninfected individuals ate infected brains. The infection, which led to death within a year, caused half of all childhood and adult deaths in New Guinea.

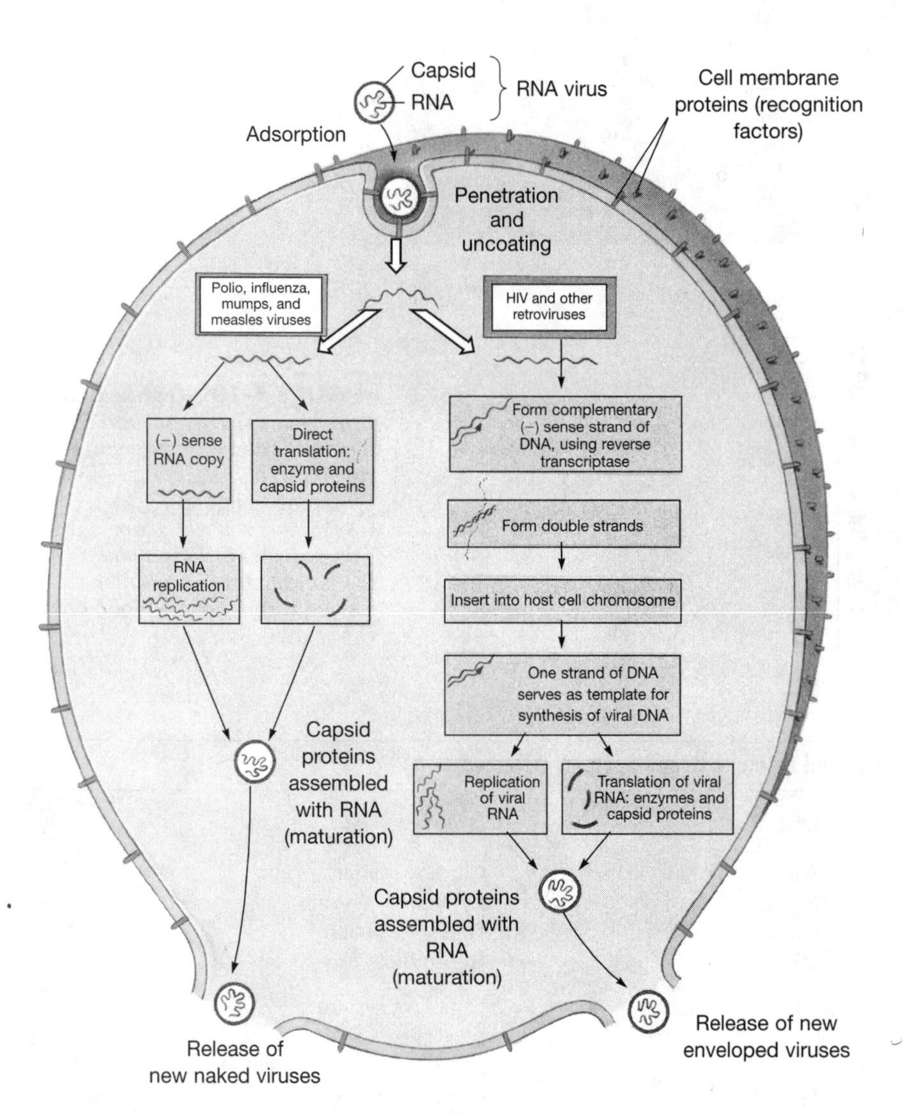

•**FIGURE A-9 Viral Replication *(continued)*. (b)** The replication of RNA viruses.

FAP Ch. 3

Other known prion diseases include *Creutzfeldt–Jakob disease* and *fatal familial insomnia.* Most cases of prion disease (kuru excepted) are the result of mutations in the normal gene that produce amyloid plaques. However, if a healthy person becomes exposed to these abnormal proteins, a prion disease will result.

Prion infection also occurs in domesticated animals. In sheep, the condition is called *scrapie;* in cows, it is called *bovine spongiform encephalopathy (BSE).* Infected cows ultimately develop an assortment of strange neurological symptoms (such as pawing at the ground and difficulty in walking), giving the condition the common name "mad cow disease." In 1995, European researchers attributed an upsurge in the incidence of Creutzfeldt–Jakob disease in humans to the consumption of meat from prion-infected cows. It remains unproven that the human infections resulted from infected beef. A number of fatal cases in England that showed brain changes similar to those of BSE, however, led to the temporary ban of British beef from the European community, the slaughter of potentially infected cows, and a change in livestock feeding practices throughout the world.

UNICELLULAR PATHOGENS AND MULTICELLULAR PARASITES

Bacteria and viruses are the best-known human pathogens, but some pathogens are eukaryotic. Examples of the most important types are included in Figure A-6c•. **Protozoa** are unicellular eukaryotes that are abundant in soil and water (Figure A-10•). They are responsible for a variety of serious human diseases, including *amoebic dysentery* and *malaria* (Table A-7). Protozoa include (1) *flagellates,* which use flagella for propulsion; (2) *amoeboids,* which include mobile,

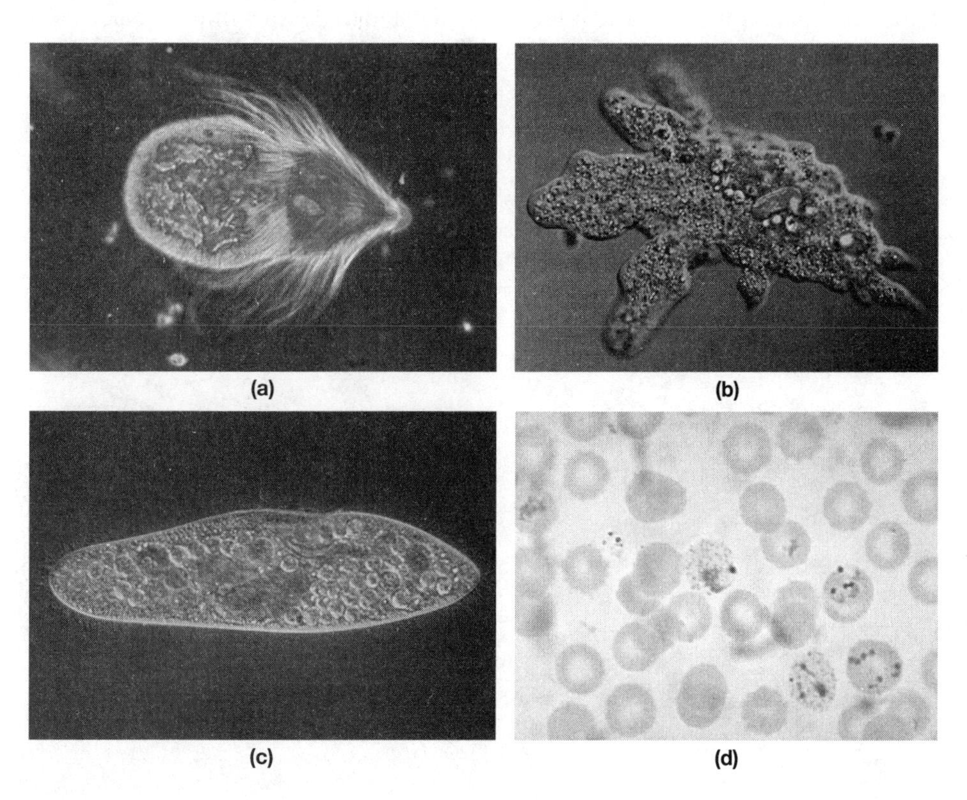

•**FIGURE A-10 Representative Protozoa.** **(a)** *Trichonympha,* a flagellate from a termite gut. **(b)** *Amoeba proteus,* a free-living form found in ponds. (LM × 310) **(c)** *Paramecium caudatum*, a free-living ciliate. **(d)** *Plasmodium vivax,* the parasite that causes malaria, stained within human blood cells. (LM × 125)

TABLE A-7 Examples of Protozoan Diseases and Primary Organ Systems Affected

Protozoa Type	Name (Genus)	Disease	Affected Organ System
Flagellates	*Trypanosoma*	African sleeping sickness	Cardiovascular system
	Leishmania	Leishmaniasis	Lymphatic system
	Giardia	Giardiasis	Digestive system
	Trichomonas	Trichomoniasis	Reproductive system
Amoeboids	*Entamoeba*	Amoebic dysentery	Digestive system
Ciliates	*Balantidium*	Dysentery	Digestive system
Sporozoans	*Plasmodium*	Malaria	Cardiovascular system
	Toxoplasma	Toxoplasmosis	Lymphatic system

amoeba-like forms that engulf their prey; (3) *ciliates,* which are covered with cilia; and (4) *sporozoans,* which are parasitic forms with complex life cycles. **Fungi** (singular, *fungus)* are eukaryotic organisms that absorb organic materials from the remains of dead cells. Mushrooms are familiar examples of very large fungi. In a *mycosis,* or fungal infection, a microscopic fungus spreads through living tissues, killing cells and absorbing nutrients. Several relatively common skin conditions (including *athlete's foot)* and a few more-serious diseases (including *histoplasmosis)* are the result of fungal infections (Table A-8).

Larger multicellular organisms (Figures A-6d and A-11•), generally referred to as *parasites,* can also invade the human body and cause diseases. The multiplication of these larger parasitic organisms in or on the body is called an **infestation**. Diseases caused by multicellular parasites are introduced in Table A-9. **Helminths** are parasitic worms that can live within the body. Helminths include **flatworms**, such as the *flukes* and *tapeworms,* and **roundworms**, or *nematodes.* These organisms, which range in size from microscopic flukes to tapeworms a meter or more in length, typically cause weakness and discomfort but do not *by themselves* kill their host. However, complications resulting from the parasitic infection, such as malnutrition, chronic bleeding, or secondary infections by bacterial or viral pathogens, can ultimately prove fatal.

TABLE A-8 Examples of Fungal Diseases and Primary Organ Systems Affected

Organism (Genus)	Disease	Affected Organ System
Aspergillus	Aspergillosis ("Farmer's lung disease")	Respiratory system
Blastomyces	Blastomycosis	Integumentary system
Histoplasma	Histoplasmosis	Respiratory system
Epidermophyton, Microsporum, and *Trichophyton*	Ringworm tinea capitis (scalp) tinea corporis (body) tinea cruris (groin) tinea unguium (nails)	Integumentary system
Candida	Candidiasis	Integumentary system
Coccidioides	Coccidioidomycosis ("San Joaquin valley fever")	Respiratory system

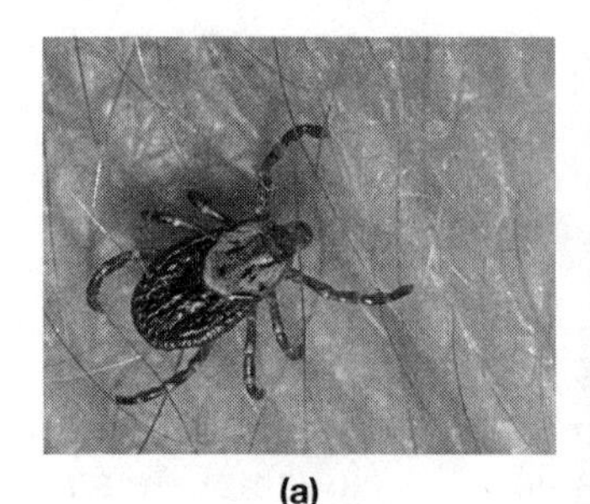

(a)

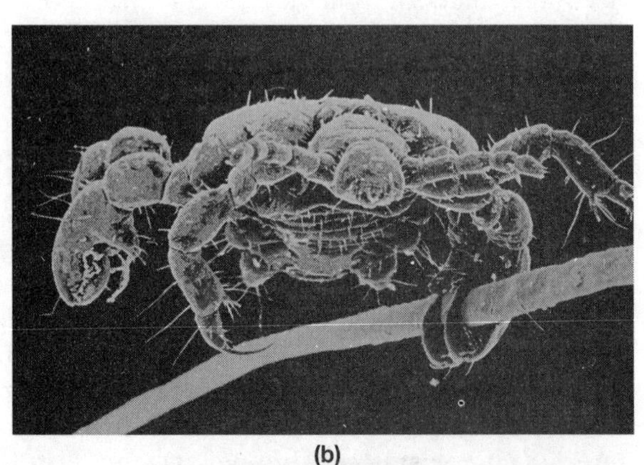

(b)

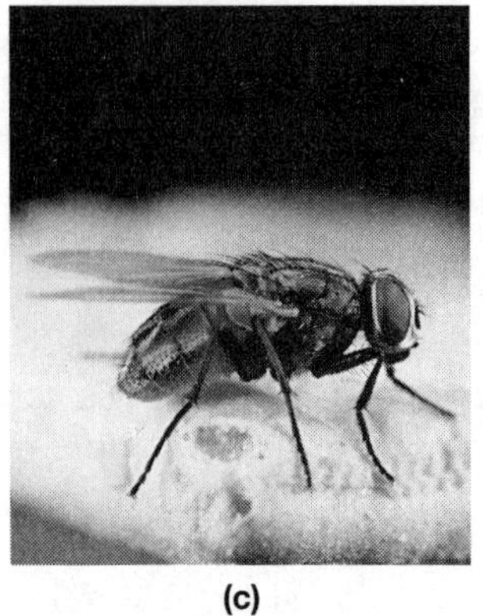

(c)

(d)

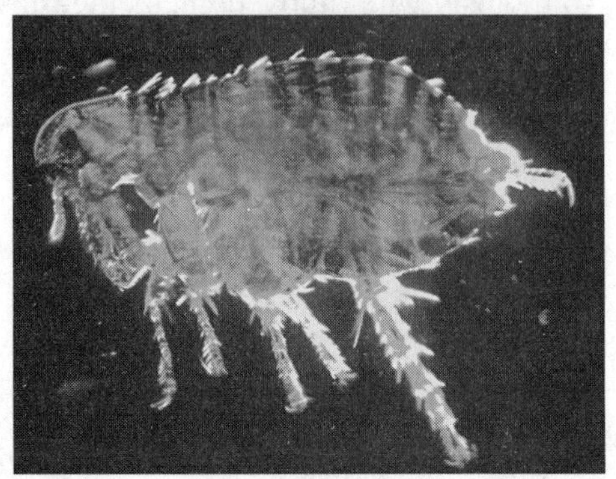

(e)

•FIGURE A-11 Representative Disease-Carrying Arthropods. **(a)** *Dermacentor andersoni,* a wood tick. **(b)** *Phthirus pubis,* a crab louse, holding onto a human pubic hair. (SEM × 55) **(c)** *Musca domestica,* the housefly, which can transport microbes on its body. (× 3) **(d)** The *Aedes* mosquito, a vector for dengue fever. **(e)** *Ctenocephalides canis,* a common flea. (× 31)

FAP Ch. 3

TABLE A-9 Examples of Diseases Caused by Multicellular Parasites and Primary Organ Systems Affected

Group	Organism	Disease/Condition	Affected Organ System
Helminths			
Roundworms	*Ascaris*	Intestinal infestation	Digestive system
	Enterobius	Pinworm infestation	Digestive system
Flatworms	*Wuchereria*	Elephantiasis	Lymphatic system
Flukes	*Fasciola, Clonorchis* (liver flukes)	Fascioliasis	Digestive system
	Schistosoma (blood fluke)	Schistosomiasis	Cardiovascular system
Tapeworms	*Taenia*	Tapeworm infestation	Digestive system
Arthropods			
Arachnids (eight legs)	Mites	Vectors of bacterial and rickettsial diseases	Various systems
	Ticks	Vectors of bacterial and rickettsial diseases	Various systems
	Spiders, scorpions	Inflammation from bites	Integumentary system
Insects (six legs)	Lice	Vectors of bacterial and rickettsial diseases	Various systems
	Human lice	Pediculosis	Integumentary system
	Mosquitoes	Vectors of bacterial and rickettsial diseases	Various systems
	Flies	Passive carriers of bacterial diseases	Various systems
	Wasps, bees	Inflammation from stings	Various systems

Arthropods make up the largest and most diverse group of animals on Earth. The major arthropods that affect humans are the *arachnids,* including scorpions, spiders, mites, and ticks, and the *insects,* such as mosquitoes, flies, lice, fleas, and bedbugs.

METHODS OF MICROANATOMY FAP *p. 65*

Over the last 50 years, our technological gadgetry has improved remarkably, enabling us to view the insides and outsides of cells in new ways. Sophisticated equipment has permitted the detailed analysis of physiological processes within cells. The basic problems facing cytologists stem from the considerable size difference between the investigator and the object of interest. Cytologists (cell biologists) and histologists (biologists who study tissues) measure intracellular structures in *micrometers* **(μm),** also known as **microns.** Although the range of cell sizes is considerable, an "average cell" is a cube roughly 10 μm × 10 μm × 10 μm. To fill a cubic millimeter, we would need a million cells. Because the human eye cannot recognize details smaller than about 0.1 mm, cytologists rely on special equipment that magnifies cells and their contents.

LIGHT MICROSCOPY

Historically, most information has been provided by **light microscopy,** a method in which a beam of light is passed through the object to be viewed. A light microscope can magnify cellular structures about 1000 times and can show details as fine as 0.25 μm. A camera can be attached to the microscope and used to produce a photograph called a **light micrograph (LM).** Unfortunately, you cannot simply pick up a cell, slap it onto a microscope slide, and take a photograph. Because individual cells are so small, you must work with large numbers of cells. Most tissues have a three-dimensional structure, and small pieces of tissue can be removed for examination. The component cells are prevented from decomposing by first exposing the tissue sample to a poison that will stop metabolic operations but will not alter cellular structures.

Even then, you still cannot look at the tissue sample through a light microscope, because a cube only 2 mm (0.078 in.) on a side will contain several million cells. You must slice the sample into thin sections. Living cells are relatively thick, and cellular contents are not transparent. Light can pass through the section only if the slices are thinner than the individual cells. Making a section that slender poses interesting technical problems. Most tissues are not very sturdy, so an attempt to slice a fresh piece will destroy the sample. (To appreciate the problem, try to slice a marshmallow into thin sections.) Thus, before you can make sections, you must embed the tissue sample in something that will make it more stable, such as wax, plastic, or epoxy. These materials will not interact with water molecules, so your sample must first be dehy-

drated (typically by immersion in 30 percent, 70 percent, 95 percent, and, finally, 100 percent alcohol). If you are embedding the sample in wax, the wax must be hot enough to melt; if you are using plastic or epoxy, the hardening process generates heat on its own.

After embedding the sample, you can section the block with a machine called a *microtome,* which uses a metal, glass, or diamond knife. For viewing by light microscopy, a typical section is about 5 μm (0.002 in.) thick. The thin sections are then placed on microscope slides. If the sample was embedded in wax, you can now remove the wax with a solvent, such as xylene. But you are not done yet. In thin sections, the cell contents are almost transparent; you cannot yet distinguish intracellular details by using an ordinary light microscope. You must first add color to the internal structures by treating the slides with special dyes called *stains.* Some stains are dissolved in water and others in alcohol. Not all types of cells pick up a given stain to the same degree, if they pick it up at all; nor do all types of cellular organelles. For example, in a sample scraped from the inside of the cheek, one stain might dye only certain types of bacteria; in a semen sample, another stain might dye only the flagella of the sperm. If you try too many stains at one time, they all run together, and you must start over. After the staining is completed, you can put coverslips over the sections (generally after you have dehydrated them again) and can see what your labors have accomplished.

Any single section can show you only a part of a cell or tissue. To reconstruct the tissue structure, you must look at a series of sections made one after the other. (See the related discussion in Chapter 1 of the text, *p. 19.*) After examining dozens or hundreds of sections, you can understand the structure of the cells and the organization of your tissue sample—or can you? Your reconstruction has left you with an understanding of what these cells look like after they have (1) died an unnatural death; (2) been dehydrated; (3) been impregnated with wax or plastic; (4) been sliced into thin sections; (5) been rehydrated, dehydrated, and stained with various chemicals; and (6) been viewed with the limitations of your equipment. A good cytologist or histologist is extremely careful, cautious, and self-critical, and much of the laboratory preparation is an art as well as a science.

ELECTRON MICROSCOPY

More-elaborate procedures can allow for the examination of finer details. In **electron microscopy,** a beam of electrons is passed through or reflected off the surface of a suitably prepared object. In **transmission electron microscopy,** the electrons pass through an ultra-thin section. Once through the section, they strike a photographic plate and produce an image known as a **transmission electron micrograph (TEM).** Transmission electron microscopy can magnify structures up to approximately 500,000 times, revealing details less than a nanometer in size. For instance, with a transmission electron microscope, you can visualize large organic molecules. In **scanning electron microscopy,** a beam of electrons reflects off the surface of an object such as a cell, a broken portion of a cell, or extracellular structures. (The surfaces are specially coated to enhance reflectivity.) After bouncing off the surface, the electrons strike a photographic plate to produce an image known as a **scanning electron micrograph (SEM).** Scanning electron microscopy can magnify structures up to only about 50,000 times but provides a three-dimensional perspective on cellular anatomy that cannot be obtained by other methods.

Such detail poses problems of its own. At the level of the light microscope, if you were to slice a large cell as you would slice a loaf of bread, you might produce 10 sections from the one cell. You could review the entire series under a light microscope in a few minutes. If you sliced the same cell for examination under an electron microscope, you would have 1000 sections, each of which could take several hours to inspect! Figure A-12a,b• compares SEM and TEM views of cells that line the intestinal tract, and Figure A-12c• shows a diagrammatic representation of the intact cell.

CELL STRUCTURE AND FUNCTION

Each cell in the body has a particular role to play in maintaining the integrity of the individual as a whole. Some conduct nerve impulses, and others manufacture hormones, build bones, or contract to produce body movements. When any of these cells malfunction, whether due to genetic abnormalities affecting enzyme function, a viral or bacterial infection, trauma, or cancer, homeostasis is threatened. This section introduces clinical and practical applications of basic principles of cellular function and discusses representative disorders resulting from problems at the cellular level of organization.

DRUGS AND THE CELL MEMBRANE FAP *p. 73*

Many clinically important drugs affect cell membranes. Although the mechanism behind the action of general anesthetics, such as *ether, chloroform,* and *nitrous oxide,* has yet to be determined, most are lipid-soluble hydrophobic molecules. The potency of an anesthetic is directly correlated with its lipid solubility. Lipid solubility may speed the drug's entry into cells and enhance its ability to block ion channels or change other properties of cell membranes. The most important clinical result is a reduction in the sensitivity and responsiveness of neurons and muscle cells.

Local anesthetics, such as *procaine* and *lidocaine,* as well as *alcohol* and *barbiturate* drugs, are also lipid-soluble. These compounds affect membrane properties by blocking sodium channels in the cell membranes

FAP Ch. 3

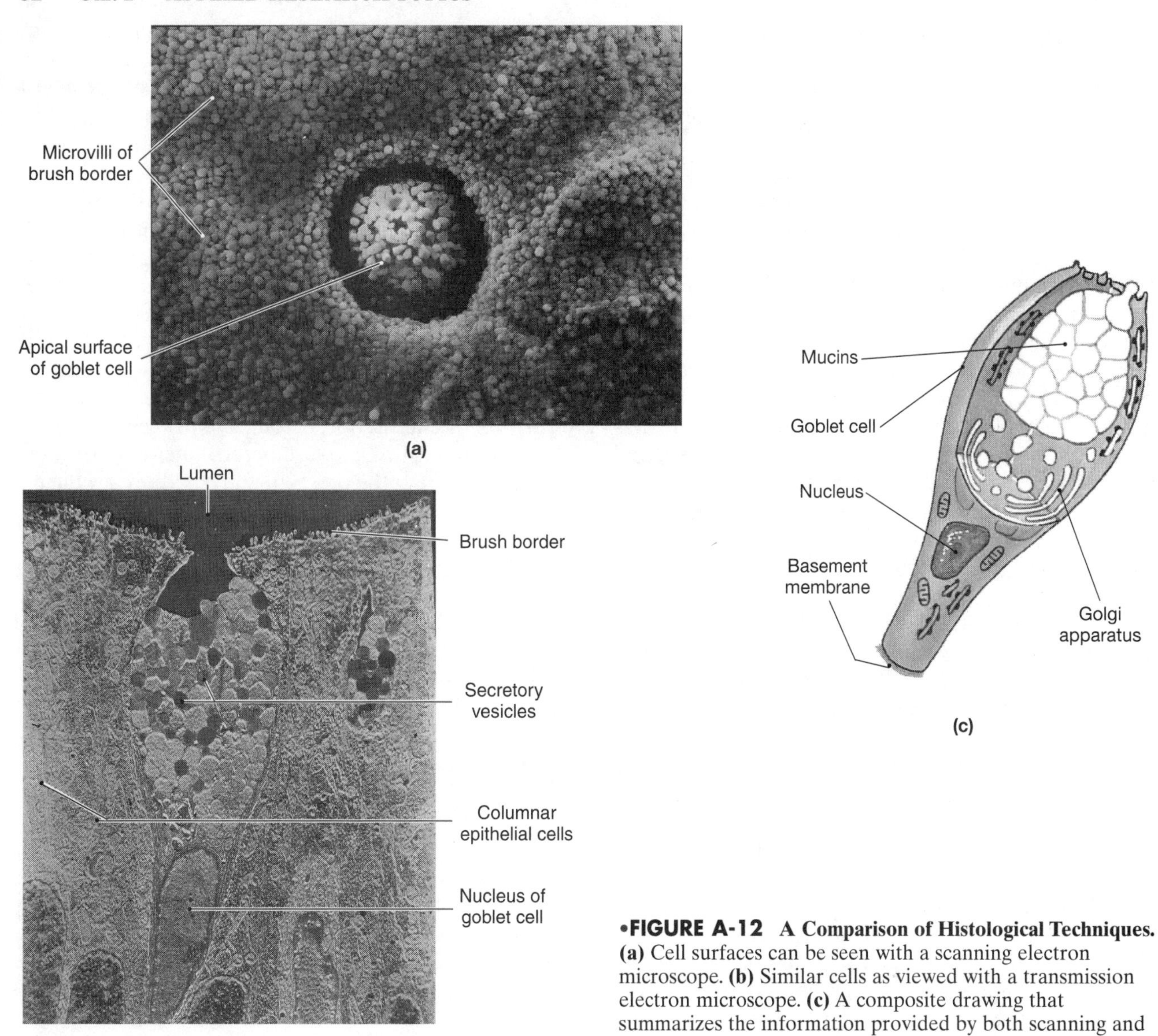

•FIGURE A-12 A Comparison of Histological Techniques. **(a)** Cell surfaces can be seen with a scanning electron microscope. **(b)** Similar cells as viewed with a transmission electron microscope. **(c)** A composite drawing that summarizes the information provided by both scanning and transmission electron microscopy.

of neurons. This blockage reduces or eliminates the responsiveness of neurons to painful (or any other) stimuli. The very powerful toxin *tetrodotoxin (TTX)* is found in some species of puffer-fish (family Tetraodontidae). Eating the internal organs of these fish causes a severe and potentially fatal form of food poisoning, marked by the disruption of normal neural and muscular activities. (Nevertheless, the flesh is considered a delicacy in Japan, where it is prepared by specially licensed chefs and served under the name *fugu.*)

Other drugs interfere with membrane receptors for hormones or chemicals that stimulate muscle or nerve cells. *Curare* is a plant extract that interferes with the chemical stimulation of muscle cell membranes. South American Indians use it to coat their hunting arrows so that wounded prey cannot run away. To prevent reflexive muscle contractions or twitches while surgery is being performed, anesthesiologists may administer a curare derivative (*d-tubocurarine* or related drugs) preoperatively to patients.

LYSOSOMAL STORAGE DISEASES FAP *p. 87*

Problems with lysosomal enzyme production cause more than 30 storage diseases that affect children. In these conditions, the lack of a specific lysosomal enzyme results in the buildup of materials normally removed and recycled by lysosomes. Eventually the cell cannot continue to function. We will consider three important examples here: Gaucher's disease, Tay-Sachs disease, and glycogen storage disease.

Gaucher's disease is caused by the buildup of *cerebrosides,* glycolipids in cell membranes. This is probably the most common type of lysosomal storage disease. The disease takes two forms: (1) an infantile form, marked by severe neurological symptoms end-

ing in death; and (2) a juvenile form, with enlargement of the spleen, anemia, pain, and relatively mild neurological symptoms. Gaucher's disease is most common among Ashkenazic Jews, at a frequency of approximately 1 in 1000 births.

Tay-Sachs disease is another hereditary disorder caused by the inability to break down glycolipids. In this case, the glycolipids are *gangliosides,* which are most abundant in neural tissue. Individuals with this condition develop seizures, blindness, dementia, and death, generally by age 3–4 years. Tay-Sachs disease is most common among the Ashkenazic Jews, at a frequency of 0.3 per 1000 births.

Glycogen storage disease (Type II) affects primarily skeletal muscle, cardiac muscle, and liver cells—the cells that synthesize and store glycogen. In this condition, the cells are unable to mobilize glycogen normally, and large numbers of insoluble glycogen granules accumulate in the cytoplasm. These granules disrupt the organization of the cytoskeleton, interfering with transport operations and the synthesis of materials. In skeletal and heart muscle cells, the buildup leads to muscular weakness and potentially fatal heart problems.

MITOCHONDRIAL DNA, DISEASE, AND EVOLUTION FAP *p. 89*

Several inheritable disorders result from abnormal mitochondrial activity. The mitochondria involved have defective enzymes that reduce their ability to generate ATP. Cells throughout the body may be affected, but symptoms involving muscle cells, neurons, and the receptor cells in the eye are most common because these cells have especially high energy demands, which are normally met by mitochondrial activity. Disorders caused by defective mitochondria are called *mitochondrial cytopathies.* In several instances, the disorders have been linked to inherited abnormalities in mitochondrial DNA. In some cases, the problem appears in one population of cells only. For example, abnormal mitochondrial DNA has been found in the motor neurons whose degeneration is responsible for *Parkinson's disease,* a neurological disorder characterized by a shuffling gait and uncontrollable tremors.

More commonly, mitochondria throughout the body are involved. Examples of conditions caused by mitochondrial dysfunction include one class of epilepsies *(myoclonic epilepsy)* and a type of blindness *(Leber's hereditary optic neuropathy).* These are inherited conditions, but the pattern of inheritance is very unusual. Although men or women may have the disease, only affected women can pass the condition on to their children. The explanation for this pattern is that the disorder results from an abnormality in the DNA of mitochondria, not in the DNA of cell nuclei. All the mitochondria in the body are produced by the replication of mitochondria present in the fertilized ovum. Few if any of those mitochondria were provided by the father; most of the mitochondria of the sperm do not remain intact after fertilization takes place. As a result, children can generally inherit these conditions only from their mother.

This brings us to an interesting concept: Virtually all your mitochondria were inherited from your mother, and hers from her mother, and so on back through time. The same is true for every other human. Now, it is known that small changes in DNA nucleotide sequences accumulate over long periods of time. Mitochondrial DNA, or *mDNA,* can therefore be used to estimate the degree of relatedness between individuals. The greater the difference between the mDNA of two individuals, the more time has passed since the lifetime of their most recent common ancestor and the more distant their relationship. On this basis, it has been estimated that all humans now alive shared a common female ancestor roughly 350,000 years ago. Appropriately, that individual has been called "Mitochondrial Eve." The existence and history of Mitochondrial Eve remain controversial.

[EXPLORE]

Additional resources, such as interactive tutorials, critical-thinking questions, clinical problems, and case studies, are available through the CD-ROM and Website for your textbook. To begin your exploration, launch the *Martini Interactive* CD-ROM or visit the Companion Website (http://www.prenhall.com/martini/fap5) and select Chapter 3.

Topics in Molecular Biology

Molecular biology is the study of the synthesis, structure, and function of macromolecules important to life, such as proteins and nucleic acids. Deciphering the genetic code and relating the intricate structure of a protein to its functions are major goals of molecular biology. Research in this area has greatly enhanced our understanding of normal functions as well as disease processes.

The field of molecular biology has revolutionized the study of medicine by uncovering a clear biochemical basis for many complex diseases. For example, in *sickle cell anemia* red blood cells undergo changes in shape that lead to blocked blood vessels and subsequent tissue damage due to oxygen starvation. This condition results when an individual carries two copies of a defective gene that determines the structure of *hemoglobin,* the oxygen-binding protein in red blood cells. The genetic defect changes just 2 of the 574 amino acids in this protein. That is enough to alter the functional properties of the hemoglobin molecule, leading to changes in the properties of the red blood cells. This type of disorder is often called a *molecular disease,* because it results from abnormalities at the molecular level of organization.

Roughly 2500 inherited disorders have now been identified, and researchers have located the defective genes responsible for cystic fibrosis, Duchenne's muscular dystrophy, and Tay-Sachs disease. Identifying the genetic defect is the vital first step toward the development of an effective gene therapy or other treatment. The treatments that are now evolving make use of the principles of *genetic engineering.*

GENETIC ENGINEERING AND GENE THERAPY

FAP *pp. 97, 788*

Once the mechanics of the genetic code were understood, everyone realized that it would be theoretically possible to change the genetic makeup of organisms—perhaps even of humans. The popular term for activities related to this goal is **genetic engineering**.

What are some of the key problems confronting genetic engineers? Genes code for proteins; the makeup of each protein is determined by the sequence of codons (nucleotide triplets) in a stretch of DNA. A human cell has 46 chromosomes, 2 meters of DNA, and roughly 10^9 triplets. If all the DNA in the human body were extracted and strung together, the resulting strand would be long enough to make several hundred round trips between Earth and the sun. Simply finding a particular gene among the 100,000 to 140,000 that each of us carries is an imposing task. Yet, before a specific gene can be modified, its location must be determined with great precision. Locating a gene involves preparing a map of the appropriate chromosome.

MAPPING THE HUMAN GENOME

Several techniques can be used to create a map of the chromosomes. **Karyotyping** (KAR-ē-ō-tī-ping; *karyon,* nucleus + *typos,* a mark) is the determination of an individual's chromosome complement. Figure A-13a• shows a set of normal human chromosomes. Each chromosome has characteristic banding patterns, and segments can be stained with special dyes. Unusual banding patterns can indicate structural abnormalities. These abnormalities are sometimes linked to specific inherited conditions, including a form of leukemia. *Down syndrome* results from the presence of an extra copy of chromosome 21 (Figure A-13b•).

In December 1993, French researchers at the Centre d'Etude de Polymorphism Humaine (CEPH, or Center for the Study of Human Polymorphism) completed the first preliminary mapping of the entire human genome. This mapping provided the landmarks and reference points needed to make more precise maps that indicate the locations of specific genes.

Mapping is useful but it is only an intermediate step on the way to the determination of the nucleotide sequence of every gene in the human genome. As of March 2000, more than 38,000 genes had been identified, with roughly 11,000 of them assigned to specific chromosomes. These numbers are a small percentage of the total, but the pace of identification and localization is increasing rapidly. The sequencing process by the Human Genome Project, which is expected to cost $2 billion and to be completed by the year 2003, will provide basic information about the location of genes on normal human chromosomes.

GENE MANIPULATION AND MODIFICATION

Suppose that the location of a defective gene has been pinpointed. Before attempting to remedy the defect in an individual, a clinician would have to determine the nature of the genetic abnormality. For example, the

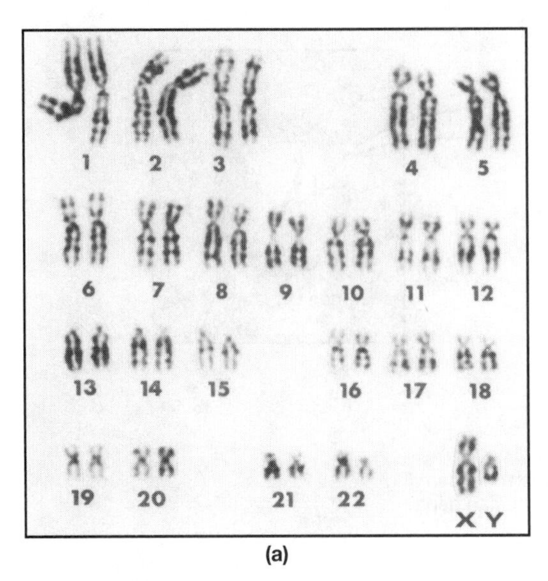

(a)

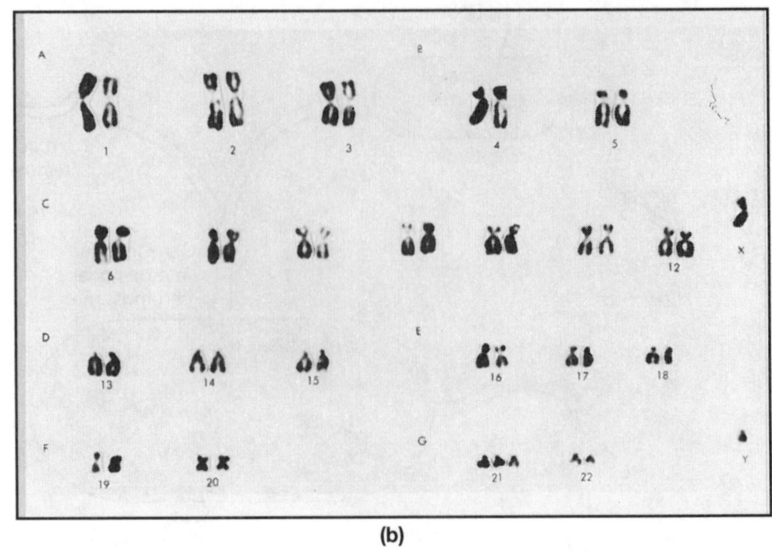

(b)

•**FIGURE A-13 Normal and Abnormal Karyotypes. (a)** A micrograph of the normal human chromosome set; the chromosomes have been arranged in this sequence for ease of comparison. **(b)** The chromosomes of an individual with Down syndrome. Notice the extra copy of chromosome 21.

gene could be inactive or overactive, or it could produce an abnormal protein. It could even be missing. Only after understanding what the problem is could the clinician try to decide how to remedy the defect. Can the gene be turned on, turned off, modified, or replaced?

What's the problem? This can be a particularly difficult question to answer. Many of the 2500 inheritable genetic disorders are classified according to general patterns of symptoms rather than any specific protein or enzyme deficiency. In some cases, the approximate location of the gene has been determined but the identity of the protein responsible for the clinical symptoms remains a mystery. A protein may have several possible abnormalities that can result in different patterns of clinical disease.

What can be done? If the gene is present but overproducing or underproducing, its activity might be controlled by introducing chemical repressors or inducers. Another approach relies on gene splicing to produce a protein that is missing or present in inadequate quantities. **Gene splicing** begins with the localization of the gene, followed by its isolation. A "healthy" copy of that gene is then spliced into the relatively simple DNA strand of a bacterium, creating *recombinant DNA (rDNA)* (Figure A-14•). Bacteria grow and reproduce rapidly under laboratory conditions, and before long a colony of identical bacteria has formed. All the members of the colony carry the introduced gene and will manufacture the corresponding protein. The protein can be extracted, concentrated, and administered to individuals who are deficient in the activity of that gene. *Hemophilia* (a deficiency of blood clotting factors) and a form of diabetes caused by insulin deficiency can be treated in this way.

Gene splicing is also used to obtain large quantities of proteins that normally are present in very small concentrations. *Interferon,* an antiviral protein, and human growth hormone are compounds now being produced commercially by means of gene-splicing technology.

The most revolutionary strategies involve "fixing" abnormal cells by giving them copies of normal genes. In general, this method poses significant targeting problems, because the gene must be introduced into the right kind of cell. For example, placing liver enzymes in fingernails would not correct a metabolic disorder. But when the target cells can be removed and isolated, as in the case of bone marrow, the technique is promising. The removal of a defective gene does not appear to be a practical approach, and the focus has been on adding genes that can take over normal functions.

In September 1990, the first gene therapy trials were initiated. The procedure was used to treat a 4-year-old girl who had *adenosine deaminase deficiency (ADA).* A rare condition, ADA affects only about 20 children worldwide each year. Without this enzyme, toxic chemicals build up in cells of the immune system. As these cells die, the body's defenses break down.

ADA results in a complex of symptoms known as *severe combined immunodeficiency disease,* or *SCID.* SCID can also be caused by other enzyme disorders affecting cells of the immune system. Symptoms include chronic respiratory infections, diarrhea, and a low resistance to viral or bacterial infections. Most children with ADA die from infections that would pose no threat to healthy children. A new drug called *PEG-ADA,* an altered form of the missing enzyme, can prolong life, but it does not cure the condition.

In the 1990 clinical trial, blood cells were collected and *lymphocytes,* the primary cells of the immune system, were removed. Short segments of DNA containing the normal gene for the production of adenosine deaminase were then inserted into the nuclei of these

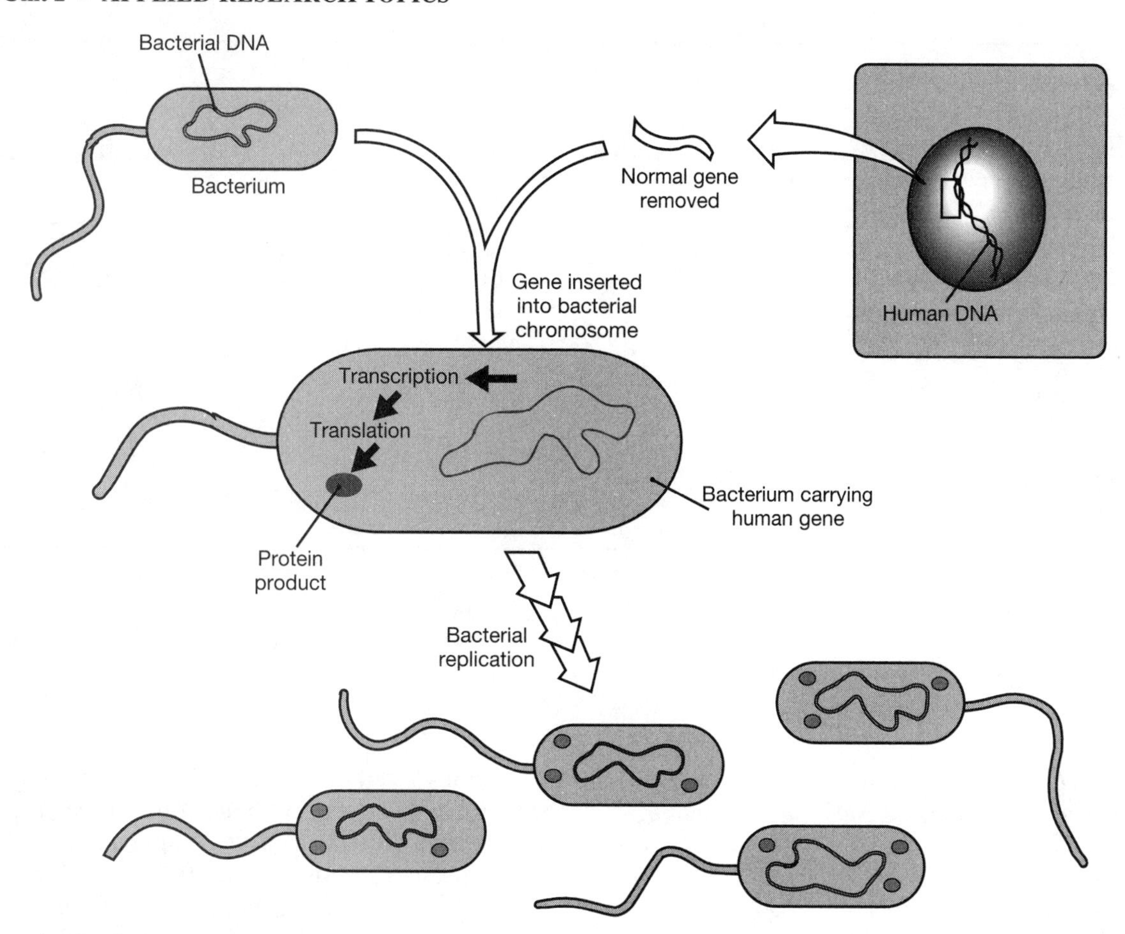

•FIGURE A-14 Gene Splicing. A gene is removed from a human cell nucleus and attached to the DNA in a bacterium, where it directs the production of a human protein. Bacterial replication creates a colony of bacteria that share the introduced gene and can yield large quantities of the protein product.

cells, and the modified lymphocytes were returned to the body. Roughly a billion modified cells were reintroduced. Over time, these modified cells divided to produce a large population of normal immune cells. The experiment was a success, and this method has been used to treat several other children with ADA. Each has regained and retained immune function, although concern remains that the engineered lymphocytes might not recognize and attack as great a variety of antigens (foreign compounds, pathogens, or toxins) as can the lymphocytes of normal individuals. The treatment must be repeated periodically (in the original case, at 3–5 month intervals). In more-recent experiments on individuals with ADA, the genes for bone marrow stem cell production were modified; these patients regained apparently normal immune function without the need for repetitive treatments.

The major limitation of gene therapy for disorders other than ADA, which involves blood cells, has been the lack of a reliable mechanism for the insertion of genes into specific cells inside the body. Viruses are an option, but potential pathogens as delivery vehicles of genetic material are a source of concern. However, the use of viruses to deliver genes to defective cells has been approved for clinical trials for the treatment of about 20 disorders. Instead of viruses, small lipid packets called *liposomes* can deliver genes to cells that line the respiratory tract; they are administered in a fine mist that can be inhaled. As is not unusual for a new method of treatment, there have been more failures than successes so far. For example, recent attempts to treat cystic fibrosis and a common form of muscular dystrophy by liposome inhalation have been unsuccessful.

These procedures attempt to relieve the symptoms of disease by inserting genes into defective somatic cells. The new genes do not change the genetic structure of reproductive cells; because oocytes and sperm retain the original genetic pattern, the genetic defect would be passed to any future generations. Researchers are much further away from practical methods of changing the genetic characteristics of reproductive cells. Mouse eggs fertilized outside the body have been treated and transplanted into the uterus of a second mouse for development. The gene added was one for a growth hormone obtained from a rat, and the large "supermouse" that resulted demonstrated that such manipulations can be performed. The possibilities of manipulating the characteristics of valuable animal stocks, such as cattle, sheep, or chickens, are quite exciting. The potential for al-

tering the genetic characteristics of humans is intimidating. Before any clinical variations on this theme are tested, our society will have to come to grips with some difficult ethical issues.

To get a sense of the kinds of problems we might have to deal with, discuss the following questions with your friends and classmates.

GENETIC ENGINEERING—QUESTIONS TO THINK ABOUT

1. Your 51-year-old father has recently been diagnosed with a hereditary disorder that affects the brain. The prognosis is poor, and the disorder currently has no cure. The physician advises that you be tested for the presence of the faulty gene.
   - Would you be tested? Would you want to know if you had the defective gene?
   - If you were tested, who should have access to the results: Your insurance company? Your children, spouse, or fiancé? Your employer?
2. *Eugenics* is the control of the hereditary characteristics of individuals to improve the species. The study of eugenics became distorted by the work of scientists under Adolf Hitler's control. Prenatal testing now permits the diagnosis of a variety of inherited disorders before birth. This information could be used to "improve the species" by selectively terminating pregnancies. Is this practice advisable or ethical?
3. Many scientists, such as Nobel laureate James Watson, consider the human genome to be the blueprint for a human. If an individual's genetic code determines every characteristic of that person, can people accused of crimes be held legally responsible for their actions?
4. Studies in the past have shown that men with an extra Y chromosome are more violent and predisposed to crime than are normal men. More-recent studies have revealed no greater tendency toward violence among males with an extra Y chromosome. A prejudice against such individuals still exists, even though the original studies are known to be seriously flawed.
   - What type of controls are needed to ensure that new information about genetic abnormalities is not released before that information is confirmed?
   - Not every person with a specific genotype will develop the same characteristics to the same degree. How can possible stereotypes be avoided when information is released?
5. You are an airline employer trying to offer your employees the least-expensive health insurance available. The insurance company requires a blood test on each potential employee to determine genetic abnormalities. You learn that a candidate for a job as a pilot has a genetic predisposition for a heart attack. Would this information affect your decision to hire that individual?

A CLOSER LOOK: CANCER FAP *pp. 101, 136*

FAP Ch. 3 Ch. 4

Twenty-five percent of all people in the United States develop cancer during their lifetime. In 1999, roughly 563,100 people in the United States died of some form of cancer, making it second only to heart disease.

CANCER CAUSES

Relatively few types of cancer are inherited; 18 hereditary types have been identified to date, including two forms of leukemia. Most cancers develop through the interaction of genetic and environmental factors, and it is difficult to separate the two completely.

Genetic Factors

Two related genetic factors are involved in the development of cancer: (1) hereditary predisposition and (2) oncogene activation.

An individual born with genes that increase the likelihood of cancer is said to have a *hereditary predisposition* for the disease. Such a person may never develop cancer, but his or her chances are higher than average. The inherited genes generally affect the abilities of tissues to metabolize toxins, control mitosis and growth, perform repairs after injury, or identify and destroy abnormal tissue cells. As a result, body cells become sensitive to local or environmental factors that would have little effect on normal tissues.

Cancers can also result from somatic-cell mutations that modify genes involved with cell growth, differentiation, or mitosis. As a result, an ordinary cell is converted into a cancer cell. The modified genes are called **oncogenes** (ON-kō-jēnz); the normal genes are called **proto-oncogenes.** *Oncogene activation* occurs by the alteration of normal somatic genes. Because these mutations do not affect reproductive cells, the cancers caused by active oncogenes are not inherited.

A proto-oncogene, like other genes, has a regulatory component that turns the gene "on" and "off" and a structural component that contains the mRNA triplets that determine protein structure. Mutations in either portion of the gene may convert it to an active oncogene. A change of just one nucleotide out of a chain of 5000 can convert a normal proto-oncogene to an active oncogene. In some cases, a viral infection can trigger the activation of an oncogene. For example, one of the human papilloma viruses appears to be responsible for many cases of cervical cancer.

More than 50 proto-oncogenes have been identified. In addition, a group of anticancer genes has been discovered. These genes, called *tumor-suppressing genes (TSGs),* or *anti-oncogenes,* suppress division and growth in normal cells. Mutations that alter TSGs make oncogene activation more likely. Such mutation has been suggested as an important factor in promoting several cancers, including several blood cell cancers, breast cancer, and ovarian cancer. Examples of important suppressor genes include the genes *p53*

FAP
Ch. 3
Ch. 4

and *p16*. Mutations affecting the *p53* gene are responsible for the majority of cancers of the colon, breast, and liver. Abnormal *p16* gene activity may be involved in as many as half of all cancer cases.

Environmental Factors

Many cancers can be directly or indirectly attributed to environmental factors called **carcinogens** (kar-SIN-ō-jenz). Carcinogens stimulate the conversion of a normal cell to a cancer cell. Some carcinogens are *mutagens* (MŪ-ta-jenz)—that is, they damage DNA strands and may cause chromosomal breakage. Radiation is a mutagen that has carcinogenic effects.

The environment contains many chemical carcinogens. Plants manufacture poisons that protect them from insects and other predators, and although their carcinogenic activities are often relatively weak, many common spices, vegetables, and beverages contain compounds that are carcinogens if consumed in large quantities. Animal tissues may also store or concentrate toxins, and hazardous compounds of many kinds can be swallowed in contaminated food. A variety of laboratory and industrial chemicals, such as coal tar derivatives and synthetic pesticides, have been shown to be carcinogenic. Cosmic radiation, X-rays, UV radiation, and other radiation sources can also cause cancer. It has been estimated that 70–80 percent of all cancers are the result of chemical or environmental factors, and 40 percent are due to a single stimulus: cigarette smoke.

Specific carcinogens will affect only those cells capable of responding to that particular physical or chemical stimulus. The responses vary because differentiation produces cell types with specific sensitivities. For example, benzene can produce a cancer of the blood; cigarette smoke, a lung cancer; and vinyl chloride, a liver cancer. Very few stimuli can produce cancers throughout the body. Radiation exposure is a notable exception. In general, cells undergoing mitosis are most likely to be vulnerable to chemical or radiational carcinogens. As a result, cancer rates are highest in epithelial tissues, where stem cell divisions occur rapidly, and lowest in nervous and muscle tissues, where divisions do not normally occur.

DETECTION AND INCIDENCE OF CANCER

Physicians who specialize in the identification and treatment of cancers are called **oncologists** (on-KOL-ō-jists; *onkos,* mass). Pathologists and oncologists classify cancers according to their cellular appearance and their sites of origin. More than a hundred kinds have been described, but broad categories are usually used to indicate the location of the primary tumor. A tumor is defined as a "new growth" resulting from uncontrolled cell division. A tumor can be *malignant* or *benign* and may *metastasize* (spread) rapidly or very slowly. Only malignant tumors are called cancers. Table A-10 summarizes information about benign and malignant tumors (cancers) associated with the major tissues of the body.

TABLE A-10 Benign and Malignant Tumors in the Major Tissue Types

Tissue	Description
Epithelia	
Carcinoma	Any cancer of epithelial origin
Adenocarcinomas	Cancers of glandular epithelia
Angiosarcomas	Cancers of endothelial cells
Mesotheliomas	Cancers of mesothelial cells
Connective tissues	
Fibromas	Benign tumors of fibroblast origin
Lipomas	Benign tumors of adipose tissue
Liposarcomas	Cancers of adipose tissue
Leukemias	Cancers of blood-forming tissues
Lymphomas	Cancers of lymphoid tissues
Chondromas	Benign tumors in cartilage
Chondrosarcomas	Cancers of cartilage
Osteomas	Benign tumors in bone
Osteosarcomas	Cancers of bone
Muscle tissues	
Myxomas	Benign muscle tumors
Myosarcomas	Cancers of skeletal muscle tissue
Cardiac sarcomas	Cancers of cardiac muscle tissue
Leiomyomas	Benign tumors of smooth muscle tissue
Leiomyosarcomas	Cancers of smooth muscle tissue
Neural tissues	
Gliomas	Cancers of neuroglial origin
Neuromas	Cancers of neuronal origin

A statistical profile of cancer incidence and survival rates in the United States is shown in Table A-11. The numbers from other countries are different. For example, *bladder cancer* is common in Egypt, *stomach cancer* in Japan, and *liver cancer* in Africa. Variations in the combination of genetic factors and dietary, infectious, and other environmental factors are thought to be responsible for these differences.

CLINICAL STAGING AND TUMOR GRADING

The detection of a cancer often begins during a routine physical examination, when the physician notices an abnormal lump or growth. Many laboratory and diagnostic tests are necessary for the correct diagnosis of cancer. Information is usually obtained by the examination of a tissue sample, or *biopsy,* typically supplemented by medical imaging and blood studies. A biopsy is one of the most significant diagnostic procedures because it permits a direct look at the tumor cells. Not only do malignant cells have an abnormal-

TABLE A-11 Cancer Incidence and Survival Rates in the United States

Site	Estimated New Cases (2000)	Estimated Deaths (2000)	5-Year Survival Rates: Diagnosis Date 1970–73	5-Year Survival Rates: Diagnosis Date 1989–94
Digestive tract				
Esophagus	12,300	12,100	4%	12%
Stomach	21,500	13,000	13%	21%
Colon and rectum	130,200	56,300	47%	62%
Respiratory tract				
Lung and bronchus	164,100	156,900	10%	14%
Urinary tract				
Kidney and renal pelvis	31,200	11,900	46%	59%*
Urinary bladder	53,200	12,200	61%	82%
Reproductive system				
Breast	184,200	41,200	68%	85%
Ovary	23,100	14,000	36%	50%
Testis	6900	300	72%	95%
Prostate gland	180,400	31,900	63%	93%
Nervous system	16,500	13,000	20%	28%*
Circulatory system	199,800	109,600	22%	41%*
Skin (melanoma only)	47,700	7700	68%	88%

Data courtesy of the American Cancer Society
*These data are for 1986–1991; not reported for 1989–1994.

ly high rate of mitosis but they are also structurally distinct from healthy body cells.

If the tissue appears cancerous, other important questions must be answered, including the following: What is the measurable size of the primary tumor? Has the tumor invaded surrounding tissues? Has the cancer already metastasized to develop secondary tumors? Are any regional lymph nodes affected? The answers to these questions are combined with observations from the physical exam, the biopsy results, and information from any imaging procedures to develop an accurate diagnosis and prognosis.

In an attempt to develop a standard system, national and international cancer organizations have developed the *TNM system* for staging (that is, identifying the stage of progression of) cancers. The letters stand for *tumor* (T) size and invasion, *lymph node* (N) involvement, and degree of *metastasis* (M):

- Tumor size is graded on a scale of 0 to 4. T0 indicates the absence of a primary tumor, and the largest dimensions and greatest amount of invasion are categorized as T4.
- Lymph nodes filter the tissue fluids from nearby capillary beds. The fluid, called *lymph,* then returns to the general lymphatic circulation. Once cancer cells have entered the lymphatic system, they can spread very quickly throughout the body. Lymph node involvement is graded on a scale of 0 to 3. A designation of N0 indicates that no lymph nodes have been invaded by cancer cells. A classification of N1 to N3 indicates the involvement of increasing numbers of lymph nodes:

  N1 indicates the involvement of a single lymph node less than 3 cm in diameter.

  N2 includes one medium-sized (3–6 cm) node or multiple nodes smaller than 6 cm.

  N3 indicates the presence of a single lymph node larger than 6 cm in diameter, whether or not other nodes are involved.
- Metastasis is graded on a scale of 0 to 1. M0 indicates that there is no evidence of metastasis, whereas M1 indicates that the cancer cells have produced secondary tumors in other portions of the body.

This grading system provides a general overview of the progression of the disease. For example, a tumor classified as T1N1M0 has a better prognosis than one classified as T4N2M1. The latter tumor would be much more difficult to treat. The grading system alone does not provide all the information needed to plan treatment, however, because different types of cancer progress in different ways. Therapies must vary

FAP
Ch. 3
Ch. 4

accordingly. Thus *leukemia,* a cancer of the blood-forming tissues, is treated differently than colon cancer. We will consider specific treatments in discussions dealing with cancers that affect individual body systems. The next section provides a general overview of the strategies used to treat cancer.

CANCER TREATMENT

It is unfortunate that the media tend to describe cancer as though it were one disease rather than many. This simplistic perspective fosters the belief that some dietary change, air ionizer, or wonder drug will be found that can prevent or cure the affliction. No single, universally effective cure for cancer is possible, because there are too many separate causes, possible mechanisms, and individual differences.

The goal of cancer treatment is to achieve remission. A tumor in **remission** either ceases to grow or decreases in size. The treatment of malignant tumors must accomplish one of these two objectives to produce remission:

1. *The surgical removal or destruction of individual tumors.* Tumors containing malignant cells can be surgically removed or destroyed by radiation, heat, or freezing. These techniques are very effective if the treatment is undertaken before extensive metastasis has occurred. For this reason, early detection is important in improving survival rates for all forms of cancer.
2. *The killing of metastasized cells throughout the body.* This is much more difficult and potentially dangerous, because healthy tissues are likely to be damaged at the same time. At present, the most widely approved treatments are chemotherapy and radiation.

Chemotherapy involves the administration of drugs that will either kill the cancerous tissues or prevent mitotic divisions. These drugs typically affect stem cells in normal tissues, and the side effects are usually unpleasant. For example, because chemotherapy slows the regeneration and maintenance of epithelia of the skin and digestive tract, patients lose their hair and experience nausea and vomiting. Several drugs are often administered simultaneously or in sequence, because, over time, cancer cells can develop a resistance to a single drug. Chemotherapy is often used in the treatment of many kinds of metastasized cancer.

Massive doses of radiation are sometimes used to treat advanced cases of *lymphoma,* a cancer of the immune system. In this rather drastic procedure, enough radiation is administered to kill all the blood-forming cells in the body. After treatment, new blood cells must be provided by a bone marrow transplant. In later sections dealing with the lymphatic system, we will discuss marrow transplants, lymphomas, and other cancers of the blood.

An understanding of molecular mechanisms and cell biology is leading to new approaches that may revolutionize cancer treatment. One approach focuses on the fact that cancer cells are ignored by the immune system. In **immunotherapy,** chemicals are administered that help the immune system recognize and attack cancer cells. More-elaborate experimental procedures involve the creation of customized antibodies by the gene-splicing techniques discussed on p. 35. The resulting antibodies are specifically designed to attack the tumor cells in each particular patient. Although this technique shows promise, it remains difficult, costly, and very labor-intensive.

A second procedure builds on the first. In **boron neutron capture therapy (BNCT),** antibodies made to attack cancer cells are labeled with an isotope of boron (B). After these antibodies are administered, the patient is irradiated with neutrons. These neutrons do not damage normal tissues. However, the boron atoms absorb neutrons and release alpha particles (two neutrons + two protons). This radiation kills the cancer cells quite effectively. Because the cancer cells absorb the radiation, healthy tissues are unaffected.

CANCER AND SURVIVAL

Advances in chemotherapy, radiation procedures, and molecular biology have produced significant improvements in the survival rates of several types of cancer. However, the improved survival rates indicated in Table A-11 reflect advances not only in therapy but also in early detection. Much of the credit goes to increased public awareness and concern about cancer. In general, the odds of survival increase markedly if the cancer is detected early, especially before it undergoes metastasis. Despite the variety of possible cancers, the American Cancer Society has identified seven "warning signs" that mean it's time to consult a physician. These are presented in Table A-12.

TABLE A-12 Seven Warning Signs of Cancer

Seven Warning Signs of Cancer
Change in bowel or bladder habits
A sore that does not heal
Unusual bleeding or discharge
Thickening or lump in breast or elsewhere
Indigestion or difficulty in swallowing
Obvious change in a wart or mole
Nagging cough or hoarseness

[EXPLORE]

Additional resources, such as interactive tutorials, critical-thinking questions, clinical problems, and case studies, are available through the CD-ROM and Website for your textbook. To begin your exploration, launch the *Martini Interactive* CD-ROM or visit the Companion Website (http://www.prenhall.com/martini/fap5) and select Chapter 3 or 4.

UNIT 3

The Body Systems: Clinical and Applied Topics

This section relates aspects of the normal anatomy and physiology of each body system to specific clinical conditions, diagnostic procedures, and other relevant topics. We will briefly review each body system in order to identify patterns among the most common disorders that affect each system. Critical-Thinking Questions about each system and Clinical Problems involving multiple systems are located on your Companion Website. These exercises will give you the chance to apply the concepts you have learned here.

The Integumentary System

The structures of the integumentary system include skin, hair, nails, and several types of exocrine glands. The integumentary system has a variety of functions, including the protection of the underlying tissues, the maintenance of body temperature, the excretion of salts and water in sweat, cutaneous sensation, and the production of vitamin D_3.

The skin is the most visible organ of the body. As a result, abnormalities are easily recognized. A bruise, for example, typically creates a swollen and discolored area where the walls of blood vessels have been damaged. Changes in skin color, skin tone, and the overall condition of the skin commonly accompany illness or disease. These changes can assist in diagnosis. For example, extensive bruising without obvious cause may indicate a blood-clotting disorder; yellowish skin and mucous membranes may signify *jaundice,* a sign that generally indicates some type of liver disorder. The general condition of the skin can also be significant. In addition to color changes, for example, changes in skin flexibility, elasticity, dryness, or sensitivity commonly follow the malfunctions of other organ systems.

EXAMINATION OF THE SKIN FAP *p. 144*

When examining a patient, dermatologists use a combination of investigative interviews ("What has been in contact with your skin lately?" or "How does it feel?") and physical examination to arrive at a diagnosis. The condition of the skin is carefully observed. Notes are made about the presence of **lesions,** which are changes in skin structure caused by disease processes. Lesions are also called **skin signs,** because they are measurable, visible abnormalities of the skin surface. Figure A-15• diagrams the most common skin signs and related disorders.

The distribution of lesions can be an important clue to the source of the problem. For example, in *shingles* (*herpes zoster*), painful vesicular eruptions on the skin follow the path of peripheral sensory nerves. A ring of slightly raised, scaly (papular) lesions is typical of fungal infections that may affect the trunk, scalp, or nails. Examples of skin infections and allergic reactions are included in Table A-13, with descriptions of the related skin signs. We consider skin lesions caused by trauma in the section titled "Trauma to the Skin" (p. 48).

Table A-13 considers signs on the skin surface, but signs involving the accessory organs of the skin can also be important. For instance:

- Nails have a characteristic shape that can change due to an underlying disorder. An example is *clubbing* of the fingernails, commonly a sign of *emphysema* or *congestive heart failure.* In these conditions, the fingertips broaden and the nails become distinctively curved.
- The condition of the hair can be an indicator of overall health. For example, depigmentation and coarseness of hair occur in the protein deficiency disease *kwashiorkor.*

DIAGNOSING SKIN CONDITIONS

Table A-14 (p. 44) introduces several major types of vascular lesions. A single vascular lesion, such as a *hematoma,* can have multiple causes. This is one of the challenges facing dermatologists; the signs may be apparent, but the underlying causes may not. Making matters more difficult, many skin disorders produce the same signs of uncomfortable sensations. For example, **pruritus** (proo-RĪ-tus), an irritating itching sensation, is an extremely common symptom associated with a variety of skin conditions. Questions about medical history, medications, possible sources of infection, environmental exposure, and other signs, such as bleeding at the gums, can be the key to making an accurate diagnosis.

Pain is another common symptom of many skin disorders. Although pain is unwelcome, cutaneous sensation is an important function of the integumentary system. Its importance is dramatically demonstrated in the condition *Hansen's disease,* or *leprosy.* Hansen's disease is caused by a bacterium that has an affinity for cool regions of the body. The bacterium destroys cutaneous nerve endings that are sensitive to touch, pain, heat, and cold. Damage to the extremities then occurs and accumulates, because the individual is no longer aware of painful stimuli. We will consider Hansen's disease further in a later section.

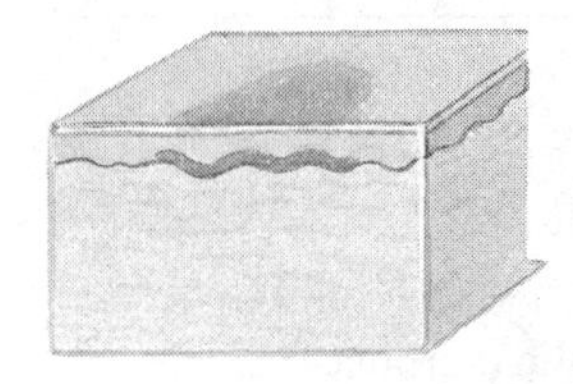

A flat **macule** is a localized change in skin color. Example: freckles

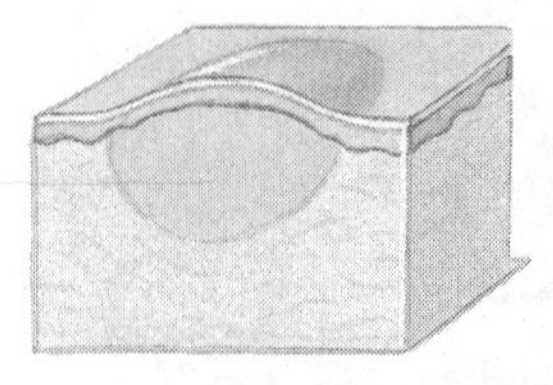

Accumulation of fluid in the papillary dermis may produce a **wheal**, a localized elevation of the overlying epidermis. Example: hives

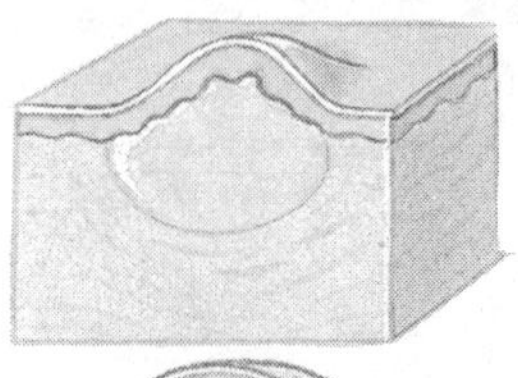

A **papule** is a solid elevated area containing epidermal and papillary dermal components. Example: mosquito or other insect bite

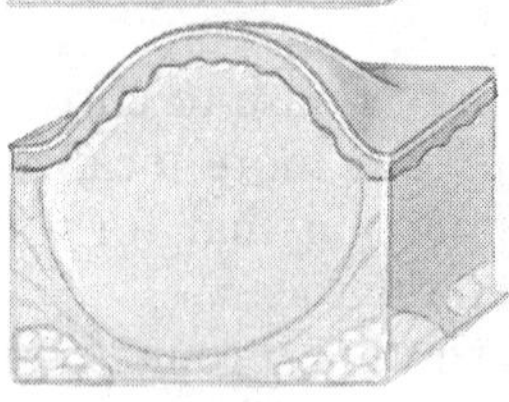

Nodules are large papules that may extend into the subcutaneous layer. Example: cyst

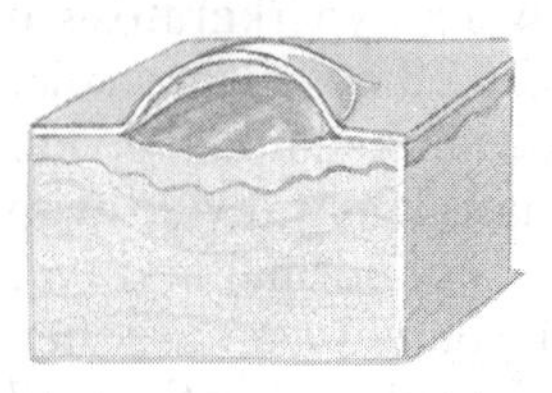

A **vesicle**, or blister, is a papule with a fluid core. A large vesicle may be called a bulla. Example: second-degree burn

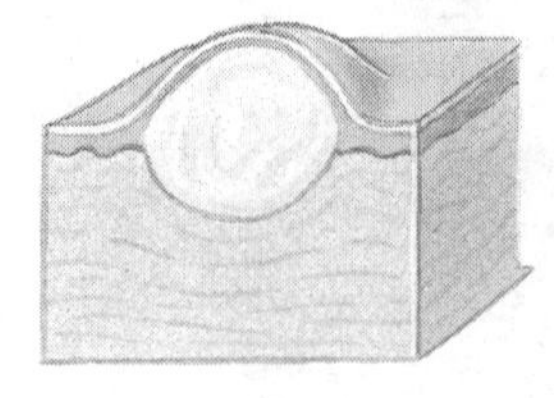

A **pustule** is a papule-sized lesion filled with pus. Example: acne pimple

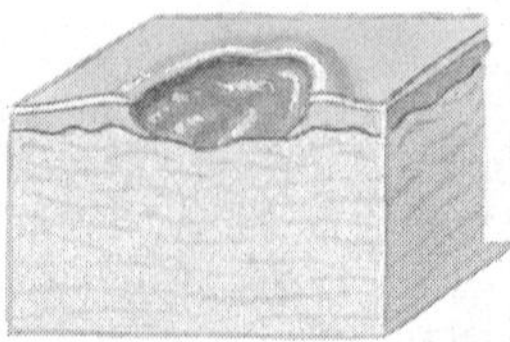

An **erosion**, or ulcer, may occur following the rupture of a vesicle or pustule. Eroded sites have lost part or all of the normal epidermis. Example: decubitis ulcer

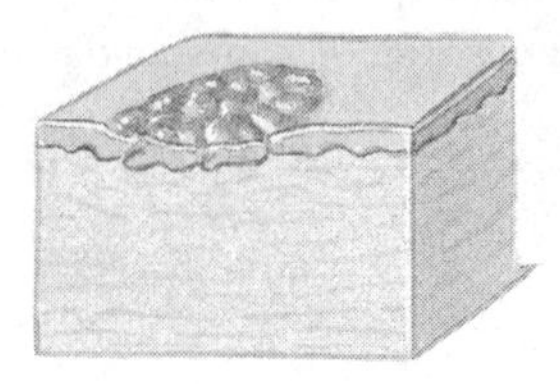

A **crust** is an accumulation of dried sebum, blood, or interstitial fluid over the surface of the epidermis. Examples: seborrheic dermatitis, scabs, impetigo

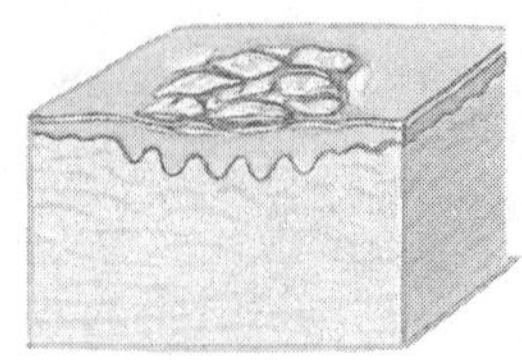

Scales form as a result of abnormal keratinization. They are thin plates of cornified cells. Example: psoriasis

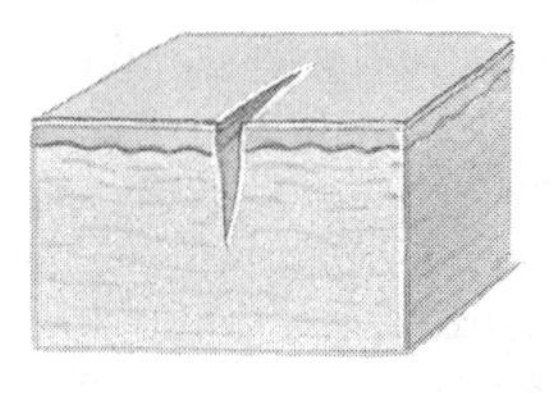

A **fissure** is a split in the integument that extends through the epidermis and into the dermis. Example: athlete's foot

•**FIGURE A-15** Skin Signs

TABLE A-13 Skin Signs of Various Disorders

Cause	Examples	Resulting Skin Lesions
Viral infections	Chickenpox	Lesions begin as macules and papules but develop into vesicles.
	Measles (rubeola)	A maculopapular rash that begins at the face and neck and spreads to the trunk and limbs
	Erythema infectiosum (fifth disease)	A maculopapular rash that begins on the cheeks (slapped-cheek appearance) and spreads to the limbs
	Herpes simplex	Raised vesicles that heal with a crust
Bacterial infections	Impetigo	Vesiculopustular lesions with exudate and yellow crusting
Fungal infections	Ringworm	An annulus (ring) of scaly papular lesions with central clearing
Parasitic infections	Scabies	Linear burrows with a small papule at one end
	Lice (pediculosis)	Dermatitis: excoriation (scratches) due to pruritus (itching)
Allergies to medication	Penicillin	Wheals (urticaria or hives)
Food allergies	Eggs, certain fruits	Wheals
Environmental allergies	Poison ivy	Vesicles

FAP
Ch. 5

TABLE A-14 Examples of Vascular Lesions

Lesion	Features	Some Possible Causes
Ecchymosis	Reddish purple, blue, or yellow bruising related to trauma	Trauma; blood-clotting disorder if bruising is abnormal; some vitamin deficiencies; thrombocytopenia; increased tendency to bruise can be due to aging or sun damage
Hematoma	Pooling of blood, forming a mass; associated with pain and swelling	Broken blood vessel
Petechiae	Small red to purple pinpoint dots appearing in clusters	Leukemia; septicemia (toxins in blood); thrombocytopenia
Erythema	Red, flushed color of skin due to dilation of blood vessels in the skin	Extensive: drug reactions Localized: burns, dermatitis

Figure A-16• is an overview of skin disorders. Diagnostic tests that may prove useful in distinguishing among them include the following:

- *The scraping of affected tissue*, a process often performed to check for fungal infections
- *The culturing of bacteria* removed from a lesion to aid in identification and to determine drug sensitivity
- *Biopsy* of affected tissue to view cell structure
- *Skin tests,* through which various types of disorders can be detected. In a skin test, a localized area of the skin is exposed to an inactivated pathogen, a portion of a pathogen, or a substance capable of producing an allergic reaction in sensitive individuals. Exposure occurs by injection or surface application. For example, in a skin test for *tuberculosis,* a small quantity of tuberculosis antigens is injected *intradermally* (*intra,* within). If the individual has been infected in the past or currently has tuberculosis, *erythema* (*erythros,* red; a change in skin color) and swelling will occur at the injection site 24–72 hours later. *Patch testing* is used to check sensitivity to *allergens,* environmental agents that can cause allergic reactions. In a patch test, the allergen is applied to the skin surface. If erythema, swelling, and/or itching develop, the individual is sensitive to that allergen.

ENVIRONMENTAL STRESS, INFLAMMATION, AND INFECTIONS OF THE SKIN

FAP *p. 147*

Disorders of Keratin Production

Not all skin signs are the result of infection, trauma, or allergy. Some skin signs are the normal response to environmental stresses. One common response is the excessive production of keratin. This process is called **hyperkeratosis** (hī-per-ker-a-TŌ-sis). The most obvious effects are easily observed as calluses and corns. Calluses are thickened patches that appear on already thick-skinned areas, such as the palms of the hands or the soles or heels of the foot, in response to chronic abrasion and distortion. Corns are more-localized areas of excessive keratin production that form in areas of thin skin on or between the toes.

In **psoriasis** (so-RĪ-a-sis), the stratum germinativum becomes unusually active, causing hyperkeratosis in specific areas, including the scalp, elbows, palms, soles, groin, or nails. Normally, an individual stem cell divides once every 20 days, but in psoriasis it may divide every day and a half. Keratinization is abnormal and typically incomplete by the time the outer layers are shed. The affected areas appear to be covered with small, silvery scales that continuously flake off. Psoriasis develops in 20–30 percent of the individuals with an inherited tendency for the condition. Roughly 5 percent of the U.S. population has psoriasis to some degree, often triggered by stress and anxiety. Most cases are painless and treatable.

Xerosis (ze-RŌ-sis), or dry skin, is a common complaint of the elderly and people who live in arid climates. Under these conditions, cell membranes in the outer layers of skin gradually deteriorate and the stratum corneum becomes more a collection of scales than a single sheet. The scaly surface is much more permeable than an intact layer of keratin, and the rate of insensible perspiration increases. In persons with severe xerosis, the rate of insensible perspiration may increase by up to 75 times.

Pressure on the skin, another form of stress, can produce *decubitis ulcers,* or *bedsores. Decubitis* means "to lie down"; an *ulcer* is a localized loss of epithelium. Decubitis ulcers form where dermal blood vessels are compressed against deeper structures such as bones or joints (FAP, *p. 151*).

Inflammation and Infections of the Skin

Inflammation is a complex process that helps defend against pathogens and injury. Because skin contains an abundance of sensory receptors, inflammation can be very painful. The various forms of **dermatitis** are dis-

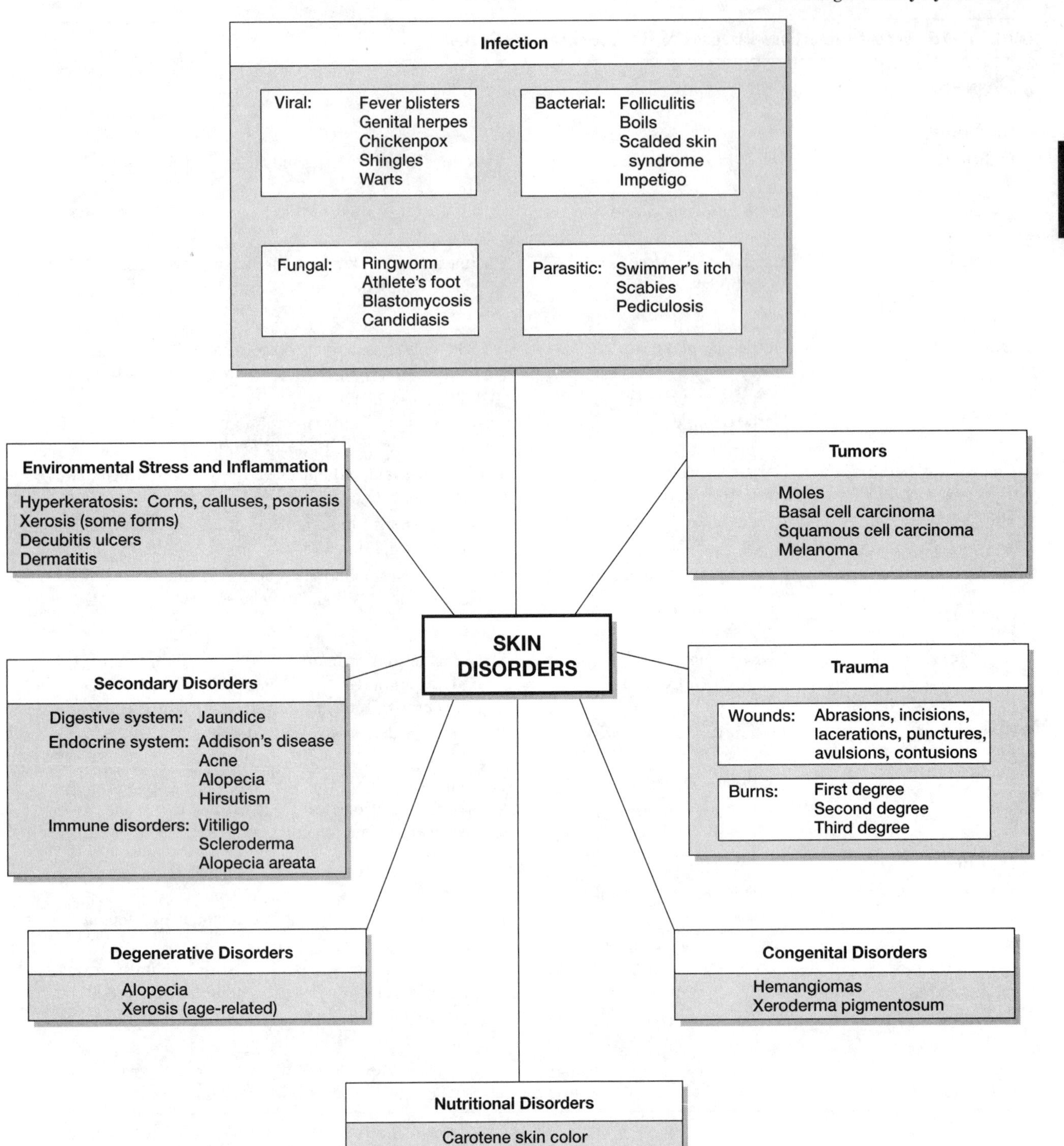

•**FIGURE A-16** **Disorders of the Integumentary System**

cussed in the text *(p. 152)*. Table A-15 summarizes information about common infections of the integumentary system.

SKIN TUMORS AND SKIN CANCERS

FAP *p. 148*

Almost everyone has several benign tumors of the skin; freckles and moles are examples (FAP, *p. 149*). Less common, but equally harmless, are benign tumors affecting dermal circulation, such as capillary hemangiomas (FAP, *p. 151*). *Skin cancers,* which are more dangerous, are the most common form of cancer.

An **actinic keratosis** is a scaly area that appears on sun-damaged skin. It is an indication that sun damage is occurring, but it is not a sign of skin cancer. In contrast, **basal cell carcinoma** is a malignant cancer

FAP Ch. 5

TABLE A-15 Common Infectious Diseases of the Integumentary System

Disease	Organism (Name)	Description
Bacteria		
Folliculitis	*Staphylococcus aureus*	Infections of hair follicles may form pimples, furuncles (boils), and abscesses.
Scalded skin syndrome	*Staphylococcus aureus*	In infants; large areas of skin blister, peel off, and leave wet, red areas.
Impetigo	*Staphylococci*, *Streptococci*, or both	Pustules form on skin, dry, and become crusts; skin pigment may not reappear after healing.
Viruses		
Oral herpes	Herpes simplex 1	Vesicles (blisters), also called cold sores, form on lips (fever blisters) and hands; vesicles disappear but may reappear at various times.
Genital herpes	Herpes simplex 2	Lesions similar to those in oral herpes form on external genitalia; vesicles disappear and reappear.
Chickenpox (*varicella*)	Herpes varicella-zoster	In children; small, red macules form vesicles, which dry and become crusts.
Shingles (*zoster*)	Herpes varicella-zoster	In adults; lesions form a pattern, usually on trunk; severe pain often follows attack.
Warts	Human papillomaviruses	Dermal warts form in the epidermis; genital warts may become malignant and may be sexually transmitted.
Fungi		
Ringworm (*tinea*)	*Epidermophyton*, *Microsporum*, and *Trichophyton*	Dry, scaly lesions form on the skin in different parts of the body: scalp *(tinea capitis)*, body *(tinea corporis)*, groin *(tinea cruris)*, foot *(tinea pedis)*, and nails *(tinea unguium)*.
Blastomycosis	*Blastomyces dermatitidis*	Pustules and abscesses form in the skin; may affect other organs.
Candidiasis	*Candida albicans*	Normal inhabitant of the human body; may infect many organs; red lesions form in skin infections; nails may also become infected.
Parasites		
Swimmer's itch	*Schistosoma* worms (flukes)	Freshwater larval stages of schistosome worms (flukes) burrow into skin and cause itching.
Scabies	*Sarcoptes scabiei* (itch mite)	Itch mite burrows and lays eggs in skin in areas between fingers and at the wrists, armpits, and genitals; entrance marked by tiny, scaly swellings that become red and itchy.
Pediculosis	*Pediculus humanus* (human body louse)	Lice infestations on body and scalp; bites produce redness, dermatitis, and itching.
	Phthirus pubis (pubic louse)	"Crabs"; lice infestation of the pubic area; their bites produce intense itching.

that originates in the stratum germinativum (basal layer) (Figure A-17a•). This form is the most common skin cancer. Roughly two-thirds of these cancers appear in body areas subjected to chronic UV exposure. Genetic factors have been identified that predispose people to this condition. **Squamous cell carcinomas** are less common but almost totally restricted to areas of sun-exposed skin. Metastasis seldom occurs in squamous cell carcinomas and virtually never in basal cell carcinomas, and most people survive these cancers. The usual treatment involves the surgical removal of the tumor, and 95 percent of patients survive 5 years or more after treatment. (This statistic, the *5-year survival rate*, is a common method of reporting long-term prognoses.)

Compared with these common and seldom life-threatening cancers, **malignant melanomas** (mel-a-NŌ-maz) (Figure A-17b•) are extremely dangerous. In this condition, cancerous melanocytes grow rapidly and metastasize through the lymphatic system. The outlook for long-term survival changes dramatically, depending on when the condition is diagnosed. If the cancer is localized, the 5-year survival rate is 99 percent; if it is widespread, the survival rate drops to 14 percent.

To detect melanoma at an early stage, you must examine your skin and you must know what to look for.

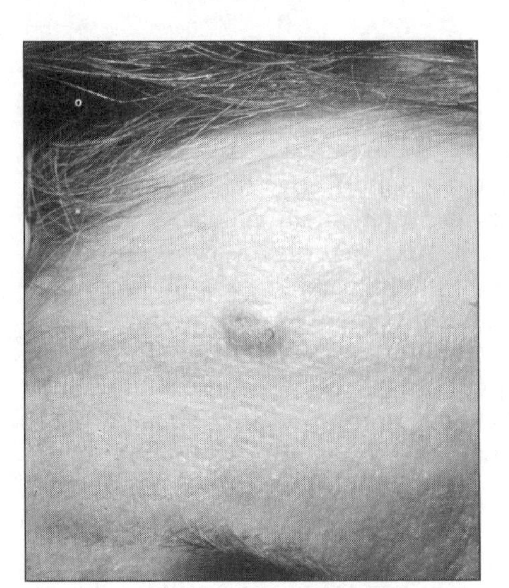

(a)

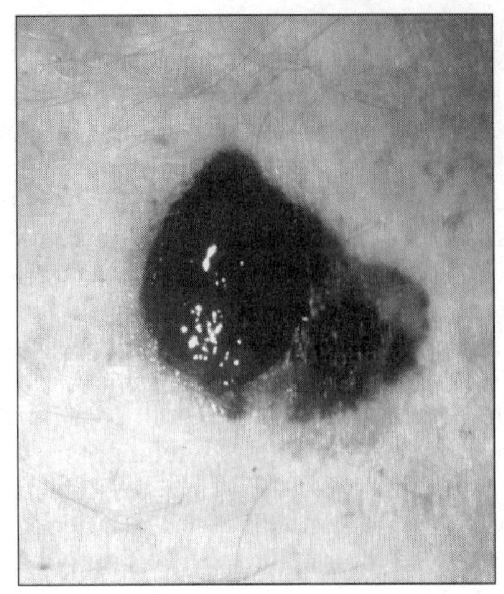

(b)

•**FIGURE A-17 Skin Cancers.** **(a)** Basal cell carcinoma. **(b)** Melanoma.

FAP Ch. 5

The mnemonic ABCD makes it easy to remember the key points of detection:

A is for *asymmetry:* Melanomas tend to be irregular in shape. Typically they are raised; they may ooze or bleed.

B is for *border:* The edges are generally unclear, irregular, and in some cases notched.

C is for *color:* Melanomas are generally mottled, with many colors (tan, brown, black, red, pink, white, and/or blue).

D is for *diameter:* A growth more than about 5 mm (0.2 in.) in diameter, or roughly the area covered by the eraser on a pencil, is dangerous.

A new experimental treatment for melanoma uses genetic engineering technology to manufacture antibodies that target the MSH (melanocyte-stimulating hormone) receptors on the surfaces of melanocytes. Melanocytes coated with these antibodies are then recognized and attacked by cells of the immune system.

Fair-skinned individuals who live in the tropics are most susceptible to all forms of skin cancer, because their melanocytes are unable to shield them from the UV radiation. Sun damage can be prevented by avoiding exposure to the sun during the middle of the day and by using a sunblock (not a tanning oil)—a practice that also delays the cosmetic problems of aging and wrinkling. *Everyone* who spends any time out in the sun should choose a broad-spectrum sunblock with a sun protection factor (SPF) of at least 15; blondes, redheads, and people with very fair skin are better off with an SPF of 20 to 30. (The risks are the same for those who spend time in a tanning salon or tanning bed.) Wearing a hat with a brim and panels to shield the neck and face provides added protection.

The use of sunscreens will be even more important as the ozone gas in the upper atmosphere is further destroyed by industrial emissions. Ozone absorbs UV radiation before it reaches Earth's surface; in doing so, ozone assists the melanocytes in preventing skin cancer. Australia, the continent that is most affected by the depletion of ozone above the South Pole (the "ozone hole"), is already reporting an increased incidence of skin cancers.

SECONDARY DISORDERS OF THE SKIN

Several diseases that have primary impacts on other systems secondarily affect the skin. For example, skin color can change as a result of disorders in both the digestive and endocrine systems. The endocrine system can also alter skin condition through its effects on hair follicles and sebaceous glands.

BALDNESS AND HIRSUTISM FAP *p. 155*

Two factors interact to determine baldness. A bald individual has a genetic susceptibility triggered by large quantities of male sex hormones. Many women carry the genetic background for baldness, but unless major hormonal abnormalities develop, as in certain endocrine tumors, nothing happens. (In some women, however, hair thins after menopause.)

Male pattern baldness affects the top of the head and forehead first, only later reducing the hair density along the sides. Thus hair follicles can be removed from the sides and implanted on the top or front of the head, temporarily delaying a receding hairline. This procedure is expensive, and not every hair transplant is successful.

Alopecia areata (al-ō-PĒ-shē-uh ar-ē-AH-ta) is a localized hair loss that can affect either gender. The cause is not known, and the severity of hair loss varies from case to case. This condition is associated with

FAP Ch. 5

several disorders of the immune system. It has also been suggested that stress can promote alopecia areata in individuals who are genetically prone to baldness.

Hairs are dead, keratinized structures, so no amount of oiling, shampooing, or dousing with kelp extracts, vitamins, or nutrients will influence the follicle buried in the dermis. Skin conditions that affect follicles can contribute to hair loss. Temporary baldness can also result from exposure to radiation or to many of the toxic (poisonous) drugs used in cancer therapy.

Untested treatments for baldness were banned by the Food and Drug Administration (FDA) in 1984. When rubbed onto the scalp, *Minoxidil,* a drug originally marketed for the control of blood pressure, appears to stimulate inactive follicles. It is available on a nonprescription basis *(Rogaine™).* Treatment involves applying a 2 percent solution to the scalp twice daily. It is most effective in preventing the progression of early hair loss, and it must be continued indefinitely.

Hirsutism (HER-soot-izm; *hirsutus,* bristly) is hair growth on women that occurs in patterns generally characteristic of men. Because considerable overlap exists between the two genders in terms of normal hair distribution, and because racial and genetic differences are significant, the precise definition is more often a matter of personal taste than of objective analysis. Age and sex hormones may play a role, because hairiness increases late in pregnancy and menopause produces a change in body hair patterns.

Severe hirsutism is commonly associated with abnormal androgen (male sex hormone) production, in either the ovaries or the adrenal glands. Unwanted follicles can be permanently "turned off" by plucking a growing hair and removing the papilla. *Electrocautery,* which destroys the follicle with a jolt of electricity, requires the services of a professional, but the results are more reliable. Patients may also be treated with drugs that reduce or prevent androgen stimulation of the follicles.

ACNE

FAP p. 156

Individuals with a genetic tendency toward acne have larger-than-average sebaceous glands. When the ducts of these glands become blocked, their secretions accumulate, inflammation develops, and bacterial infection can occur. The condition generally surfaces at puberty, when sex hormone production accelerates and stimulates the sebaceous glands. Their secretory output may be further encouraged by anxiety, stress, physical exertion, certain foods, or drugs.

The visible signs of acne are called **comedos** (ko-MĒ-dōz). Closed comedos ("whiteheads") contain accumulated, stagnant secretions. Open comedos ("blackheads") contain more-solid material that has been invaded by bacteria. Although neither condition indicates the presence of dirt in the pores, washing may help keep superficial oiliness down.

Acne generally fades after sex hormone concentrations stabilize. Topical antibiotics; vitamin A derivatives, such as *Retin-A™;* or peeling agents may help reduce inflammation and minimize scarring. In cases of severe acne, the most effective treatment normally involves the discouragement of bacteria by the administration of antibiotic drugs. However, oral antibiotic therapy has risks, including the development of antibiotic-resistant bacteria, so this therapy is not used unless other treatment methods have failed. Truly dramatic improvements in severe cases have been obtained with the prescription drug *Accutane™.* This compound is structurally similar to vitamin A, and it reduces oil gland activity on a long-term basis. A number of minor side effects, such as dry skin rashes, have been reported; these apparently disappear when the treatment ends. However, the use of Accutane during the first month of pregnancy carries a 25-times-normal risk of inducing birth defects, so women must avoid pregnancy while using this drug.

TRAUMA TO THE SKIN

FAP p. 158

Trauma is a physical injury caused by pressure, impact, distortion, or other mechanical force. Trauma to the skin is very common, and a number of terms are used to describe it. Such injuries generally affect all components of the integument, and each type of wound presents a different series of problems to clinicians attempting to limit damage and promote healing.

An **open wound** is an injury producing a break in the epithelium. The major categories of open wounds are illustrated in Figure A-18•. **Abrasions** are the result of scraping against a solid object (Figure A-18a•). Bleeding may be slight, but a considerable area may be open to invasion by microorganisms. **Incisions** are linear cuts produced by sharp objects (Figure A-18b•). Bleeding can be severe if deep vessels are damaged. The bleeding may help flush the wound, and closing the incision with bandages or stitches can limit the area open to infection while healing is underway. A **laceration** is a jagged, irregular tear in the surface produced by solid impact or by an irregular object (Figure A-18c•). Tissue damage is extensive, and repositioning the opposing sides of the injury may be difficult. Despite the bleeding that generally occurs, lacerations are prone to infection. **Punctures** result when slender, pointed objects pierce the epithelium (Figure A-18d•). Little bleeding results, and any microbes delivered under the epithelium in the process are likely to find conditions to their liking. In an **avulsion**, chunks of tissue are torn away by brute force in, for example, an auto accident or an explosion (Figure A-18e•). Bleeding may be considerable, and even more serious internal damage may be present.

Closed wounds can affect any internal tissue, but because the epithelium is intact, the likelihood of infection is reduced. A **contusion** is a bruise causing bleeding in the dermis. "Black and blue" marks are fa-

FAP
Ch. 5

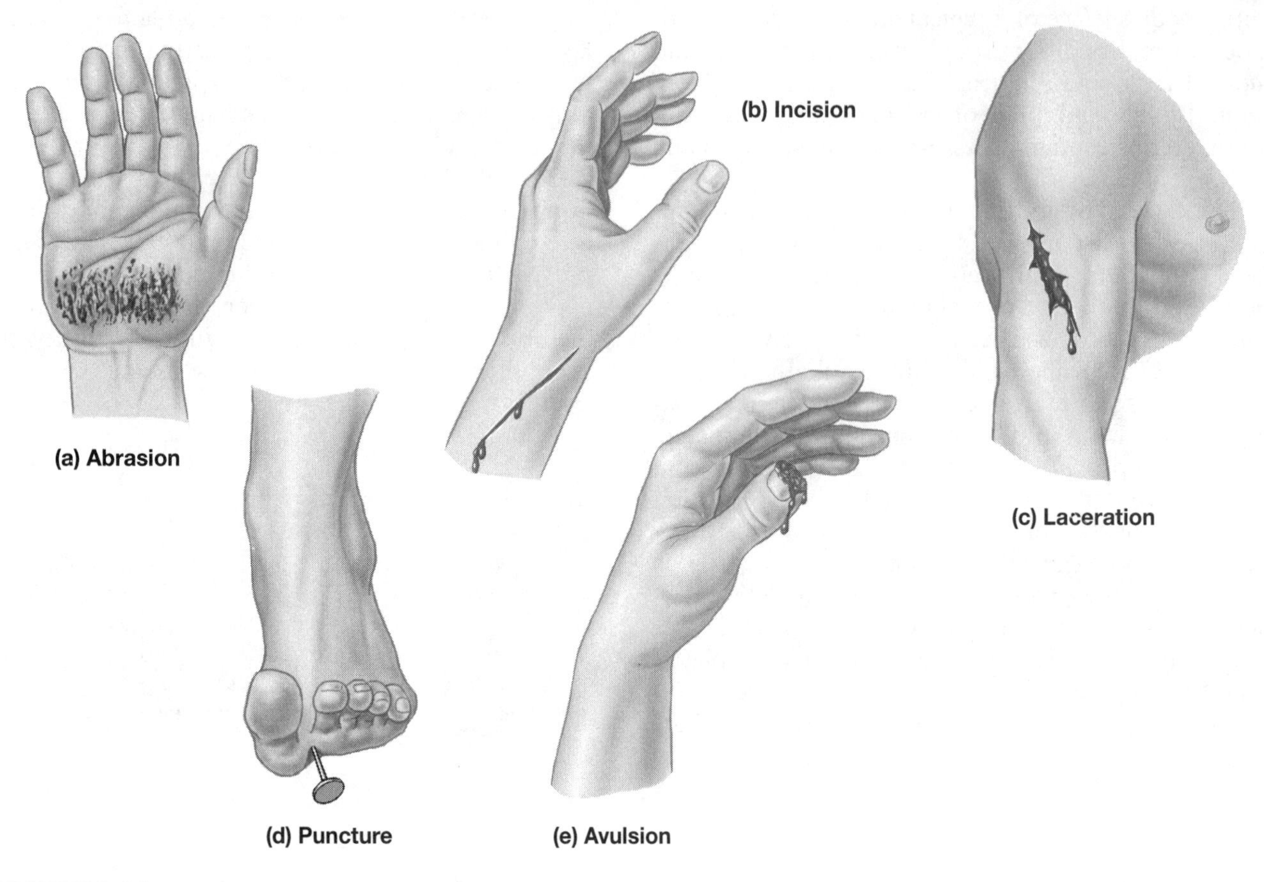

•**FIGURE A-18** Major Types of Open Wounds

miliar examples of contusions; most are not dangerous, but contusions of the head, such as "black eyes," may be accompanied by life-threatening intracranial bleeding. Closed wounds affecting internal organs and organ systems are almost always serious threats to life.

Skin can regenerate effectively even after considerable damage; skin repairs are discussed in the text *(p. 158).* Burns, a type of trauma that can affect the epidermis, dermis, and deeper tissues, are also discussed in the text *(p. 162).*

TREATMENT OF SKIN CONDITIONS

Topical **anti-inflammatory drugs,** such as the steroid *hydrocortisone,* can be used to reduce the redness and itching that accompany a variety of skin conditions. Some systemic (injected or swallowed) drugs may also be helpful; *aspirin* is a familiar systemic drug with anti-inflammatory properties. Isolated growths, such as skin tumors or warts, can be surgically removed. Alternatively, abnormal cells can be destroyed by electrical currents **(electrosurgery)** or by freezing **(cryosurgery).** Ultraviolet radiation may help conditions such as acne or psoriasis, whereas the use of a sunscreen or a sunblock is important in controlling outbreaks caused by sensitivity to the sun.

SYNTHETIC SKIN FAP *p. 162*

Skin grafts involve covering areas where skin has been completely destroyed or lost. Pieces of undamaged skin from other areas of the body are used. If the damaged area is large, there may not be enough normal skin available for grafting. *Epidermal culturing* can produce a new epithelial layer to cover a burn site. The cells in a 3-cm^2 epidermal sample from a burn patient can be cultured in a controlled environment that contains epidermal growth factors, fibroblast growth factors, and other stimulatory chemicals. The number of cells doubles in the first 18 hours. The cells are then separated, and culturing continues. After 3 days in culture, the number of cells has increased by more than 16 times. After a week, the number has increased by over 300 times. The cells are now in small clusters, and they have begun forming layers. Germinative cells are on the bottom, attached to the glass of the culturing vessel. The more superficial layers of cells roughly resemble those of normal epidermis, although keratin production does not occur. This artificially produced epidermis can now be transplanted to cover an injury site. The larger the area that must be covered, the longer the culturing process continues. After 3 or 4 weeks of culturing, the cells obtained in the original sample can provide enough epidermis to cover the

entire body surface of a typical adult. The absence of a dermis makes this graft epidermis less flexible and more fragile than normal—a particular problem around joints and areas of friction. Also, these skin grafts do not contain accessory organs, such as hairs or sweat glands.

The success of grafting is limited by the contamination of the wound site by bacteria while the epidermis is being cultured. The contamination is prevented by the use of a skin *allograft.* In this procedure, skin from a frozen cadaver is removed and placed over the wound as a temporary method of sealing the surface. Before the immune system of the patient attacks it, the graft will be partially or completely removed to provide a binding site for the epidermal transplant. After grafting, the complete reorganization and repair of the dermis and epidermis at the injury site takes approximately 5 years.

A second new procedure provides a model for dermal repairs that takes the place of normal tissue. A special synthetic skin is used. The imitation has a plastic (Silastic®) "epidermis" and a dermis composed of collagen fibers and ground cartilage. The collagen fibers are taken from cow skin, and the cartilage is taken from sharks. Over time, fibroblasts migrate among the collagen fibers and gradually replace the model framework with their own. The Silastic epidermis is intended only as a temporary cover that will be replaced by either skin grafts or a cultured epidermal layer.

Several other procedures have been approved or are awaiting FDA approval. All involve the application of a temporary dermal layer that can guide the repair process. The most interesting approach, awaiting FDA approval, involves the application of a dermal layer created by culturing fibroblasts obtained from the foreskins of circumcised newborn male infants.

[EXPLORE]

Additional resources, such as interactive tutorials, critical-thinking questions, clinical problems, and case studies, are available through the CD-ROM and Website for your textbook. To begin your exploration, launch the *Martini Interactive* CD-ROM or visit the Companion Website (http://www.prenhall.com/martini/fap5) and select Chapter 5.

The Skeletal System

The skeletal framework of the body is composed of at least 206 bones and the associated tendons, ligaments, and cartilages. The skeletal system has a variety of important functions, including the support of soft tissues, blood cell production, mineral and lipid storage, and, through its relationships with the muscular system, the support and movement of the body as a whole. Skeletal system disorders can thus affect many other systems. The skeletal system is in turn influenced by the activities of other systems. For example, weakness or paralysis of skeletal muscles will lead to a weakening of the associated bones.

Although the bones you study in the lab may seem to be rigid and permanent structures, the living skeleton is dynamic and undergoes continuous remodeling. The remodeling process involves bone deposition by osteoblasts and *osteolysis,* or dissolution of bone matrix, by osteoclasts. As indicated in Figure A-19•, the net result of the remodeling varies with the following five factors:

1. *The age of the individual.* During development, bone deposition occurs faster than bone resorption; as the amount of bone increases, the skeleton grows. At maturity, bone deposition and resorption are in balance. As the aging process continues, the rate of bone deposition declines and the bones become less massive. This gradual weakening, called *osteopenia,* begins at age 30–40 and may ultimately progress to osteoporosis (p. 57).
2. *The applied physical stresses.* Heavily stressed bones become thicker and stronger, and lightly stressed bones become thinner and weaker. Skeletal weakness

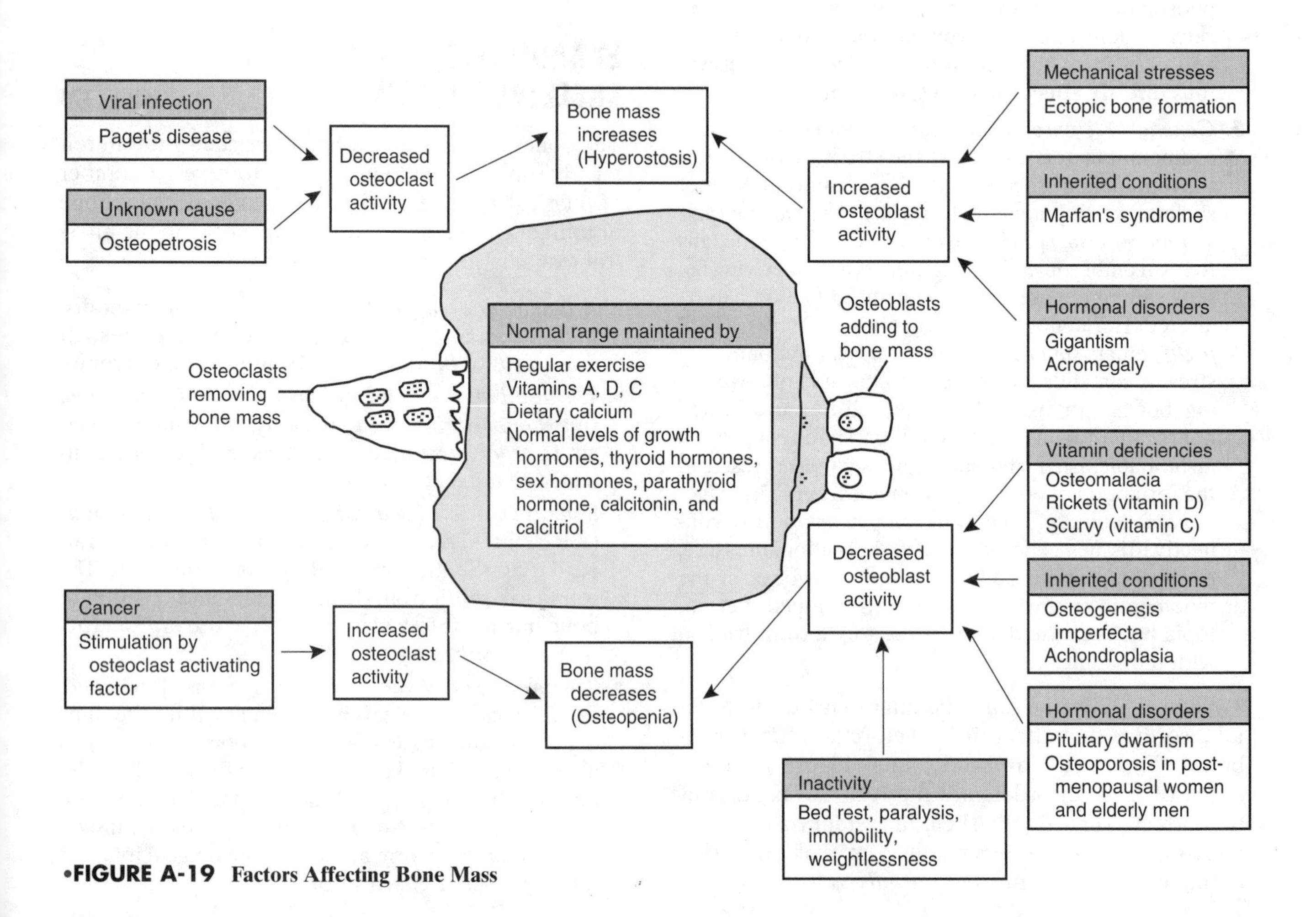

•**FIGURE A-19** **Factors Affecting Bone Mass**

can therefore result from muscular disorders, such as *myasthenia gravis* (p. 67) and the *muscular dystrophies* (p. 66), and from conditions that affect motor neurons of the central nervous system, such as spinal cord injuries (p. 76), *demyelination disorders* (p. 74), and *multiple sclerosis* (p. 76).

3. *Circulating hormone levels.* Changing levels of growth hormone (GH), androgens and estrogens, thyroid hormones, parathyroid hormone, and calcitonin increase or decrease the rate of mineral deposition in bone. As a result, many disorders of the endocrine system affect the skeletal system. For example:

FAP Ch. 6

   - Conditions affecting the skin, liver, or kidneys can interfere with calcitriol production.
   - Thyroid and parathyroid disorders can alter levels of thyroid hormones, parathyroid hormone, and calcitonin.
   - Pituitary gland disorders and liver disorders can affect the production of both GH and somatomedin.
   - Reproductive system disorders can alter circulating levels of androgens and estrogens.

   We will describe many of these conditions in the section dealing with the endocrine system.

4. *Rates of calcium and phosphate absorption and excretion.* For bone mass to remain constant, the rate of calcium and phosphate excretion, primarily at the kidneys, must be balanced by the rate of calcium and phosphate absorption at the digestive tract. Dietary calcium deficiencies or problems that reduce calcium and phosphate absorption at the digestive tract will thus directly affect the skeletal system.

5. *Genetic or environmental factors.* Genetic or environmental factors can affect the structure of bone or the remodeling process. A number of abnormalities of skeletal development, such as *Marfan's syndrome* and *achondroplasia* (p. 55), are inherited. When bone fails to form embryonically in certain areas, underlying tissues can be exposed and associated function can be altered. This abnormality occurs in a *cleft palate* (FAP *p. 205)* and in *spina bifida* (FAP *p. 214*). Environmental stresses can alter the shape and contours of developing bones. For example, some cultures use lashed boards to form an infant's skull to a shape considered fashionable. Environmental forces can also result in the formation of bone in unusual locations. These *heterotopic bones* (p. 54) may develop in a variety of connective tissues exposed to chronic friction, pressure, or mechanical stress. For example, cowboys in the nineteenth century sometimes developed heterotopic bones in the dermis of the thigh, from friction with the saddle.

Figure A-20• diagrams the major classes of skeletal disorders that affect the structure and function of bones (Figure A-20a•) and joints (Figure A-20b•). Some of these disorders are the result of conditions that affect primarily the skeletal system *(osteosarcoma, osteomyelitis);* others result from problems originating in other systems *(acromegaly, rickets).*

Traumatic injuries, such as fractures or dislocations, and infections also damage cartilages, tendons, and ligaments. A somewhat different array of conditions affects the soft tissues of the bone marrow. Red bone marrow contains the stem cells for red blood cells, white blood cells, and platelets. Blood diseases characterized by blood cell overproduction (*leukemia, polycythemia,* pp. 116, 108) or underproduction (several *anemias,* p. 115) result in bone marrow abnormalities.

THE SYMPTOMS OF BONE AND JOINT DISORDERS

A common symptom of a skeletal system disorder is pain. Bone pain and joint pain are common symptoms associated with many bone disorders. As a result, the presence of pain does not provide much help in identifying a specific bone or joint disorder. A person may be able to tolerate chronic, aching bone or joint pain and therefore not seek medical assistance until more-definitive symptoms appear. By then the condition is relatively advanced. For example, a symptom that may require immediate attention is a *pathologic fracture.* Pathologic fractures are the result of weakening of the skeleton by disease processes, such as *osteosarcoma* (a bone cancer). These fractures can be caused by physical stresses that are easily tolerated by normal bones.

EXAMINATION OF THE SKELETAL SYSTEM

FAP *p. 175*

The bones of the skeleton cannot be seen without relatively sophisticated equipment. However, a number of physical signs can assist in the diagnosis of a bone or joint disorder. Important factors noted in the physical examination include the following:

- *A limitation of movement or stiffness.* Many joint disorders, such as the various forms of arthritis, restrict movement or produce stiffness at one or more joints.
- *The distribution of joint involvement and inflammation.* In a *monoarthritic* condition, only one joint is affected. In a *polyarthritic* condition, several joints are affected simultaneously.
- *Sounds associated with joint movement. Bony crepitus* (KREP-i-tus) is a crackling or grating sound generated during the movement of an abnormal joint. The sound can result from the movement and collision of bone fragments following a joint fracture or from friction at an arthritic joint.
- *The presence of abnormal bone deposits.* Thickened, raised areas of bone develop around fracture sites during the repair process. Abnormal bone deposits can also develop around the joints in the fingers. These deposits are called *nodules.* When palpated, nodules are solid and painless. Nodules, which can restrict movement, commonly form at the interphalangeal joints of the fingers in osteoarthritis.

FAP Ch. 6

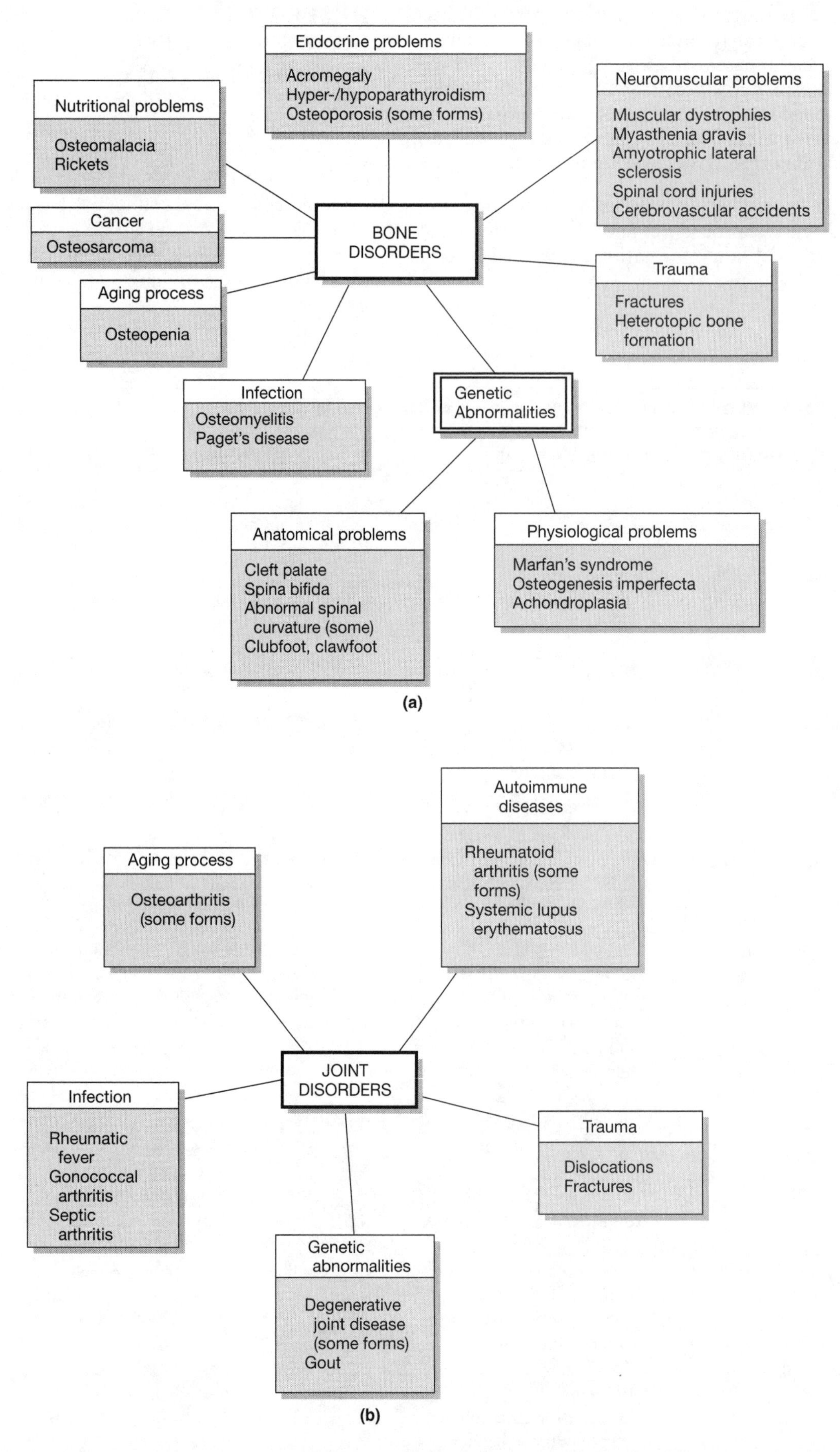

•**FIGURE A-20 An Overview of Disorders of the Skeletal System. (a)** Bone disorders. **(b)** Joint disorders.

- *Abnormal posture.* Bone disorders that affect the vertebral column can result in abnormal posture. This result is most apparent when the condition alters the normal spinal curvature. Examples include *kyphosis, lordosis,* and *scoliosis* (p. 57). A condition involving an intervertebral joint, such as a herniated disc, will also produce abnormal posture and movement.

Table A-16 summarizes descriptions of the most important diagnostic procedures and laboratory tests that can be used to obtain information about the status of the skeletal system.

FAP Ch. 6

HETEROTOPIC BONE FORMATION FAP *p. 175*

Heterotopic (*hetero,* different + *topos,* place), or **ectopic** (*ektos,* outside), **bones** are bones that develop in unusual places. Such bones dramatically demonstrate the adaptability of connective tissues. Physical or chemical events can stimulate the development of osteoblasts in normal connective tissues. For example, sesamoid bones develop within tendons near points of friction and pressure. Bone can also form within a large blood clot at an injury site or within portions of

TABLE A-16 Examples of Tests Used in the Diagnosis of Bone and Joint Disorders

Diagnostic Procedure	Method and Result	Representative Uses
X-ray of bone and joint	Standard X-ray; film sheet with radiodense tissues in white on a black background	Detects fractures, tumors, dislocations, reduction in bone mass, and bone infections (osteomyelitis)
Bone scans	Injected radiolabeled phosphate accumulates in bones, and radiation emitted is converted into an image.	Especially useful in diagnosis of metastatic bone cancer; detects fractures, early infections, and degenerative bone diseases
Arthrocentesis	Insertion of a needle into joint for aspiration of synovial fluid	See section on analysis of synovial fluid (below).
Arthroscopy	Insertion of fiber-optic tubing into a joint cavity; attached camera displays joint interior.	Detects abnormalities of the menisci and ligaments; useful in differential diagnosis of joint disorders
MRI	Standard MRI (FAP *p. 23*), produces computer generated images.	Detects bone and soft tissue abnormalities
Laboratory Test	**Normal Values in Blood Plasma or Serum**	**Significance of Abnormal Values**
Alkaline phosphatase	Adults: 30–85 mIU/ml Children: 60–300 mIU/ml (higher values occur during adolescent bone growth)	Elevated levels occur in adults due to abnormal osteoblast activity; elevated levels occur in bone cancer, Paget's disease, and multiple myeloma.
Calcium	Adults: 8.5–10.5 mg/dl Children 8.5–11.5 mg/dl	Elevated levels occur in bone cancers and multiple fractures and in persons who are immobilized for a prolonged period.
Phosphorus	Adults: 2.3–4.7 mg/dl Children: 4.0–7.0 mg/dl	Typically elevated when calcium levels are low; elevated levels occur in acromegaly, parathyroid disorders, and bone tumors.
Uric acid	Adult males: 3.5–8.0 mg/dl Adult females: 2.8–6.8 mg/dl	Elevated levels occur with gout, which develops when uric acid crystals (products of purine metabolism) build up in a joint.
Rheumatoid factor	Adults: A negative result is normal; 1:20–1:80 titer or higher is one criterion for rheumatoid arthritis.	About 75% of people diagnosed with rheumatoid arthritis have a positive test for this factor; liver and collagen diseases and systemic lupus erythematosus may give positive results.
Synovial fluid analysis	WBC: $<200/mm^3$ RBC: none Glucose: <10 mg/dl below serum glucose levels Protein: 1.0–3.0 g/dl No uric acid crystals	Elevated white blood cell count suggests bacterial infection; mild elevation indicates inflammatory process; decreased glucose levels and decreased viscosity of fluid indicates inflammation of joint; uric acid crystals indicate gout.

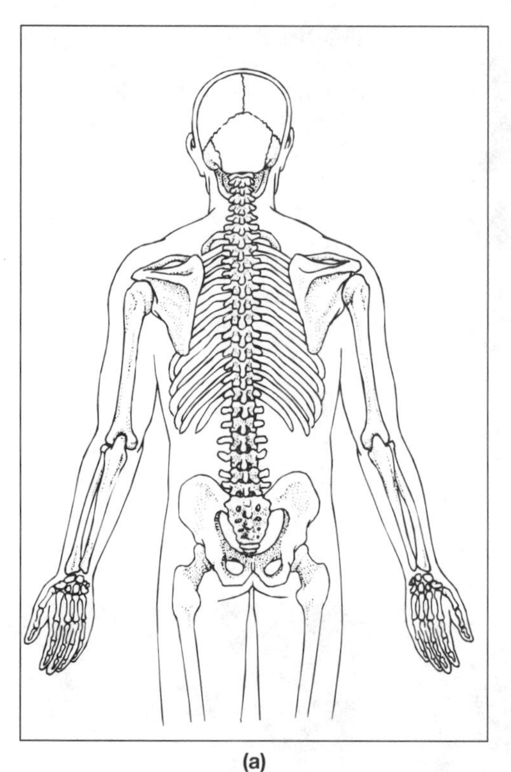

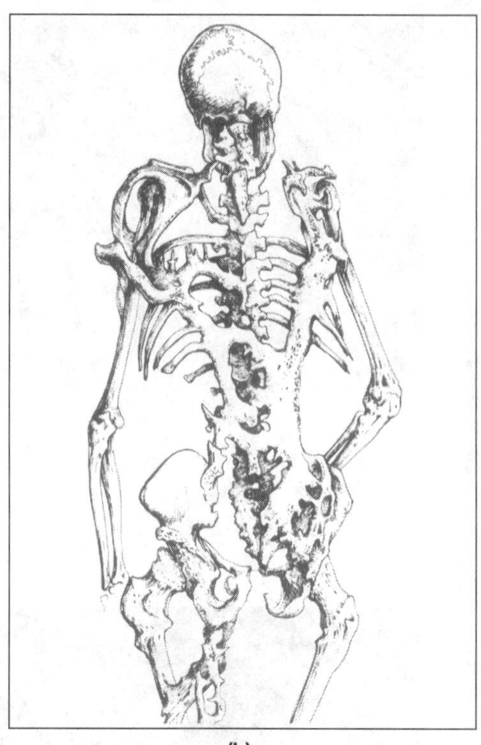

•FIGURE A-21 Heterotopic Bone Formation. (a) The skeleton of a healthy adult male, posterior view. [1974, L. B. Halstead Wykerman Publications, Ltd. (London)] **(b)** The skeleton of an adult male with advanced myositis ossificans.

the dermis subjected to chronic abuse. Other triggers include foreign chemicals and problems that affect calcium excretion and storage.

Almost any connective tissues can be affected. Ossification within a tendon or around joints can painfully interfere with movement. Bones can also form within the kidneys, between skeletal muscles, in the pericardium, in the walls of arteries, and around the eyes.

Myositis ossificans (mī-ō-SĪ-tis os-SIF-i-kanz) involves the deposition of bone around skeletal muscles. A muscle injury can trigger a minor case. Severe cases have no known cause, but they certainly provide the most dramatic demonstrations of heterotopic bone formation. If the process does not reverse itself, the muscles of the back, neck, and upper limbs will gradually be replaced by bone. The extent of the conversion can be seen in Figure A-21•. Figure A-21a• shows the skeleton of a healthy adult male; Figure A-21b• shows the skeleton of a 39-year-old man with advanced myositis ossificans. Several of the vertebrae have fused into a solid mass, and major muscles of the back, shoulders, and hips have undergone extensive ossification. Compare this figure with Figures 11-3b and 11-14a in the text *(pp. 319, 336)* to identify the specific muscles.

CONGENITAL DISORDERS OF THE SKELETON

FAP *p. 181*

Several inherited conditions result in abnormal bone formation. Three examples are osteogenesis imperfecta, Marfan's syndrome, and achondroplasia.

Osteogenesis imperfecta, appearing in roughly 1 individual in 20,000, affects the organization of collagen fibers. Osteoblast function is impaired, growth is abnormal, and in severe forms the bones are very fragile, leading to progressive skeletal deformation and repeated fractures. Fibroblast activity is also affected, and the ligaments and tendons can be very "loose," permitting excessive movement at the joints.

Marfan's syndrome is also linked to defective connective tissue structure. Extremely long and slender limbs, the most obvious physical indication of this disorder, result from excessive cartilage formation at the epiphyseal cartilages (Figure A-22a•).

Achondroplasia (ā-kon-drō-PLĀ-sē-uh) also results from abnormal epiphyseal activity. The child's epiphyseal cartilages grow unusually slowly, and the adult has short, stocky limbs (Figure A-22b•). Although other skeletal abnormalities occur, the trunk is normal in size, and sexual and mental development remain unaffected. The adult is an *achondroplastic dwarf.*

In **osteomalacia** (os-tē-ō-ma-LĀ-shē-uh; *malakia,* softness) the size of the skeletal elements does not change, but their mineral content decreases, softening the bones. The osteoblasts work hard, but the matrix doesn't accumulate enough calcium salts. This condition, *rickets,* occurs in adults or children whose diet contains inadequate levels of calcium or vitamin D_3.

The excessive formation of bone is termed **hyperostosis** (hī-per-os-TŌ-sis). In osteopetrosis (os-tē-ō-pe-TRŌ-sis; *petros,* stone), the total mass of the skeleton gradually increases as a result of a decrease in osteoclast activity. Remodeling stops, and the shapes of the

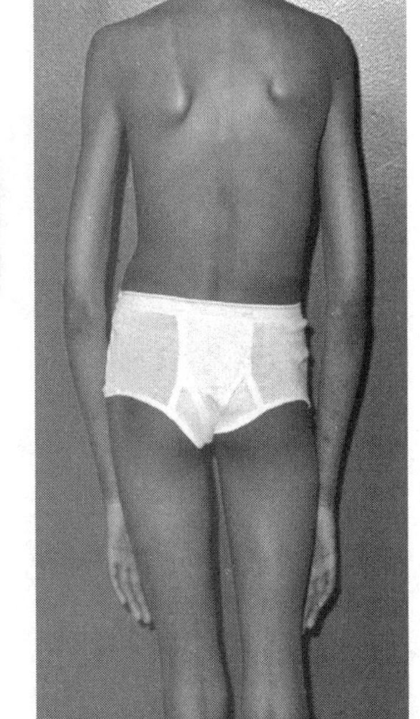

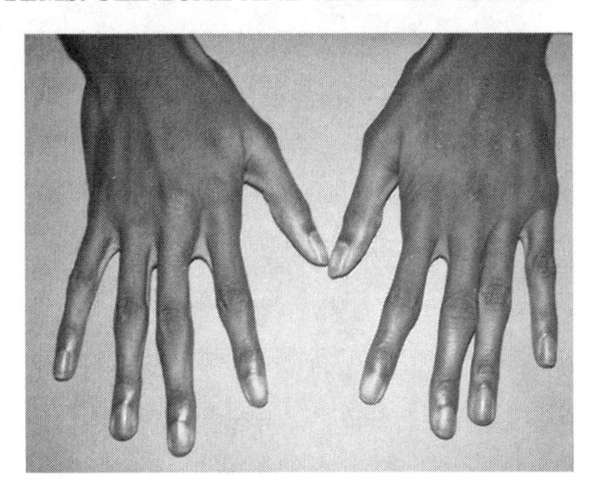

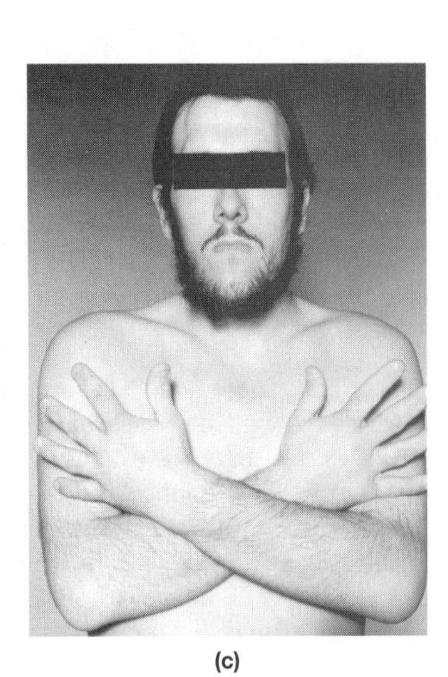

(a) (b) (c)

•**FIGURE A-22 Disorders of Bone Formation. (a)** Marfan's syndrome (posterior view and hands). **(b)** Achondroplasia. **(c)** Acromegaly.

bones gradually change. Osteopetrosis in children produces a variety of skeletal deformities. The primary cause of this relatively rare condition is unknown.

In **acromegaly** (*akron,* extremity + *megale,* great) an excessive amount of growth hormone is released after puberty, when most of the epiphyseal cartilages have already closed. Cartilages and small bones respond to the hormone, however, resulting in abnormal growth at the hands, feet, lower jaw, skull, and clavicle (Figure A-22c•).

STIMULATION OF BONE GROWTH AND REPAIR FAP *p. 185*

Despite the considerable capacity for bone repair, every fracture does not heal as expected. A *delayed union* is a repair that proceeds more slowly than anticipated. *Nonunion,* or no repair, can occur as a result of complicating infection, continued movement, or other factors preventing complete callus formation.

Several techniques can induce bone repair. Surgical bone grafting is the most common treatment for nonunion. This method immobilizes the bone fragments and provides a bony model for the repair process. Dead bone or bone fragments can be used; alternatively, living bone from another site, such as the iliac crest or part of a rib, can be inserted. Surgeons can insert a shaped patch, made by mixing crushed bone and water, as an alternative to bone grafting. Bone transplants from cadavers have been used, but unless these bones are thoroughly sterilized, blood-borne diseases, including AIDS, can be transmitted. The calcium carbonate skeleton of corals has been sterilized and used as well.

Another approach involves the stimulation of osteoblast activity by strong electrical fields at the injury

site. This procedure has been used to promote bone growth after fractures have refused to heal normally. Wires may be inserted into the skin, implanted in the adjacent bone, or wrapped around a cast. The overall success rate of about 80 percent is truly impressive.

One experimental method of inducing bone repair involves the mixing of bone marrow cells into a soft matrix of bone collagen and ceramic. This combination is used like a putty at the fracture site. Mesenchymal cells in the marrow divide, producing chondrocytes that create a cartilaginous patch; the patch is later converted to bone by periosteal cells. A second experimental procedure uses a genetically engineered protein to stimulate the conversion of osteoprogenitor cells into active osteoblasts. Although results in animal experimentation have been promising, the latter technique has not yet been approved for human trials.

DEGENERATIVE DISORDERS OF THE SKELETAL SYSTEM

OSTEOPOROSIS AND AGE-RELATED SKELETAL ABNORMALITIES FAP *p. 187*

Osteoporosis (os-tē-ō-por-ō-sis; *porosus,* porous) is a condition that produces a reduction in bone mass sufficient to compromise normal function. The distinction between the "normal" osteopenia of aging and the clinical condition of osteoporosis is a matter of degree. Current estimates indicate that 29 percent of women between the ages of 45 and 79 can be considered osteoporotic. The increase in incidence after menopause has been linked to a decrease in the production of estrogens (female sex hormones). The incidence of osteoporosis in men of the same age is estimated at 18 percent.

The excessive fragility of the bones commonly leads to breakage, and subsequent healing is impaired. Vertebrae may collapse, distorting the vertebral articulations and putting pressure on spinal nerves. Supplemental estrogens, dietary changes to elevate calcium levels in blood, exercise that stresses bones and stimulates osteoblast activity, and calcitonin administration by nasal spray appear to slow but not prevent the development of osteoporosis. The inhibition of osteoclast activity by drugs called *bisphosphonates,* such as *Fosamax®*, can reduce the risk of spine and hip fractures in elderly women.

Osteoporosis can also develop as a secondary effect of many cancers. Cancers of the bone marrow, breast, or other tissues release a chemical known as *osteoclast-activating factor.* This compound increases both the number and activity of osteoclasts and produces a severe osteoporosis.

Infectious diseases that affect the skeletal system become more common as individuals age. In part, this fact reflects the higher incidence of fractures, combined with slower healing and the reduction of immune defenses.

Osteomyelitis (os-tē-ō-mī-e-LĪ-tis; *myelos,* marrow) is a painful and destructive bone infection generally caused by bacteria. This condition, most common in people over 50 years of age, can lead to dangerous systemic infections. A virus appears to be responsible for **Paget's disease**, also known as **osteitis deformans** (os-tē-Ī-tis de-FOR-manz). This condition may affect up to 10 percent of the population over 70. Osteoclast activity accelerates, producing areas of acute osteoporosis, and osteoblasts produce abnormal matrix proteins. The result is a gradual deformation of the skeleton. Bisphosphonate treatment may slow progression of the disease, as in osteoporosis, by reducing osteoclast activity.

FAP Ch. 7

KYPHOSIS, LORDOSIS, AND SCOLIOSIS FAP *p. 212*

In **kyphosis** (kī-FŌ-sis), the normal thoracic curvature becomes exaggerated posteriorly, producing a "round-back" appearance (Figure A-23a•). This condition can be caused by (1) osteoporosis with compression fractures affecting the anterior portions of vertebral bodies, (2) chronic contractions in muscles that insert on the vertebrae, or (3) abnormal vertebral growth.

In **lordosis** (lor-DŌ-sis), or "swayback," both the abdomen and buttocks protrude abnormally (Figure A-23b•). The cause is an anterior exaggeration of the lumbar curvature.

Scoliosis (skō-lē-Ō-sis) is an abnormal lateral curvature of the spine (Figure A-23c•). This lateral deviation can occur in one or more of the movable vertebrae. Scoliosis is the most common distortion of the spinal curvature. This condition results from developmental problems, such as incomplete vertebral formation, or from muscular paralysis affecting one side of the back. In four out of five cases, the structural or functional cause of the abnormal spinal curvature is impossible to determine. Scoliosis generally appears in girls during adolescence, when periods of growth are most rapid. Treatment may consist of a combination of exercises and braces that offer limited if any benefit. Severe cases can be treated through surgical straightening with implanted metal rods or cables.

PROBLEMS WITH THE ANKLE AND FOOT FAP *p. 240*

The ankle and foot are subjected to a variety of stresses during normal daily activities. In a *sprain,* a ligament is stretched to the point at which some of the collagen fibers are torn. The ligament remains functional, and the structure of the joint is not affected. The most common cause of a **sprained ankle** is a forceful inversion of the foot that stretches the lateral ligament. An ice pack is generally required to reduce swelling.

FAP Ch. 8

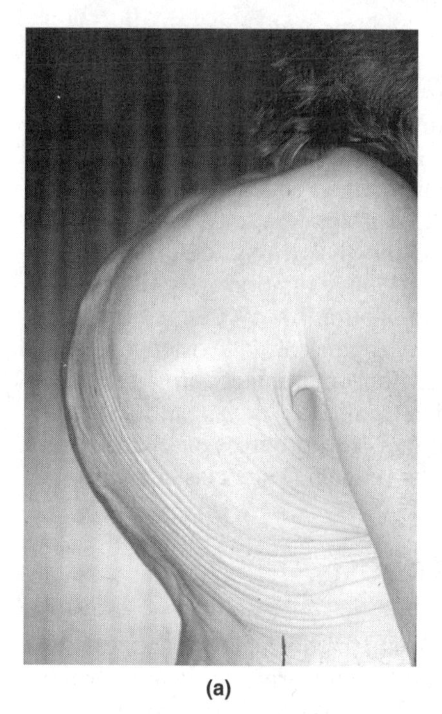
(a)

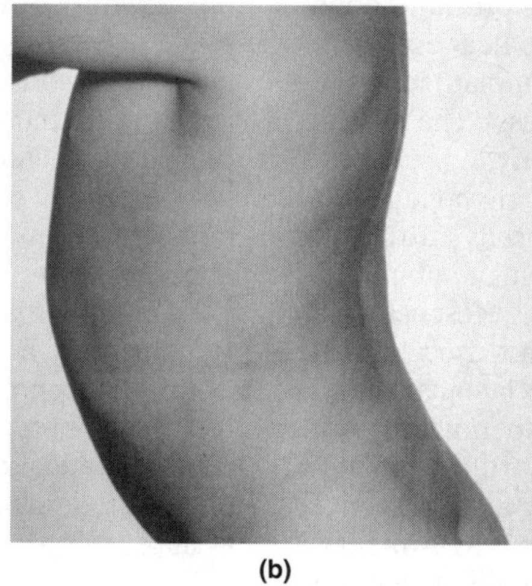
(b)

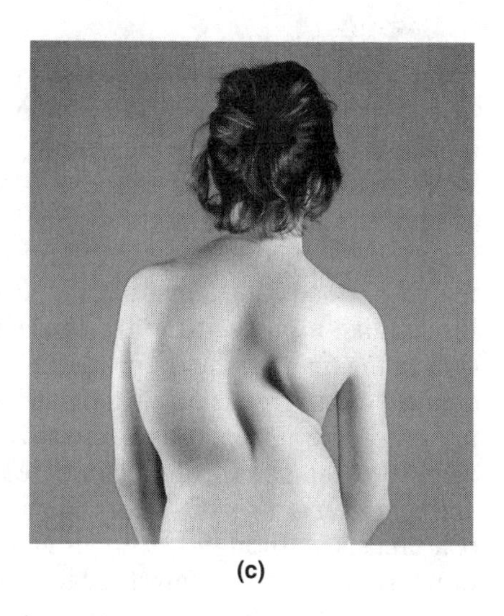
(c)

•**FIGURE A-23** **Spinal Deformities.** **(a)** Kyphosis. **(b)** Lordosis. **(c)** Scoliosis.

With rest and support, the ankle should heal in about 3 weeks.

In more-serious incidents, the entire ligament can be torn apart, or the connection between the ligament and the lateral malleolus can be so strong that the bone breaks instead of the ligament. In general, a broken bone heals more quickly and effectively than does a completely torn ligament. A dislocation typically accompanies such injuries.

In a **dancer's fracture,** the proximal portion of the fifth metatarsal is broken. Most such cases occur while the body weight is being supported by the longitudinal arch of the foot. A sudden shift in weight from the medial portion of the arch to the lateral, less elastic border breaks the fifth metatarsal close to its distal articulation.

Individuals with abnormal arch development are most likely to suffer metatarsal injuries. Someone with *flat feet* loses or never develops the longitudinal arch. "Fallen arches" develop as tendons and ligaments stretch and become less elastic. Obese individuals or those who must constantly stand or walk on the job are likely candidates. Children have very mobile articulations and elastic ligaments, so they commonly have flexible flat feet. Their feet look flat only while they are standing, and the arch appears when they stand on their toes or sit down. In most cases, the condition disappears as growth continues.

Clawfeet are also produced by muscular abnormalities. In individuals with a clawfoot, the median longitudinal arch becomes exaggerated because the plantar flexors overpower the dorsiflexors. Muscle cramps or nerve paralysis is responsible. The condition tends to develop in adults and gets progressively worse with age.

Congenital talipes equinus, or *clubfoot,* is an inherited developmental abnormality discussed in the text (FAP *p. 240*).

A MATTER OF PERSPECTIVE FAP *p. 241*

Several professions focus on the structure and function of the human skeleton. Each specialty has a different perspective, with its own techniques, traditions, and biases. People seeking medical advice may find it difficult to choose the most appropriate specialist. Unfortunately, misinformation and misconceptions have polarized opinions regarding the relative merits of the various specialists.

An **orthopedist,** or **orthopedic surgeon,** is a physician who specializes in surgery restoring the structure and function of bones and articulations. An orthopedist might typically be involved in the reconstruction of a limb after a severe fracture, the replacement of an arthritic hip, knee surgery, and so forth. Subspecialties focus on particular regions, such as the hands. An orthopedist is trained to recognize the underlying condition when an injury is not the primary cause of a particular problem. As a physician, an orthopedist must complete (1) a 4-year college program, (2) a 4-year medical or osteopathic school, (3) an internship, (4) a residency program, and (5) national boards (examinations) and then be licensed by the state in which he or she practices medicine. It can take up to 12 or 13 years of training after high school to complete all the requirements.

Osteopaths (D.O.) have a slightly different perspective than that of medical doctors (M.D.). Osteopaths must complete undergraduate college and an osteopathic medical school, serve an internship, and pass national and state examinations. Additional residency training may also be required. A century ago, when osteopathy was founded, drug therapies for the treatment of diseases were unreliable, and osteopaths put their trust in rest, cleanliness, and proper nutrition, in some cases accompanied by manipulation of the skeleton. Osteopaths today in general place a greater emphasis than do orthopedists on preventive medicine, such as the maintenance of normal structural relationships, proper nutrition, exercise, and a healthy environment. In terms of evaluating and treating bone or joint conditions, there is no substantive difference today between osteopaths and medical doctors other than the amount of skeletal manipulation prescribed by osteopaths. Both osteopaths and medical doctors can go through residency training to become orthopedists.

The viewpoint of a **chiropractor** (D.C.) differs considerably from that of either orthopedists or osteopaths. In its original form, chiropractic proposed that all forms of disease result from abnormal functioning of the nervous system. This rather sweeping premise was modified as physiological principles became better understood. *Straight chiropractors* continue to place primary emphasis on vertebral alignment as the source of spinal nerve compression that causes peripheral musculoskeletal symptoms. These practitioners use modern radiological techniques to evaluate vertebral positions but generally avoid the use of drugs. Treatment involves physical manipulation of the vertebral column to bring the vertebral elements into proper alignment.

Other classes of chiropractors may have widely varying approaches and techniques. Many claim that spinal manipulation alone can cure conditions totally unrelated to the skeletal, muscular, or nervous system. There is no known mechanism that would lend plausibility to those claims, nor is there valid supporting evidence. (Valid evidence must be testable, unbiased, and repeatable—criteria discussed in the section on the scientific method, pp. 2–4.)

Because so much variability exists among the practitioners, generalizations are difficult if not impossible to make. However, chiropractors in every state are licensed. For example, to be licensed in Hawaii an applicant needs more than 2 years of undergraduate credits (60 semester hours), a diploma from 1 of 16 accredited 4-year chiropractic colleges, and completion of a national examination.

Podiatrists (D.P.M.), or *chiropodists,* are specialists in foot structure and related surgery. They commonly treat sports injuries, design supports for feet affected by trauma or developmental problems, and perform limited surgical procedures. Podiatrists must be graduates of an accredited podiatry college and may be required to pass national and state examinations.

Where you take your aching bones is your choice, but the way the practitioner views your problem will obviously affect your treatment. Be wary of any individual who claims to have the answer to every question. A competent practitioner knows his or her personal and professional limitations and is willing to refer a patient to another specialist when appropriate.

DIAGNOSING AND TREATING DISC PROBLEMS

FAP p. 257

FAP Ch. 9

The most common sites for disc problems are between vertebrae C_5 and C_6, L_4 and L_5, and L_5 and S_1. A clinician can generally determine the location of the injured disc by noting the distribution of abnormal sensations. For example, someone with a herniated disc at L_4–L_5 will experience pain in the hip, the groin, the posterior and lateral surfaces of the thigh, the lateral surface of the calf, and the superior surface of the foot. A herniation at L_5–S_1 produces pain in the buttocks, the posterior thigh, the posterior calf, and the sole of the foot. Most lumbar disc problems can be treated successfully with some combination of rest, back braces, painkillers, and physical therapy. Surgery to relieve the symptoms is required in only about 10 percent of cases involving lumbar disc herniation. The primary method of treatment involves removing the offending disc, and, if necessary, fusing the vertebral bodies together to prevent relative movement. Accessing the disc requires that the laminae of the nearest vertebral arch be removed. For this reason, the procedure is known as *laminectomy* (la-mi-NEK-to-mē).

In rare cases in which the herniated portion of the disc does not extend well into the vertebral foramen, portions of the disc can be removed by a small tool that is guided to the site by radiological imaging. This procedure is faster and easier than a laminectomy.

ARTHROSCOPIC SURGERY AND THE DIAGNOSIS OF JOINT INJURIES

FAP p. 263

An **arthroscope** uses fiber optics to permit the exploration of a joint without major surgery. Optical fibers are thin threads of glass or plastic that conduct light. The fibers can be bent around corners, so they can be introduced into a knee or other joint and moved around, enabling the physician to see what is going on inside the joint. If necessary, the apparatus can be modified to perform surgical modification of the joint at the same time. This procedure, called **arthroscopic surgery,** has greatly simplified the treatment of knee and other joint injuries. Figure A-24a• is an arthroscopic view of the interior of an injured knee, showing a damaged meniscus. Small pieces of cartilage can be removed and the meniscus surgically trimmed. A total **meniscectomy,** the removal of the affected cartilage, is generally

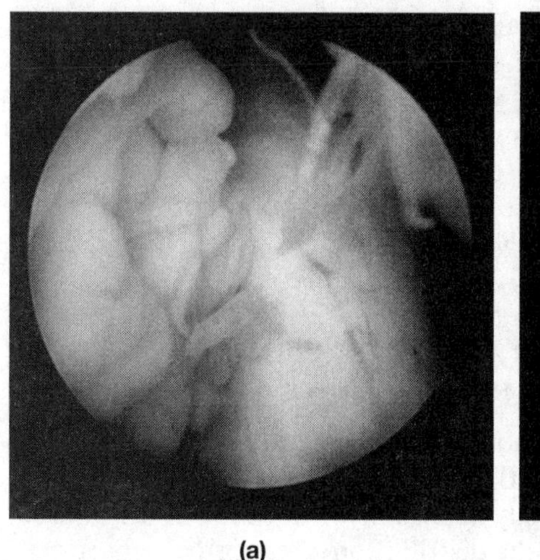

(a)

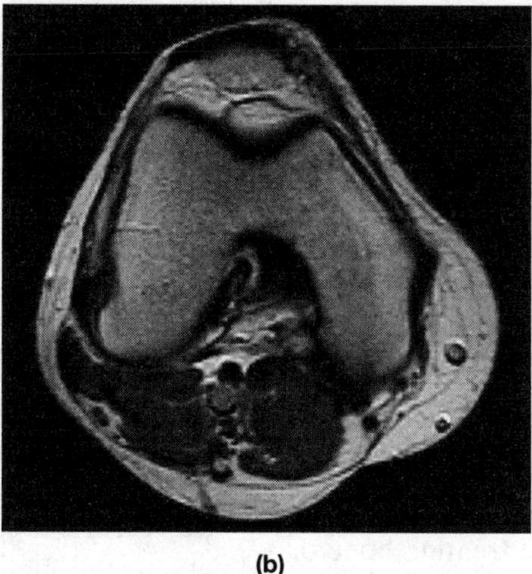

(b)

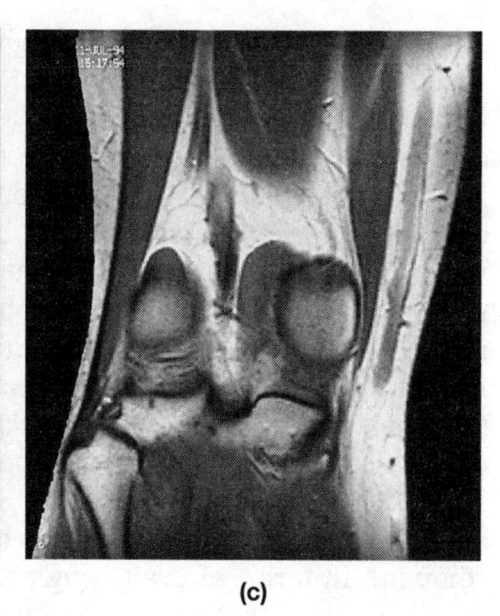

(c)

•**FIGURE A-24 Arthroscopy and MRI Scans of the Knee. (a)** An arthroscopic view of a damaged knee, showing the torn edge of an injured meniscus. **(b)** A transverse and **(c)** a frontal MRI scan of the knee joint. For identification of the structures visualized in these images, see *Scans 5–7.*

avoided, because it leaves the joint prone to develop degenerative joint disease. New tissue-culturing techniques may someday permit the replacement of the meniscus or even the articular cartilage.

An arthroscope cannot show the physician soft tissue details outside the joint cavity, and repeated arthroscopy eventually leads to the formation of scar tissue and other joint problems. Magnetic resonance imaging (MRI) is a cost-effective and noninvasive method of viewing and examining, without injury, soft tissues around the joint. Figures A-24b and A-24c• are MRI views of the knee joint. Notice the image clarity and the soft tissue details visible in these scans.

RHEUMATISM, ARTHRITIS, AND SYNOVIAL FUNCTION FAP *p. 265*

Rheumatism (ROO-ma-tizm) is a general term that indicates pain and stiffness affecting the skeletal system, the muscular system, or both. There are several major forms of rheumatism. **Arthritis** (ar-THRĪ-tis) includes all the rheumatic diseases that affect synovial joints. Arthritis always involves damage to the articular cartilages, but the specific cause varies. For example, arthritis can result from bacterial or viral infection, injury to the joint, metabolic problems, or autoimmune disorders.

Proper synovial function requires healthy articular cartilages. When an articular cartilage has been damaged, the matrix begins to break down; the exposed cartilage changes from a slick, smooth gliding surface to a rough feltwork of bristly collagen fibers. This feltwork drastically increases friction, damaging the cartilage further and producing pain. Eventually the central area of the articular cartilage may completely disappear, exposing the underlying bone.

Fibroblasts are attracted to areas of friction, and they begin tying the opposing bones together with a network of collagen fibers. This network may later be converted to bone, locking the articulating elements into position. Such a bony fusion, called **ankylosis** (an-ke-LŌ-sis), eliminates the friction, but only by the drastic remedy of making movement impossible.

The diseases of arthritis are usually considered as either degenerative or inflammatory. **Degenerative diseases** begin at the articular cartilages, and modification of the underlying bone and inflammation of the joint occur secondarily. **Inflammatory diseases** start with the inflammation of synovial tissues, and damage later spreads to the articular surfaces. We will consider one example of each type.

Osteoarthritis (os-tē-o-ar-THRĪ-tis), also known as **degenerative arthritis** or **degenerative joint disease** (DJD), generally affects older individuals. In the U.S. population, 25 percent of women and 15 percent of men over 60 years of age show signs of this disease. The condition seems to result from cumulative wear and tear on the joint surfaces. Some individuals, however, may have a genetic predisposition to develop osteoarthritis; researchers have isolated a gene linked to the disease. This gene codes for an abnormal form of collagen that differs from the normal protein in only 1 of its 1000 amino acids.

Rheumatoid arthritis is an inflammatory condition that affects roughly 2.5 percent of the adult U.S. population. The cause is uncertain, although allergies, bacteria, viruses, and genetic factors have all been proposed. The synovial membrane becomes swollen and inflamed, a condition known as **synovitis** (sī-nō-VĪ-tis). The cartilaginous matrix begins to break down, and the process accelerates as dying cartilage cells release lysosomal enzymes.

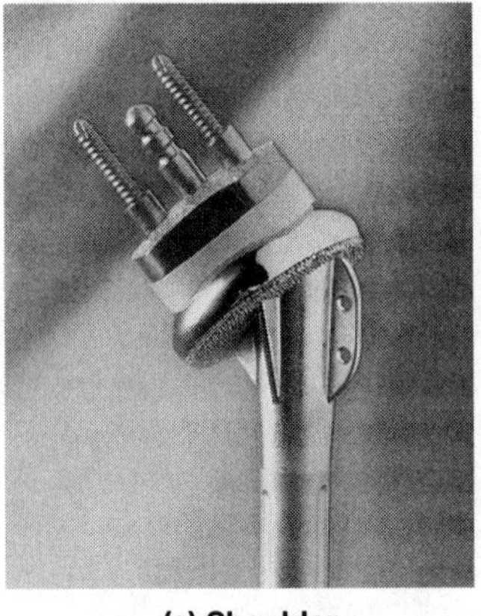
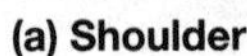
(a) Shoulder

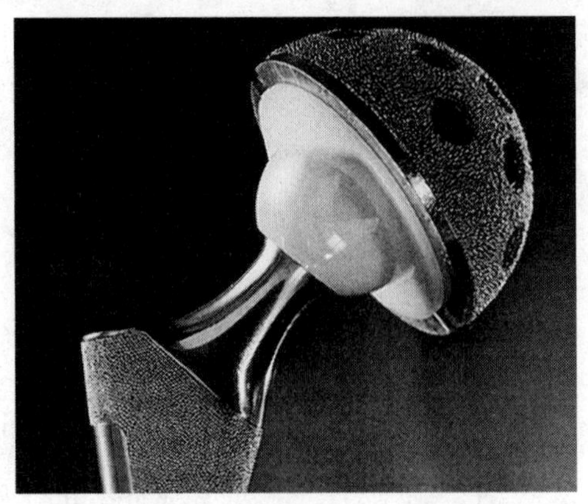
(b) Hip

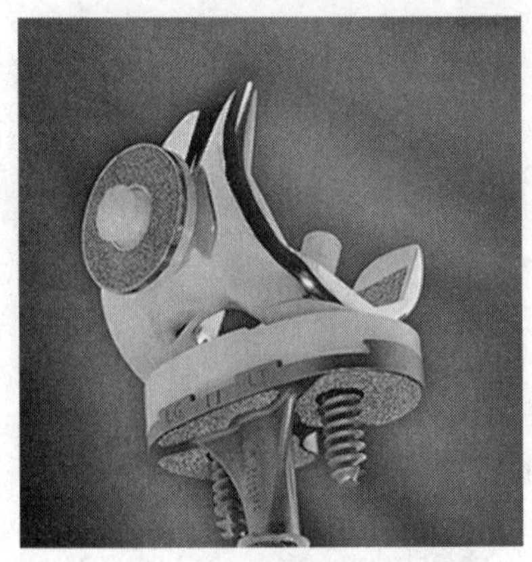
(c) Knee

FAP Ch. 9

•**FIGURE A-25** **Artificial Joints**

Advanced stages of inflammatory and degenerative forms of arthritis produce an inflammation that spreads into the surrounding area. Ankylosis, common in the past when complete rest was routinely prescribed for arthritis patients, is rarely seen today. Regular exercise, physical therapy, drugs that reduce inflammation (such as aspirin), and even immune suppressants (such as *methotrexate*) may slow the progression of the disease. Nutritional supplements such as glucosamine and chondroitin sulfate and other compounds sold to "help repair cartilage" are broken down by digestive enzymes and have no effect on chondrocyte activity. However, some people report less pain while taking these supplements. Surgical procedures can realign or redesign the affected joint. In extreme cases involving the hip, knee, elbow, or shoulder, the defective joint can be replaced by an artificial one. Prosthetic (artificial) joints, such as those shown in Figure A-25•, are weaker than natural ones, but elderly people seldom stress them to their limits.

HIP FRACTURES, AGING, AND PROFESSIONAL ATHLETES FAP *p. 233*

Today, two very different groups of people suffer hip fractures: (1) individuals over age 60, whose bones have been weakened by osteoporosis, and (2) young, healthy professional athletes, who subject their hips to extreme forces. When the injury is severe, the vascular supply to the joint is damaged. As a result, two problems can gradually develop:

1. *Avascular necrosis.* The mineral deposits in the bone of the pelvis and femur are turned over very rapidly, and osteocytes have high energy demands. A reduction in blood flow first injures and then kills the osteocytes. When bone maintenance stops in the affected region, the matrix begins to break down. This process is called *avascular necrosis.*
2. *Degeneration of articular cartilages.* The chondrocytes in the articular cartilages absorb nutrients from the synovial fluid, which circulates around the joint cavity as the bones change position. A fracture of the femoral neck is generally followed by joint immobility and poor circulation to the synovial membrane. The combination results in a gradual deterioration of the articular cartilages of the femur and acetabulum.

In recent years the frequency of hip fractures and fracture-dislocations among young, healthy professional athletes has increased dramatically. The dislocations tear blood vessels in the articular capsule and along the femoral neck, where the capsular fibers attach. The result is typically avascular necrosis and the degeneration of the articular cartilages at the hip.

In a "total hip" replacement, the damaged portion of the femur is removed, and an artificial femoral head and neck are attached by a spike that extends into the marrow cavity of the shaft. Special cement is used to anchor it in place and to attach a new articular surface to the acetabulum. Joint replacement eliminates the pain and restores full range of motion. However, prosthetic joints, such as those shown in Figure A-25•, are weaker than natural ones. They are generally implanted in older people who seldom stress them to their limits. Whether an artificial hip can tolerate the stresses of professional sports is an open question. It is also an important question, because the frequency of such injuries is increasing.

[EXPLORE]

Additional resources, such as interactive tutorials, critical-thinking questions, clinical problems, and case studies, are available through the CD-ROM and Website for your textbook. To begin your exploration, launch the *Martini Interactive* CD-ROM or visit the Companion Website (http://www.prenhall.com/martini/fap5) and select Chapter 6, 7, 8, or 9.

The Muscular System

The muscular system includes more than 700 skeletal muscles that are directly or indirectly attached to the skeleton by tendons or aponeuroses. The muscular system produces movement as the contractions of skeletal muscles pull on the attached bones. Muscular activity does not always result in movement, however; it can also help stabilize skeletal elements and prevent movement. Skeletal muscles are also important in guarding entrances or exits of internal passageways, such as those of the digestive, respiratory, urinary, or reproductive systems, and in generating heat to maintain a stable body temperature.

Skeletal muscles contract only under the command of the nervous system. For this reason, clinical observation of muscular activity may provide indirect information about the nervous system as well as direct information about the muscular system. The assessment of facial expressions, posture, speech, and gait can be an important part of a physical examination. Classical signs of muscle disorders include the following:

- *Gower's sign* is a distinctive method of standing up from a sitting or lying position on the floor. This sign is characteristic of young children with *muscular dystrophy* (p. 66). They move from a sitting position to a standing position by pushing the trunk off the floor with the hands and then moving the hands to the knees. The hands are then used as braces to force the body into the standing position. This extra support is necessary, because the pelvic muscles are too weak to swing the weight of the trunk over the legs.
- *Ptosis* is a drooping of the upper eyelid. It may be seen in *myasthenia gravis* (p. 67), *botulism* (p. 67), and *myotonic dystrophy* (p. 67) or it may follow damage to the cranial nerve (N III) that innervates the *levator palpebrae superioris muscle* of the eyelid.
- A *muscle mass,* an abnormal dense region within a muscle, is sometimes seen or felt in a skeletal muscle. A muscle mass results from torn muscle tissue; a hematoma; a parasitic infection, such as *trichinosis* (p. 64); or bone deposition, as in *myositis ossificans* (p. 55).
- Abnormal contractions may indicate problems with the muscle tissue or its innervation. *Muscle spasticity* exists when a muscle has excessive muscle tone. A *muscle spasm* is a sudden, strong, and painful involuntary contraction.
- *Muscle flaccidity* exists when the relaxed skeletal muscle appears soft and loose and its contractions are very weak or absent.
- *Muscle atrophy* is skeletal muscle deterioration, or *wasting,* due to disuse, immobility, or interference with the normal innervation.
- Abnormal patterns of muscle movement, such as *tics, choreiform movements,* or *tremors,* and muscular paralysis are generally caused by nervous system disorders. We will describe these movements further in sections dealing with abnormal nervous system function.

SIGNS AND SYMPTOMS OF MUSCULAR SYSTEM DISORDERS

Two common symptoms of muscular disorders are pain and weakness in the affected skeletal muscles. The potential causes of muscle pain include the following:

- *Muscle trauma.* Examples of traumatic injuries to a skeletal muscle include a laceration, a deep bruise or crushing injury, a muscle tear, and a damaged tendon.
- *Muscle infection.* Skeletal muscles can be infected by viruses, as in some forms of *myositis* (muscle inflammation), or colonized by parasitic worms, such as those responsible for *trichinosis* (p. 64). These types of infections generally produce pain that is restricted to the involved muscles. Diffuse muscle pain can develop in the course of other infectious diseases, such as influenza.
- *Related problems with the skeletal system.* Muscle pain can result from skeletal problems, such as arthritis (p. 60) or a sprained ligament near the point of muscle origin or insertion.
- *Problems with the nervous system.* Muscle pain can be related to the inflammation of sensory neurons or the stimulation of pain pathways in the central nervous system (CNS).

Muscle strength can be evaluated by applying an opposite force against a specific action. For example, an examiner might exert a gentle extending force on a patient's forearm while asking the patient to flex the elbow. Because the muscular and nervous systems are so closely interrelated, a single symptom, such as muscle weakness, can have a variety of causes (Figure A-26•). Muscle weakness can also develop as

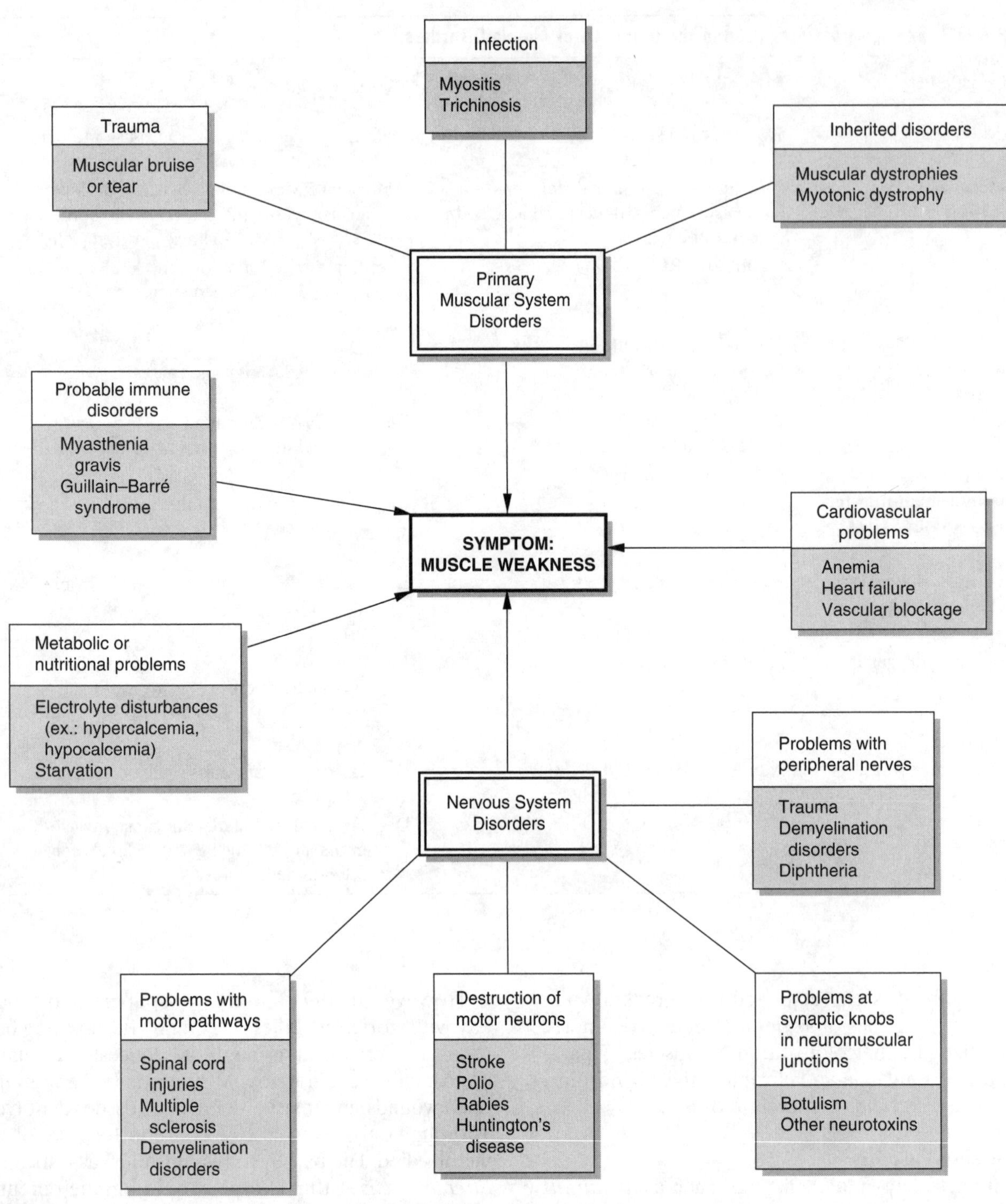

•FIGURE A-26 Potential Causes of Muscle Weakness

FAP Ch. 10

a consequence of a condition that affects the entire body, such as anemia or starvation.

DIAGNOSTIC TESTS

Table A-17 provides information about representative diagnostic procedures for muscular system abnormalities. Figure A-27• (p. 64) provides an overview of muscular system disorders.

DISORDERS OF THE MUSCULAR SYSTEM

DISRUPTION OF NORMAL MUSCLE ORGANIZATION

FAP p. 272

A variety of disorders are characterized by a disruption in the structural organization of skeletal muscles. We will consider only a few representative examples:

TABLE A-17 Examples of Tests Used in the Diagnosis of Muscle Disorders

Diagnostic Procedure	Method and Result	Representative Uses
Muscle biopsy	Removal of a small amount of affected muscle tissue	Determines muscle disease; also used to detect cyst formation or larvae to diagnose trichinosis
Electromyography (EMG)	Insertion of a probe that transmits measurements of electrical activity in contracting muscles	Abnormal EMG readings occur in disorders such as myasthenia gravis, amyotrophic lateral sclerosis (ALS), and muscular dystrophy.
MRI	Standard MRI	Useful in the detection of muscle diseases and associated soft tissue damage
Laboratory Test	**Normal Values in Blood Plasma or Serum**	**Significance of Abnormal Values**
Aldolase	Adults: less than 8 U/l 22–59 mU/l (SI units)	Elevated levels occur in muscular dystrophy but not in myasthenia gravis or multiple sclerosis.
Aspartate aminotransferase (AST or SGOT)	Adults: 7–40 U/l Children: 15–55 U/l	Elevated levels occur in some muscle diseases and can occur after exercise.
Creatine phosphokinase (CPK or CK)	Adults: 30–180 IU/l	Elevated levels occur in muscular dystrophy and myositis.
Lactate dehydrogenase (LDH)	Adults: 100–190 U/l	Elevated levels occur in some muscle diseases and in lactic acidosis.
Electrolytes		
Potassium	Adults: 3.5–5.0 mEq/l	Decreased levels of potassium can cause muscle weakness.
Calcium	Adults: 8.5–10.5 mg/dl	Decreased calcium levels can cause muscle tremors and tetany; increased levels can cause muscle flaccidity.

FAP Ch. 10

necrotizing fasciitis, characterized by a breakdown in the connective tissues of skeletal muscles; trichinosis, characterized by the colonization of muscles by parasites; and fibromyalgia and chronic fatigue syndrome, two muscle disorders of uncertain origin.

Necrotizing Fasciitis

Several bacteria produce enzymes such as *hyaluronidase* or *cysteine protease.* Hyaluronidase breaks down hyaluronic acid and disassembles the associated proteoglycans. Cysteine protease breaks down connective tissue proteins. The bacteria that produce these enzymes are dangerous, because they can spread rapidly by liquefying the matrix and dissolving the intercellular cement that holds epithelial cells together. The *streptococci* are one group of bacteria that secrete both of these enzymes. *Streptococcus A* bacteria are involved in many human diseases, most notably "strep throat," an infection of the pharynx. In most cases, the immune response is sufficient to contain and ultimately defeat these bacteria before extensive tissue damage has occurred.

However, in 1994 tabloid newspapers had a field day with stories of "killer bugs" and "flesh-eating bacteria" that terrorized residents of Gloucester, England. The details were horrific: Minor cuts became major open wounds, and interior connective tissues dissolved. Although only seven cases were reported, five of the victims died. The pathogen responsible was a strain of *Streptococcus A* that overpowered immune defenses and swiftly invaded and destroyed soft tissues. Moreover, the pathogens eroded their way along the fascial wrapping that covers skeletal muscles and other organs. This condition is **necrotizing fasciitis.** In most cases, myositis also occurs, followed by degeneration of the muscle tissue. Some form of soft-tissue invasion occurs 75–150 times annually in the United States; this number is 5–10 percent of the invasive *Streptococcus A* wound infections reported each year.

Trichinosis

Trichinosis (trik-i-NŌ-sis; *trichos,* hair + *nosos,* disease) results from infection by the parasitic nematode worm *Trichinella spiralis.* Symptoms include diarrhea, weak-

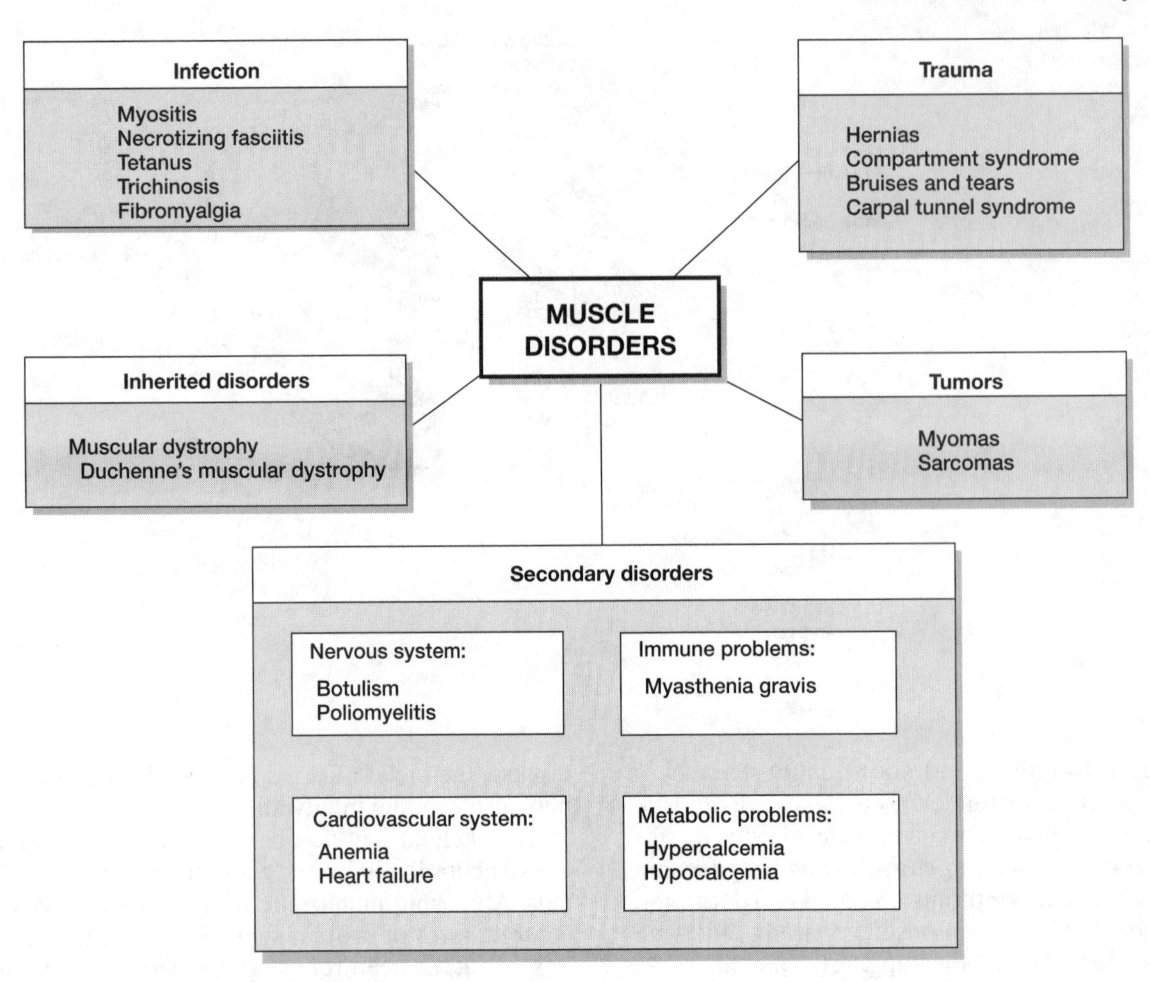

•**FIGURE A-27 Disorders of the Muscular System**

FAP Ch. 10

ness, and muscle pain and are caused by the invasion of skeletal muscle tissue by larval worms, which create small pockets within the perimysium and endomysium (Figure A-28•). Muscles of the tongue, eyes, diaphragm, chest, and legs are most often affected.

Larvae are common in the flesh of pigs, horses, dogs, and other mammals. The larvae are killed when the meat is cooked; people are most often exposed by eating undercooked infected pork. Once eaten, the larvae mature within the human intestinal tract, where they mate and produce eggs. The new generation of larvae then migrates through the body tissues to reach the muscles, where they complete their early development. The migration and subsequent settling produce a generalized achiness, muscle and joint pain, and swelling in infected tissues. An estimated 1.5 million Americans carry *Trichinella* in their muscles, and up to 300,000 new infections occur each year. The mortality rate for people who have symptoms severe enough to require treatment is approximately 1 percent.

Fibromyalgia and Chronic Fatigue Syndrome

Fibromyalgia (*-algia,* pain) is a disorder that has formally been recognized only since the mid-1980s. Although first described in the early 1800s, the condition is still somewhat controversial, because the reported symptoms cannot be linked to any anatomical or physiological abnormalities. However, physicians now recognize a distinctive pattern of symptoms that warrant consideration as a clinical entity.

Fibromyalgia may be the most common musculoskeletal disorder affecting women under 40 years of age; as many as 6 million individuals in the United States have this condition. Symptoms include chronic aches, pain, and stiffness, plus multiple tender points at specific, characteristic locations. The four most common tender points are (1) the medial surface of the knee, (2) distal to the lateral epicondyle of the humerus, (3) near the external occipital crest of the skull, and (4) at the junction between the second rib and its costal cartilage. An additional clinical criterion is that the pains and stiffness cannot be explained by other mechanisms. Individuals with this condition frequently report chronic fatigue; they feel tired on awakening and often complain of waking up repeatedly during the night. Fibromyalgia may also be associated with other conditions, such as irritable bowel syndrome (p. 167) or chronic depression.

Most of these symptoms could be attributed to other problems. For example, chronic depression can

•**FIGURE A-28** **The Life Cycle of *Trichinella spiralis***

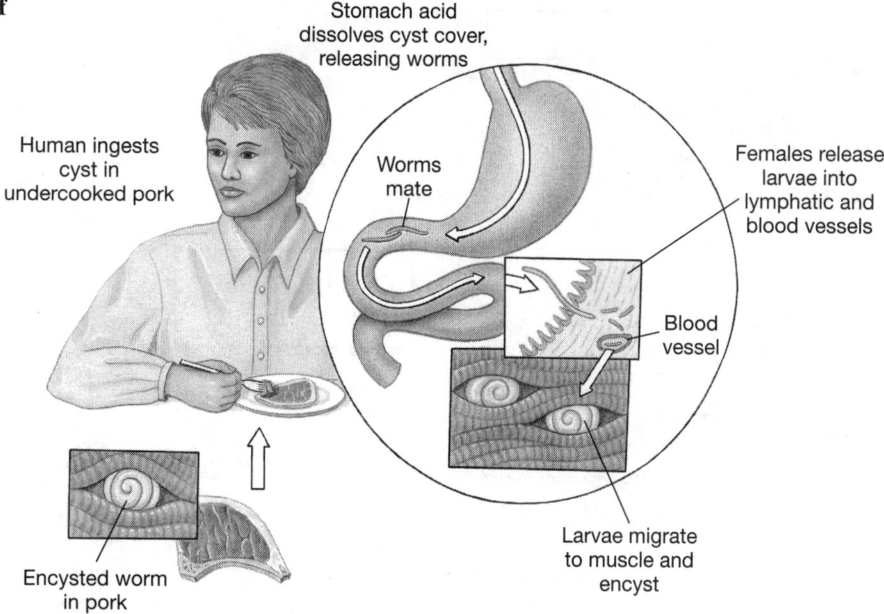
Stomach acid dissolves cyst cover, releasing worms
Human ingests cyst in undercooked pork
Worms mate
Females release larvae into lymphatic and blood vessels
Blood vessel
Larvae migrate to muscle and encyst
Encysted worm in pork

FAP Ch. 10

by itself lead to fatigue and poor-quality sleep. As a result, the pattern of tender points is the diagnostic key to fibromyalgia. This symptom clearly distinguishes fibromyalgia from **chronic fatigue syndrome** (CFS). The current symptoms accepted as a definition of CFS include (1) sudden onset, generally following a viral infection; (2) disabling fatigue; (3) muscle weakness and pain; (4) sleep disturbance; (5) fever; and (6) enlargement of cervical lymph nodes. Roughly twice as many women as men are diagnosed with CFS.

Attempts to link either fibromyalgia or CFS to a viral infection or to some physical or psychological trauma have not been successful, and the causes remain mysterious. For both conditions, treatment at present is limited to relieving symptoms when possible. For example, anti-inflammatory medications may help relieve pain; antidepressant medications may improve sleep patterns and reduce depression; and exercise programs may help maintain normal range of motion.

The Muscular Dystrophies FAP *p. 277*

The **muscular dystrophies** (DIS-trō-fē-z) are inherited diseases that produce progressive muscle weakness and deterioration. One of the most common and best understood conditions is **Duchenne's muscular dystrophy (DMD).** This form of muscular dystrophy appears in childhood, commonly between the ages of 3 and 7. The condition generally affects only males; the incidence is roughly 30 per 100,000 male births. A progressive muscular weakness develops, and the child typically requires a wheelchair by age 12. Most individuals die before age 20 due to respiratory paralysis or cardiac problems. Skeletal muscles are primarily affected, although for some reason the facial muscles continue to function normally. In later stages of the disease, the facial muscles and cardiac muscle tissue may also become involved.

The skeletal muscle fibers in a person with DMD are structurally different from those of other individuals. Abnormal membrane permeability, cholesterol content, rates of protein synthesis, and enzyme composition have been reported. Individuals with DMD also lack the protein *dystrophin,* found in normal muscle fibers. Dystrophin is a large protein attached to the inner surface of the sarcolemma, near the triads. Although the functions of this protein remain uncertain, dystrophin is suspected to play a role in the regulation of calcium ion channels in the sarcolemma. In children with DMD, calcium channels remain open for an extended period, and calcium levels rise in the sarcoplasm to the point at which key proteins denature. The muscle fiber then degenerates. Researchers have recently identified and cloned the gene for dystrophin production; that gene is located on the X chromosome. Rats with DMD have been cured by the insertion of this gene into their muscle fibers, a technique that may soon be used to treat humans.

The inheritance of DMD is sex-linked: Women carrying the defective genes are unaffected, but each of their male children will have a 50 percent chance of developing DMD. Now that the specific location of the gene has been identified, it is possible to determine whether a woman is carrying the defective gene. It is also possible to use an innovative prenatal test to determine whether a fetus has inherited the condition. In this procedure, a small sample of fluid is collected from the membranous sac that surrounds the fetus. This fluid contains fetal cells called *amniocytes,* which are collected and cultivated in the laboratory. Researchers then insert a gene called *MyoD,* which triggers the differentiation of the amniocytes into

skeletal muscle fibers. These cells can then be tested not only for the signs of muscular dystrophy but also for indications of other inherited muscular disorders.

Attempts have been made to give children with DMD injections that contain donated normal myoblasts, in the hope that these myoblasts will fuse with developing muscle fibers and will provide normal dystrophin-production genes. Results have been inconclusive. Animal studies indicate that symptoms of DMD develop when fewer than 20 percent of the muscle fibers contain normal dystrophin. In one recent study that used muscle biopsies, 1–10 percent of the muscle fibers of treated children contained normal dystrophin. Efforts to improve the delivery method and increase the treatment's effects are under way.

Myotonic Dystrophy **Myotonic dystrophy** is a form of muscular dystrophy that occurs in the United States at an incidence of 13.5 per 100,000 population. Symptoms may develop in infancy, but more commonly they develop after puberty. As with other forms of muscular dystrophy, adults developing myotonic dystrophy experience a gradual reduction of muscle strength and control. Problems with other systems, especially the cardiovascular and digestive systems, typically develop. There is no effective treatment.

The inheritance of myotonic dystrophy is unusual, because children of an individual with myotonic dystrophy commonly develop symptoms more severe than those of the parent. The increased severity of the condition appears to be related to the presence of multiple copies of a specific gene on chromosome 19. For some reason, the nucleotide sequence of that gene gets repeated several times, and the number can increase from generation to generation. This phenomenon has been called a "genetic stutter." The greater the number of copies, the more severe the symptoms. It is not known why the stutter develops or how the genetic duplication affects the severity of the condition. Evidence indicates that the extra nucleotides in some way interfere with the transcription of an adjacent gene involved with the control of muscle tone.

PROBLEMS WITH THE CONTROL OF MUSCLE ACTIVITY FAP *pp. 281, 298*

Another group of disorders interferes with normal neuromuscular communication by affecting either the nerve's ability to issue commands or the muscle's ability to respond.

Botulism

Botulinus (bot-ū-LĪ-nus) **toxin** prevents the release of acetylcholine (ACh) at synaptic terminals. It thus produces a severe and potentially fatal paralysis of skeletal muscles. A case of botulinus poisoning is called **botulism.**[1] The toxin is produced by the bacterium *Clostridium botulinum,* which does not need oxygen to grow and reproduce. Because this bacterium can live quite well in a sealed can or jar, most cases of botulism are linked to improper canning or storing procedures, followed by failure to cook the food adequately before it is eaten. Canned tuna or beets, smoked fish, and cold soups are the foods most commonly linked to botulism. Boiling for a half-hour destroys both the toxin and the bacteria.

Symptoms generally begin 12–36 hours after a contaminated meal is eaten. The initial symptoms are typically disturbances in vision, such as double vision or a painful sensitivity to bright light. These symptoms are followed by other sensory and motor problems, including blurred speech and an inability to stand or walk. Roughly half of botulism patients experience intense nausea and vomiting. These symptoms persist for days to weeks, followed by a gradual recovery; some patients are still in recovery after a year.

The major risk of botulinus poisoning is respiratory paralysis and death by suffocation. Treatment is supportive: bed rest, observation, and, if necessary, the use of a mechanical respirator. In severe cases, an antitoxin and drugs that promote the release of ACh, such as *guanidine hydrochloride,* may be administered. The overall mortality rate in the United States is about 10 percent.

FAP Ch. 10

Myasthenia Gravis

Myasthenia gravis (mī-as-THĒ-nē-uh GRA-vis) is characterized by a general muscular weakness that tends to be most pronounced in the muscles of the arms, head, and chest. The first symptom is generally a weakness of the eye muscles and drooping eyelids. Facial muscles are commonly weak as well, and the individual develops a peculiar smile known as the "myasthenic snarl." As the disease progresses, weakness of the pharynx leads to problems with chewing and swallowing, and holding the head upright becomes difficult.

The muscles of the upper chest and upper extremities are next to be affected. All the voluntary muscles of the body may ultimately be involved. Severe myasthenia gravis produces respiratory paralysis, with a mortality rate of 5–10 percent. However, the disease does not always progress to such a life-threatening stage. For example, roughly 20 percent of people with the disease experience no symptoms other than eye problems.

The condition results from a decrease in the number of ACh receptors on the motor end plate. Before the remaining receptors can be stimulated enough to trigger a strong contraction, the ACh molecules are destroyed by cholinesterase. As a result, muscular weakness develops.

The primary cause of myasthenia gravis appears to be a malfunction of the immune system. Roughly

[1] This disorder was described more than 200 years ago by German physicians who treated patients poisoned by dining on contaminated sausages. *Botulus* is the Latin word for sausage.

70 percent of individuals with myasthenia gravis have an abnormal thymus, an organ involved with the maintenance of normal immune function. In myasthenia gravis, the immune response attacks the ACh receptors of the motor end plate as if they were foreign proteins. For unknown reasons, 1.5 times as many women as men are affected. The typical age at onset is 20–30 for women, versus over 60 for men. Estimates of the incidence of this disease in the United States range from 2 to 10 cases per 100,000 population.

One approach to therapy involves the administration of drugs, such as *neostigmine,* that are termed **cholinesterase inhibitors.** As their name implies, these compounds are enzyme inhibitors; they tie up the active sites at which cholinesterase normally binds ACh. With cholinesterase activity reduced, the concentration of ACh at the synapse can rise enough to stimulate the surviving receptors and produce muscle contraction. Corticosteroid therapy is typically beneficial, as is the surgical removal of the thymus.

FAP Ch. 10

Polio

Because skeletal muscles depend on their motor neurons for stimulation, disorders that affect the nervous system can have an indirect effect on the muscular system. **Polio** is caused by the *poliovirus,* a virus that does not produce clinical symptoms in roughly 95 percent of infected individuals. The virus produces variable symptoms in the remaining 5 percent. Some individuals develop a nonspecific illness resembling the flu. Other individuals develop a brief *meningitis* (p. 76), an inflammation of the protective membranes that surround the CNS. In still another group of people, the virus attacks somatic motor neurons in the CNS.

In this third form of the disease, the individual develops a fever 7–14 days after infection. The fever subsides but recurs roughly a week later, accompanied by muscle pain, cramping, and paralysis of one or more limbs. Respiratory paralysis may also occur, and the mortality rate of this form of polio is 2–5 percent of children and 15–30 percent of adults. If the individual survives, some degree of recovery generally occurs over a period of up to 6 months.

For unknown reasons, the survivors of paralytic polio may develop progressive muscular weakness 20–30 years after the initial infection. This *postpolio syndrome* is characterized by fatigue, muscle pain and weakness, and, in some cases, muscle atrophy. There is no treatment for this condition, although rest seems to help.

Polio has been almost completely eliminated from the U.S. population due to a successful immunization program. In 1954, 18,000 new cases occurred in the United States; there were 8 in 1976, and none since 1994. The World Health Organization reports that polio has been eradicated from the entire Western Hemisphere. Unfortunately, many parents today refuse to immunize their children against the poliovirus, because they assume that the disease has been "conquered." Failure to immunize is a mistake, because (1) there is still *no cure* for polio, (2) the virus remains in the environment in many areas of the world, and (3) up to 38 percent of children ages 1–4 years have not been immunized. A major epidemic could therefore develop very quickly if the virus were brought into the United States from another part of the world.

For years, the vaccine that has been used is the oral Sabin vaccine, preferred for its ease of administration and better immune stimulation than the injectable vaccine. However, unlike the injected vaccine, the oral vaccine carries a 1 in 1 million risk that the immunized person will develop polio. In 1996 the Centers for Disease Control and Prevention (CDC) recommended the use of either a combination of injected and oral vaccines or the injected vaccine alone.

HERNIAS FAP *p. 335*

When the abdominal muscles contract forcefully, pressure in the abdominopelvic cavity can increase dramatically. That pressure is applied to internal organs. If the individual exhales at the same time, the pressure is relieved because the diaphragm can move upward as the lungs collapse. But during vigorous isometric exercises or when lifting a weight while holding your breath, pressure in the abdominopelvic cavity can rise to 106 kg/cm^2 (1500 lb/in.^{2}), roughly 100 times the normal pressure. A pressure that high can cause a variety of problems, including hernias. A **hernia** develops when a visceral organ protrudes abnormally through an opening in a muscular wall or partition. There are many types of hernias; we will consider only *inguinal* (groin) *hernias* and *diaphragmatic hernias* here.

Late in the development of males, the testes descend into the scrotum by passing through the abdominal wall at the **inguinal canals.** In adult males, the sperm ducts and associated blood vessels penetrate the abdominal musculature at the inguinal canals as the *spermatic cords,* on their way to the abdominal reproductive organs. In an inguinal hernia, the inguinal canal enlarges and the abdominal contents, such as a portion of the greater omentum, small intestine, or (more rarely) urinary bladder, are forced into the inguinal canal (Figure A-29•). If the herniated structures become trapped or twisted, surgery may be required to prevent serious complications. Inguinal hernias are not always caused by unusually high abdominal pressures. Injuries to the abdomen, or inherited weakness or distensibility of the canal, can have the same effect.

The esophagus and major blood vessels pass through an opening in the diaphragm, the muscle that separates the thoracic and abdominopelvic cavities. In a **diaphragmatic hernia,** also called a *hiatal hernia* (hī-Ā-tal; *hiatus,* a gap or opening), abdominal organs

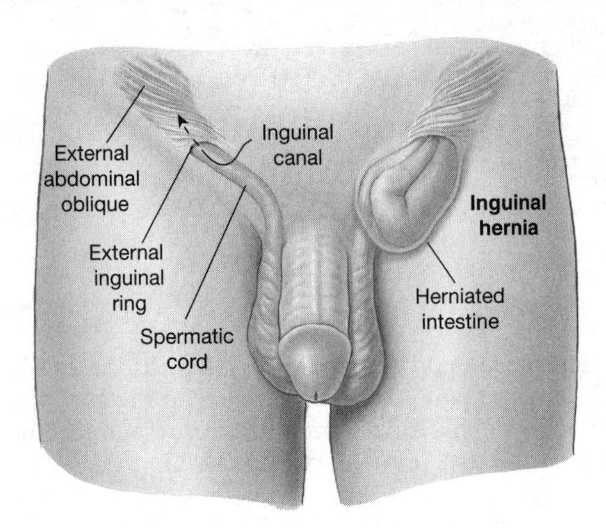

•**FIGURE A-29** **An Inguinal Hernia**

slide into the thoracic cavity, typically through the *esophageal hiatus,* the opening used by the esophagus. The severity of the condition depends on the location and size of the herniated organ or organs. Hiatal hernias are very common, and most go unnoticed. Radiologists see them in about 30 percent of individuals whose upper gastrointestinal tracts are examined with barium contrast techniques. When clinical complications develop, they generally occur because abdominal organs that have pushed into the thoracic cavity are exerting pressure on structures or organs there. As is the case with inguinal hernias, a diaphragmatic hernia can result from congenital factors or from an injury that weakens or tears the diaphragm.

SPORTS INJURIES

FAP *p. 339*

Sports injuries affect amateurs and professionals alike. A 5-year study of college football players indicated that 73.5 percent experienced mild injuries, 21.5 percent moderate injuries, and 11.6 percent severe injuries during their playing careers. Contact sports are not the only activities that show a significant injury rate; a study of 1650 joggers running at least 27 miles a week reported 1819 injuries in a single year.

Muscles and bones respond to increased use by enlarging and strengthening. Poorly conditioned individuals are therefore more likely than people in good condition to subject their bones and muscles to intolerable stresses. Training is also important in minimizing the use of antagonistic muscle groups and in keeping joint movements within the intended ranges of motion. Planned warm-up exercises before athletic events stimulate circulation, improve muscular performance and control, and help prevent injuries to muscles, joints, and ligaments. Stretching exercises stimulate blood flow to muscles and help keep ligaments and joint capsules supple. Such conditioning extends the range of motion and prevents sprains and strains when sudden loads are applied.

Dietary planning can also be important in preventing injuries to muscles during endurance events, such as marathon running. Emphasis has commonly been placed on the importance of carbohydrates, leading to the practice of "carbohydrate loading" before a marathon. But while operating within aerobic limits, muscles also utilize amino acids extensively, so an adequate diet must include both carbohydrates and proteins.

Improved playing conditions, equipment, and regulations also play a role in reducing the incidence of sports injuries. Jogging shoes, ankle or knee braces, helmets, mouth guards, and body padding are examples of equipment that can be effective. The substantial penalties now earned for personal fouls in contact sports have reduced the numbers of neck and knee injuries.

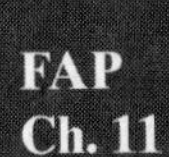

Several injuries common to those engaged in active sports can also affect nonathletes, although the primary causes may be very different. A partial listing of activity-related conditions includes the following:

- *Bone bruise:* bleeding within the periosteum of a bone
- *Bursitis:* an inflammation of the bursae at joints
- *Muscle cramps:* prolonged, involuntary, and painful muscular contractions
- *Sprains:* tears or breaks in ligaments or tendons
- *Strains:* tears in muscles
- *Stress fractures:* cracks or breaks in bones subjected to repeated stresses or trauma
- *Tendinitis:* an inflammation of the connective tissue surrounding a tendon

We have discussed many of these conditions in related sections of the text.

Finally, many sports injuries would be prevented if people who engage in regular exercise would use common sense and recognize their personal limitations. It can be argued that some athletic events, such as the ultramarathon, place such excessive stresses on the cardiovascular, muscular, respiratory, and urinary systems that these events cannot be recommended, even for athletes in peak condition.

[EXPLORE]

Additional resources, such as interactive tutorials, critical-thinking questions, clinical problems, and case studies, are available through the CD-ROM and Website for your textbook. To begin your exploration, launch the *Martini Interactive* CD-ROM or visit the Companion Website (http://www.prenhall.com/martini/fap5) and select Chapter 10 or 11.

The Nervous System

The nervous system is a highly complex and interconnected network of neurons and supporting neuroglia. Neural tissue is extremely delicate, and the characteristics of the extracellular environment must be kept within narrow homeostatic limits. When homeostatic regulatory mechanisms break down under the stress of genetic or environmental factors, infection, or trauma, symptoms of neurological disorders appear.

Literally hundreds of disorders affect the nervous system. A *neurological examination* attempts to trace the source of the problem through evaluation of the sensory, motor, behavioral, and cognitive functions of the nervous system. Figure A-30• introduces several major categories of nervous system disorders. We will discuss many of these examples in the sections that follow. Table A-18 summarizes representative infectious diseases of the nervous system.

THE SYMPTOMS OF NEUROLOGICAL DISORDERS

The nervous system has varied and complex functions, and the symptoms of neurological disorders are equally diverse. However, a few symptoms accompany a wide variety of disorders:

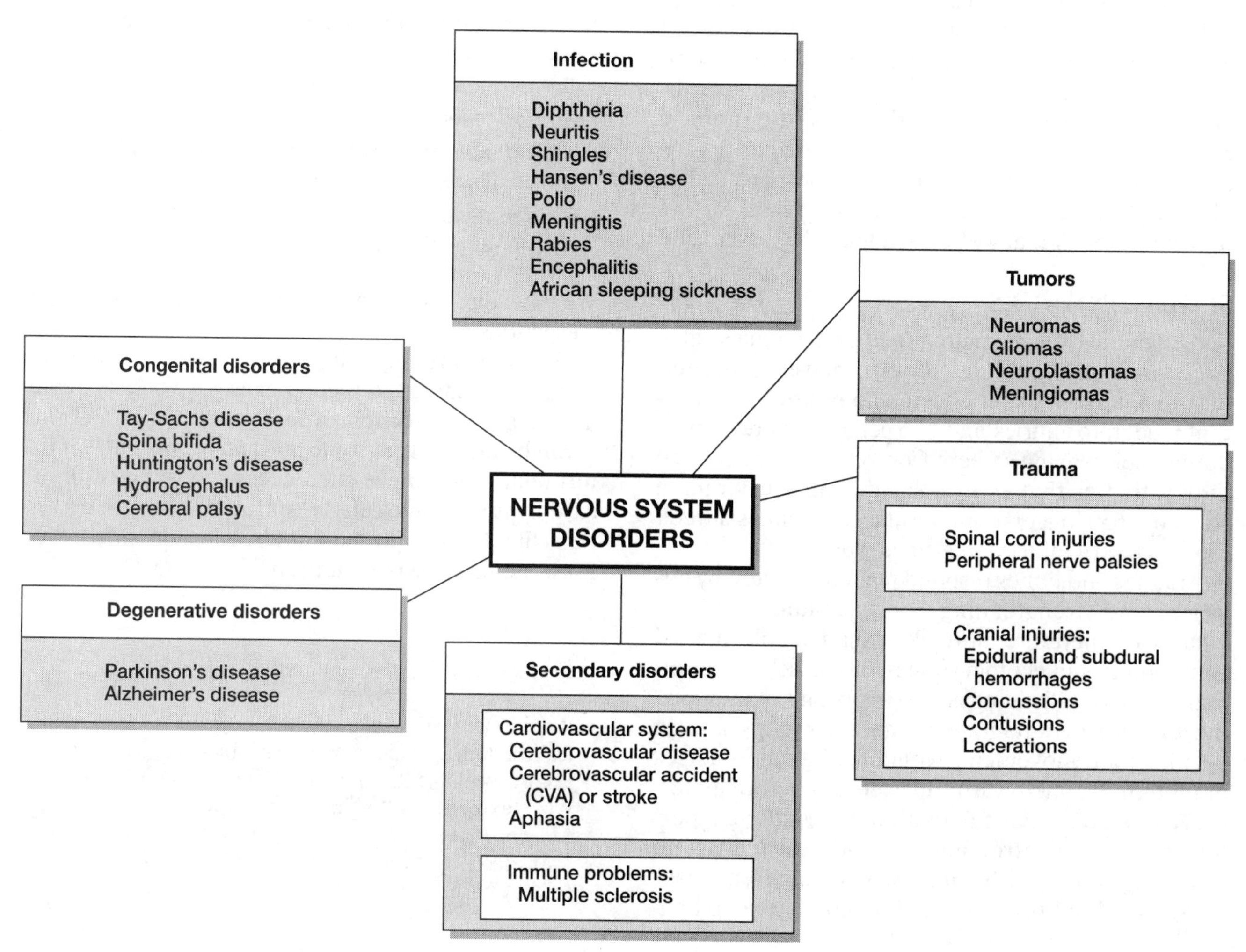

•**FIGURE A-30** **Nervous System Disorders**

TABLE A-18 Examples of Infectious Diseases of the Nervous System

Disease	Organism(s)	Description
Bacterial Diseases		
Hansen's disease (leprosy)	*Mycobacterium leprae*	Progresses slowly; invades nerves and produces sensory loss and motor paralysis; cartilage and bone may degenerate.
Bacterial meningitis		Inflammation of the spinal or cranial meninges
	Haemophilus influenzae	*Haemophilus* meningitis; usually infects children (age 2 months–5 years); vaccine available
	Neisseria meningitidis	Meningococcal meningitis; usually infects children and adults (age 5–40 years); treatment with antibiotics
	Streptococcus pneumoniae	Streptoccocal meningitis; usually infects adults over age 40; high mortality rate (40%)
Brain abscesses	Various bacteria	Infection increases in size and compresses the brain.
Viral Diseases		
Poliomyelitis	Polioviruses	Polio has different forms; only one attacks motor neurons, leading to paralysis of limbs and muscle atrophy. Vaccine is available.
Rabies	Rabies virus	Virus invades the central nervous system through peripheral nerves. Untreated cases are fatal; treatment involves rabies antitoxin.
Encephalitis	Various encephalitis viruses	Inflammation of the brain; fever and headache; no vaccine is available. Transmission occurs by mosquitoes. Eastern equine encephalitis is most lethal (50–75% mortality rate).
Parasitic Diseases		
African sleeping sickness	*Trypanosoma brucei*	Caused by a flagellated protozoan; infection occurs through bite of tsetse fly; infects blood, lymph nodes, and then nervous system. Symptoms include headache, tiredness, weakness, and paralysis, before coma and death; no vaccine is available.

FAP Ch. 12

- *Headache.* Roughly 90 percent of headaches are *tension headaches,* which are thought to be due to muscle tension, or *migraine headaches,* which have both neurological and circulatory origins (see p. 72). Neither of these conditions is life-threatening.
- *Muscle weakness.* Muscle weakness can have an underlying neurological basis, as we noted in the section "Signs and Symptoms of Muscular System Disorders" (see Figure A-26•, p. 63). The examiner must determine the origin of the symptom before treatment can be prescribed. Myopathies (muscle disease) must be differentiated from neurological diseases such as demyelinating disorders, neuromuscular junction dysfunction, and peripheral nerve damage.
- *Paresthesias.* Loss of feeling, numbness, or tingling sensations may develop after damage to (1) a sensory nerve (cranial or spinal nerve) or (2) sensory pathways in the central nervous system (CNS). The effects can be temporary or permanent. For example, a *pressure palsy* may last a few minutes, whereas the paresthesia that develops distal to an area of severe spinal cord damage will probably be permanent (p. 78).

THE NEUROLOGICAL EXAMINATION

FAP *p. 361*

During a physical examination, information about the nervous system is obtained indirectly by assessing sensory, motor, and intellectual functions. Examples of factors noted in the physical examination include:

- *State of consciousness.* There are many levels of consciousness, ranging from unconscious and incapable of being aroused, to fully alert and attentive, to hyperexcitable. We introduce the names assigned to the various levels of consciousness in a later section (p. 85).
- *Reflex activity.* The general state of the nervous system, and especially the state of peripheral sensory and motor innervation, can be checked by testing specific reflexes (p. 79). For example, the *knee-jerk reflex* will not be normal if damage has occurred in associated segments of the lumbar spinal cord, their spinal nerve roots, or the peripheral nerves involved in the reflex. Musculoskeletal abnormalities also affect reflex responses, but such abnormalities can usually be detected on further examination.

- *Abnormal speech patterns.* Normal speech involves intellectual processing, motor coordination at the speech centers of the brain, precise respiratory control, the regulation of tension in the vocal cords, and the adjustment of the musculature of the palate and face. Problems with the selection, production, or use of words commonly follow damage to the cerebral hemispheres, as in a stroke (p. 128).
- *Abnormal motor patterns.* An individual's posture, balance, and mode of walking, or *gait,* are useful indicators of the level of motor coordination. Clinicians also ask about abnormal involuntary movements that may indicate a *seizure,* a temporary disorder of cerebral function (p. 84).

A number of diagnostic procedures and laboratory tests can be used to obtain additional information about the status of the nervous system. Table A-19 summarizes information about these procedures.

FAP Ch. 12

DISORDERS OF THE NERVOUS SYSTEM

HEADACHES

Almost everyone has experienced a **headache** at one time or another. Diagnosis and treatment pose a number of problems, primarily because, as we noted earlier, headaches can be produced by a wide variety of conditions. The most common causes of headache are either vascular or muscular problems.

Most headaches do not merit a visit to a neurologist. The vast majority of headaches are associated with muscle tension, such as tight neck muscles, but a variety of other factors can also be responsible. For example, headaches may develop due to one of the following problems:

- *CNS problems,* such as infections (meningitis, encephalitis, rabies) or brain tumors
- *Trauma,* such as a blow to the head (p. 48)
- *Cardiovascular disorders,* such as a *stroke* (p. 128)
- *Metabolic disturbances,* such as low blood sugar
- *Related muscle tension,* such as stiff neck or temporomandibular joint (TMJ) syndrome

Migraine headaches affect roughly 5 percent of the population. An individual with a *classic migraine* experiences visual or other sensory signals that an attack is imminent. The headache pain may then be accompanied by disturbances in vision or somatic sensation, extreme anxiety, nausea, or disorientation. The symptoms generally persist for several hours. A *common migraine* typically lacks warning signs.

Evidence indicates that migraine headaches begin at a portion of the mesencephalon known as the *dorsal raphe.* Electrical stimulation of the dorsal raphe can produce changes in cerebral blood flow; several drugs with anti-migraine action inhibit neurons there. The most effective drugs stimulate a class of serotonin receptors that are abundant in the dorsal raphe.

The trigger for **tension headaches** probably involves a combination of factors, but sustained contractions of the neck and facial muscles are most commonly implicated. Tension headaches can last for days or can occur daily over longer periods of time, typically without the throbbing, pulsing sensations characteristic of migraine headaches. Instead, the person may complain of a feeling of pressure or vise-like compression. Some tension headaches do not involve the muscles but accompany severe depression or anxiety.

AXOPLASMIC TRANSPORT AND DISEASE FAP p. 365

With a soft flutter of wings, dark shapes drop from the sky onto the backs of grazing cattle. Each shape is a small bat whose scientific name, *Desmodus rotundus,* is less familiar than the popular term *vampire bat.* Vampire bats inhabit tropical and semitropical areas of North, Central, and South America. They range from the Texas coast to Chile and southern Brazil. These rather aggressive animals are true vampires, subsisting on a diet of fresh blood. Over the next hour, every bat in the flight—which may number in the hundreds—will consume about 65 ml of blood through small slashes in the skin of their prey.

As unpleasant as this blood collection may sound, it is not the blood loss that is the primary cause for concern. The major problem is that these bats can be carriers for the rabies virus. **Rabies** is an acute disease of the central nervous system. The rabies virus can infect any mammal, wild or domestic. With few exceptions, the result is death within 3 weeks. For unknown reasons, bats can survive rabies infection for an indefinite period. As a result, an infected bat can serve as a carrier for the disease. Because many bat species, including vampire bats, form dense colonies, a single infected individual can spread the disease through the entire colony.

Rabies is generally transmitted to people through the bite of a rabid animal. About 15,000 cases of rabies occur each year worldwide, the majority of them the result of dog bites. Only about five of those cases, however, are diagnosed in the United States, where most dogs and cats are vaccinated against rabies. In the United States, most cases of rabies are caused instead by the bites of raccoons, foxes, skunks, or bats.

Although these bites generally involve peripheral sites, such as the hand or foot, the symptoms are caused by CNS damage. The virus present at the injury site is absorbed by the synaptic knobs of peripheral nerves in the region. It then gets a free ride to the CNS, courtesy of retrograde flow. During the first few days after exposure, the individual may experience

TABLE A-19 Representative Diagnostic and Laboratory Tests for Nervous System Disorders and Their Common Uses

Diagnostic Procedure	Method and Result	Representative Uses
Lumbar puncture (spinal tap)	Needle aspiration of CSF from the subarachnoid space in the lumbar area of the spinal cord	See Analysis of CSF (below) for diagnostic uses.
Skull X-ray	Standard X-ray	Detects fractures and possible sinus involvement or other bony abnormalities
Electroencephalography (EEG)	Electrodes placed on the scalp detect electrical activity of the brain, and EEG produces graphic record.	Detects abnormalities in frequency and amplitude of brain waves due to cranial trauma or neurological disorders such as seizures
Computerized tomography (CT) scan of the brain	Standard CT; contrast media are commonly used.	Detects tumors; cerebrovascular abnormalities, such as aneurysms (weakened areas in vessel walls); scars; strokes; or areas of edema
Cerebral angiography and **digital subtraction angiography (DSA)**	Dye is injected into an artery of the neck, and the movement of the dye is observed via serial X-rays; DSA transfers dye-location information to a computer for image enhancement.	Detects abnormalities in the cerebral vessels, such as aneurysms or blockages
Positron emission tomography (PET) scan	Radiolabeled compounds injected into bloodstream accumulate at specific areas of the brain; the radiation emitted is monitored by a computer that generates a reconstructed image.	Determines blood flow to specific regions of the brain; detects focal points of brain metabolic activity; may help diagnose mental illness; also useful in the diagnosis of Parkinson's disease and Alzheimer's disease
Magnetic resonance imaging (MRI)	Standard MRI; contrast media are commonly used to enhance visualization.	Detects brain tumors, hemorrhaging, edema, spinal cord injury, and other structural abnormalities

Laboratory Test	Normal Values	Significance of Abnormal Values
Analysis of CSF		
Pressure of CSF	<200 cm H_2O	Pressure higher than 200 cm H_2O is considered abnormal, possibly indicating hemorrhaging, brain tumor, blood clots around the brain, or infection.
Color of CSF	Clear and colorless	Increased cloudiness suggests hemorrhage, faulty puncture technique, or infection.
Glucose in CSF	50–75 mg/dl	Decreased levels occur when CNS tumors or infections are present.
Protein in CSF	15–45 mg/dl	Elevated levels occur in some infectious processes, such as meningitis and encephalitis. Protein levels may be elevated during inflammatory processes or brain tumor formation.
Cells present in CSF	No RBCs present; WBC count should be less than 5 per mm^3.	RBCs appear with subarachnoid hemorrhage; increased number of neutrophils occurs in bacterial infections, such as bacterial meningitis; increase in lymphocyte numbers occurs in viral meningitis; cancer cells from a brain tumor may be shed into CSF.
Culture of CSF	Organism causing infectious process in brain or spinal cord can be cultured for identification and determination of sensitivity to antibiotics.	Culture can determine causative agent in meningitis or brain abscess.

headache, fever, muscle pain, nausea, and vomiting. The individual then enters a phase marked by extreme excitability, hallucinations, muscle spasms, and disorientation. Swallowing becomes difficult, accounting for an early name for rabies, *hydrophobia* (literally, fear of water). The accumulation of saliva makes the individual appear to be "foaming at the mouth." Coma and death soon follow.

Preventive treatment was developed by Louis Pasteur in the late 1800s. It must begin almost immediately after exposure and consists of injections that contain antibodies against the rabies virus followed by a series of vaccinations against rabies. This postexposure treatment may not be sufficient after a massive infection, which can lead to death in as little as 4 days. Individuals such as veterinarians or field biologists who are at high risk of exposure commonly take a preexposure series of vaccinations. These injections bolster the immune defenses and improve the effectiveness of the postexposure treatment. Without preventive treatment, rabies infection in humans is always fatal.

FAP Ch. 12

Rabies is perhaps the most dramatic example of a clinical condition directly related to axoplasmic transport. However, many toxins, including heavy metals, some pathogenic bacteria, and other viruses, use this mechanism to enter the CNS.

DEMYELINATION DISORDERS FAP *p. 370*

Demyelination disorders are linked by a common symptom: the destruction of myelinated axons in the CNS and peripheral nervous system (PNS). The mechanism responsible for this loss differs in each of these disorders. We will consider only the major categories of demyelination disorders:

- *Heavy metal poisoning.* Chronic exposure to heavy metal ions, such as arsenic, lead, or mercury, can lead to damage of neuroglia and to demyelination. As demyelination occurs, the affected axons deteriorate, and the condition becomes irreversible. Historians note several interesting examples of heavy metal poisoning with widespread impact. For example, lead contamination of drinking water has been cited as one factor in the decline of the Roman Empire. In the seventeenth century, the great physicist Sir Isaac Newton is thought to have suffered several episodes of physical illness and mental instability brought on by his use of mercury in chemical experiments. Well into the nineteenth century, mercury used in the preparation of felt presented a serious occupational hazard for those employed in the manufacture of stylish hats. Over time, mercury absorbed through the skin and across the lungs accumulated in the CNS, producing neurological damage that affected both physical and mental function. (This effect is the source of the expression "mad as a hatter.") More recently, Japanese fishermen working in Minamata Bay, Japan, collected and consumed seafood contaminated with mercury discharged from a nearby chemical plant. Levels of mercury in their systems gradually rose to the point at which clinical symptoms appeared in hundreds of people. Making matters worse, pregnant women exposed to mercury had babies with severe, crippling birth defects.
- *Diphtheria.* **Diphtheria** (dif-THĒ-rē-uh; *diphthera,* leather + *-ia,* disease) is a disease that results from a bacterial infection of the respiratory tract. In addition to restricting airflow and sometimes damaging the respiratory surfaces, the bacteria produce a powerful toxin that injures the kidneys and adrenal glands, among other tissues. In the nervous system, diphtheria toxin damages Schwann cells and destroys myelin sheaths in the PNS. This demyelination leads to sensory and motor problems that can ultimately produce a fatal paralysis. The toxin also affects cardiac muscle cells, and heart enlargement and heart failure may occur. The fatality rate for untreated cases ranges from 35 to 90 percent, depending on the site of infection and the subspecies of bacterium. Because an effective vaccine exists, cases are relatively rare in countries with adequate health care. Russia experienced an epidemic in the 1990s when vaccines were not widely available.
- *Multiple sclerosis.* **Multiple sclerosis** (skler-Ō-sis; *sklerosis,* hardness), or **MS,** is a disease characterized by recurrent incidents of demyelination that affects axons in the optic nerve, brain, and spinal cord. Common symptoms include partial loss of vision and problems with speech, balance, and general motor coordination, including bowel and urinary bladder control. The time between incidents and the degree of recovery vary from case to case. In about one-third of all cases, the disorder is progressive, and each incident leaves a greater degree of functional impairment. The average age at the first attack is 30–40 years; the incidence among women is 1.5 times that among men. Corticosteroid injections and interferon have slowed the progression of the disease. We will discuss MS further in a later section (p. 76).
- *Guillain–Barré syndrome.* **Guillain–Barré syndrome** is characterized by a progressive but reversible demyelination. Symptoms initially involve weakness of the legs, which spreads rapidly to muscles of the trunk and arms; in severe cases, even breathing is affected. These symptoms generally increase in intensity for 1–2 weeks before subsiding. The mortality rate is low (under 5 percent), but some permanent loss of motor function can occur. The cause is unknown, but because roughly two-thirds of Guillain–Barré patients develop symptoms within 2 months after they have experienced a viral infection, it is suspected that the condition may result from a malfunction of the immune system. (We consider the mechanism involved in Chapter 22 of the text; see FAP *p. 788.*)

NEUROTOXINS IN SEAFOOD FAP *p. 382*

Several forms of human poisoning result from the ingestion of seafood that contains neurotoxins, poisons that primarily affect neurons. **Tetrodotoxin** (te-TRŌ-dō-tok-sin), or **TTX,** is found in the liver, gonads, and blood of certain Pacific puffer fish species, and a related compound is found in the skin glands of some sala-

manders. Tetrodotoxin selectively blocks voltage-regulated sodium ion channels, effectively preventing nerve cell activity. The usual result is death from paralysis of the respiratory muscles. Despite the risks, the Japanese consider the puffer fish a delicacy, served under the name *fugu.* Specially licensed chefs prepare this dish, carefully removing the potentially toxic organs. Nevertheless, a mild tingling and sense of intoxication are considered desirable, and several people die each year as a result of improper preparation of fugu.

Saxitoxin (sak-si-TOK-sin), or **STX,** can have a similarly lethal effect. Saxitoxin and related poisons are produced by several species of marine microorganisms. When these organisms undergo a population explosion, they color the surface waters, producing a "red tide." Eating seafood that has become contaminated by feeding on the toxic microbes can result in symptoms of **paralytic shellfish poisoning** (**PSP;** from clams, mussels, or oysters) or **ciguatoxin** (sēg-wa-TOK-sin; *cigua*, a sea snail) (**CTX;** from fish). Mild cases produce symptoms of paresthesias and a curious reversal of hot-versus-cold sensations. Severe cases, which are relatively rare, result in coma and death due to respiratory paralysis.

NEUROACTIVE DRUGS FAP *p. 390*

The text discussion on FAP *p. 390* introduced the concept that many drugs and toxins work by interfering with normal synaptic function. Table A-20 provides information about specific chemical compounds that affect ACh activity. A comparable range of drugs is available to target other types of chemical synapses.

TAY-SACHS DISEASE FAP *p. 398*

Tay-Sachs disease is a genetic abnormality that involves the metabolism of *gangliosides,* important components of neuron cell membranes. It is a *lysosomal storage disease* caused by abnormal lysosome activity. Individuals with this condition lack the enzyme needed to break down one particular ganglioside, which accumulates in the lysosomes of CNS neurons and causes them to deteriorate. Affected infants seem normal at birth, but neurological problems begin to appear within 6 months. The progression of symptoms typically includes muscle weakness, blindness, seizures, and death, generally before age 4 years. No effective treatment exists, but prospective parents can be tested to determine whether they carry the gene responsible

TABLE A-20 Drugs Affecting Acetylcholine (ACh) Activity at Synapses

Drug	Mechanism	Effects
Hemicholinium	Blocks ACh synthesis	Produces symptoms of synaptic fatigue
***Botulinus* toxin**	Blocks ACh release directly	Paralyzes voluntary muscles; produced by bacteria; responsible for a deadly type of food poisoning
Barbiturates	Decrease rate of ACh release	Produce muscle weakness; depress CNS activity; administered as sedatives and anesthetics
Procaine (Novocain®)	Reduces membrane permeability to sodium	Prevents stimulation of sensory neurons; used as a local anesthetic
Tetrodotoxin (TTX) Saxitoxin (STX) Ciguatoxin (CTX)	Blocks sodium ion channels	Eliminates production of action potentials; produced by some marine organisms during normal metabolic activity
Neostigmine	Prevents ACh inactivation by cholinesterase	Produces sustained contraction of skeletal muscles; other effects on cardiac muscle, smooth muscle, and glands; used to treat myasthenia gravis and to counteract overdoses of tubocurarine
Insecticides (malathion, parathion, etc.) and nerve gases	As above	As above
***d*-tubocurarine**	Prevents ACh binding to postsynaptic receptor sites	Paralyzes voluntary muscles; curare produced by South American plant
Nicotine	Binds to ACh receptor sites	Low doses facilitate voluntary muscles; high doses cause paralysis; an active ingredient in cigarette smoke; addictive for most people
Succinylcholine	Reduces sensitivity to ACh	Paralyzes voluntary muscles; used to produce temporary muscular relaxation during surgery
Atropine	Competes with ACh for binding sites on postsynaptic membrane	Reduces heart rate, smooth muscle activity; decreases salivation; dilates pupils; produces skeletal muscle weakness at high doses; produced by deadly nightshade plant

for this condition. The disorder is most prevalent in Ashkenazic Jews of Eastern European origin. A prenatal test is available to detect this condition in a fetus.

MENINGITIS FAP p. 409

The warm, dark, nutrient-rich environment of the cranial or spinal meninges provides ideal conditions for a variety of bacteria and viruses. Microorganisms that cause meningitis include bacteria associated with middle ear and sinus infections; with pneumonia, streptococcal ("strep"), staphylococcal ("staph"), or meningococcal infections; and with tuberculosis. These pathogens may gain access to the meninges by traveling within blood vessels or by entering at sites of vertebral or cranial injury. Headache, chills, high fever, disorientation, and rapid heart and respiratory rates appear as higher centers are affected. Without treatment, delirium, coma, convulsions, and death may follow within hours.

FAP Ch. 13

The most common clinical assessment involves checking for a "stiff neck" by asking the patient to touch the chin to the chest. Meningitis affecting the cervical portion of the spinal cord results in a marked increase in the muscle tone of the extensor muscles of the neck. So many motor units become activated that flexion of the neck becomes painfully difficult, if not impossible.

The mortality rate for viral and bacterial forms of meningitis ranges from 1 to more than 50 percent, depending on the type of virus or bacteria, the age and health of the individual, and other factors. There is no effective treatment for viral meningitis, but bacterial meningitis can be combated with antibiotics and the maintenance of proper fluid balance and electrolyte balance. The incidence of the most common form of bacterial meningitis, caused by *Haemophilus influenzae,* has been dramatically reduced by immunization.

SPINAL ANESTHESIA FAP p. 410

Injecting a local anesthetic around a nerve produces a temporary blockage of nerve function. This procedure can be done peripherally, as when skin lacerations are sewn up, or at sites around the spinal cord to obtain more widespread anesthetic effects. Although an *epidural block,* the injection of an anesthetic into the epidural space, has the advantage of affecting only the spinal nerves in the immediate area of the injection, epidural anesthesia can be difficult to achieve in the upper cervical, midthoracic, and lumbar regions, where the epidural space is extremely narrow. **Caudal anesthesia** involves the introduction of anesthetics into the epidural space of the sacrum. Injection at this site paralyzes and anesthetizes lower abdominal and perineal structures. Caudal anesthesia is widely used to control pain during labor and delivery.

Local anesthetics can also be introduced into the subarachnoid space of the spinal cord. However, the effects spread as the circulation and diffusion of cerebrospinal fluid (CSF) distribute the anesthetic along the spinal cord. Problems with overdosing are seldom serious, because the need to control the patient's position during administration can limit the distribution of the drug to some degree. Moreover, because the diaphragmatic breathing muscles are controlled by upper cervical spinal nerves, respiration continues even if all thoracic and abdominal segments have been paralyzed.

MULTIPLE SCLEROSIS FAP p. 413

Multiple sclerosis (MS), introduced in the discussion of demyelination disorders, is a disease that produces muscular paralysis and sensory losses through demyelination. The initial symptoms appear as the result of myelin degeneration within the white matter of the lateral and posterior columns of the spinal cord or along tracts within the brain. For example, spinal cord involvement can produce weakness, tingling sensations, and a loss of "position sense" for the limbs. During subsequent attacks, the effects become more widespread. The cumulative sensory and motor losses may eventually lead to a generalized muscular paralysis.

Recent evidence suggests that this condition may be linked to a defect in the immune system that causes it to attack myelin sheaths. Individuals with MS have lymphocytes that do not respond normally to foreign proteins. Because several viral proteins have amino acid sequences similar to those of normal myelin, it has been proposed that MS results from a case of mistaken identity. For unknown reasons, MS appears to be associated with cold and temperate climates. It has been suggested that individuals who develop MS may have an inherited susceptibility to a virus and that this susceptibility is exaggerated by environmental conditions. The yearly incidence in the United States averages around 50 cases for every 100,000 people in the population. Improvement has been noted in some MS patients treated with *interferon,* a peptide secreted by cells of the immune system, and recently corticosteroid treatment has been linked to a slowdown in the progression of MS.

SPINAL CORD INJURIES AND EXPERIMENTAL TREATMENTS FAP p. 413

At the outset, any severe injury to the spinal cord produces a period of sensory and motor paralysis termed **spinal shock.** The skeletal muscles become weak; neither somatic nor visceral reflexes function; and the brain no longer receives sensations of touch, pain, heat, or cold. The location and severity of the injury determine how long these symptoms persist and how completely the individual recovers.

Violent jolts, such as those associated with blows or gunshot wounds near the spinal cord, can cause **spinal concussion** without visibly damaging the spinal cord. Spinal concussion produces a period of spinal shock, but the symptoms are only temporary and recovery may be complete in a matter of hours. More-serious

injuries, such as vertebral fractures, generally involve physical damage to the spinal cord. In a **spinal contusion,** hemorrhages occur in the meninges, pressure rises in the cerebrospinal fluid, and the white matter of the spinal cord may degenerate at the site of injury. Partial recovery over a period of weeks may leave some functional losses. Recovery from a **spinal laceration** by vertebral fragments or other foreign bodies tends to be far slower and less complete. **Spinal compression** occurs when the spinal cord becomes squeezed or distorted within the vertebral canal. In a **spinal transection,** the spinal cord is completely severed. At present, surgical procedures cannot repair a severed spinal cord.

Many spinal cord injuries involve some combination of compression, laceration, contusion, and partial transection. Relieving pressure and stabilizing the affected area through surgery (such as *spinal fusion,* the immobilization of adjacent vertebrae) may prevent further damage and allow the injured spinal cord to recover as much as possible.

Two avenues of research are being pursued, one biological and the other electronic.

Biological Methods

A major biological line of investigation involves the introduction of stem cells and the biochemical control of nerve growth and regeneration.

Treated with embryonic stem cells at the injury site 9 days after a crushing injury to the spine, laboratory rats recovered some limb mobility and strength. Oligodendrocytes, astrocytes, and functional neurons developed at the injury site. Neural stem cells have also been proposed for the treatment of strokes, Parkinson's disease, and Alzheimer's disease. The recent discovery that the adult brain contains populations of inactive stem cells has opened a new line of investigation: What are the factors that activate resident stem cells?

Neurons are influenced by a combination of growth promoters and growth inhibitors. Damaged myelin sheaths apparently release an inhibitory factor that slows the repair process. Researchers have made an antibody, *IN-1,* that inactivates the inhibitory factor released in the damaged spinal cords of rats. The treatment stimulates repairs, even in severed spinal cords. Because gray matter has no myelin, the chances of regrowth of severed axons are enhanced if the proximal cut end is guided into the gray matter rather than being left in the white matter. Researchers are attempting to devise a suitable means of accomplishing this rerouting.

A partial listing of compounds known to affect nerve growth and regneration includes **nerve growth factor** (NGF), **brain-derived neurotrophic factor** (BDNF), **neurotrophin-3** (NT-3), **neurotrophin-4** (NT-4), **glial growth factor, glial maturation factor, ciliary neurotrophic factor,** and **growth-associated protein 43** (GAP-43). Many of these factors have now been synthesized by means of gene-splicing techniques, and sufficient quantities are available to permit their use in experiments on humans and other mammals. Initial results are promising, and these factors in various combinations are being evaluated for treatment of CNS injuries and the chronic degeneration seen in Alzheimer's disease and Parkinson's disease.

Electronic Methods

Several research teams are experimenting with the use of computers to stimulate specific muscles and muscle groups electrically. The technique is called *functional electrical stimulation,* or FES. This approach commonly involves the implantation of a network of wires beneath the skin with their tips in skeletal muscle tissue. The wires are connected to a small computer worn at the waist. The wires deliver minute electrical stimuli to the muscles, depolarizing their membranes and causing contractions. With this equipment and lightweight braces, quadriplegics have walked several hundred yards and paraplegics several thousand. The *Parastep*™ system, which uses a microcomputer controller, is now undergoing clinical trials.

FAP Ch. 13

Equally impressive results have been obtained by using a network of wires woven into the fabric of close-fitting garments. This network provides the necessary stimulation without the complications and maintenance problems that accompany implanted wires. A paraplegic woman in a set of electronic "hot pants" completed several miles of the 1985 Honolulu Marathon, and more recently another paraplegic woman walked down the aisle at her own wedding.

Such technological solutions can provide only a degree of motor control without accompanying sensation. Everyone would prefer a biological procedure that would restore the functional integrity of the nervous system. For now, however, computer-assisted programs such as FES can improve the quality of life for thousands of paralyzed individuals. The 1995 horseback-riding injury of actor Christopher Reeve has brought publicity and research funds to this area.

SHINGLES AND HANSEN'S DISEASE

FAP p. 416

In **shingles,** or *herpes zoster,* the *herpes varicella-zoster* virus attacks neurons within the dorsal roots of spinal nerves and sensory ganglia of cranial nerves. This disorder produces a painful rash whose distribution corresponds to that of the affected sensory nerves (Figure A-31•). Shingles develops in adults who were first exposed to the virus as children. The initial infection produces symptoms known as chickenpox. After this encounter, the virus remains dormant within neurons of the anterior gray horns of the spinal cord. It is not known what triggers the reactivation of this pathogen. Fortunately for those affected, attacks of shingles generally heal and leave behind only unpleasant memories.

Most people suffer only a single episode of shingles in their adult lives. However, the problem can recur in

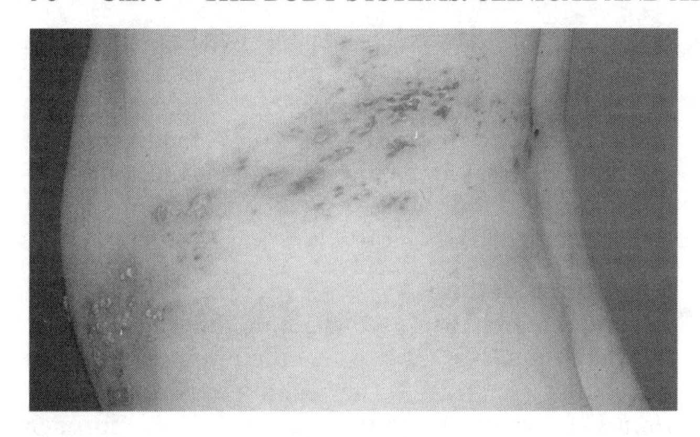

•**FIGURE A-31** **Shingles.** The left side of a person with shingles. The skin eruptions follow the distribution of dermatomal innervation.

people with weakened immune systems, including those with AIDS or some forms of cancer. Treatment for shingles typically involves large doses of the antiviral drug *acyclovir (Zovirax®)*.

FAP Ch. 13

The condition traditionally called **leprosy,** now more commonly known as **Hansen's disease,** is an infectious disease caused by a bacterium, *Mycobacterium leprae.* It is a disease that progresses slowly, and symptoms may not appear for up to 30 years after infection. The bacterium invades peripheral nerves, especially those in the skin, producing initial sensory losses. Over time, motor paralysis develops, and the combination of sensory and motor loss can lead to recurring injuries and infections. The eyes, nose, hands, and feet may develop deformities as a result of neglected injuries (Figure A-32a,b•). This disease has several forms; peripheral nerves are always affected, but some forms also involve extensive lesions of the skin and mucous membranes.

Only about 5 percent of those exposed to *Mycobacterium leprae* develop symptoms; people living in the tropics are at greatest risk. There are about 2000 cases in the United States and an estimated 12–20 million cases worldwide. If detected before deformities occur, the disease can generally be treated successfully with drugs such as rifampin and dapsone. Treated individuals are not infectious, and the practice of confining "lepers" in isolated compounds has been discontinued.

PERIPHERAL NEUROPATHIES FAP *p. 416*

Peripheral neuropathies, or peripheral nerve palsies, are characterized by regional losses of sensory and motor function as a result of nerve trauma or compression. **Brachial palsies** result from injuries to the brachial plexus or its branches. **Crural palsies** involve the nerves of the lumbosacral plexus.

Palsies appear for several reasons. The *pressure palsies* are especially interesting; a familiar but mild example is the experience of having an arm or leg "fall asleep." The limb becomes numb, and afterward an uncomfortable "pins-and-needles" sensation, or **paresthesia,** accompanies the return to normal function.

These incidents are seldom clinically significant, but they provide graphic examples of the effects of more-serious palsies that can last for days to months. In **radial nerve palsy,** pressure on the back of the arm interrupts the function of the radial nerve, so the extensors of the wrist and fingers are paralyzed. This condition is also known as "Saturday night palsy," because falling asleep on a couch with your arm over the seat back (or beneath someone's head) can produce the right combination of pressures. Students may also be familiar with **ulnar palsy,** which can result from prolonged contact between an elbow and a desk. The ring and little fingers lose sensation, and the fingers cannot be adducted. We considered *carpal tunnel syndrome,* a neuropathy resulting from compression of the *median nerve* at the wrist, in the text *(pp. 233, 344)*.

Men who carry large wallets in their hip pockets may develop symptoms of **sciatic compression** after they drive or sit in one position for extended periods. As nerve function declines, the individuals notice some

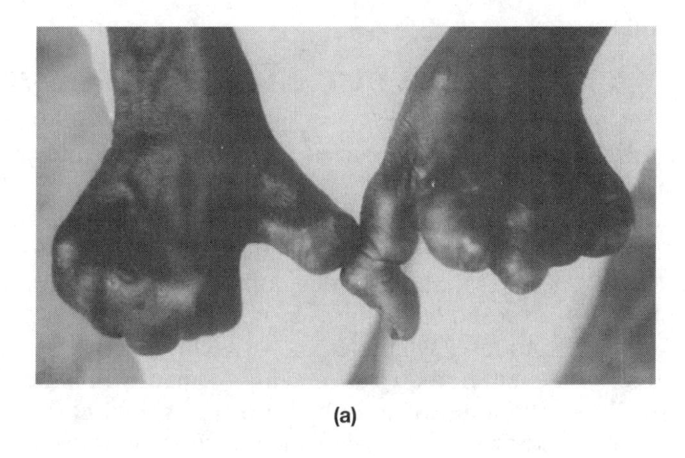

(a)

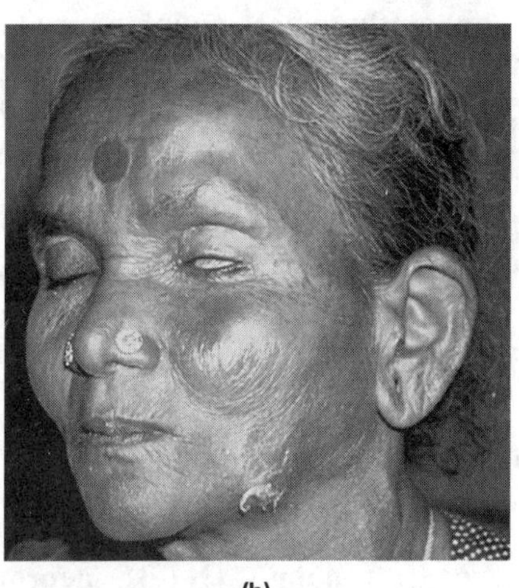

(b)

•**FIGURE A-32** **Hansen's Disease.** **(a)** The distal limbs are gradually deformed as untreated Hansen's disease progresses. **(b)** This disease also affects facial features, typically starting with degenerative changes around the eyes and at the nose and ears.

lumbar or gluteal pain, numbness along the back of the leg, and weakness in the leg muscles. Similar symptoms result from the compression of nerve roots that form the sciatic nerve by a distorted lumbar intervertebral disc. This condition is termed **sciatica,** and one or both lower limbs are be affected, depending on the site of compression. Finally, sitting with your legs crossed can produce symptoms of a **peroneal palsy.** Sensory losses from the top of the foot and side of the leg are accompanied by a decreased ability to dorsiflex or evert the foot.

DAMAGE AND REPAIR OF PERIPHERAL NERVES FAP *p. 416*

If a peripheral axon is damaged but not displaced, normal function may eventually return as the cut stump grows across the injury site away from the soma and along its former path. The mechanics of this process were described at the close of Chapter 12 in the text. For normal function to be restored, several things must happen. The severed ends must be relatively close together (1–2 mm, or 0.04–0.08 in.); they must remain in proper alignment; and there must be no physical obstacles between them, such as the collagen fibers of scar tissue. These conditions can be created in the laboratory, using experimental animals and individual axons or small fascicles. But in accidental injuries to peripheral nerves, the edges are likely to be jagged; intervening segments may be lost entirely, and elastic contraction in the surrounding connective tissues may pull the cut ends apart and misalign them.

Until recently, the surgical response would involve trimming the injured nerve ends, neatly sewing them together, and hoping for the best. This procedure was typically unsuccessful, in part because scalpels do not produce a smoothly cut surface and because the thousands of broken axons would never be perfectly aligned. Moreover, axons are not highly elastic, so if a large segment of the nerve was removed, crushed, or otherwise destroyed, there would be no way to bring the intact ends close enough to permit regeneration. In such instances, a **nerve graft** could be inserted, using a section from some other, less important peripheral nerve. The functional results were even less likely to be wholly satisfactory, because the growing axonal tips had to find their way across not one but two gaps, and the chances for successful alignment were proportionately smaller. Nevertheless, any return of function was better than none!

Research now focuses on the physical and biochemical control of nerve regeneration, using the growth factors introduced earlier (p. 77). Another promising strategy is the use of a synthetic sleeve to guide nerve growth. The sleeve is a tube with an outer layer of silicone around an inner layer of cowhide collagen bound to the proteoglycans from shark cartilage. Using this sleeve as a guide, axons can grow across gaps as large as 20 mm (0.79 in.). The procedure has yet to be tried on humans, and functional restoration in other mammals is generally incomplete because proper alignment does not always occur.

REFLEXES AND DIAGNOSTIC TESTING FAP *p. 430*

Many reflexes can be assessed through careful observation and the use of simple tools. The procedures are easy to perform, and the results can provide valuable information about damage to the spinal cord or spinal nerves. By testing a series of spinal and cranial reflexes, a physician can assess the function of sensory pathways and motor centers throughout the spinal cord and brain.

Neurologists test many reflexes; only a few are so generally useful that physicians make them part of a standard physical examination. These reflexes are shown in Figure A-33•.

The *patellar reflex* (Figure 13-17, FAP *p. 426*), *ankle jerk* (Figure A-33a•), *biceps reflex* (Figure A-33b•), and *triceps reflex* (Figure A-33c•) are stretch reflexes controlled by specific segments of the spinal cord. Testing these reflexes provides information about the corresponding spinal segments. For example, a normal patellar, or knee jerk, reflex indicates that spinal nerves and spinal segments L_2–L_4 are undamaged. The Babinski sign (Figure 13-20, FAP *p. 430*) is normally absent in adults because descending commands inhibit the response. Spinal cord injury or strokes may reduce this inhibition, resulting in reemergence of the Babinski sign.

The *abdominal reflex* (Figure A-33d•) is normally present in adults. In this reflex, a light stroking of the skin produces a reflexive twitch in the abdominal muscles that moves the navel toward the stimulus. This reflex is facilitated by descending commands; it disappears after descending tracts have been damaged.

Abnormal Reflex Activity

In **hyporeflexia,** normal reflexes are weak but apparent, especially with reinforcement (FAP *p. 430*). In **areflexia** (ā-rē-FLEK-sē-uh; *a-*, without), normal reflexes fail to appear, even with reinforcement. Hyporeflexia or areflexia may indicate temporary or permanent damage to skeletal muscles, dorsal or ventral nerve roots, spinal nerves, the spinal cord, or the brain.

Hyperreflexia occurs when higher centers maintain a high degree of facilitation along the spinal cord. Under these conditions, reflexes are easily triggered and the responses may be grossly exaggerated. This effect also results from spinal cord compression or diseases that target higher centers or descending tracts. One potential result of hyperreflexia is the appearance of alternating contractions in opposing muscles. When one muscle contracts, it stimulates the stretch receptors in the opposing muscle. The stretch reflex

FAP Ch. 13

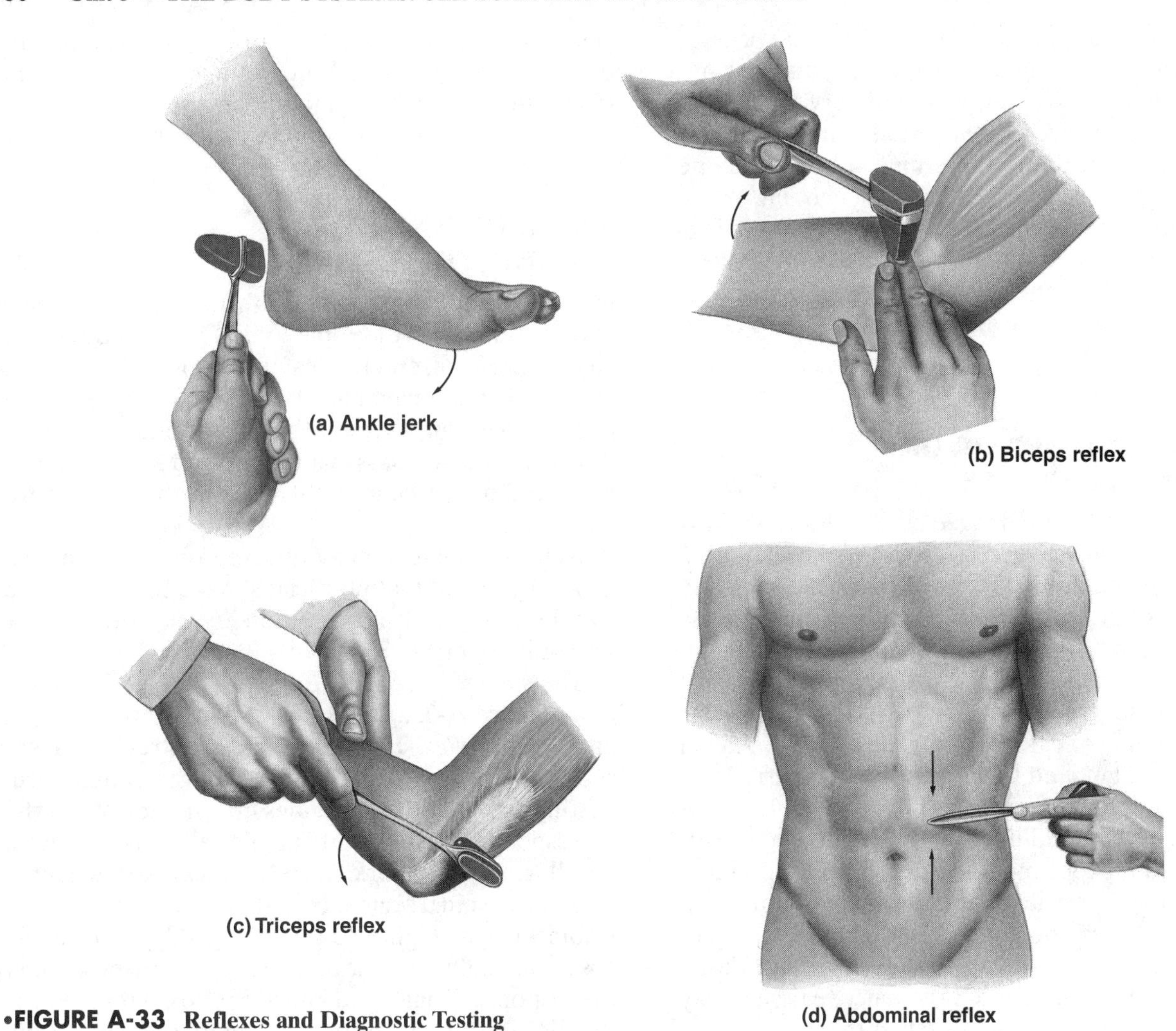

•**FIGURE A-33** **Reflexes and Diagnostic Testing**

then triggers a contraction in that muscle, and this contraction stretches receptors in the original muscle. This self-perpetuating sequence, which can be repeated indefinitely, is called **clonus** (KLŌ-nus). In a hyperreflexive person, a tap on the patellar tendon will set up a cycle of kicks rather than just one or two.

A more extreme hyperreflexia develops if the motor neurons of the spinal cord lose contact with higher centers. In many cases, after a severe spinal injury, the individual first experiences a temporary period of areflexia known as spinal shock (p. 76). When the reflexes return, they respond in an exaggerated fashion, even to mild stimuli. For example, the lightest touch on the skin surface may produce a massive withdrawal reflex. The reflex contractions may occur in a series of intense muscle spasms strong enough to break bones. In the **mass reflex,** the entire spinal cord becomes hyperactive for several minutes, issuing exaggerated skeletal muscle and visceral motor commands.

CRANIAL TRAUMA FAP *p. 441*

Cranial trauma is a head injury resulting from harsh contact with another object. Head injuries account for over half the deaths attributed to trauma. Roughly 8 million cases of cranial trauma arise each year in the United States; over a million of these involve intracranial hemorrhaging, concussion, contusion, or laceration of the brain. We presented the characteristics of spinal concussion, contusion, and laceration on p. 48; comparable descriptions are applied to brain injuries.

Concussions typically accompany even minor head injuries. A concussion involves a temporary loss of consciousness and some degree of amnesia (p. 85). Physicians examine concussed individuals quite closely and may X-ray the skull to check for skull fractures or cranial bleeding. Mild concussions produce a brief interruption of consciousness and little memory loss. Severe concussions produce extended periods of unconsciousness and abnormal neurological functions. Severe concussions are typically associated with **contusions** (bruises) or **lacerations** (tears) of the brain tissue; the possibilities for recovery vary with the areas affected. Extensive damage to the reticular formation can produce a permanent state of unconsciousness, whereas damage to the lower brain stem generally proves fatal.

APHASIA FAP *p. 449*

Aphasia is a disorder affecting the ability to speak or read. *Global aphasia,* discussed in the text, results from extensive damage to the general interpretive area or to the associated sensory tracts. There are several other forms of aphasia. **Major motor aphasia** can develop after a brief period of global aphasia. This condition is extremely frustrating for the individual, who can understand language and knows how to respond but lacks the motor control necessary to produce the right combinations of sounds. It is also known as *nonfluent,* or *expressive, aphasia.* In *fluent,* or *receptive, aphasia,* the person does not understand what is heard and does not make sense while speaking. The individual words and sounds come easily, but they convey no meaning.

Lesser degrees of aphasia commonly follow a minor stroke. There is no initial period of global aphasia, and the individual can understand spoken and written words. The problems encountered with speaking or writing gradually fade. Many individuals with minor aphasia recover completely.

THE CEREBRAL NUCLEI AND PARKINSON'S DISEASE FAP *p. 460*

The cerebral nuclei contain two discrete populations of neurons. One group stimulates motor neurons by releasing acetylcholine (ACh), and the other inhibits motor neurons by releasing gamma aminobutyric acid (GABA). Under normal conditions, the excitatory neurons remain inactive and the descending tracts are primarily responsible for inhibiting motor neuron activity. If the descending tracts are severed in an accident, the loss of inhibitory control leads to a generalized state of muscular contraction known as **decerebrate rigidity.**

The excitatory neurons are quiet because they are continuously exposed to the inhibitory effects of the neurotransmitter dopamine. This compound is manufactured by neurons in the substantia nigra and is carried by axoplasmic transport to synapses in the cerebral nuclei. If the ascending tract or the dopamine-producing neurons are damaged, this inhibition is lost, and the excitatory neurons become increasingly active. This increased activity produces the motor symptoms of **Parkinson's disease,** or *paralysis agitans.*

Parkinson's disease is characterized by a pronounced increase in muscle tone. Voluntary movements become hesitant and jerky, because a movement cannot occur until one muscle group manages to overpower its antagonists. This condition is called **spasticity.** Individuals with Parkinson's disease exhibit spasticity during voluntary movement and a continual **tremor** when at rest. A tremor represents a tug of war between antagonistic muscle groups that produces a background shaking of the limbs, in this case at a frequency of 4–6 cycles per second. Individuals with Parkinson's disease also have difficulty starting voluntary movements. Even changing one's facial expression requires intense concentration, and the individual acquires a blank, static expression. Finally, the positioning and preparatory adjustments normally performed automatically no longer occur. Every aspect of each movement must be voluntarily controlled, and the extra effort requires intense concentration that can prove tiring and extremely frustrating. Other symptoms include rigid posture and a slow, shuffling walk. In the late stages of this condition, other CNS effects, such as depression, hallucinations, and dementia, commonly appear.

Providing the cerebral nuclei with dopamine can significantly reduce the symptoms in two-thirds of Parkinson's patients. Intravenous dopamine injection is not effective, because the molecule cannot cross the blood–brain barrier. The most common procedure involves the oral administration of the drug L-DOPA (levodopa), a compound that crosses the capillaries and is then converted to dopamine. Unfortunately, it appears that with repeated treatment, the capillaries become less permeable to L-DOPA, so the required dosage must increase. The effectiveness of L-DOPA can be increased by giving it in combination with other drugs, such as *amantadine* or *bromocriptine.* Amantadine accelerates dopamine release at synaptic terminals, and bromocriptine, a dopamine agonist, stimulates dopamine receptors on postsynaptic membranes. The drug *deprenyl* appears to slow the progression of the disease in some patients by affecting the biochemical pathways of dopamine removal.

Surgery to control Parkinson's symptoms focuses on the destruction of areas within the cerebral nuclei or thalamus to control the motor symptoms of tremor and rigidity. The high success rate of drug therapy has greatly reduced the number of surgical procedures. Recent attempts to transplant tissues that produce dopamine or related compounds into the cerebral nuclei have met with limited success. Variable results have been obtained with the transplantation of tissue from the adrenal glands. The transplantation of fetal brain tissue into adult brains has been more successful, with several research groups reporting improvement in patients' motor skills up to 3 years after surgery. In the future, the insertion or activation of neural stem cells may become an important treatment method.

Most individuals with Parkinson's disease are elderly. The champion boxer Muhammad Ali developed symptoms relatively early in life (in his 40s), possibly as a result of repeated cranial trauma during his career. Since 1983, an increasing number of young people have developed this condition. In that year, a drug appeared on the streets rumored to be "synthetic heroin." In addition to the compound that produced the "high" sought by users, the drug contained several contaminants, including a complex molecule with the abbreviated name **MPTP.** This accidental byproduct of the synthetic process destroys neurons of the substantia nigra, eliminating the manufacture and transport of dopamine to the cerebral nuclei. As a result

of exposure to this drug, approximately 200 young, healthy adults have developed symptoms of severe Parkinson's disease. Why MPTP targets these particular neurons, and not all the CNS neurons that produce dopamine, remains a mystery.

CEREBELLAR DYSFUNCTION FAP p. 462

Cerebellar function can be altered permanently by trauma or a stroke or temporarily by drugs such as alcohol. These alterations can produce disturbances in motor control. In severe ataxia, balance problems are so great that the individual cannot sit or stand upright. Less-severe conditions cause an obvious unsteadiness and irregular patterns of movement. The individual typically watches his or her feet to see where they are going and controls ongoing movements by intense concentration and voluntary effort. Reaching for something becomes a major exertion, because the only information available must be gathered by sight or touch while the movement is taking place. Without the cerebellar ability to adjust movements while they are occurring, the individual becomes unable to anticipate the time course of a movement. Most commonly, a reaching movement ends with the hand overshooting the target. This inability to anticipate and stop a movement precisely is called **dysmetria** (dis-MET-rē-uh; *dys-*, bad + *metron,* measure). In attempting to correct the situation, the person usually overshoots again in the opposite direction, and then again. The hand oscillates back and forth until either the object can be grasped or the attempt is abandoned. This oscillatory movement is known as an **intention tremor.**

FAP Ch. 14

Clinicians check for ataxia by watching an individual walk in a straight line; the usual test for dysmetria involves touching the tip of the index finger to the tip of the nose. Because many drugs impair cerebellar performance, the same tests are used by police officers to check drivers suspected of driving while under the influence of alcohol or other drugs.

Analysis of Gait and Balance

To check gait and balance, the examiner typically asks the person to walk a line, first with the eyes open and then with the eyes closed. This procedure checks how much the individual is relying on visual information to fine-tune motor functions. If the gait is normal with the eyes open but abnormal and unsteady with the eyes closed, there are problems with the balance pathways and perhaps the cerebellum as well. Heel-to-toe walking on a straight line will magnify any gait abnormalities or loss of balance sensations.

While the subject is walking, the examiner also watches how the heels and toes are placed and how the arms swing back and forth. The pattern of limb movement during walking is normally regulated by the cerebral nuclei. Problems with these nuclei, as in Parkinson's disease, will upset the pace and rhythm of these movements.

The *Romberg test* is used to check balance and equilibrium sensations. A *positive Romberg sign* exists if the individual cannot maintain balance with the feet together and eyes closed. A positive Romberg sign may be exhibited by persons with brain damage, multiple sclerosis (p. 76), peripheral neuropathies (p. 78), or several vestibular disorders.

TIC DOULOUREUX FAP P. 468

Tic douloureux, or **trigeminal neuralgia,** affects 1 individual out of every 25,000. Sufferers complain of severe, almost totally debilitating pain that arrives with a sudden, shocking intensity and then disappears. In most cases, only one side of the face is involved, and the pain is along the sensory path of the maxillary and mandibular branches of the trigeminal nerve. This condition generally affects adults over 40 years of age; the cause in most cases is unknown. The possibility that a tumor exists along the course of the nerve is usually evaluated by an MRI scan. The pain can often be controlled by drug therapy, but surgical procedures may eventually be required. The goal of the surgery is to destroy the afferent nerves that carry the pain sensations. This goal can be attempted by actually cutting the nerves, a procedure called a **rhizotomy** (*rhiza-,* root), or by injecting chemicals such as alcohol or phenol into the nerves at the foramina ovale and rotundum. The sensory fibers can also be destroyed by inserting an electrode and cauterizing the sensory nerve trunks as they leave the semilunar ganglion.

CRANIAL REFLEXES AND CRANIAL NERVE TESTS FAP p. 474

Cranial reflexes are reflex arcs that involve the sensory and motor fibers of cranial nerves. Table A-21 lists representative examples of cranial reflexes and their functions. These reflexes are clinically important because they provide a quick and easy method for observing the condition of cranial nerves and specific nuclei and tracts in the brain. Thus, they help localize the site of damage or disease.

Cranial somatic reflexes are seldom more complex than the somatic reflexes of the spinal cord. Table A-21 indicates the normal functions of four somatic reflexes: the *corneal reflex,* the *tympanic reflex,* the *auditory reflexes,* and the *vestibulo-ocular reflexes*. In many instances, these reflexes are used to check for damage to the cranial nerves or processing centers involved. The brain stem contains many reflex centers that control visceral motor activity. Many of these reflex centers are located in the medulla oblongata, and they can direct very complex visceral motor responses to stimuli. These reflexes are essential to the control of respiratory, digestive, and cardiovascular functions.

A variety of tests are used to monitor the condition of specific cranial nerves. For example:

TABLE A-21 Cranial Reflexes

Reflex	Stimulus	Afferents	Central Synapse	Efferents	Response
Somatic reflexes					
Corneal reflex	Contact with corneal surface	N V	Motor nucleus for N VII	N VII	Blinking of eyelids
Tympanic reflex	Loud noise	N VIII	Inferior colliculus	N VII	Reduced movement of auditory ossicles
Auditory reflexes	Loud noise	N VIII	Motor nuclei of brain stem and spinal cord	N III, IV, VI, VII, X, and cervical nerves	Eye and/or head movements triggered by sudden sounds
Vestibulo-ocular reflexes	Rotation of head	N VIII	Motor nuclei controlling eye muscles	N III, IV, VI	Opposite movement of eyes to stabilize field of vision
Visceral reflexes					
Direct light reflex	Light striking photoreceptors	N II	Superior colliculus	N III	Constriction of ipsilateral pupil
Consensual light reflex	Light striking photoreceptors	N II	Superior colliculus	N III	Constriction of contralateral pupil

FAP Ch. 14

- The olfactory nerve (N I) is assessed by asking the subject to distinguish among various odors.
- Cranial nerves II, III, IV, and VI are assessed while the vision and movement of the eyes are checked. First, the person is asked to hold the head still and track the movement of the examiner's finger with the eyes. For the eyes to track the finger, the oculomotor muscles and cranial nerves must be functioning normally. For example, if the person cannot track with the right eye a finger that is moving from left to right, the right lateral rectus muscle or N VI on the right side may be damaged.
- Cranial nerve V, which provides motor innervation to the muscles of mastication, can be checked by asking the person to clench the teeth. The jaw muscles are then palpated. If motor components of N V on one side are damaged, the muscles on that side will be weak or flaccid. Sensory components of N V can be tested by lightly touching areas of the forehead and side of the face.
- The facial nerve (N VII) is checked by watching the muscles of facial expression or asking the person to perform particular facial movements. Wrinkling the forehead, raising the eyebrows, pursing the lips, and smiling are controlled by the facial nerve. If a branch of N VII has been damaged, the affected side will show muscle weakness or drooping. For example, the corner of the mouth may sag and fail to curve upward when the person smiles. Special sensory components of N VII can be checked by placing solutions known to stimulate taste receptors on the anterior third of the tongue.
- The glossopharyngeal and vagus nerves (N IX and N X) can be evaluated by watching the person swallow something. Examination of the soft palate arches and uvula for normal movement is important.
- The accessory nerve (N XI) can be checked by asking the person to shrug the shoulders. Atrophy of the sternocleidomastoid or trapezius muscles may also indicate problems with the accessory nerve.
- The hypoglossal nerve (N XII) can be checked by having the person extend the tongue and move it from side to side.

AMYOTROPHIC LATERAL SCLEROSIS

FAP *p. 492*

Demyelinating disorders affect both sensory and motor neurons, producing losses in sensation and motor control. **Amyotrophic lateral sclerosis (ALS)** is a progressive disease that affects specifically motor neurons, leaving sensory neurons intact. As a result, individuals with ALS experience a loss of motor control but have no loss of sensation in the affected regions. Motor neurons throughout the CNS are destroyed. Neurons involved with the innervation of skeletal muscles are the primary targets. Neurons involved with sensory information and intellectual functions are unaffected.

Symptoms of ALS generally do not appear until the individual is over age 40. It occurs at an incidence of 3–5 cases per 100,000 population worldwide. The disorder is somewhat more common among males than among females. The pattern of symptoms varies with the specific motor neurons involved. When motor neurons in the cerebral hemispheres of the brain are the first to be affected, the individual experiences

difficulty in performing voluntary movements and has exaggerated stretch reflexes. If motor neurons in other portions of the brain and the spinal cord are targeted, the individual experiences weakness, initially in one limb but gradually spreading to other limbs and ultimately the trunk. When the motor neurons innervating skeletal muscles degenerate, a loss of muscle tone occurs. Over time, the skeletal muscles atrophy. The disease progresses rapidly, and the average survival after diagnosis is just 3–5 years. Because intellectual functions remain unimpaired, a person with ALS remains alert and aware throughout the course of the disease. This is one of the most disturbing aspects of the condition.

The primary cause of ALS is uncertain; only 5–10 percent of ALS cases appear to have a genetic basis. At the cellular level, it appears that the underlying problem lies at the postsynaptic membranes of motor neurons. It has been suggested that an abnormal receptor complex for the neurotransmitter *glutamate* in some way leads to the buildup of free radicals, such as NO (nitric oxide), that ultimately kill the neuron. In a recent clinical study, treatment with *riluzole,* an experimental drug that blocks glutamate receptors, has shown promise in extending the life of ALS patients. The Food and Drug Administration (FDA) has approved this drug for clinical use.

FAP Ch. 15

SEIZURES AND EPILEPSIES FAP *p. 493*

A *seizure* is a temporary disorder of cerebral function, accompanied by abnormal, involuntary movements; unusual sensations; or inappropriate behavior. The individual may lose consciousness for the duration of the attack. There are many types of seizures. The terms **epilepsy** and *seizure disorder* refer to more than 40 conditions characterized by a recurring pattern of seizures over extended periods. In roughly 75 percent of patients, no obvious cause can be determined.

Seizures of all kinds are usually accompanied by a marked change in the pattern of electrical activity that is monitored in an electroencephalogram. The change begins in one portion of the cerebral cortex but may thereafter spread to adjacent regions, potentially involving the entire cortical surface. The neurons at the site of origin of the change are abnormally sensitive. When they become active, they may facilitate and subsequently stimulate adjacent neurons. As a result, the abnormal electrical activity can spread across the entire cerebral cortex.

The extent of the cortical involvement determines the nature of the observed symptoms. A **focal seizure** affects a relatively restricted cortical area, producing sensory or motor symptoms or both. The individual generally remains conscious throughout the attack. If the seizure occurs within a portion of the primary motor cortex, the activation of pyramidal cells will produce uncontrollable movements. The muscles affected or the specific sensations experienced provide an indication of the precise region involved. In a **temporal lobe seizure,** the disturbance spreads to the sensory cortex and association areas, so the individual may also experience unusual memories, sights, smells, or sounds. Involvement of the limbic system can also produce sudden emotional changes. The individual may lose consciousness at some point during the seizure.

Convulsive seizures are associated with uncontrolled muscle contractions. In a **generalized seizure,** the entire cortical surface is involved. Generalized seizures range from prolonged, major events to brief, almost unnoticed incidents. We will consider only two examples here, *grand mal* and *petit mal seizures.*

Some epileptic attacks involve powerful, uncoordinated muscular contractions that affect the face, eyes, and limbs. These are symptoms of a **grand mal seizure.** During a grand mal attack, the cortical activation begins at a single focus and then spreads across the entire surface. There may be no warning, but some individuals experience a vague apprehension or awareness that a seizure is about to begin. A sudden loss of consciousness follows, and the individual drops to the floor as major muscle groups go into tonic contraction. The body remains rigid for several seconds before a rhythmic series of contractions occurs in the limb muscles. Incontinence may occur. After the attack subsides, the individual may appear disoriented or sleep for several hours. Muscles or bones subjected to extreme stresses may be damaged, and the person will probably be sore for days after the incident.

Petit mal seizures are very brief (less than 10 seconds in duration) and involve few motor abnormalities. Typically, the individual loses consciousness suddenly, with no warning. It is as if an internal switch were thrown and the conscious mind turned off. Because the individual is "not there" for brief periods during petit mal attacks, the incidents are known as *absence seizures.* During the seizure, small motor activities, such as fluttering of the eyelids or trembling of the hands, may occur.

Petit mal seizures generally begin between ages 6 and 14. They can occur hundreds of times a day, so the child lives each day in small segments separated by blank periods. The individual is aware of brief losses of consciousness that occur without warning but, due to embarrassment, may not seek help. He or she becomes extremely anxious about the timing of future attacks. However, the motor signs are so minor that they tend to go unnoticed by family members, and the psychological stress caused by this condition is in many cases overlooked. The initial diagnosis is typically made during counseling for learning problems. (You have probably taken an exam after you have missed 1 or 2 lectures out of 20. Imagine taking an exam after you have missed every third minute of every lecture.)

Both petit mal and grand mal forms of epilepsy can be treated with barbiturates or other anticonvulsive drugs, such as *phenytoin sodium (Dilantin®)* or *valproic acid (Depakene®).*

AMNESIA

FAP p. 496

Amnesia, or memory loss, occurs suddenly or progressively, and recovery is complete, partial, or nonexistent, depending on the nature of the problem. In **retrograde amnesia** (*retro-*, behind), the individual loses memories of past events. Some degree of retrograde amnesia commonly follows a head injury; after a car wreck or a fall, many victims are unable to remember the moments preceding the accident. In **anterograde amnesia** (*antero-*, ahead), an individual may be unable to store additional memories, but earlier memories are intact and accessible. The problem appears to involve an inability to generate long-term memories. At least two drugs—*diazepam (Valium®)* and *Halcion®*—have been known to cause brief periods of anterograde amnesia. Brain injuries can cause more-prolonged memory problems. A person with permanent anterograde amnesia lives in surroundings that are always new. Magazines can be read, chuckled over, and reread a few minutes later with equal pleasure, as if they had never been seen before. Clinicians must introduce themselves at every meeting, even if they have been treating the patient for years.

Post-traumatic amnesia (PTA) commonly develops after a head injury. The duration of the amnesia varies with the severity of the injury. This condition combines the characteristics of retrograde and anterograde amnesias; the individual can neither remember the past nor consolidate memories of the present.

ALTERED STATES

FAP p. 496

A fine line may separate normal from abnormal states of awareness, and many variations in these states are clinically significant. The state of **delirium** involves wild oscillations in the level of wakefulness. The individual has little or no grasp of reality and often experiences hallucinations. When capable of communication, the person seems restless, confused, and unable to deal with situations and events. Delirium typically develops (1) in individuals with very high fevers or some metabolic abnormalities; (2) in patients experiencing withdrawal from addictive drugs, such as alcohol; (3) after hallucinogenic drugs, such as LSD, are taken; (4) as a result of brain tumors affecting the temporal lobes; and (5) in the late stages of some infectious diseases, including syphilis and AIDS. The term **dementia** implies a more stable, chronic state characterized by deficits in memory, spatial orientation, language, or personality. We shall discuss *senile dementia,* a form of dementia that may develop in the elderly, in a later section.

Table A-22 indicates the entire range of conscious and unconscious states, ranging from *delirium* through *coma.* Assessing the level of consciousness in an awake person involves noting the person's general alertness, patterns of speech, the content of that speech (as an indication of ongoing thought processes), and general motor abilities. Orientation to three concepts—Who are you? What day is it? Where are you?—is a basic part of the *mental status* assessment. In an unconscious person, the *pupillary reflex* provides important information about the status of the brain stem. When light is shined into one eye, the pupils of both eyes should constrict. This reflex is coordinated by the superior colliculus in the mesencephalon. If the pupils are unreactive, or *fixed,* serious brain damage has occurred.

A person's level of consciousness is typically reported in terms of the *Glasgow scale,* a classification system given in Table A-23. When the rating is completed, a tally is taken. Someone with a total score of 7 or less is probably comatose. Low scores in more than one section indicate that the condition is relatively severe. A comatose individual with a score of 3–5 has probably suffered irreversible brain damage.

FAP Ch. 15

TABLE A-22 States of Consciousness

Level or State	Description
Conscious states	
Normal consciousness	Aware of self and external environment; well-oriented; responsive
Delirium	Disorientation, restlessness, confusion, hallucinations, and agitation, alternating with other conscious states
Confusion	Reduced awareness; easily distracted; easily startled by sensory stimuli; person alternates between drowsiness and excitability; resembles minor form of delirium state
Dementia	Difficulties with spatial orientation, memory, language; changes in personality
Somnolence	Extreme drowsiness, but response to stimuli is normal
Chronic vegetative state	Conscious but unresponsive; no evidence of cortical function
Unconscious states	
Asleep	Can be aroused by normal stimuli (light touch, sound, etc.)
Stupor	Can be aroused by extreme and/or repeated stimuli
Coma	Cannot be aroused and does not respond to stimuli (coma states can be further subdivided according to the effect on reflex responses to stimuli)

TABLE A-23 The Glasgow Scale

Area or Aspect Assessed	Response	Score
Motor abilities	None	1
	Decerebrate rigidity (individual is supine with extension of knees, ankles in plantar flexion; arms adducted and elbows extended; forearms pronated; finger joints flexed)	2
	Decorticate rigidity (individual is supine with elbows, wrists, and fingers flexed and pressed against the chest; lower limbs rigidly extended with ankles in plantar flexion)	3
	Withdrawal reflex to stimulus	4
	Ability to pinpoint painful stimulus	5
	Ability to move body parts according to verbal request	6
Eyes	No reaction	1
	Opens eyes with painful stimulus	2
	Opens eyes upon verbal request	3
	Opens eyes spontaneously	4
Verbal ability	None	1
	Sounds emitted are not understandable	2
	Words used are not appropriate	3
	Able to speak but is not oriented	4
	Converses and is oriented	5

FAP Ch. 15

SLEEP DISORDERS FAP p. 497

Sleep disorders include abnormal patterns of REM or deep sleep, variations in the time of onset or the time devoted to sleeping, and unusual behaviors performed while sleeping. Sleep disorders of one kind or another are very common, affecting an estimated 25 percent of the U.S. population at any time.

Many clinical conditions affect sleep patterns or are exaggerated by the autonomic changes that accompany the various stages of sleep. Major clinical categories of sleep disorders include *parasomnias, insomnias, hypersomnias,* and *sleep apnea.*

Parasomnias are abnormal behaviors performed during sleep. Sleepwalking, sleep talking, teeth-grinding, and so forth commonly involve slow-wave sleep, when the skeletal muscles are not maximally inhibited. Parasomnias can have psychological rather than physiological origins.

Insomnia is characterized by shortened sleeping periods, difficulty in getting to sleep, or early awakening with an inability to return to sleep. This is the most common sleep disorder; an estimated one-third of adults have insomnia. Temporary insomnia may accompany stress, such as family arguments or other crises. Insomnia of longer duration can result from chronic depression, illness, or drug abuse. Treatment varies with the primary cause of the insomnia. For example, the treatment of mild, temporary insomnia may involve an exercise program and a reduction of caffeine intake. Severe insomnia can be temporarily treated with drugs, usually *benzodiazepines* such as *temazepam, flurazepam,* or *triazolam.* The treatment of underlying depression may also prove helpful.

Hypersomnia involves extremely long periods of otherwise normal sleep. The individual may sleep until noon and nap before dinner, despite an early retirement in the evening. These conditions have physiological or psychological origins, and successful treatment may involve drug therapy or counseling or both. Two important examples of hypersomnias are *narcolepsy* and *sleep apnea.*

Roughly 0.2–0.3 percent of the population has **narcolepsy,** characterized by dropping off to sleep at inappropriate times. The sleep lasts only a few minutes and is preceded by a period of muscle weakness. Any exciting stimulus, such as laughter or other strong emotion, may trigger the attack. (Interestingly, this condition is inherited and occurs in other mammals; narcoleptic dogs will keel over into a sound sleep when presented with their favorite treats.) In some instances, the sleep period is followed by a brief period of amnesia. Drugs that block REM sleep, such as methylphenidate *(Ritalin®),* may reduce or eliminate narcoleptic attacks.

In **sleep apnea,** a sleeping individual stops breathing for short periods of time, probably due to a dis-

turbance or abnormality in the respiratory control mechanism or blockage of the upper airway. The problem can be exaggerated by various drugs (including alcohol), obesity, hypertension, tonsillitis, and other medical conditions that affect either the CNS or cause intermittent obstruction of the respiratory passageways. Roughly half those who experience sleep apnea show simultaneous cardiac arrhythmias, but the link between the two is not understood. Treatment focuses primarily on alleviating potential causes; for example, overweight individuals are encouraged to lose weight, and enlarged tonsils are removed. In some cases, air under pressure is administered through the nose to keep the upper airway open. This treatment is called *continuous positive airway pressure (CPAP)*.

HUNTINGTON'S DISEASE FAP *p. 498*

Huntington's disease is an inherited disease marked by a progressive deterioration of mental abilities. About 25,000 Americans have this condition. In Huntington's disease, the cerebral nuclei and frontal lobes of the cerebral cortex show degenerative changes. The basic problem is the destruction of ACh-secreting and GABA-secreting neurons in the cerebral nuclei. The cause of this deterioration is not known. The first signs of the disease generally appear in early adulthood. As you would expect in view of the areas affected, the symptoms involve difficulties in performing voluntary and involuntary patterns of movement and a gradual decline in intellectual abilities that eventually lead to dementia and death.

Screening tests can now detect the presence of the gene that causes Huntington's disease, which is an autosomal dominant gene located on chromosome 4. In people with Huntington's disease, a gene of uncertain function contains a variable number of repetitions of the nucleotide sequence CAG. This DNA segment appears to be unstable, and the number of repetitions can change from generation to generation. The duplication or deletion is thought to occur during gamete formation. The more repetitions, the earlier in life the symptoms appear and the more severe the symptoms. The link between the multiple copies of the CAG nucleotide and the disorder has yet to be understood. There is no effective treatment. The children of a person with Huntington's disease have a 50 percent chance of receiving the gene and developing the disease.

PHARMACOLOGY AND DRUG ABUSE FAP *p. 498*

The drug industry certainly qualifies as "big business." Yearly sales approach $10 billion, and each year well over $1 billion is spent on advertising campaigns. If you watch television, you are presented with a dazzling array of advertisements for alcoholic beverages, cold medicines, diet pills, headache remedies, anxiety relievers, and sleep promoters. Although roughly one-third of the drugs sold affect the CNS, few consumers have any understanding of how these drugs exert their effects. Despite an overwhelming interest in medicinal and "recreational" (typically illicit) drugs, we seldom encounter accurate information about their mechanisms of action or the associated hazards. Unfounded and inaccurate rumors are therefore quite common; the concept that "natural" drugs are safer or more effective than "artificial" (synthetic) drugs is an example of a dangerous but popular misconception.

The major categories of drugs that affect CNS function are presented in Table A-24. Considerable overlap exists among these categories. For example, any drug that causes heavy sedation will also depress the perception of pain and alter mood at the same time. Well-known prescription, nonprescription, and illicit drugs are included in this table where appropriate. Some of the nonprescription drugs, most of the prescription drugs, and all of the illegal CNS-active drugs are prone to abuse because tolerance and addiction occur.

FAP Ch. 15

Sedatives lower the general level of CNS activity, reduce anxiety, and have a calming effect. **Hypnotic drugs** further depress activity, producing drowsiness and promoting sleep. All levels of CNS activity are reduced, so the effects can range from a mild relaxation to a general anesthesia, depending on the drug and the dosage administered. Sedatives and hypnotic drugs are prescribed more often than any others; almost half of the CNS-active drugs sold are sedatives or hypnotics.

Analgesics, the second most popular group, provide relief from pain. Some, like aspirin, act in the periphery by reducing the source of the painful stimulation. (Aspirin slows the release of prostaglandins, which promote inflammation and stimulate pain receptors.) Others, especially the drugs structurally related to opium, target the CNS processing of pain sensations. The compounds bind to receptors on the surfaces of CNS neurons, notably in the cerebral nuclei, the limbic system, thalamus, hypothalamus, midbrain, medulla oblongata, and spinal cord. The drugs traditionally administered, such as morphine, mimic the activity and structure of endorphins already present in the CNS.

Psychotropics are used to produce changes in mood and emotional state. Certain forms of **depression** have been shown to result from the inadequate production of norepinephrine, dopamine, or serotonin in key nuclei of the brain. Overexcitement, or **mania,** has been correlated with an overproduction of these compounds. Enhancing or blocking the production of these transmitters can often provide the mental stabilization needed to alleviate conditions such as these.

Antipsychotics reduce the hallucinations and behavioral or emotional extremes that characterize the severe mental disorders known as **psychoses,** but they leave other mental functions relatively intact.

TABLE A-24 A Simple Classification of Drugs That Affect the CNS

Category (Common name)	Actions	Examples	
		Prescription	Nonprescription
Sedatives and hypnotics ("downers")	Depress CNS activity; may promote sleep, reduce anxiety, create calm; those used to depress seizures are considered "anticonvulsants"; can be addictive	Barbiturates (phenobarbital, Nembutal®, amytal); benzodiazepines (Librium®, Valium®)	Sominex®, Nytol®, Benadryl®, alcohol
Analgesics ("pain killers")	Relieve pain at source or along CNS pathways	Opiates (morphine, Demerol®, codeine)	Aspirin, Tylenol®, ibuprofen
Psychotropics ("mood changers")	Alter CNS function and change mental state and/or mood; many are addictive.	Antipsychotics (chlorpromazine); cocaine; anti-depressants (imipramine); anti-anxiety drugs (Librium, Valium); mood stabilizers (lithium)	Caffeine, alcohol, kava, St. John's wort
Anticonvulsants	Inhibit spread of cortical stimulation	Dilantin®; also sedative-hypnotics	
Stimulants	Facilitate CNS activity; addictive; typically affect cardiovascular system	Xanthines (caffeine); amphetamines	Diet pills, Sudafed®, Actifed®, caffeine

FAP Ch. 15

Phenothiazines, notably *chlorpromazine (Thorazine®),* were initially used to control nausea and vomiting and to promote sedation and relaxation. However, they ultimately proved more useful in controlling psychotic behavior.

The term *tranquilizer* was first used for the early antipsychotic drugs, such as *Reserpine®*. Later, the category was expanded to include "minor tranquilizers," such as *Valium®*, whose effects more closely resemble those of sedatives. Eventually, it became apparent that "tranquilization" was more useful as a descriptive term than as a classification for specific drugs. For example, a tranquilizing effect can be produced by sedatives (alcohol, Valium), hypnotics (barbiturates), and antipsychotics.

The first **anticonvulsants,** used to control seizures, were sedatives such as phenobarbital. Over time, however, patients required larger and larger doses to achieve the same level of effect. This phenomenon, called **tolerance,** commonly appears when any CNS-active drug is administered. *Dilantin®* and related drugs have powerful anticonvulsant effects without producing tolerance.

Stimulants facilitate activity in the CNS. Few stimulants are used clinically, as they may trigger convulsions or hallucinations. Several of the ingredients in coffee and tea are CNS stimulants, caffeine being the most familiar example. **Amphetamines** prompt the release of norepinephrine, stimulating the respiratory and cardiovascular control centers and elevating muscle tone to the point at which tremors may begin. Amphetamines also stimulate dopamine and serotonin release in the CNS and facilitate cranial and spinal reflexes.

Drug abuse has a very hazy definition; it implies that the individual voluntarily uses a drug in some inappropriate manner. Any drug use that violates medical advice, prevalent social mores, or common law can be included within this definition. **Drug addiction** refers to an overwhelming dependence and compulsion to use a specific drug, despite the medical or legal risks involved. If use is stopped or prevented, the individual will typically suffer physical and psychological symptoms of **withdrawal.** The physiological foundations of drug dependence and addiction vary with the compound considered. In general, the mechanisms are poorly understood.

The dangers inherent in recreational drug use have been repeatedly overlooked or ignored. Until recently, the abuse of common addictive drugs such as alcohol and nicotine was considered relatively normal, or at least excusable. The risks of alcohol abuse can hardly be overstated; it has been estimated that 40 percent of young men have problems with alcohol consumption, and 7 percent of the adult population show signs of alcohol abuse or addiction. Long-term abuse increases the risks of diabetes, liver and kidney disorders, cancer, cardiovascular disease, and digestive system malfunctions. In addition, CNS disturbances commonly lead to accidents, due to poor motor control, and violent behaviors, due to interference with normal emotional balance and analytical function. Medications used for other problems, such as those for hypertension or even antibiotics, can also have CNS effects, ranging from drowsiness to hallucinations or seizures. These side effects remind us of the importance of overall homeostasis and our inability to separate the mind from the body.

ALZHEIMER'S DISEASE FAP *p. 499*

In its characteristic form, Alzheimer's disease produces a gradual deterioration of mental organization. Affected individuals lose memories, verbal and reading skills, and emotional control. Initial symptoms are subtle—moodiness, irritability, depression, and a general lack of energy. These symptoms are often ignored, overlooked, or dismissed. Elderly relatives are viewed as eccentric or irascible and are humored whenever possible.

As the condition progresses, however, it becomes more difficult to ignore or accommodate. An individual with Alzheimer's disease has difficulty making decisions, even minor ones. Mistakes—sometimes dangerous ones—are made, through either bad judgment or forgetfulness. For example, the person might light the gas burner, place a pot on the stove top, and go into the living room. Two hours later, the pot, still on the stove, melts and starts a fire.

As memory losses continue, the problems become more severe. The individual may forget relatives, his or her home address, or how to use the telephone. The memory loss commonly starts with an inability to store long-term memories, followed by the loss of recently stored memories, and eventually the loss of basic long-term memories, such as the sound of the individual's own name. The loss of memory affects both intellectual and motor abilities, and a person with severe Alzheimer's disease has difficulty in performing even the simplest motor tasks. Although by this time victims are relatively unconcerned about their mental state or motor abilities, the condition can have devastating emotional effects on the immediate family.

Individuals with Alzheimer's disease show a pronounced decrease in the number of cortical neurons, especially in the frontal and temporal lobes. This loss is correlated with inadequate ACh production in the *nucleus basalis* of the cerebrum. Axons leaving this region project throughout the cerebral cortex; when ACh production declines, cortical function deteriorates.

Most cases of Alzheimer's disease are associated with unusually large concentrations of *neurofibrillary tangles* and *plaques* in the nucleus basalis, hippocampus, and parahippocampal gyrus. The tangles are intracellular masses of abnormal microtubular proteins. The plaques are extracellular masses that form around a core that consists of an abnormal protein called **beta-amyloid.** Beta-amyloid is also found in other tissues, including the skin, blood vessels, subcutaneous layer, and intestines. Because this protein appears in small quantities in the blood and cerebrospinal fluid of many Alzheimer's patients, a screening test is now being developed to detect the condition before mental deterioration becomes pronounced. Familial cases of Alzheimer's disease are associated with mutations on either chromosome 21 or a small region of chromosome 14. A number of experimental protocols are undergoing clinical trials, but as yet there is no effective treatment for this condition.

HYPERSENSITIVITY AND SYMPATHETIC FUNCTION FAP *p. 510*

Two interesting clinical conditions result from the disruption of normal sympathetic functions. In **Horner's syndrome,** the sympathetic postganglionic innervation to one side of the face becomes interrupted. This interruption may be the result of an injury, a tumor, or some progressive condition such as multiple sclerosis. The affected side of the face becomes flushed as vascular tone decreases. Sweating stops in the affected region, and the pupil on that side becomes markedly constricted. Other symptoms include a drooping eyelid and an apparent retreat of the eye into the orbit.

Raynaud's disease most commonly affects young women. In this condition, the sympathetic system orders excessive peripheral vasoconstriction. The fingers, toes, ears, and nose may become deprived of their normal circulatory supply and take on pale white or blue coloration. The symptoms may spread to adjacent areas as the disorder progresses. Attempts to block the vasoconstriction with drugs have met with variable success. A regional **sympathectomy** (sim-path-EK-tō-mē) may be performed, cutting the fibers that provide sympathetic innervation to the affected area, but results are typically disappointing.

Under normal conditions, sympathetic tone provides the effectors with a background level of stimulation. After the elimination of sympathetic innervation, peripheral effectors may become extremely sensitive to norepinephrine and epinephrine. This hypersensitivity can produce extreme alterations in vascular tone and other functions after stimulation of the adrenal medulla. If the sympathectomy involves cutting the postganglionic fibers, hypersensitivity to circulating norepinephrine and epinephrine may eliminate the beneficial effects. The prognosis improves if the preganglionic fibers are transected, because the ganglionic neurons will continue to release small quantities of neurotransmitter across the neuromuscular or neuroglandular junctions. This release keeps the peripheral effectors from becoming hypersensitive.

PHARMACOLOGY AND THE AUTONOMIC NERVOUS SYSTEM FAP *p. 522*

The treatment of many clinical conditions involves the manipulation of autonomic function. As we noted in the text *(p. 522), mimetic drugs* or *blocking agents* can be administered to counteract or reduce symptoms of abnormal autonomic nervous system (ANS) activity. Table A-25 relates important mimetic drugs and blocking agents to specific autonomic activities.

Mimetic drugs have advantages over the neurotransmitters or hormones whose actions they simulate. Whereas neurotransmitters must be administered into the bloodstream by injection and their duration of action is limited, mimetic drugs can be administered in a variety of ways, and their actions can be sustained.

TABLE A-25 Drugs and the ANS

Drug	Mechanism	Action	Clinical Uses
Sympathomimetic			
Phenylephrine (Neosynephrine®)	Stimulates α_1 receptors	Elevates blood pressure; stimulates smooth muscle	As a topical nasal decongestant and to elevate low blood pressure
Clonidine	Stimulates α_2 receptors	Lowers blood pressure	Treatment of high blood pressure
Isoproterenol	Stimulates ß receptors	Stimulates heart rate; dilates respiratory passages	Treatment of respiratory disorders and as a cardiac stimulant during cardiac resuscitation
Dobutamine	Stimulates β_1 receptors	Stimulates cardiac output	Treatment of heart failure or heart attack
Albuterol, terbutaline	Stimulate β_2 receptors	Dilate respiratory passages	Treatment of asthma or premature labor
Ephedrine/ pseudoephedrine	Stimulates norepinephrine release at neuromuscular and neuroglandular junctions	Similar to effects of epinephrine	As a nasal decongestant and to elevate blood pressure or dilate respiratory passageways
Sympathetic blocking agents			
Prazosin (Minipress®)	Blocks α_1 receptors	Lowers blood pressure	Treatment of high blood pressure
Propranolol (Inderal®)	Blocks β_1 and β_2 receptors	Reduces metabolic activity in cardiac muscle but may constrict respiratory passageways; slows heart rate	Treatment of high blood pressure; used to reduce heart rate and force of contraction in heart disease
Metoprolol	Blocks β_1 receptors	Reduces metabolic activity in cardiac muscle	Similar to those of Inderal but has less of an effect on respiratory muscles
Parasympathomimetic			
Physostigmine, neostigmine	Block action of acetylcholinesterase	Increase ACh concentrations at parasympathetic neuromuscular and neuroglandular junctions	Stimulate digestive tract and smooth muscles of urinary bladder
Pilocarpine	Stimulates muscarinic receptors	Similar to effects of ACh	Applied topically to cornea of eye to cause pupillary contraction; treatment of glaucoma
Parasympathetic blocking agents			
Atropine, related drugs	Block muscarinic receptors	Inhibit parasympathetic activity	Treatment of diarrhea; used to dilate pupils; raise heart rate, reduce secretions of digestive and respiratory tracts prior to surgery, treat accidental exposure to anticholinesterase drugs, such as pesticides or military nerve gases

FAP Ch. 16

For example, sympathomimetic drugs can be applied topically, by spray, by inhalation, or in drops. They can also be injected to elevate blood pressure and improve cardiac performance after severe blood loss or heart muscle failure. Sympathomimetic drugs are used to treat a variety of disorders; we will consider only representative examples here.

Phenylephrine stimulates alpha receptors, causing a constriction of peripheral vessels and elevating blood pressure. It is prescribed in some cases of low blood pressure. Two other sympathomimetic drugs, *ephedrine* and *pseudoephedrine,* stimulate beta receptors. These drugs form the basis of several cold remedies (such as *Actifed*® and *Sudafed*®) that reduce nasal congestion and open respiratory passages through their effects on β_2 receptors. Undesirable side effects, such as jitteriness, anxiety, sleeplessness, and high blood pressure, result from the facilitation of CNS pathways and the stimulation of peripheral α and β_1 receptors. Recently these drugs have been abused by amphetamine ad-

dicts unable to obtain amphetamines. Apparently, these addicts are deliberately seeking the side effects that other people find distressing and objectionable.

Sympathomimetic drugs that selectively target β_1 receptors, such as *dobutamine,* are especially valuable in increasing heart rate and blood pressure. They are commonly prescribed to improve the performance of a failing heart. Sympathomimetic drugs targeting β_2 receptors, such as *albuterol* or *terbutaline,* have been developed to treat the bronchial constriction that accompanies asthma attacks.

Sympathetic blocking agents that prevent a normal response to neurotransmitters or sympathomimetic drugs include alpha-blockers and beta-blockers. **Alpha-blockers** eliminate the peripheral vasoconstriction that accompanies sympathetic stimulation. The alpha-blockers include *prazosin,* which selectively targets α_1 receptors and is used to reduce high blood pressure. The drug can reduce constriction of the prostatic urethra, so it is used to treat restricted urine flow in *(benign) prostatic hypertrophy* (FAP *p. 978*). **Beta-blockers** are effective and clinically useful for treating chronic high blood pressure and for reducing the risk of death after a heart attack. In general, beta-blockers decrease the heart rate and force of contraction, reducing the strain on the heart and simultaneously lowering peripheral blood pressure. *Propranolol* and *metoprolol* are two of the most popular beta-blockers currently on the market. Propranolol affects both β_1 and β_2 receptors, so persons with asthma may experience difficulties in breathing as their respiratory passageways constrict. Metoprolol targets β_1 receptors almost exclusively, leaving the respiratory smooth muscles relatively unaffected.

Parasympathomimetic drugs may be used to increase activity along the digestive tract and encourage defecation and urination. *Physostigmine* and *neostigmine* are important parasympathomimetic drugs that work by blocking the action of acetylcholinesterase. Because this enzyme is rendered inoperative, levels of ACh within the synapses climb and parasympathetic activity is enhanced. In Chapter 12 we mentioned the blocking agent *d-tubocurarine,* which blocks neuromuscular transmission (FAP *p. 390*). The administration of physostigmine or neostigmine can counteract the paralytic effects of this drug.

Parasympathetic blocking agents, such as *atropine,* target the muscarinic receptors at neuromuscular and neuroglandular junctions. These drugs have diverse effects, but they are typically used to control the diarrhea and cramps associated with various forms of food poisoning. The drug *Lomotil®,* known as the "traveler's friend," can provide temporary relief from diarrhea for the duration of a plane flight home. Among its other effects, atropine causes an elevation of the heart rate due to a loss of parasympathetic tone. *Scopolamine* has similar effects on peripheral tissues, but it has greater influence on the CNS. Its most useful effects are promoting drowsiness, reducing nausea, and relieving anxiety. As a result, scopolamine is commonly given to a patient who is being prepared for surgery, prior to the administration of the anesthetic agent. (Scopolamine is also administered transdermally to control nausea, as we noted in the text on *pp. 150* and *563*.)

A CLOSER LOOK: TRANSDUCTION

FAP *p. 528*

Transduction is the process that converts a stimulus into a series of action potentials in a sensory neuron. Those action potentials are intrepreted by the CNS as a sensation. Transduction can be divided into three steps:

STEP 1: *An arriving stimulus changes the transmembrane potential of the receptor membrane.* The nature of the interaction between stimulus and receptor determines the receptor's specificity. Regardless of the nature of the interaction, however, the result is always the same: The transmembrane potential of the receptor cell changes. The change in the transmembrane potential that accompanies receptor stimulation is called a **receptor potential.**

The receptor potential may be a depolarization or a hyperpolarization. It is a graded potential change: the stronger the stimulus, the larger the receptor potential.

STEP 2: *The receptor potential directly or indirectly affects a sensory neuron.* A membrane depolarization that leads to an action potential in a sensory neuron is called a generator potential. The typical receptors for the general senses are the dendrites of sensory neurons (Figure A-34a•). Receptor potentials in the dendrites spread to and summate at the initial segment, producing a generator potential that can be large enough to trigger an action potential in the axon. When a sensory neuron acts as a receptor, the terms *receptor potential* and *generator potential* can be used interchangeably.

Sensations of taste, hearing, equilibrium, and vision are provided by specialized receptor cells that communicate with sensory neurons across chemical synapses. The receptor cells develop graded receptor potentials in response to stimulation, and the change in membrane potential alters the rate of neurotransmitter release at the synapse. In the case of taste receptors, stimulation triggers the release of neurotransmitters that depolarize the membrane of the sensory neuron (Figure A-34b•). The larger the receptor potential, the more neurotransmitter released and the greater the depolarization. When specialized receptor cells are involved, the receptor potential and the generator potential are distinct in both location and timing. The receptor potential appears in the receptor cell when the stimulus arrives. The generator potential develops later, in the sensory neuron, following the arrival of neurotransmitter at the postsynaptic membrane.

STEP 3: *Action potentials travel to the CNS along an afferent fiber.* When a generator potential appears, action potentials develop in the afferent fiber. For reasons that

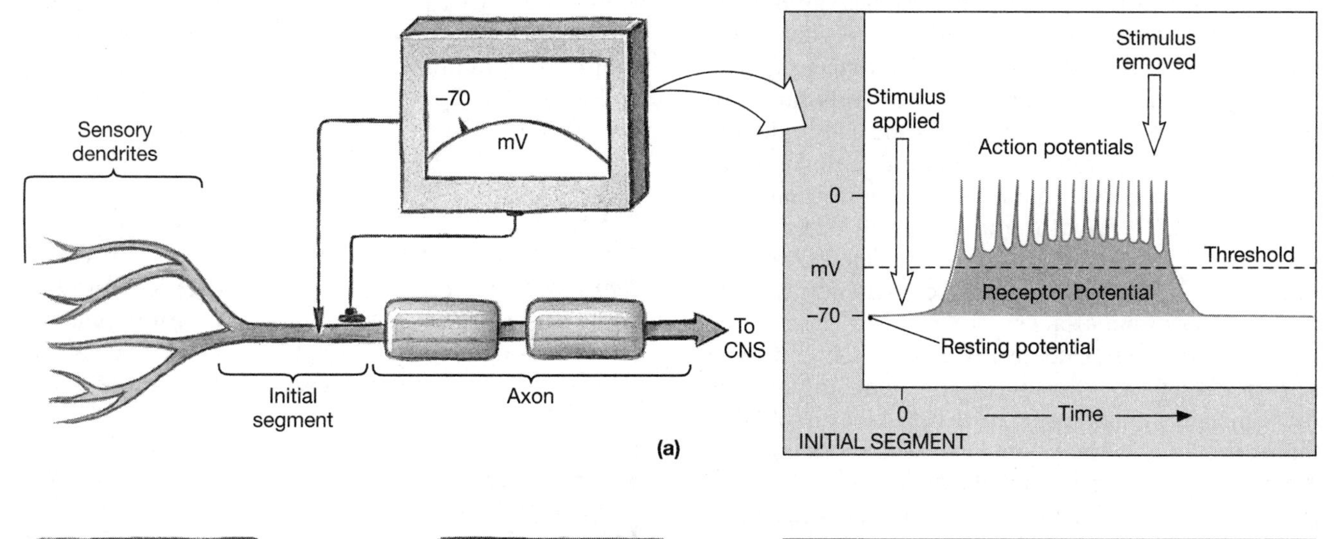

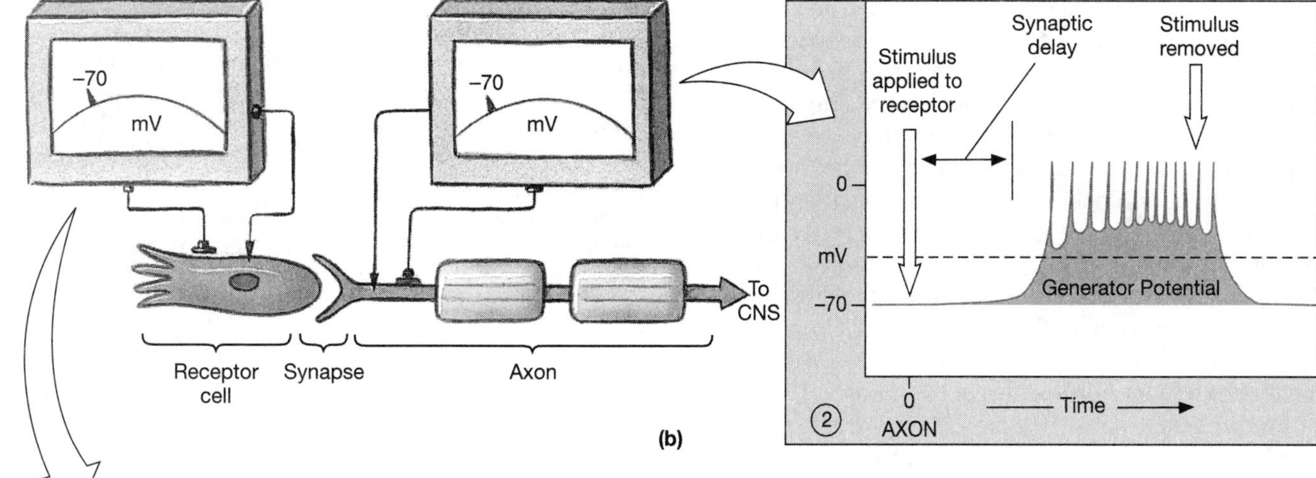

FAP Ch. 17

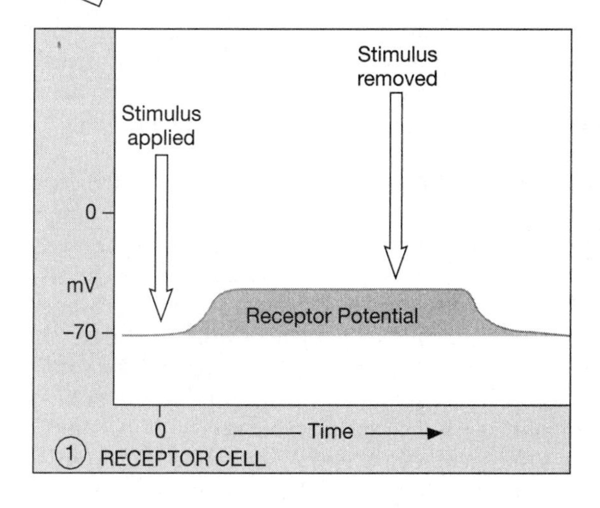

•**FIGURE A-34 Transduction. (a)** When a sensory neuron acts as a receptor, a stimulus that depolarizes the dendrites may bring the initial segment of the axon to threshold. The receptor and the neuron are the same cell, so the receptor potential is a generator potential. **(b)** In the special senses of taste, equilibrium, hearing, and vision, the receptor cells are specialized cells that communicate with neurons across chemical synapses. ① The receptor cell shows a receptor potential in response to stimulation. In this example, the receptor potential is a depolarization that accelerates neurotransmitter release, and ② the neurotransmitter produces a generator potential in the postsynaptic membrane.

we discussed in the text in Chapter 12, the greater the depolarization produced by the generator potential, the higher the frequency of action potentials in the afferent fiber. The arriving information is then processed and interpreted by the CNS at the conscious and subconscious levels.

ANALYZING SENSORY DISORDERS FAP p. 528

A recurring theme of the text is that an understanding of how a system works enables you to predict how things might go wrong. You are already familiar with the organization and physiology of sensory systems,

and we discussed some of the most important clinical problems in clinical comments on the preceding pages. Placing the entire array into categories provides an excellent example of a strategy that can be used to analyze any system in the body.

Every sensory system contains peripheral receptors, afferent fibers, ascending tracts, nuclei, and areas of the cerebral cortex. Any malfunction affecting the system must involve one of those components. Any clinical diagnosis requires you to seek answers to a series of yes-or-no questions, eliminating one possibility at a time until the nature of the problem becomes apparent.

A CLOSER LOOK: PAIN MECHANISMS, PATHWAYS, AND CONTROL FAP p. 531

The sensory neurons that bring pain sensations into the CNS release *glutamate* and *Substance P* as neurotransmitters. These two neurotransmitters have very different but complementary effects. Glutamate produces an immediate depolarization of the postsynaptic membrane by binding to *AMPA* receptors,[1] which open channels that permit Na^+ entry and K^+ departure from the cytoplasm. The excitatory effects are restricted to the postsynaptic cell, because glutamate is rapidly reabsorbed by the synaptic knobs. The binding to AMPA receptors stimulates the interneuron, and pain impulses ascend to the thalamus within the spinothalamic tracts.

When the interneuron depolarizes under glutamate stimulation, the depolarization makes another membrane receptor, *NMDA,* available for glutamate binding. The binding process triggers the opening of calcium ion channels in the membrane, and the entry of Ca^{2+} leads to the activation of second messengers with varied effects on the neuron. The net result is that the stimulated neurons become strongly facilitated, and pain sensitivity increases.

Substance P released by sensory neurons has a more widespread stimulatory effect. This neurotransmitter diffuses through the gray matter of the dorsal gray horn of the spinal cord, where it affects large numbers of interneurons. Substance P binds to membrane receptors and is brought into the cytoplasm by means of receptor-mediated endocytosis. The result is the stimulation of some interneurons, producing sensations of pain, and the facilitation of many other interneurons involved with pain pathways.

The ascending pain sensations are widely distributed. Most ascending fibers travel within the lateral spinothalamic tracts for projection to the primary sensory cortex. The thalamus also relays pain sensations to the cingulate gyrus, an emotional center of the limbic system. Some of the ascending fibers do not reach the thalamus but synapse in the reticular formation or hypothalamus instead. Pain can thus influence CNS activities at both the conscious and subconscious levels.

Endorphins and enkephalins are neuromodulators whose release inhibits activity along pain pathways in the brain. These compounds, structurally similar to morphine, are found in the limbic system, hypothalamus, and reticular formation. The pain centers in these areas also use Substance P as a neurotransmitter. Endorphins bind to the presynaptic membrane and prevent the release of Substance P, thereby reducing the conscious perception of pain, although the painful stimulus remains.

Due to the facilitation that results from glutamate and Substance P release, the level of pain experienced (especially chronic pain) can be out of proportion to the amount of painful stimuli or the apparent tissue damage. This effect may be one reason why people differ so widely in their perception of the pain associated with childbirth, headaches, or back pain. This facilitation is also presumed to play a role in phantom limb pain; the sensory neurons may be inactive, but the hyperexcitable interneurons may continue to generate pain sensations. An interesting piece of evidence comes from clinical research: When general anesthesia shuts down the cerebral cortex and prevents conscious perception of pain, the interneurons of the spinal cord are not anesthetized. As a result, they become facilitated, and this effect can exaggerate postoperative pain. In one recent study, roughly 20 percent of patients who had a limb amputated under general anesthesia experienced phantom limb pain. However, when spinal facilitation was prevented, none of the patients experienced that pain.

Acute and Chronic Pain

Pain management poses a number of problems for clinicians. Painful sensations can result from tissue damage or sensory nerve irritation; it can originate where it is perceived, be referred from another location, or represent a false signal generated along the sensory pathway. The treatment differs in each case, and an accurate diagnosis is an essential first step. Acute pain is the result of tissue injury; the cause is apparent, and local treatment of the injury is typically effective in relieving the pain. The most effective solution is to stop the damage, end the stimulation, and suppress the painful sensations at the injury site. Pain sensations are suppressed when topical or locally injected anesthetics inactivate nociceptors in the immediate area. Analgesic drugs can also be administered. They work in many different ways; we will consider only a few examples here.

[1]AMPA receptors and the other NMDA receptors discussed were originally named after administered compounds that would bind to them. AMPA receptors bind **a**lpha-amino-3-hydroxy-5-**m**ethyl-4-isoxazole **p**ropionic **a**cid; NMDA receptors bind **N-m**ethyl-**D-a**spartate, a synthetic molecule related to glutamate.

Tissue injury results in damage to cell membranes. A fatty acid called arachidonic acid escapes from injured membranes. In interstitial fluid, an enzyme called *cyclo-oxygenase* converts arachidonic acid molecules to prostaglandins; it is these prostaglandins that stimulate nociceptors in the area. Aspirin, ibuprofen, and related analgesics reduce inflammation and suppress pain by blocking the action of cyclo-oxygenase.

Chronic pain is more difficult to categorize and treat. It includes (1) pain from an injury that persists after tissue structure has been repaired; (2) pain from a chronic disease, such as cancer; and (3) pain without an apparent cause. Chronic pain in part reflects permanent facilitation of the pain pathways and the creation of a reverberating "pain memory." Complex psychological and physiological components are also involved. For example, many chronic pain patients develop a tolerance for pain medications, and insomnia and depression are common complaints. Chronic pain can be helped by antidepressants, which affect neurotransmitter levels. Counseling may help the person focus attention outward rather than inward; the outward focus can lessen the perceived level of pain and reduce the amount of pain medication required. Curiously, developing a second, acute source of pain, such as a herpes-zoster infection, can reduce the perception of preexisting chronic pain.

FAP Ch. 17

In some cases, chronic pain and severe acute pain can be suppressed by inhibition of the central pain pathway. Analgesics related to morphine reduce pain by mimicking the action of endorphins. The perception of pain may be altered, although the pain remains. For example, patients on morphine report being aware of painful sensations, but they are not affected by them. Surgical steps can be taken to control severe pain; for instance, (1) the sensory innervation of an area can be destroyed by an electrical current, (2) the dorsal roots carrying the painful sensations can be cut (a *rhizotomy*), (3) the ascending tracts in the spinal cord can be severed (a *tractotomy*), or (4) thalamic or limbic centers can be stimulated or destroyed. These options, listed in order of increasing degree of effect, surgical complexity, and associated risk, are used only when other methods of pain control have failed to provide relief.

In the Chinese technique of acupuncture to control pain, fine needles are inserted at specific locations and are either heated or twirled by the therapist. Several theories have been proposed to account for the positive effects, but none is widely accepted or proven. It has been suggested that the pain relief may follow endorphin release. It is not known how acupuncture stimulates endorphin release; the acupuncture points do not correspond to the distribution of any of the major peripheral nerves.

Many other aspects of pain generation and control remain a mystery. Up to 30 percent of patients who receive a nonfunctional medication subsequently experience a significant reduction in pain. It has been suggested that this *placebo effect* results from endorphin release triggered by the expectation of pain relief. Although the medication has no direct effect, the indirect effect is quite significant and complicates the evaluation of analgesic medications.

ASSESSMENT OF TACTILE SENSITIVITIES

FAP *p. 533*

Regional sensitivity to light touch can be checked by gentle contact with a fingertip or a slender wisp of cotton. The **two-point discrimination test** provides a more detailed sensory map of tactile receptors. Two fine points of a bent paper clip or another object are applied to the skin surface simultaneously. The subject then describes the contact. When the points fall within a single receptive field, the individual will report only one point of contact. A normal individual loses two-point discrimination at 1 mm (0.04 in.) on the surface of the tongue, at 2–3 mm (0.08–0.12 in.) on the lips, at 3–5 mm (0.12–0.20 in.) on the backs of the hands and feet, and at 4–7 cm (1.6–2.75 in.) over the general body surface.

Vibration receptors are tested by applying the base of a tuning fork to the skin. Damage to an individual spinal nerve produces insensitivity to vibration along the paths of the related sensory nerves. If the sensory loss results from spinal cord damage, the injury site can typically be located by walking the tuning fork down the spinal column, resting its base on the vertebral spines.

Descriptive terms are used to indicate the degree of sensitivity in the area considered. *Anesthesia* implies a total loss of sensation; the individual cannot perceive touch, pressure, pain, or temperature sensations from that area. *Hypesthesia* is a reduction in sensitivity, and *paresthesia* is the presence of abnormal sensations, such as the pins-and-needles sensation when an arm or leg "falls asleep" due to pressure on a peripheral nerve. (We discussed several types of *pressure palsies* that produce temporary paresthesia on p. 78.)

OTITIS MEDIA AND MASTOIDITIS

FAP *p. 558*

Acute otitis media is an infection of the middle ear, most commonly of bacterial origin. It typically affects infants and children and is occasionally seen in adults. The pathogens tend to gain access via the auditory tube, generally during an upper respiratory infection. As the pathogens multiply in the middle ear cavity, white blood cells rush to the site and pus accumulates. Eventually, the tympanic membrane may rupture, producing a characteristic oozing from the external auditory canal. The bacteria can generally be controlled by antibiotics, the pain reduced by analgesics, and the stagnant mucus and swelling reduced by deconges-

tants. In the United States, it is rare for otitis media to progress to the stage at which rupture of the tympanic membrane occurs.

Serous otitis media (SOM) involves the accumulation of clear, thick, glue-like fluid in the middle ear. It can follow acute otitis media or can result from chronic nasal infection and allergies. This condition causes hearing loss; affected toddlers may have delayed speech development. Treatment involves decongestants, antihistamines, and, in severe cases, drainage of the middle ear cavity into the external auditory canal through an implanted shunt. As toddlers grow, the auditory tube enlarges, allowing better drainage during upper respiratory infections, so both forms of otitis media become less common.

Otitis media is extremely common in developing countries where medical care and antibiotics are not readily available. Many children and adults in these countries have *chronic otitis media,* a condition characterized by chronic or recurring bouts of middle ear infection. This condition produces scarring or perforation of the tympanic membrane, which leads to some degree of hearing loss. Resulting damage to the inner ear or the auditory ossicles can further reduce hearing.

If the pathogens leave the middle ear and invade the air cells within the mastoid process, **mastoiditis** develops. The connecting passageways are very narrow; as the infection progresses, the individual experiences severe earaches, fever, and swelling behind the ear in addition to symptoms of otitis media. The major risk of mastoiditis is the spread of the infection to the brain via the connective tissue sheath of the facial nerve (N VII). Prompt and powerful antibiotic therapy is needed. If the problem remains, the person may have to undergo *mastoidectomy* (the opening and drainage of the mastoid sinuses).

VERTIGO AND MÉNIÈRE'S DISEASE FAP *p. 564*

The term **vertigo** describes an inappropriate sense of motion, usually a spinning sensation. This meaning distinguishes it from "dizziness," a sensation of light-headedness and disorientation that typically precedes a fainting spell. Vertigo can result from abnormal conditions in or stimulation of the inner ear or from problems elsewhere along the sensory pathway that carries equilibrium sensations. It commonly accompanies CNS infections, and many people experience vertigo when they have a high fever.

Any event that sets endolymph in motion can stimulate the equilibrium receptors and produce vertigo. Placing an ice pack in contact with the area over the temporal bone or flushing the external auditory canal with cold water may chill the endolymph in the outermost portions of the labyrinth and establish a temperature-related circulation of fluid. A mild and temporary vertigo is the result. The consumption of excessive quantities of alcohol or exposure to certain drugs can also produce vertigo by changing the composition of the endolymph or disturbing the hair cells of the inner ear.

Other causes of vertigo include viral infection of the vestibular nerve and damage to the vestibular nucleus or its tracts. Acute vertigo can also result from damage caused by abnormal endolymph production, as in Ménière's disease. Probably the most common cause of vertigo is *motion sickness,* which we discuss in the text (FAP *p. 563*).

In **Ménière's disease,** the distortion of the membranous labyrinth of the inner ear by high fluid pressures ruptures the membranous wall and mixes endolymph and perilymph. The receptors in the vestibule and semicircular canals then become highly stimulated. The individual may be unable to start a voluntary movement because he or she is experiencing intense spinning or rolling sensations. In addition to the vertigo, the person may "hear" unusual sounds as the cochlear receptors are activated.

TESTING AND TREATING HEARING DEFICITS FAP *p. 571*

In the most common hearing test, an individual listens to sounds of varying frequency and intensity generated at irregular intervals. A record is kept of the responses, and the graphed record, or **audiogram,** is compared with that of an individual with normal hearing (Figure A-35•). **Bone conduction tests** are used to discriminate between conductive deafness and nerve deafness. If you put your fingers in your ears and talk quietly, you can still hear yourself because the bones of the skull conduct the sound waves to the cochlea, bypassing the middle ear. In a bone conduction test, the physician places a vibrating tuning fork against the skull. If the subject hears the sound of the tuning fork when it is in contact with the skull but not when it is held next to the entrance to the external auditory canal, the problem must lie with the external or middle ear. If the subject remains unresponsive to both stimuli, the problem must be at the receptors or along the auditory pathway.

Several effective treatments exist for conductive deafness. A hearing aid overcomes the loss in sensitivity by increasing the intensity of stimulation. Surgery may repair the tympanic membrane or free damaged or immobilized auditory ossicles. Artificial ossicles may also be implanted if the original ones are damaged beyond repair.

Few treatments are available for nerve deafness. Mild conditions may be overcome by the use of a hearing aid if some functional hair cells remain. In a **cochlear implant,** a small, battery-powered device is inserted beneath the skin behind the mastoid process. Small wires run through the round window to reach the cochlear nerve; the implant "hears" a sound and

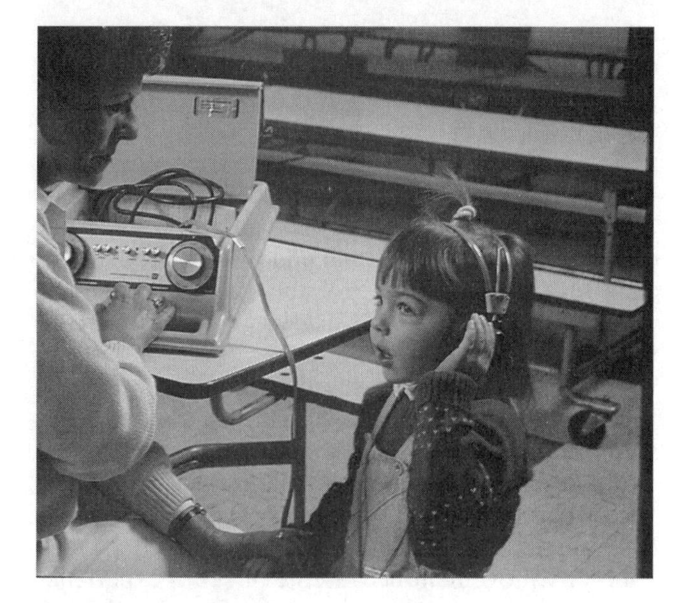

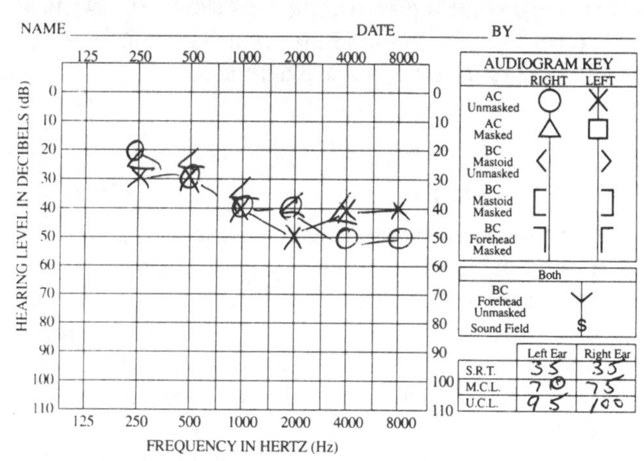

•**FIGURE A-35** An Audiogram

stimulates the nerve directly. Increasing the number of wires and varying their implantation sites make it possible to create a number of different frequency sensations. Those sensations do not approximate normal hearing—there is as yet no way to target the specific afferent fibers responsible for the perception of a particular sound. Instead, a random assortment of afferent fibers are stimulated, and the individual must learn to recognize the meaning and probable origin of the perceived sound. Although this technique is obviously not perfect, improvements in electrode placement and computer processing are being made. Each year, more and more severely deaf children and adults benefit from this treatment.

A new approach involves inducing the regeneration of hair cells of the organ of Corti. Researchers working with mammals other than humans have been able to induce hair cell regeneration both in cultured hair cells and in live animals. This is a very exciting area of research, and there is hope that it may ultimately lead to an effective treatment for human nerve deafness.

[EXPLORE]

Additional resources, such as interactive tutorials, critical-thinking questions, clinical problems, and case studies, are available through the CD-ROM and Website for your textbook. To begin your exploration, launch the *Martini Interactive* CD-ROM or visit the Companion Website (http://www.prenhall.com/martini/fap5) and select Chapter 12, 13, 14, 15, 16, or 17.

The Endocrine System

The endocrine system provides long-term regulation and adjustment of homeostatic mechanisms and a variety of body functions. For example, the endocrine system is responsible for the regulation of fluid and electrolyte balance, cell and tissue metabolism, growth and development, and reproductive functions. The endocrine system also assists the nervous system in responding to stressful stimuli.

The endocrine system is composed of nine major endocrine glands and several other organs, such as the heart and kidneys, that have other important functions. The hormones secreted by these endocrine organs are distributed by the circulatory system to target tissues throughout the body. Each hormone affects a specific set of target tissues that may differ from that of other hormones. The selectivity is based on the presence or absence of hormone-specific receptors in the target cell's cell membrane, cytoplasm, or nucleus.

A CLASSIFICATION OF ENDOCRINE DISORDERS

Homeostatic regulation of circulating hormone levels primarily involves negative feedback control mechanisms. The feedback loop involves an interplay between the endocrine organ and its target tissues. An endocrine gland may release a particular hormone in response to one of three types of stimuli:

1. *Some hormones are released in response to variations in the concentrations of specific substances in body fluids.* Parathyroid hormone, for example, is released when calcium levels decline.
2. *Some hormones are released only when the gland cells receive hormonal instructions from other endocrine organs.* For example, the rate of production and release of triiodothyronine (T_3) and tetraiodothyronine (T_4, thyroxine) by the thyroid gland is controlled by thyroid-stimulating hormone (TSH) from the anterior pituitary gland. The secretion of TSH is in turn regulated by the release of thyrotropin-releasing hormone (TRH) from the hypothalamus.
3. *Some hormones are released in response to neural stimulation.* The release of epinephrine and norepinephrine from the adrenal medullae during sympathetic activation is an example.

Endocrine disorders can therefore develop due to abnormalities in the endocrine gland, the endocrine or neural regulatory mechanisms, or the target tissues. Figure A-36• provides an overview of the major classes of endocrine disorders. In the discussion that follows, we will use the thyroid gland as an example, because the text introduces major types of thyroid gland disorders. These *primary disorders* may result in overproduction *(hypersecretion)* or underproduction *(hyposecretion)* of hormones. For example, clinicians may categorize a thyroid disorder as *primary hyperthyroidism* or *primary hypothyroidism* if the problem originates within the thyroid gland.

DISORDERS DUE TO ENDOCRINE GLAND ABNORMALITIES

Most endocrine disorders are the result of problems within the endocrine gland itself. Causes of hyposecretion include the following:

- *Metabolic factors.* Hyposecretion may result from a deficiency in some key substrate needed to synthesize that hormone. For example, hypothyroidism can be caused by inadequate dietary iodine levels or by exposure to drugs that inhibit iodine transport or utilization at the thyroid gland.
- *Physical damage.* Any condition that interrupts the normal circulatory supply or that physically damages the endocrine cells may become inactive immediately or after an initial surge of hormone release. If the damage is severe, the gland can become permanently inactive. For instance, temporary or permanent hypothyroidism can result from the infection or inflammation of the gland (*thyroiditis*), from the interruption of normal blood flow, or from exposure to radiation as part of treatment for cancer of the thyroid gland or adjacent tissues. The thyroid gland can also be damaged in an *autoimmune disorder* that results in the production of antibodies that attack and destroy normal follicle cells.
- *Congenital disorders.* An individual may be unable to produce normal amounts of a particular hormone because (1) the gland itself is too small, (2) the required enzymes are abnormal, (3) the receptors that trigger secretion are relatively insensitive because a mutation affects G-protein structure, or (4) the gland cells lack the receptors normally involved in stimulating secretory activity.

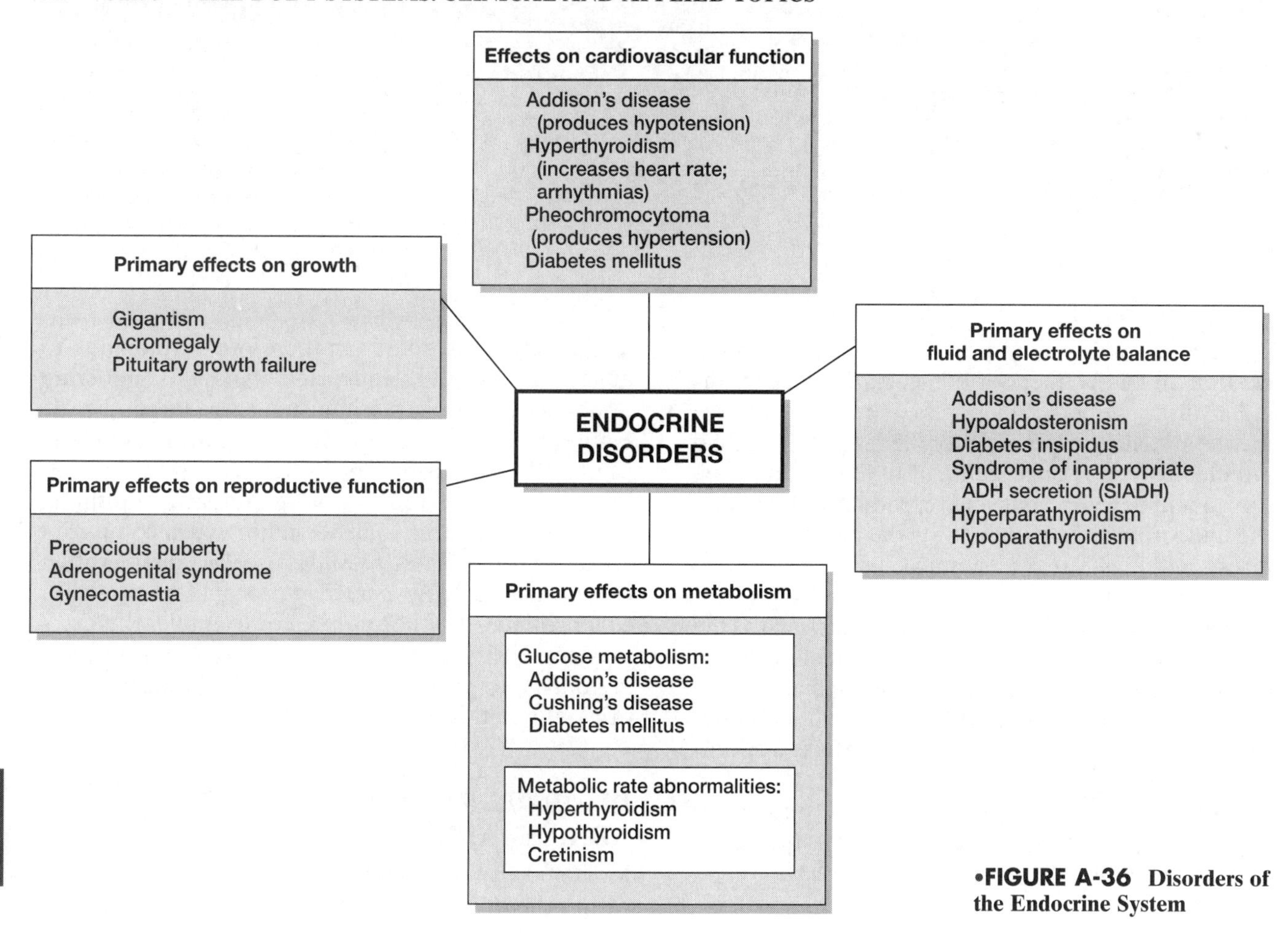

FAP Ch. 18

•**FIGURE A-36** **Disorders of the Endocrine System**

DISORDERS DUE TO ENDOCRINE OR NEURAL REGULATORY MECHANISM ABNORMALITIES

Endocrine disorders can instead result from problems with other endocrine organs involved in the negative feedback control mechanism. For example:

- *Secondary hypothyroidism* can be caused by inadequate TSH production at the pituitary gland or by inadequate TRH secretion at the hypothalamus.
- *Secondary hyperthyroidism* can be caused by excessive TRH or TSH production. Secondary hyperthyroidism may develop in individuals with tumors of the pituitary gland.

DISORDERS DUE TO TARGET TISSUE ABNORMALITIES

Endocrine abnormalities can also be caused by the presence of abnormal hormonal receptors in target tissues. In this case, the gland involved and the regulatory mechanisms are normal but the peripheral cells are unable to respond to the circulating hormone. The best example of this type of abnormality is *Type 2 diabetes,* in which peripheral cells do not respond normally to insulin.

THE SYMPTOMS OF ENDOCRINE DISORDERS

Knowledge of the individual endocrine organs and their functions makes it possible to predict the symptoms of specific endocrine disorders. For example, thyroid hormones increase basal metabolic rate, body heat production, perspiration, restlessness, and heart rate. An elevated metabolic rate, increased body temperature, weight loss, nervousness, excessive perspiration, and an increased or irregular heartbeat are common symptoms of hyperthyroidism. Conversely, a low metabolic rate, decreased body temperature, weight gain, lethargy, dry skin, and a reduced heart rate typically accompany hypothyroidism. The symptoms associated with over- and underproduction of major hormones are summarized in Table 18-9 in FAP *(p. 617).*

THE DIAGNOSIS OF ENDOCRINE DISORDERS

The first step in the diagnosis of an endocrine disorder is the physical examination. Several disorders produce characteristic physical signs that reflect abnormal hormone activities. A few examples were introduced in the text:

- *Cushing's disease* results from an oversecretion of glucocorticoids by the adrenal cortex. As the condition progresses, the normal pattern of fat distribution in the body shifts. Adipose tissue accumulates in the abdominal area, the lower cervical area (causing a "buffalo hump"), and the face (a "moonface"), but the extremities become relatively thin.
- *Acromegaly* results from the oversecretion of growth hormone in adults. In this condition, the facial features become distorted due to excessive cartilage growth, and the lower jaw protrudes, a sign known as *prognathism.* The hands and feet also become enlarged.
- *Adrenogenital syndrome* results from the oversecretion of androgens by the adrenal glands in females. Hair growth patterns change, and *hirsutism* (p. 49) develops.
- Hypothyroidism can produce a distinctively enlarged thyroid gland, or *goiter.*
- Hyperthyroidism can produce protrusion of the eyes, or *exophthalmos.*

These signs are very useful, but many other signs and symptoms related to endocrine disorders are less definitive. For example, *polyuria,* or increased urine production, can result from hyposecretion of ADH *(diabetes insipidus)* or a form of *diabetes mellitus;* a symptom such as hypertension (high blood pressure) can be caused by a variety of cardiovascular or endocrine problems. In these instances, many diagnostic decisions are based on blood tests, which can confirm the presence of an endocrine disorder by detecting abnormal levels of circulating hormones, followed by tests that determine whether the primary cause of the problem lies with the endocrine gland, the regulatory mechanism(s), or the target tissues. Table A-26 provides an overview of important blood tests and other tests used in the diagnosis of endocrine disorders.

DISORDERS OF THE ENDOCRINE SYSTEM

THYROID GLAND DISORDERS FAP *p. 597*

Hypothyroidism typically results from some problem that involves the thyroid gland rather than a problem with the pituitary production of TSH. In primary hypothyroidism, TSH levels are elevated, because the pituitary gland attempts to stimulate thyroid activity, but levels of T_3 and T_4 are depressed. One form of hypothyroidism results from a mutation that affects the structure of the G-proteins at TSH receptors. The structural change reduces the receptors' sensitivity to TSH and depresses thyroid activity. Treatment of chronic hypothyroidism, such as the hypothyroidism that follows radiation exposure, generally involves the administration of synthetic thyroid hormones (thyroxine) to maintain normal blood concentrations.

A **goiter** is an enlargement of the thyroid gland. The enlargement generally indicates increased thyroid follicle size, but thyroxine release may be increased or decreased, depending on the cause of the goiter. Most goiters develop when the thyroid gland is unable to synthesize and release adequate amounts of thyroid hormones. Under continuing TSH stimulation, thyroglobulin production accelerates and the thyroid follicles enlarge. One type of goiter occurs if the thyroid fails to obtain enough iodine to meet its synthetic requirements. (This condition is now rare in the United States, because iodized table salt is widely available.) Administering iodine may not solve the problem entirely; the sudden availability of iodine can produce symptoms of hyperthyroidism as the stored thyroglobulin becomes activated. The usual therapy involves the administration of thyroxine, which has a negative feedback effect on the hypothalamus and pituitary gland, thus inhibiting the production of TSH. Over time, the resting thyroid may return to its normal size.

Hyperthyroidism, also known as *thyrotoxicosis,* occurs when thyroid hormones are produced in excessive quantities. In **Graves' disease,** excessive thyroid activity leads to goiter and the symptoms of hyperthyroidism. Protrusion of the eyes, or **exophthalmos** (eks-ahf-THAL-mōs), may also appear for unknown reasons. Graves' disease has a genetic autoimmune basis and affects many more women than it does men. Treatment may involve the use of antithyroid drugs, the surgical removal of portions of the glandular mass, or the destruction of part of the gland by exposure to radioactive iodine.

FAP Ch. 18

Hyperthyroidism may also result from inflammation or, rarely, thyroid tumors. In extreme cases, the individual's metabolic processes accelerate out of control. During a *thyrotoxic crisis,* or "thyroid storm," the individual experiences a dangerously high fever, a rapid heart rate, and the malfunctioning of a variety of physiological systems.

DISORDERS OF PARATHYROID FUNCTION FAP *p. 598*

When the parathyroid glands secrete inadequate or excessive amounts of parathyroid hormone, calcium concentrations move outside normal homeostatic limits. **Hypoparathyroidism** with hypocalcemia can develop after neck surgery, especially a thyroidectomy, if the blood supply to the parathyroid glands is restricted. In many other cases, the primary cause of the condition is uncertain. Parathyroid hormone (PTH) is extremely costly; because supplies are very limited, PTH administration is not used to treat this condition, despite its probable effectiveness. As an alternative, a dietary combination of vitamin D and calcium can be used to elevate body fluid calcium concentrations. (As we noted in Chapter 6 of the text, vitamin D stimulates the absorption of calcium ions across the lining of the digestive tract.)

TABLE A-26 Representative Diagnostic Procedures for Disorders of the Endocrine System

Diagnostic Procedure	Method and Result	Representative Uses
Pituitary Gland		
Wrist and hand X-ray film	Standard X-rays of epiphyseal cartilages for estimation of "bone age," based on the time of closure of epiphyseal cartilages	Compares a child's bone age and chronological age; a bone age greater than 2 years behind the chronological age suggests possible growth hormone deficiency with hypopituitarism or pituitary growth failure.
X-ray study of sella turcica	Standard X-ray of the sella turcica, which houses the pituitary gland	Determines the size of the pituitary gland; detects pituitary tumors
CT scan of pituitary gland	Standard cross-sectional CT; contrast media may be used.	
MRI of pituitary gland	Standard MRI	
Thyroid Gland		
Thyroid scanning	Radionuclide given by mouth accumulates in the thyroid, giving off radiation captured to create an image of the thyroid.	Determines size, shape, and abnormalities of the thyroid gland; detects presence of nodules that may be tumors or hyperactive or hypoactive areas; determines cause of a mass in the neck
Ultrasound examination of thyroid	Sound waves reflected off internal structures are used to generate a computer image.	Detects thyroid cysts or tumors, enlarged lymph nodes, or abnormalities in the shape or size of the thyroid gland
Radioactive iodine uptake test (RAIU)	Radioactive iodine is ingested and trapped by the thyroid; detector determines the amount of radioiodine taken up over a period of time.	Determines hyperactivity or hypoactivity of the thyroid gland
Parathyroid Glands		
Ultrasound examination of parathyroid glands	Standard ultrasound	Determines structural abnormalities of the parathyroid gland, such as enlargement
Adrenal Glands		
Ultrasound of adrenal gland	Standard ultrasound	Determines abnormalities in adrenal gland size or shape; detects pheochromocytoma
CT scan of adrenal gland	Standard cross-sectional CT	Determines abnormalities in adrenal gland size or shape
Adrenal angiography	Injection of radiopaque dye for examination of the vascular supply to the adrenal gland	Detects tumors and hyperplasia
Laboratory Test	**Normal Values in Blood Plasma or Serum**	**Significance of Abnormal Values**
Pituitary Gland		
Growth hormone (GH)	>10 ng/ml	<10 ng/ml on stimulation for GH suggests deficiency (found in hypopituitarism and pituitary growth failure).
Plasma ACTH	Adults, morning: 20–80 pg/ml Adults, late afternoon: 10–40 pg/ml	Elevated levels of ACTH could indicate stress, pituitary tumor, hyperpituitarism, or Addison's disease with a compensatory elevation of ACTH; decreased levels suggest hypopituitarism, Cushing's disease, or carcinoma of the adrenal gland.
Serum TSH	Adults: <3 ng/ml	Elevated levels indicate hyperpituitarism or primary hypothyroidism; decreased levels suggest hypopituitarism or hyperthyroidism.
Serum LH	Premenopausal females: 3–30 mIU/ml Females, midcycle: 30–100 mIU/ml	Elevated levels occur with pituitary tumors and hyperpituitarism; decreased levels occur in hypopituitarism and adrenal tumors.

FAP Ch. 18

TABLE A-26 Representative Diagnostic Procedures for Disorders of the Endocrine System *(continued)*

Laboratory Test	Normal Values in Blood Plasma or Serum	Significance of Abnormal Values
Thyroid Gland		
Free serum T_4	Adults: 1.0–2.3 ng/dl	Elevated levels occur in hyperthyroidism; decreased levels occur in hypothyroidism.
Calcitonin (plasma)	Adult males: <40 pg/ml Adult females: <20 pg/ml	Elevated levels occur in carcinoma of the thyroid gland.
Thyroid autoantibody test	Negative to 1:20	High titers occur with thyroid tumors and in autoimmune disorders such as Graves' disease.
TSH stimulation test	Patients are given TSH via injection; thyroid gland is monitored for the normal increased function by checking T_4 levels.	Persons with a hypoactive thyroid who have increased T_4 levels after TSH administration have a pituitary problem secondary hypothyroidism; persons whose thyroid does not respond to TSH have primary hypothyroidism.
Parathyroid Glands		
Serum parathyroid hormone (PTH)	Adults: PTH-N 400–900 pg/ml PTH-C 200–600 pg/ml	Increased levels occur in hyperparathyroidism and hypercalcemia; decreased levels occur in hypoparathyroidism.
Serum phosphorus	Adults: 3–4.5 μg/dl	Increased levels occur in hypoparathyroidism; decreased levels occur in hyperparathyroidism.
Serum calcium	Adults: 8.5–10.5 mg/dl	Decreased levels occur in hypoparathyroidism; increased levels occur in hyperparathyroidism.
Adrenal Glands		
Plasma ACTH	Adults, morning: 20–80 pg/ml Adults, late afternoon: 10–40 pg/ml	Increased levels occur in stress, hypofunction of adrenal glands, and pituitary hyperactivity; decreased levels occur with Cushing's disease, carcinoma of adrenal gland, and intake of steroids such as prednisone or cortisone.
Plasma cortisol	Adults, morning: 5–23 μg/dl Adults, afternoon: 3–13 μg/dl	Increased levels occur in adrenal hyperactivity, Cushing's disease, stress, and steroid use; decreased levels occur in Addison's disease and pituitary hypofunction.
Serum aldosterone	Adults: 4–30 ng/dl supine: decreased <1 ng/dl elevated >9 ng/dl	Increased levels occur in dehydration, hyperactivity of adrenal glands, and hyponatremia; decreased levels occur in hypernatremia and adrenal hypoactivity.
Serum sodium	Adults: 135–145 mEq/l	Increased levels occur in dehydration; decreased levels occur with adrenocortical insufficiency.
Serum potassium	Adults: 3.5–5.0 mEq/l	Increased levels occur with hypoactivity of adrenal glands and hypoaldosteronism; decreased levels occur with hyperactivity of adrenal glands and aldosteronism.
Urine hydroxycorticosteroid (detects breakdown products of cortisol)	Adults: 2–12 mg/24 hour	Increased levels occur in Cushing's disease; decreased levels occur in Addison's disease.
Urine ketosteroids (detects metabolites of androgens)	Adults: 5–25 mg/24 hour	Increased levels occur in adrenal hyperactivity and hyperpituitarism; decreased levels occur in adrenal hypoactivity and hypopituitarism.
Pancreas		
Glucose tolerance test (monitors serum glucose following glucose ingestion by a patient who has fasted for 12 hours)	Adults: Results after ingestion of 75–100 g of glucose: 0 hour 75–115 mg/dl 1/2 hour <160 mg/dl 1 hour <170 mg/dl 2 hours <140 mg/dl	Increased levels occur in diabetes mellitus, Cushing's disease, alcoholism, and infections; decreased levels occur in hyperinsulinism and hypoactivity of adrenal glands.
Serum insulin	Adults: 5–25 μU/ml	Increased levels occur in early Type 2 diabetes and obesity; decreased levels occur in Type 1 diabetes.
Glycosylated hemoglobin (Alc)	Adults: 4–7%	Increased levels occur in diabetes mellitus.

In **hyperparathyroidism,** calcium concentrations become abnormally high. Calcium salts in the skeleton are mobilized, and bones are weakened. On X-rays the bones have a light, airy appearance, because the dense calcium salts no longer dominate the tissue. Central nervous system (CNS) function is depressed, thinking slows, memory is impaired, and the individual often experiences emotional swings and depression. Nausea and vomiting occur, and in severe cases the individual becomes comatose. Muscle function deteriorates, and skeletal muscles become weak. Other tissues are typically affected as calcium salts crystallize in joints, tendons, and the dermis; calcium deposits may produce masses called *kidney stones,* which block filtration and conduction passages in the kidneys.

Hyperparathyroidism most commonly results from a tumor of the parathyroid gland. Treatment involves the surgical removal of the overactive tissue. Fortunately, humans have four parathyroid glands, and the secretion of even a portion of one gland can maintain normal calcium concentrations.

FAP Ch. 18

DISORDERS OF THE ADRENAL CORTEX FAP *pp. 601, 602*

Clinical problems related to the adrenal gland vary with the adrenal zone involved. The conditions may result from changes in the functional capabilities of the adrenal cells (primary conditions) or disorders that affect the regulatory mechanisms (secondary conditions).

In **hypoaldosteronism,** the zona glomerulosa fails to produce enough aldosterone, generally either as an early sign of adrenal insufficiency or because the kidneys are not releasing adequate amounts of renin. Low aldosterone levels lead to excessive losses of water and sodium ions at the kidneys, and the water loss in turn leads to low blood volume and a fall in blood pressure. The resulting changes in electrolyte concentrations, including *hyperkalemia* (high extracellular K^+ levels), affect transmembrane potentials, eventually causing dysfunctions in neural and muscular tissues.

Hypersecretion of aldosterone results in **aldosteronism,** or *hyperaldosteronism.* Under continued aldosterone stimulation, the kidneys retain sodium ions in exchange for potassium ions that are lost in urine. Hypertension and hypokalemia occur as extracellular potassium levels decline, increasing the concentration gradient for potassium ions across cell membranes. This increase leads to an acceleration in the rate of potassium diffusion out of the cells and into interstitial fluids. The reduction in intracellular and extracellular potassium levels eventually interferes with the function of excitable membranes, especially cardiac muscle cells and neurons, and kidney cells.

Addison's disease results from inadequate stimulation of the zona fasciculata by adrenocorticotropic hormone (ACTH) or from the inability of the adrenal cells to synthesize the necessary hormones, generally due to autoimmune problems or infection. Affected individuals become weak and lose weight owing to a combination of appetite loss, hypotension, and hypovolemia. They cannot adequately mobilize energy reserves, and their blood glucose concentrations fall sharply within hours after a meal. Stresses cannot be tolerated, and a minor infection or injury can lead to a sharp and fatal decline in blood pressure. A particularly interesting symptom is the increased production of the pigment melanin in the skin. The ACTH molecule and the melanocyte-stimulating hormone (MSH) molecule are similar in structure, and at high concentrations ACTH stimulates the MSH receptors on melanocytes. President John F. Kennedy had this disorder.

Cushing's disease results from the overproduction of glucocorticoids. The symptoms resemble those of a protracted and exaggerated response to stress. (The stress response is discussed in the text on *p. 612.*) Glucose metabolism is suppressed, lipid reserves are mobilized, and peripheral proteins are broken down. Lipids and amino acids are mobilized in excess of the existing demand. The energy reserves are shuffled around, and the distribution of body fat changes. Adipose tissues in the cheeks and around the base of the neck become enlarged at the expense of other areas, producing a "moon-faced" appearance. The demand for amino acids falls most heavily on the skeletal muscles, which respond by breaking down their contractile proteins. This response reduces muscular power and endurance. The skin becomes thin and may develop *stria,* or stretch marks.

If the primary cause is ACTH oversecretion at the anterior pituitary gland, the most common source is a pituitary *adenoma* (a benign tumor of glandular origin). Microsurgery can be performed through the sphenoid bone to remove the adenomatous tissue. Some oncology centers use pituitary radiation rather than surgery. Several pharmacological therapies act at the hypothalamus, rather than at the pituitary gland, to prevent the release of corticotropin-releasing hormone (CRH). The drugs used are serotonin antagonists, gamma aminobutyric acid (GABA) transaminase inhibitors, or dopamine agonists. Alternatively, a bilateral adrenalectomy (the removal of the adrenal glands) can be performed, but further complications may arise as the adenoma enlarges. Cushing's disease also results from the production of ACTH outside the pituitary gland; for example, the condition may develop with one form of lung cancer (oat cell carcinoma). The removal of the causative tumor in some cases relieves the symptoms.

The chronic administration of large doses of steroids is required in some cases to treat severe asthma and some cancers and to prevent organ transplant rejection. Prolonged use of such large doses can produce symptoms similar to those of Cushing's disease, but such treatment is usually avoided.

DISORDERS OF THE ADRENAL MEDULLAE

FAP p. 603

The overproduction of epinephrine by the adrenal medullae may reflect chronic sympathetic activation. A **pheochromocytoma** (fē-ō-krō-mō-sī-TŌ-muh) is a tumor that produces catecholamines in massive quantities. The tumor generally develops within an adrenal medulla but may also involve other sympathetic ganglia. The most dangerous symptoms are rapid and irregular heartbeat and high blood pressure; other symptoms include uneasiness, sweating, blurred vision, and headaches. This condition is rare, and surgical removal of the tumor is the most effective treatment.

LIGHT AND BEHAVIOR

FAP p. 603

Exposure to sunlight can do more than induce a tan or promote the formation of vitamin D_3. Evidence indicates that daily light/dark cycles have widespread effects on the central nervous system, with melatonin playing a key role. Several studies have indicated that residents of temperate and higher latitudes in the Northern Hemisphere undergo seasonal changes in mood and activity patterns. These people feel most energetic from June through September, and they experience relatively low spirits from December through March. (The opposite situation occurs in the Southern Hemisphere, where winter and summer are reversed relative to the Northern Hemisphere.) The degree of seasonal variation differs from individual to individual: Some people display no symptoms; other people are affected so severely that they seek medical attention. The observed symptoms are called **seasonal affective disorder (SAD).** Individuals with SAD experience depression and lethargy and have difficulty concentrating. They tend to sleep for long periods, perhaps 10 hours or more a day. They may also go on eating binges and crave carbohydrates.

Melatonin secretion appears to be regulated by exposure to sunlight, not simply by exposure to light. Normal interior lights are apparently not strong enough or do not release the right mixture of light wavelengths to depress melatonin production. Because many people spend very little time outdoors in the winter, melatonin production increases then; the depression, lethargy, and concentration problems appear to be linked to elevated melatonin levels in blood. In experiments, comparable symptoms can be produced in a healthy individual by an injection of melatonin.

Many individuals with SAD are successfully treated by exposure to sun lamps that produce full-spectrum light. Experiments are underway to define exactly how intense the light must be and to determine the minimal effective time of exposure. Some people have been using melatonin or MSH obtained (in varying doses and purity) from health-food stores to treat SAD, insomnia, and jet lag. Because the health-food market is unregulated and few if any controlled studies have been done, it remains unclear whether melatonin and MSH are truly effective therapies.

DIABETES MELLITUS

FAP p. 607

Insulin-Dependent Diabetes Mellitus

The primary cause of **insulin-dependent diabetes mellitus (IDDM),** or **Type 1 diabetes,** is inadequate insulin production by the beta cells of the pancreatic islets. Glucose transport in most cells cannot occur in the absence of insulin. When insulin concentrations decline, cells can no longer absorb glucose; tissues remain glucose-starved despite the presence of adequate or even excessive amounts of glucose in the bloodstream.

After a meal rich in glucose, blood glucose concentrations may become so elevated that the kidney cells cannot reclaim all the glucose molecules that enter the urine. The high urinary concentration of glucose limits the ability of the kidneys to conserve water, so the individual urinates frequently and may become dehydrated. The chronic dehydration leads to disturbances of neural function (blurred vision, tingling sensations, disorientation, fatigue) and muscle weakness.

Despite high blood concentrations, glucose cannot enter endocrine tissues, and the endocrine system responds as if glucose were in short supply. Alpha cells release glucagon, and glucocorticoid production accelerates. Peripheral tissues then break down lipids and proteins to obtain the energy needed to continue functioning. The breakdown of large numbers of fatty acids promotes the generation of molecules called **ketone bodies.** These small molecules are metabolic acids whose accumulation in large numbers can cause a dangerous reduction in blood pH. This condition, called **ketoacidosis,** commonly triggers vomiting. In severe cases, it can precede a fatal coma.

If the individual survives (an impossibility without insulin therapy), long-term treatment involves a combination of dietary control and the administration of insulin, either by injection or by infusion, using an **insulin pump.** The treatment is complicated by the fact that tissue glucose demands vary with food eaten, physical activity, emotional state, stress, and other factors that are hard to assess or predict. Dietary control, including the regulation of the type of food, time of meals, and amount consumed, can help reduce oscillations in blood glucose levels. Modern insulin pumps can be programmed by the user to compensate for changes in activity and eating patterns. However, it remains difficult to maintain stable and normal blood glucose levels over long periods of time, even with an insulin pump. Blood glucose levels are checked to determine the amount and type of insulin needed. At present, small blood samples must be taken 2 or more times per day.

Since 1990, pancreas transplants have been used to treat diabetes in the United States. The procedure is generally limited to gravely ill patients already undergoing kidney transplantation. The graft success rate over

5 years is roughly 50 percent, and the procedure is controversial. Pancreatic islet transplantation has also been attempted but with variable success.

A more promising approach is the use of a **biohybrid artificial pancreas.** This procedure has been used to treat Type 1 diabetes in dogs. Islet cells can be cultured in the laboratory and inserted within an artificial membrane. The membrane contains pores that allow fluid movement but prevent interactions between the islet cells and white blood cells that would reject them as foreign. The islet cells monitor the blood glucose concentration and secrete insulin or glucagon as needed. The biohybrid artificial pancreas can be located almost anywhere that has an adequate blood supply; in human trials, it is inserted under the skin of the abdomen. In 1994, a human patient became independent of insulin after the insertion of artificially encapsulated human islet cells.

Type 1 diabetes most commonly appears in individuals under 40 years of age. Because it typically appears in childhood, it has been called **juvenile-onset diabetes.** Roughly 80 percent of people with this type of diabetes have circulating antibodies that target the surfaces of beta cells. The disease may therefore be an *autoimmune disorder,* a condition that results when the immune system attacks normal body cells. (Possible mechanisms responsible for autoimmune disorders are discussed in Chapter 22 of the text.) Consequently, attempts have been made to prevent the appearance of Type 1 diabetes with *azathioprine (Imuran®),* a drug that suppresses the immune system. This treatment is dangerous, however, because compromising immune function increases the risk of acquiring serious infections or of developing cancer. Type 1 diabetes is complex and probably reflects a combination of genetic and environmental factors. To date, genes associated with the development of Type 1 diabetes have been localized to chromosomes 6, 11, and 18.

FAP Ch. 18

Non-Insulin-Dependent Diabetes Mellitus

Non-insulin-dependent diabetes mellitus (NIDDM), or **Type 2 diabetes,** typically affects obese individuals over 40 years of age. Because of the age factor, this condition is also called **maturity-onset diabetes.** Type 2 diabetes is far more common than Type 1 diabetes, occurring in an estimated 6.6 percent of the U.S. population. Some 500,000 new cases are diagnosed each year in the United States alone, and 90 percent of them involve obese individuals.

In Type 2 diabetes, insulin levels are normal or elevated but peripheral tissues no longer respond normally. Treatment consists of weight loss and dietary restrictions that may elevate insulin production and tissue response. The drug *metformin (Glucophage®)* lowers plasma glucose concentrations, primarily by reducing glucose synthesis and release at the liver. The use of metformin in combination with other drugs that affect glucose metabolism promises to improve the quality of life for many Type 2 diabetes patients.

The standard testing procedures for Type 2 diabetes check for primary diabetic signs: high fasting blood glucose concentrations, the appearance of glucose in urine, and an inability to reduce elevated glucose levels. The latter sign is examined by means of a *glucose tolerance test.* Blood concentrations are monitored after a fasting subject consumes 75–100 g (2.6–3.2 oz) of glucose. In an individual who is not diabetic, the glucose will enter the bloodstream, insulin production will rise, and peripheral tissues will absorb the glucose so rapidly that blood concentrations will remain relatively normal. Without adequate insulin production, glucose concentrations skyrocket to more than twice normal levels.

[EXPLORE]

Additional resources, such as interactive tutorials, critical-thinking questions, clinical problems, and case studies, are available through the CD-ROM and Website for your textbook. To begin your exploration, launch the *Martini Interactive* CD-ROM or visit the Companion Website (http://www.prenhall.com/martini/fap5) and select Chapter 18.

The Cardiovascular System

The components of the cardiovascular system include the blood, heart, and blood vessels. Blood flows through a network of thousands of miles of vessels in the body, transporting nutrients, gases, wastes, hormones, and ions and redistributing the heat generated by active tissues. The exchange of materials between blood and peripheral tissues occurs across the walls of tiny capillaries that are situated between the arterial and venous systems. The total capillary surface area for exchange is truly enormous, averaging about 6300 square meters—about half the area of a football field.

Because the cardiovascular system plays a key role in supporting all other systems, disorders of this system will affect virtually every cell in the body. One method of organizing the many potential disorders involving the cardiovascular system is by the nature of the primary problem, whether it affects the blood, the heart, or the vascular network. Figure A-37• provides an introductory overview of major blood disorders

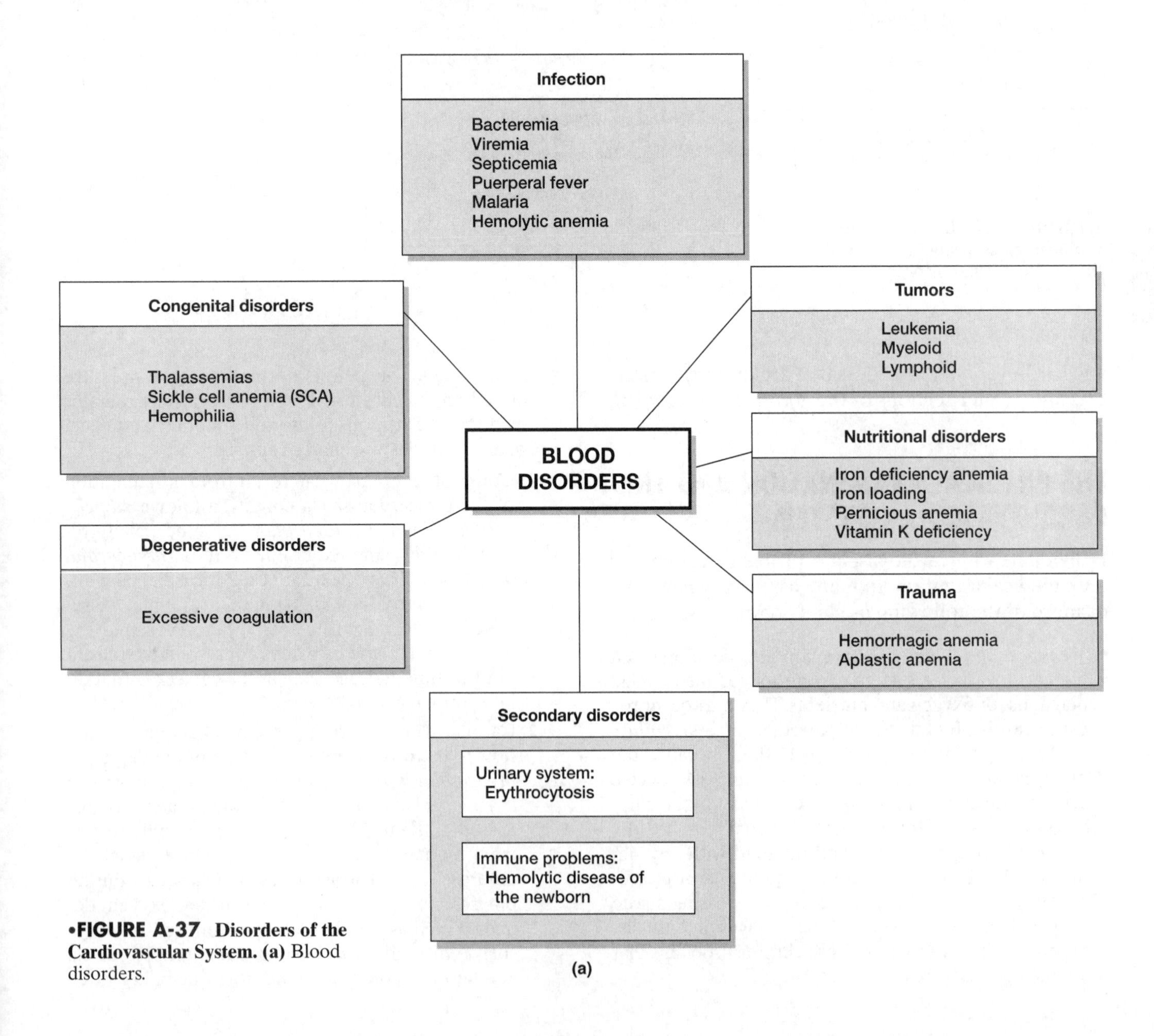

•**FIGURE A-37 Disorders of the Cardiovascular System. (a)** Blood disorders.

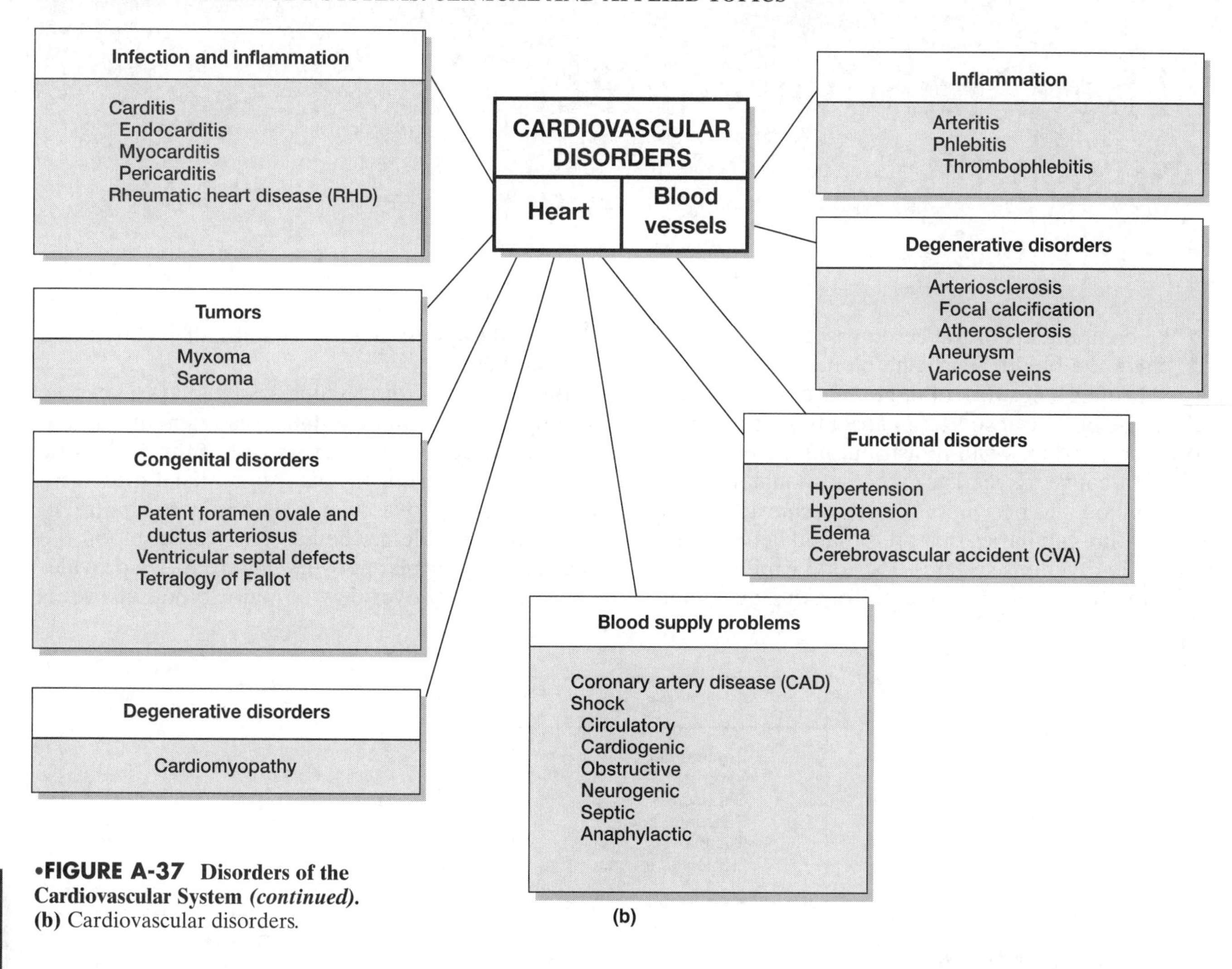

•**FIGURE A-37** **Disorders of the Cardiovascular System** ***(continued).*** **(b)** Cardiovascular disorders.

FAP Ch. 19

and cardiovascular disorders that are discussed in the text and in later sections of the *Applications Manual.*

THE PHYSICAL EXAMINATION AND THE CARDIOVASCULAR SYSTEM

Individuals with cardiovascular problems commonly seek medical attention. In many cases, they have one or more of the following as chief complaints:

- *Weakness and fatigue.* These symptoms develop when the cardiovascular system can no longer meet tissue demands for oxygen and nutrients. They may occur because cardiac function is impaired, as in *heart failure* (p. 123) or *cardiomyopathy* (p. 118), or because the blood is unable to carry normal amounts of oxygen, as in the various forms of *anemia* (p. 115). In the early stages of these conditions, the individual feels healthy at rest but becomes weak and fatigued with any significant degree of exertion because the cardiovascular system cannot keep pace with the rising tissue oxygen demands. In more-advanced stages of these disorders, weakness and fatigue are chronic problems that continue, even at rest.
- *Cardiac pain.* This pain is normally a deep pressure pain felt in the substernal region; it typically radiates down the left arm or up into the shoulder and neck. Cardiac pain has two major causes:
  1. Constant severe pain can result from inflammation of the pericardial sac, a condition known as *pericarditis.* This *pericardial pain* can superficially resemble the pain experienced in a *myocardial infarction (MI),* or heart attack. Pericardial pain differs from the pain of an MI in that (a) it may be relieved by leaning forward; (b) a fever may be present; and (c) the pain does not respond to the administration of drugs, such as *nitroglycerin,* that dilate coronary blood vessels.
  2. Cardiac pain also results from inadequate blood flow to the myocardium. This type of pain is called *myocardial ischemic pain.* Ischemic pain occurs in *angina pectoris* and in an MI. Angina pectoris, discussed in the text *(p. 667),* most commonly results from the narrowing of coronary blood vessels by atherosclerosis. The associated pain appears during physical exertion, when myocardial oxygen demands increase, and the pain is relieved by drugs such as nitroglycerin, which dilate coronary vessels and improve coronary blood flow. The pain associ-

ated with an MI is commonly felt as a heavy weight or a constriction of the chest. The pain of an MI is also distinctive because (a) it is not linked to exertion; (b) it is persistent and is not relieved by rest, nitroglycerin, or other coronary vasodilators; and (c) nausea, vomiting, and sweating may occur during the attack.

- *Palpitations.* Palpitations are a person's perception of an altered heart rate. The individual may complain of the heart "skipping a beat" or "racing." The most likely cause of palpitations is an abnormal pattern of cardiac activity known as an *arrhythmia.* The detection and analysis of arrhythmias are considered in a later section (p. 121).
- *Pain on movement.* Individuals with advanced atherosclerosis may experience pain in the extremities during exercise. The pain may become so severe that the person is unwilling or unable to walk or perform other common activities. The underlying problem is the constriction or partial occlusion of major arteries, such as the external iliac arteries to the lower limbs, by plaque formation.

These are only a few of the many symptoms that can be caused by cardiovascular disorders. In addition, the individual may notice the appearance of characteristic signs of underlying cardiovascular problems. A partial listing of important signs includes the following:

- *Edema* is an increase of fluid in tissues that occurs when (a) the pumping efficiency of the heart is decreased, (b) the plasma protein content of the blood is reduced, or (c) venous pressures are abnormally high. The tissues of the extremities are most commonly affected, and individuals experience swollen feet, ankles, and legs. When edema is so severe that pressing on the affected area leaves an indentation, the sign is called *pitting edema.* Edema is discussed in Chapters 21 and 22 of the text *(pp. 710, 756).*
- Breathlessness, or *dyspnea,* occurs when cardiac output is inadequate for tissue oxygen demands. Dyspnea may also occur with *pulmonary edema,* a buildup of fluid within the alveoli of the lungs. Pulmonary edema and dyspnea are typically associated with *congestive heart failure (CHF)* (p. 124).
- *Varicose veins* are dilated superficial veins that are visible at the skin surface. This condition, which develops when venous valves malfunction, can be caused or exaggerated by increased systemic venous pressures. Varicose veins are considered further on p. 125.
- There may be characteristic and distinctive changes in skin coloration. For example, *pallor* is the lack of normal red or pinkish color in the skin of a Caucasian or the conjunctiva and oral mucosa of darker-skinned people. Pallor accompanies many forms of anemia but can also be the result of inadequate cardiac output, shock (p. 127), or circulatory collapse. *Cyanosis* is the bluish color of the skin or mucous membranes that occurs when tissues are deficient in oxygen. Cyanosis generally results from cardiovascular or respiratory disorders.
- *Vascular skin lesions* were introduced in the discussion of skin disorders on p. 42. Characteristic vascular lesions can occur in clotting disorders (p. 116) and in *leukemia* (p. 115). For example, abnormal bruising may be the result of a disorder that affects the clotting system, platelet production, or vessel structure. *Petechiae,* which appear as purple spots on the skin surface, are typically seen in individuals with certain types of leukemia or other diseases associated with low platelet counts.

DIAGNOSTIC PROCEDURES

In many cases, the initial detection of a cardiovascular disorder occurs during the assessment stage of a physical examination:

1. When the vital signs are taken, the pulse is checked for strength, rate, and rhythm. Weak or irregular heart beats will commonly be noticed at this time.
2. The blood pressure is monitored with a stethoscope and blood pressure cuff (sphygmomanometer). Unusually high or low readings can alert the examiner to potential problems with cardiac or vascular function. However, a diagnosis of hypertension (p. 126) is usually not made on the basis of a single reading but after several readings over a period of time.
3. The heart sounds are monitored by auscultation with a stethoscope:
   - Cardiac rate and rhythm can be checked and arrhythmias detected.
   - Abnormal heart sounds, or *murmurs,* may indicate problems with atrioventricular or semilunar valves. Murmurs are noted in relation to their location in the heart (as determined by the position of the stethoscope on the chest wall), the time of occurrence in the cardiac cycle, whether the sound is low-pitched or high-pitched, and whether variations in intensity are present.
   - Nothing is usually heard during auscultation of normal vessels of the circulatory system. *Bruits* are the sounds that result from turbulent blood flow, in most cases around an obstruction within a vessel. Bruits are typically heard where large atherosclerotic plaques have formed.

Functional abnormalities of the heart and blood vessels can commonly be detected through physical assessment and the recognition of characteristic signs and symptoms. The structural basis of these problems is generally determined by the use of scans and X-rays and by monitoring electrical activity in the heart. For problems with a hematological basis, laboratory tests performed on blood samples normally provide the information necessary to reach a preliminary diagnosis. Table A-27 summarizes information about representative diagnostic procedures used to evaluate the health of the cardiovascular system.

TABLE A-27a Examples of Tests Used in Diagnosing Cardiovascular Disorders

Diagnostic Procedure	Method and Result	Representative Uses
Electrocardiography (ECG/EKG)	Electrodes placed on the chest detect electrical activity of cardiac muscle during a cardiac cycle; information is transmitted to a monitor for graphic recording.	Heart rate can be determined through study of the ECG; abnormal wave patterns may occur with cardiac irregularities such as myocardial infarction, chamber hypertrophy, or arrhythmias, such as premature ventricular contractions (PVCs), and defects in the conduction system.
Echocardiography	Standard ultrasound examination of the heart	Detects structural abnormalities of the heart; useful in determination of valve function, chamber size, vessel size, and ejection fraction
Transesophageal echocardiography	Use of an ultrasound probe attached to an endoscope and inserted into the esophagus	Provides enhanced view of posterior and inferior heart chambers
Phonocardiography	Heart sounds are monitored and graphically recorded.	Detects murmurs and abnormal heart sounds
Exercise stress test	ECG, blood pressure, and heart rate are monitored during exercise on treadmill.	Coronary artery blockage is suspected if ECG patterns and/or echocardiography results change.
Chest X-ray	Film sheet with radiodense tissues in white on negative image	Determines abnormal shape or size of the heart and abnormalities of the aorta and great vessels; detects fluid buildup in lungs
Cardiac catheterization (coronary arteriography)	For the study of the left side of the heart, a catheter is inserted into the femoral or brachial artery; for the study of the right side, the catheter is inserted into the femoral or subclavian vein; the catheter passes along the vessels to the heart. Contrast dye may be injected for better contrast as X-rays are taken.	Detects blockages and spasms in coronary arteries; also helps evaluate congenital defects and ventricular hypertrophy and determine the severity of a valvular defect; used to monitor volumes and pressures within the chambers
MRI of heart	Noninvasive images of coronary arteries; limited by heart motion and imaging speed	Detects coronary artery calcification, indicating atherosclerotic plaques and blockages
Positron emission tomography (PET)	Radionuclides are injected into the bloodstream; accumulation occurs at certain areas of the heart. A computer image for viewing is produced; seldom used clinically.	Detects tissue areas with a reduced metabolic rate due to an infarction and determines the extent of damage; also determines blood flow to the heart through the coronary arteries; may detect ischemia
Pericardiocentesis	Needle aspiration of fluid from the pericardial sac for analysis; therapeutic for cardiac tamponade.	See Pericardial fluid analysis, Table A-27b.
Doppler ultrasound	Transducer is placed over the vessel to be examined, and the echoes are analyzed by computer to provide information on blood velocity and flow direction.	Detects venous occlusion; determines cardiac efficiency; monitors fetal circulation and blood flow in umbilical vessels
Venography	Radiopaque dye is injected into a peripheral vein of a limb, and an X-ray study is done to detect venous occlusions.	Detects venous thrombosis or phlebitis

FAP Ch. 19

DISORDERS OF THE CARDIOVASCULAR SYSTEM

POLYCYTHEMIA FAP p. 628

An elevated hematocrit with a normal blood volume constitutes **polycythemia** (po-lē-sī-THĒ-mē-uh). There are several types of polycythemia. We will consider *erythrocytosis* (e-rith-rō-sī-TŌ-sis), a polycythemia that affects only red blood cells (RBCs), later in this section. **Polycythemia vera** ("true polycythemia") results from an increase in the numbers of all blood cells. The hematocrit may reach 80–90, at which point the tissues become oxygen-starved because RBCs are blocking the smaller vessels. This condition seldom strikes young people; most cases involve patients ages 60–80 years. Several treatment options are available, but none cures the condition. The cause of polycythemia vera is unknown, although some evidence links it to radiation exposure.

TABLE A-27b Laboratory Tests Useful in Diagnosing Cardiovascular Disorders

Laboratory Test	Normal Values in Blood Plasma or Serum	Significance of Abnormal Values
Creatine phosphokinase (CPK, CK)	0–36 U/l	Prolonged elevated levels of CK indicate myocardial damage; levels usually do not rise until 6–12 hours following myocardial infarction (MI).
Isoenzymes of CPK (CK-MM) (CK-MB) (CK-BB)	Varies with method; the normal CK-MB is 0–6% of the total.	CK-MM helps diagnose muscle disease. CK-MB provides information about damaged cardiac muscle; percentage rises to 60%. Levels rise within 6–12 hours following MI. CK-BB helps diagnose brain infarct or stroke.
Aspartate aminotransferase (AST or SGOT)	Adults: 7–45 U/l	Elevated in congestive heart failure due to liver damage
Lactate dehydrogenase (LDH)	Adults: 45–90 U/l	Elevated in roughly 40% of patients with congestive heart failure; elevation of LDH occurs 24–72 hours after MI.
Serum electrolytes		
Sodium	Adults: 135–145 mEq/l	Decreased levels occur with heart failure, hypotension; increased levels can occur with essential hypertension.
Potassium	Adults: 3.5–5.0 mEq/l	Decreased levels occur when certain diuretics are taken; increased levels occur with kidney failure.
Serum lipoproteins	HDL: 29–77 mg/dl LDL: 60–160 mg/dl	Higher risk of coronary heart disease occurs when LDL values are >130 mg/dl and HDL is <35 mg/dl.
Serum cholesterol	Adults: 150–240 mg/dl (desirable level <200 mg/dl)	When >240 mg/dl, high risk for atherosclerosis and coronary artery disease; elevated levels occur in familial hypercholesterolemia.
Serum triglycerides	Adults: 10–160 mg/dl (adults >50 years old may have higher values)	Elevated with acute MI and poorly controlled diabetes mellitus and some AIDS treatments
Pericardial fluid analysis		Fluid is analyzed for appearance, number, and types of blood cells; protein and glucose levels; and the presence of infectious organisms.

FAP Ch. 19

THALASSEMIA

FAP *p. 631*

The thalassemias are a diverse group of inherited blood disorders caused by the inability to produce adequate amounts of alpha or beta chains of hemoglobin. A specific condition is categorized as an **alpha-thalassemia** or a **beta-thalassemia**, depending on whether the alpha or beta hemoglobin chains are affected. Normal individuals inherit two copies of alpha-chain genes from each parent, and alpha-thalassemia develops when one or more of these genes are missing or inactive. The severity of the symptoms varies with the number of normal alpha-chain genes that remain functional. For example, an individual with three normal alpha-chain genes will not develop symptoms but can be a carrier, passing the defect to the next generation. A child whose parents are both carriers is likely to develop a more severe form of the disease:

- Individuals with two copies, rather than four copies, of the normal alpha-chain gene (one from each parent) have somewhat impaired hemoglobin synthesis. This condition is known as *alpha-thalassemia trait.* The RBCs are small and contain less than the normal quantity of hemoglobin. About 2 percent of African Americans and many Southeast Asians have alpha-thalassemia trait.
- Individuals with only one copy of the normal alpha-chain gene have very small *(microcytic)* RBCs that are relatively fragile.
- Most individuals with no functional copies of the normal alpha-chain gene die shortly after birth, because the hemoglobin synthesized cannot bind and transport oxygen normally. The incidence of fatal alpha-thalassemia is highest among Southeast Asians.

Each person inherits only one gene for the beta hemoglobin chain from each parent. These genes contain several possible mutations, and beta-thalassemia can take a variety of forms. If an individual does not receive a copy of a normal, functioning gene from either parent, the condition of **beta-thalassemia major,** or *Cooley's disease,* develops. Symptoms of this condition include severe anemia with microcytosis and a

TABLE A-27c Representative Hematology Studies for Diagnosing Blood Disorders

Laboratory Test	Normal Values in Blood Plasma or Serum	Significance of Abnormal Values
Complete blood count		
RBC count	Adult males: 4.7–6.1 million/mm^3 Adult females: 4.0–5.0 million/mm^3	Increased RBC count, Hb, and Hct occur in polycythemia vera, congenital heart disease, and events that induce hypoxia, such as moving to a high altitude.
Hemoglobin (Hb, Hgb)	Adult males: 14–18 g/dl Adult females: 12–16 g/dl	Decreased RBC count, Hb, and Hct occur with hemorrhage and the various forms of anemia, including hemolytic, iron deficiency, vitamin B_{12} deficiency, sickle cell anemia, and thalassemia or as a result of bone marrow failure or leukemia.
Hematocrit (Hct)	Adult males: 42–52 Adult females: 37–47	See sections on RBC count and Hemoglobin.
RBC indices		
Mean corpuscular volume (MCV) (measure of average volume of a single RBC)	Adults: 80–98 µl^3	Increased MCV and MCH occur in types of macrocytic anemia, including vitamin B_{12} deficiency.
Mean corpuscular hemoglobin (MCH) (measure of average amount of Hb per RBC)	Adults: 27–31 pg	Decreased MCV and MCH occur in types of microcytic anemia, such as iron deficiency anemia and thalassemia
Mean corpuscular hemoglobin concentration (MCHC)	Adults: 32–36% (derived by dividing the total Hb concentration number by the Hct value)	Decreased levels (hypochromic erythrocytes) suggest iron deficiency anemia or thalassemia.
WBC count	Adults: 5000–10,000/mm^3	Increased levels occur with chronic and acute infections, tissue death (MI, burns), leukemia, parasitic diseases, and stress; decreased levels occur in aplastic and pernicious anemias, overwhelming bacterial infection (sepsis), and viral infections.
Differential WBC count	Neutrophils: 50–70%	Increased levels occur in acute bacterial infection, myelocytic leukemia, rheumatoid arthritis, and stress; decreased levels occur in aplastic and pernicious anemia, viral infections, radiation treatment, and with some medications.
	Lymphocytes: 20–30%	Increased levels occur in lymphocytic leukemia, infectious mononucleosis, and viral infections; decreased levels occur in radiation treatment, AIDS, and corticosteroid therapy.
	Monocytes: 2–8%	Increased levels occur in chronic inflammation, viral infections, and tuberculosis; decreased levels occur in aplastic anemia and corticosteroid therapy.
	Eosinophils: 2–4%	Increased levels occur in allergies, parasitic infections, and some autoimmune disorders; decreased levels occur with steroid therapy.
	Basophils: 0.5–1%	Increased levels occur in inflammatory processes and during healing; decreased levels occur in hypersensitivity reactions.
Platelet count	Adults: 150,000–400,000/mm^3	Increased count can cause vascular thrombosis and occurs in polycythemia vera; decreased levels can result in spontaneous bleeding and occurs in different types of anemia and in some leukemias.
Bleeding time (amount of time for bleeding to stop after a small incision is made in skin)	3–7 minutes	Prolongation occurs in patients with decreased platelet count, anticoagulant therapy, aspirin ingestion, leukemia, or clotting factor deficiencies.

TABLE A-27c Representative Hematology Studies for Diagnosing Blood Disorders *(continued)*

Laboratory Test	Normal Values in Blood Plasma or Serum	Significance of Abnormal Values
Factors assay (coagulation factors I, II, V, VIII, IX, X, XI, XII)	Measured for their hemostatic activity	Decreased activity of the coagulation factors will result in defective clot formation; deficiencies can be caused by liver disease or vitamin K deficiencies.
Plasma fibrinogen (Factor I)	200–400 mg/dl	Elevated values can occur in inflammatory conditions and acute infections and also with medications such as birth control pills; decreased levels occur in liver disease, leukemia, and disseminated intravascular coagulation (DIC).
Partial thrombo-plastin time (PTT)	35–50 seconds	Prolonged by heparin
Plasma prothrombin time (PT)	Within 2 seconds of control (control should be 11–15 seconds)	Elevated levels can occur after trauma, MI, or infection. Decreased levels occur in DIC and liver disease and by medications such as streptokinase or coumadin.
Hemoglobin electrophoresis		
Hemoglobin A	Adults: within 95–98% of the total Hb	
Hemoglobin F	<2% of the total Hb after age 2 (newborn has 50–80% HbF)	Elevated levels after 6 months of age suggest thalassemia.
Hemoglobin S	0% of the total Hb	Present in sickle cell anemia and the sickle cell trait
Serum bilirubin	Adults: 0.1–1.2 mg/dl	Increased levels occur with hemolysis, liver disease (hepatitis), liver cancer, and gallstone obstruction of bile ducts.
Erythrocyte sedimentation rate (ESR) (measure of rate at which erythrocytes normal saline)	Adult males: 1–15 mm/h Adult females: 1–20 mm/h	Increased by disease processes that increase the protein concentration in the plasma, including inflammatory conditions, infections, and cancer; not useful in diagnosis of specific disorders
Bone marrow aspiration biopsy (involves laboratory examination of shape and cell size of erythrocytes, leukocytes, and megakaryocytes)	For the evaluation of hematopoiesis or the presence of tumor cells or infection	Increased RBC precursors with polycythemia vera; increased WBC precursors with leukemia; radiation therapy or chemotherapy can cause a decrease in all cell populations.

FAP Ch. 19

low hematocrit (below 20); and enlargement of the spleen, liver, heart, and areas of red bone marrow. Potential treatments for persons with severe symptoms include transfusions, splenectomy (to slow the rate of RBC recycling), and bone marrow transplantation. Beta-thalassemia minor, or beta-thalassemia trait, seldom produces clinical symptoms. An individual with this condition has one normal gene for the beta hemoglobin chain, and the rate of hemoglobin synthesis is depressed by roughly 15 percent. This decrease does not affect the RBCs' functional abilities, however, so no treatment is necessary. Blood counts from individuals with alpha-thalassemia or beta-thalassemia are similar to those from individuals who have iron deficiency anemia, and there is a risk of inappropriate treatment with iron supplements.

TRANSFUSIONS AND SYNTHETIC BLOOD

FAP *pp. 631, 637*

In a **transfusion,** blood components are provided to an individual whose blood volume has been reduced or whose blood is deficient in some way. The donated blood is obtained under sterile conditions, treated to prevent clotting and to stabilize the RBCs, and refrigerated. Transfusions of whole blood are most commonly used to restore blood volume after massive blood loss. In an **exchange transfusion,** an individual's

entire blood volume is drained off and simultaneously replaced with whole blood from another source.

Chilled whole blood remains usable for about 3.5 weeks. For longer storage, the blood must be *fractionated.* The RBCs are separated from the plasma, treated with a special antifreeze solution, and then frozen. The plasma can then be stored chilled, frozen, or freeze-dried. This procedure permits the long-term storage of rare blood types that might not otherwise be available for emergency use.

Fractionated blood has many uses. **Packed red blood cells** (PRBCs), with most of the plasma removed, are preferred for cases of anemia, in which the volume of blood may be close to normal but has below-normal oxygen-carrying capabilities. Plasma can be administered when massive fluid losses are occurring, such as after a severe burn. Plasma samples can be further fractionated to yield albumins, immunoglobulins, clotting factors, and plasma enzymes, each of which can be administered separately. White blood cells and platelets can also be collected, sorted, and stored for subsequent transfusion.

Some 12 million pints of blood are transfused each year in the United States alone, and the demand for blood or blood components commonly exceeds the supply. Moreover, many people are concerned about the risk of infection with hepatitis viruses or with HIV (the virus that causes AIDS) from contaminated blood. For these reasons, transfusion practices have changed during recent years. In general, fewer units of blood are now administered. Blood donors and collected blood are screened for hepatitis and HIV infections. There has also been an increase in **autologous transfusion,** in which blood is removed from an individual, stored, and later transfused back into the same individual when needed—for example, after a surgical procedure. Genetically engineered erythropoietin (EPO) can also be administered to help the patient's body restore its full complement of RBCs more quickly than it could without the hormone. Moreover, new technology permits the reuse of blood "lost" during surgery. The blood is collected and filtered, the platelets are removed, and the remainder of the blood is reinfused into the patient.

FAP Ch. 19

Shortages of blood and anxieties over the safety of existing stockpiles persist. In addition, some people are unable or unwilling to accept transfusions for medical or religious reasons. Thus there has been widespread interest in the development of synthetic blood components. Genetic engineering techniques are being used to address this problem. For example:

- It is now possible to synthesize one of the subunits of normal human hemoglobin, which can then be introduced into the bloodstream to increase oxygen transport and total blood volume. The hemoglobin molecules can be attached to inert carrier molecules that will prevent their filtration and loss at the kidneys.
- Hemoglobin can now be obtained from nonhuman sources. Technical progress has been rapid, but practical success has been limited. Genetically modified pigs and sheep can provide human hemoglobin and even human blood cells, and despite several remaining obstacles, this approach may eventually provide a safe and reliable source of blood for blood banks.

Whole blood substitutes are highly experimental solutions still undergoing clinical evaluation. In addition to the osmotic agents in plasma expanders (FAP *p. 627*), these solutions contain small clusters of synthetic molecules built of carbon and fluorine atoms. The mixtures, known as **perfluorochemical** (PFC) **emulsions,** can carry roughly 70 percent of the oxygen of whole blood. Animals have been kept alive after an exchange transfusion that completely replaced their circulating blood with a PFC emulsion. These PFC emulsions have the same advantages of other plasma expanders, plus the added benefits of transporting oxygen. Because no RBCs are involved, the PFC emulsions can carry oxygen to regions whose capillaries have been partially blocked by fatty deposits or blood clots. Unfortunately, PFCs do not absorb oxygen as effectively as normal blood does. To ensure that the PFCs deliver adequate oxygen to peripheral tissues, the individual must breathe air rich in oxygen, usually through an oxygen mask. In addition, phagocytes appear to engulf the PFC clusters. These problems have limited the use of PFC emulsions in humans. However, one PFC emulsion, *Fluosol®,* has been used to enhance oxygen delivery to cardiac muscle during heart surgery.

Another approach involves the manufacture of miniature erythrocytes by enclosing small bundles of hemoglobin in a lipid membrane. These **neohematocytes** (nē-ō-he-MA-tō-sīts) are spherical, with a diameter of less than 1 μm, and they can easily pass through narrow or partially blocked vessels. The major problem with this technique is that phagocytes treat neohematocytes as fragments of normal erythrocytes, so they remain in the bloodstream for only about 5 hours.

SICKLE CELL ANEMIA FAP *p. 631*

Sickle cell anemia (SCA) results from the production of an abnormal form of hemoglobin. The beta chains are involved, and the abnormal subunit is called *hemoglobin S.* Today sickle cell anemia affects 60,000–80,000 African Americans, or approximately 0.14 percent of the African-American population.

An individual with sickle cell anemia carries two copies of the abnormal gene—one from each parent. If only one copy is present, the individual has a *sickling trait.* One African American in 12 carries the sickling trait. Although it is now known that the genes are present in persons of Mediterranean, Middle Eastern, and East Indian ancestry, statistics on the in-

cidence of sickling trait and SCA in these groups are as yet unavailable.

Sickled RBCs may get stuck in small capillaries and obstruct blood flow. Symptoms of sickle cell anemia include pain and damage to a variety of organs and systems, depending on the location of the obstructions. In addition, the trapped RBCs eventually die and break down, producing a characteristic hemolytic anemia. Transfusions of normal blood can temporarily prevent additional complications, and experimental drugs can control or reduce sickling. *Hydroxyurea* is an anticancer drug that stimulates the production of fetal hemoglobin, a slightly different form of hemoglobin produced during development. This drug is effective but has toxic side effects (not surprising in an anticancer drug). The food additive **butyrate,** found in butter and other foods, appears to be even more effective in promoting the synthesis of fetal hemoglobin. In clinical trials, it has been effective in treating sickle cell anemia and other conditions caused by abnormal hemoglobin structure, such as beta-thalassemia.

In an individual with the sickling trait, most of the hemoglobin is of the normal form, and the RBCs function normally. But the presence of the abnormal hemoglobin gives the individual the ability to resist the parasitic infection that causes **malaria,** a mosquito-borne illness. The parasites that cause malaria enter the bloodstream when an individual is bitten by an infected mosquito. The microorganisms then invade, and reproduce within, the RBCs. But when they enter the RBC of a person with the sickling trait, the cell responds by sickling. Either the sickling itself kills the parasite or the sickling attracts the attention of a phagocyte that engulfs the RBC and kills the parasite. In either event, the individual remains unaffected by the malaria, whereas individuals without the sickling trait sicken and commonly die of that disease. Genetic studies indicate that the sickling mutation has evolved at least five times at different locations in Africa and India—regions where malaria poses a serious health problem.

BILIRUBIN TESTS AND JAUNDICE

FAP *p. 632*

When hemoglobin is broken down, the heme units (minus the iron) are converted to bilirubin. Normal serum bilirubin concentrations range from 0.1 to 1.2 mg/dl. Of that amount, roughly 85 percent will be removed by the liver. Several clinical conditions are characterized by an increase in the total plasma bilirubin concentration. In such conditions, bilirubin diffuses into peripheral tissues, giving them a yellow color that is most apparent in the skin and over the sclera of the eyes. This combination of signs (yellow skin and eyes) is called **jaundice** (JAWN-dis).

Jaundice can have many causes, but blood tests that determine the concentration of different forms of bilirubin can provide useful diagnostic clues. For example, **hemolytic jaundice** results from the destruction of large numbers of RBCs. When this occurs, phagocytes release massive quantities of one form of bilirubin (unconjugated) into the blood. Because the liver cells then accelerate the secretion of bilirubin in the bile, the blood concentration of another form of bilirubin (conjugated) does not increase proportionately. A blood test from a person with hemolytic jaundice would reveal (1) elevated total bilirubin, (2) high concentrations of unconjugated bilirubin, and (3) a conjugated bilirubin contribution of much less than 15 percent of the total bilirubin concentration.

These results are quite different from those seen in **obstructive jaundice.** In this condition, the ducts that remove bile from the liver are constricted or blocked. Liver cells cannot get rid of conjugated bilirubin, and large quantities diffuse into the blood. In this case, diagnostic tests would show (1) elevated total bilirubin, (2) an unconjugated bilirubin contribution of much less than 85 percent of the total bilirubin concentration, and (3) high concentrations of conjugated bilirubin.

IRON DEFICIENCIES AND EXCESSES

FAP *p. 633*

If dietary supplies of iron are inadequate, hemoglobin production slows down and symptoms of *iron deficiency anemia* appear. This form of anemia can also be caused by any condition that produces a blood loss, because the iron in the lost blood cannot be recycled. As the RBCs are replaced, iron reserves must be mobilized for use in the synthesis of new hemoglobin molecules. If those reserves are exhausted or if dietary sources are inadequate, iron deficiency results. In iron deficiency anemia, the RBCs cannot synthesize enough functional hemoglobin, and they are unusually small when they enter the bloodstream. The hematocrit declines, and the hemoglobin content and oxygen-carrying capacity of the blood are substantially reduced. Symptoms include weakness and fatigue.

FAP Ch. 19

Women are especially dependent on a normal dietary supply of iron, because their iron reserves are smaller than those of men. A healthy male's body has about 3.5 g of iron in the ionic form Fe^{2+}. Of that amount, 2.5 g is bound to the hemoglobin of circulating RBCs, and the rest is stored in the liver and bone marrow. In women, the total body iron content averages 2.4 g, with roughly 1.9 g incorporated into RBCs. Thus a woman's iron reserves consist of only 0.5 g, half that of a typical man. Moreover, the monthly menstrual cycles of premenopausal adult women result in blood losses that further stress iron reserves. When the demand for iron increases out of proportion with dietary supplies, iron deficiency develops. An estimated 20 percent of menstruating women in the United States show signs of iron deficiency. Pregnancy also stresses iron reserves, because the woman must

provide the iron needed to produce both maternal and fetal erythrocytes.

Good dietary sources of iron include liver, red meats, kidney beans, egg yolks, spinach, and carrots. Iron supplements can help prevent iron deficiency, but too much iron can be as dangerous as too little. Iron absorption across the digestive tract normally keeps pace with physiological demands. When the diet contains abnormally high concentrations of iron, or when hereditary factors increase the rate of absorption, the excess iron gets stored in peripheral tissues. This storage is called *iron loading.* Eventually, cells begin to malfunction as massive iron deposits accumulate in the cytoplasm. For example, iron deposits in pancreatic cells can lead to diabetes mellitus; deposits in cardiac muscle cells lead to abnormal heart contractions and heart failure. (Evidence suggests that iron deposits in the heart caused by the overconsumption of red meats may contribute to heart disease.) Liver cells become nonfunctional, and liver cancers may develop.

Comparable symptoms of iron loading may follow repeated transfusions of whole blood, because each unit of whole blood contains roughly 250 mg of iron. For example, as we noted previously, the severe forms of *thalassemia* result from a genetic inability to produce adequate amounts of one of the four chains of hemoglobin. Erythrocyte production and survival are reduced, and so is the blood's oxygen-carrying capacity. Most individuals with severe untreated thalassemia die in their twenties, but not because of the anemia. These patients are treated for severe anemia with frequent blood transfusions, which prolong life, but the excessive iron loading eventually leads to fatal heart problems.

FAP Ch. 19

ERYTHROCYTOSIS AND BLOOD DOPING FAP *p. 634*

In **erythrocytosis** (e-rith-rō-sī-TŌ-sis), the blood contains abnormally large numbers of RBCs. Erythrocytosis generally results from the massive release of erythropoietin (EPO) by tissues, especially the kidneys, deprived of oxygen. After moving to high altitudes, people typically experience erythrocytosis, because a lungful of air at that altitude contains fewer oxygen molecules than it would at sea level. For reasons discussed in Chapter 23 of the main text the amount of oxygen carried by an average RBC is thus reduced at high altitudes. Increasing the number of RBCs compensates for this reduction; although each RBC is carrying less oxygen than it would at sea level, more RBCs are in circulation. The hematocrit of mountaineers training at altitudes of 10,000–12,000 feet may become as high as 65.

An individual whose heart or lungs are functioning inadequately may also develop erythrocytosis. For example, this condition is commonly seen in cases of heart failure and emphysema, two conditions we will discuss in later sections. Whether the blood fails to circulate efficiently or the lungs do not deliver enough oxygen to the blood, peripheral tissues remain oxygen-poor despite the rising hematocrit. Having a higher concentration of RBCs increases the oxygen-carrying capacity of the blood, but it also makes the blood thicker and harder to push around the circulatory system. The work load on the heart increases, making a bad situation even worse.

The practice of **blood doping** has occurred among competitive athletes involved with endurance sports, such as cycling. The procedure entails removing whole blood from the athlete in the weeks before an event. The packed red cells are separated from the plasma and stored. By the time of the race, the competitor's bone marrow will have replaced the lost blood. Immediately before the event, the packed red cells are reinfused, increasing the hematocrit. The objective is to elevate the oxygen-carrying capacity of the blood and thereby increase endurance. The consequence is that the athlete's heart is placed under a tremendous strain. The long-term effects are unknown, but the practice obviously carries a significant risk; it has recently been banned in amateur sports. Attempts to circumvent this rule by the use of administered EPO in 1992–1993 resulted in the tragic deaths of 18 European cyclists.

BLOOD TESTS AND RBCS FAP *p. 634*

Several common blood tests are used to assess circulating RBCs (Table A-27c):

Reticulocyte Count

Reticulocytes are immature RBCs that are still synthesizing hemoglobin. Most reticulocytes remain in the bone marrow until they complete their maturation, but some enter the bloodstream. Reticulocytes normally account for about 0.8 percent of the erythrocyte population. Values above 1.5 percent or below 0.5 percent indicate that something is wrong with the rates of RBC survival or maturation.

Hematocrit (Hct)

The hematocrit value is the percentage of whole blood occupied by cells. The hematocrit of a normal adult averages 46 for men and 42 for women, with ranges of 42–52 for men and 37–47 for women.

Hemoglobin Concentration (Hb)

This test determines the amount of hemoglobin in the blood, expressed in grams per deciliter (g/dl). Normal ranges are 14–18 g/dl for males and 12–16 g/dl for females. The differences in hemoglobin concentration reflect the differences in hematocrit. For both genders, a single RBC contains 27–33 picograms (pg) of hemoglobin.

RBC Count

Calculations of the RBC count, the number of RBCs per microliter of blood, are based on the hematocrit and hemoglobin content and can be used to get more

information about the condition of the RBCs. Values typically reported in blood tests include:

- *Mean corpuscular volume (MCV),* the average volume of an individual RBC, in cubic micrometers (μm^3). The MCV is calculated by dividing the hematocrit by the RBC count, using the formula

$$\text{MCV} = \frac{\text{Hct}}{\text{RBC count (in millions)}} \times 10$$

This is a "cookbook" method that takes advantage of the fact that the hematocrit closely approximates the relative volume of RBCs in any unit sample of whole blood. Normal values for the MCV range from 80 to 98. For a representative hematocrit of 46 and an RBC count of 5.2 million, the mean corpuscular volume is

$$\text{MCV} = \frac{46}{5.2} \times 10 = 88.5\ \mu m^3$$

Cells of normal size are said to be **normocytic,** whereas larger-than-normal or smaller-than-normal RBCs are called **macrocytic** or **microcytic,** respectively.

- *Mean corpuscular hemoglobin concentration (MCHC),* the average amount of hemoglobin in a single RBC, expressed in picograms. Normal values range from 27 to 31 pg. The MCHC is calculated as

$$\text{MCHC} = \frac{\text{Hb}}{\text{RBC count (in millions)}} \times 10$$

RBCs containing normal amounts of hemoglobin are said to be **normochromic,** whereas the names **hyperchromic** and **hypochromic** indicate higher-than-normal or lower-than-normal hemoglobin content, respectively.

Anemia exists whenever the oxygen-carrying capacity of the blood is reduced, diminishing the delivery of oxygen to peripheral tissues. Such a reduction causes a variety of symptoms, including premature muscle fatigue, weakness, lethargy, and a lack of energy. Anemia may exist because the hematocrit is abnormally low or because the amount of hemoglobin in the RBCs is reduced. Standard laboratory tests can be used to differentiate among the various forms of anemia on the basis of the number, size, shape, and hemoglobin content of RBCs. As an example, Table A-28 shows how this information can be used to distinguish among four major types of anemia:

1. **Hemorrhagic anemia** results from severe blood loss. Erythrocytes are of normal size; each contains a normal amount of hemoglobin, and reticulocytes are present in normal concentrations, at least initially. Blood tests would therefore show a low hematocrit and low hemoglobin, but the MCV, MCHC, and reticulocyte counts would be normal.
2. In **aplastic** (ā-PLAS-tik) **anemia,** the bone marrow fails to produce new RBCs. The 1986 nuclear accident in Chernobyl (of the former USSR) caused a number of cases of aplastic anemia. The condition is fatal unless surviving stem cells repopulate the marrow or a bone marrow transplant is performed. In aplastic anemia, the circulating RBCs are normal in all respects, but because new RBCs are not being produced, the RBC count, Hct, Hb, and reticulocyte count are low.
3. In **iron deficiency anemia,** normal hemoglobin synthesis cannot occur because iron reserves are inadequate (p. 113). Because developing RBCs cannot make functional hemoglobin, they are unusually small. A blood test shows a low hematocrit, low hemoglobin content, low MCV, and low MCHC but generally a normal reticulocyte count. An estimated 60 million women worldwide have iron deficiency anemia.
4. In **pernicious** (per-NISH-us) **anemia,** normal RBC maturation ceases because the supply of vitamin B_{12} is inadequate. Erythrocyte production declines, and the RBCs are abnormally large and may develop a variety of bizarre shapes. Blood tests from a person with pernicious anemia indicate a low hematocrit with a high MCV and a normal or low reticulocyte count.
5. In **hemolytic anemia**, RBCs are breaking down in the bloodstream. The hematocrit and hemoglobin concentration are low, and the reticulocyte count is sky high. The individual RBCs are generally normal in size and hemoglobin content. In chronic cases, reticulocytes enter the bloodstream prematurely and many of the RBCs are smaller than normal.

FAP Ch. 19

THE LEUKEMIAS FAP p. 641

Leukemias characterized by the presence of abnormal granulocytes or other cells of the bone marrow are called **myeloid;** leukemias that involve abnormal lymphocytes are termed **lymphoid.** The first symptoms

TABLE A-28 RBC Tests and Anemias

Anemia Type	Hct	Hb	Reticulocyte Count	MCV	MCHC
Hemorrhagic (acute)	Low	Low	Normal or high	Normal	Normal
Aplastic	Low	Low	Very low	Normal	Normal
Iron deficiency	Low	Low	Normal or low	Low	Low
Pernicious	Low	Low	Very low	High	Normal or low
Hemolytic	Low	Low	Very high (3 × normal)	Normal or low	Normal

appear as immature and abnormal white blood cells (WBCs) enter the bloodstream. As the number of white blood cells increases, they travel through the circulation, invading tissues and organs throughout the body.

These cells are extremely active, and they require abnormally large amounts of energy. As in other cancers, described in Chapter 4 of the text and elsewhere in this manual (p. 37), invading leukemic cells gradually replace the normal cells, especially in the bone marrow. Red blood cell, normal WBC, and platelet formation declines, with resulting anemia, infection, and impaired blood clotting. Untreated leukemias are invariably fatal.

Leukemias are classified as *acute* (short and severe) or *chronic* (prolonged). Acute leukemias are linked to radiation exposure, hereditary susceptibility, viral infections, or unknown causes. Chronic leukemias are related to chromosomal abnormalities or immune system malfunctions. Survival in untreated acute leukemia averages about 3 months; individuals with chronic leukemia may survive for years.

Effective treatments exist for some forms of leukemia but not for others. For example, when acute lymphoid leukemia is detected early, 85–90 percent of patients can be held in remission for 5 years or longer, but only 10–15 percent of patients with acute myeloid leukemia survive 5 years or more. The yearly mortality rate for leukemia (all types) in the United States has not declined appreciably in the past 30 years, remaining at about 6.8 per 100,000 population. However, new treatments are being developed that show promise when used to combat specific forms of leukemia. For example, the administration of *gamma-interferon,* a hormone of the immune system, has been very effective in treating hairy cell leukemia and chronic myeloid leukemia.

FAP Ch. 19

One option for treating acute leukemias is to perform a bone marrow transplant. In this procedure, massive chemotherapy or radiation treatment is given, enough to kill all the cancerous cells. Unfortunately, this treatment also destroys the patient's blood cells and stem cells in the bone marrow and other blood-forming tissues. The individual then receives an infusion of healthy bone marrow cells that repopulate the blood and marrow tissues.

If the bone marrow is extracted from another person (a **heterologous marrow transplant**), care must be taken to ensure that the blood types and tissue types are compatible (see Chapters 19 and 22 of the text). If they are not, the new lymphocytes may attack the patient's tissues, with potentially fatal results. Best results are obtained when the donor is a close relative. In an **autologous marrow transplant,** bone marrow is removed from the patient, cleansed of cancer cells, and reintroduced after radiation or chemotherapy treatment. Although this method produces fewer complications, the preparation and cleansing of the marrow are technically difficult and a recurrence of leukemia is more likely.

Bone marrow transplants are also performed to treat patients whose bone marrow has been destroyed by toxic chemicals or radiation. For example, heterologous transplants were used successfully in the former USSR to treat survivors of the Chernobyl nuclear reactor accident in 1986.

A CLOSER LOOK: THE CLOTTING SYSTEM FAP *p. 646*

Figure A-38a•, the clotting cascade, presents the interactions of the intrinsic and extrinsic pathways and on the roles of the coagulation factors. Figure A-38b• presents the fibrinolytic system, which is activated as thrombin forms. This linked pathway provides a mechanism for the gradual removal of a blood clot during tissue repair.

TESTING THE CLOTTING SYSTEM FAP *p. 649*

Several clinical tests check the efficiency of the clotting system (Table A-27c).

Bleeding Time

This test measures the time it takes a small skin wound to seal. There are variations on this procedure, with normal values ranging from 3 to 7 minutes. Aspirin prolongs bleeding time by affecting platelet function and suppressing the extrinsic pathway.

Coagulation Time

In this test, a sample of whole blood stands under controlled conditions until a visible clot has formed. Normal values range from 3 to 15 minutes. The test has several potential sources of error and so is not very accurate. Nevertheless, it is the simplest test that can be performed on a blood sample. More-sophisticated tests begin by adding citrate ions to the sample. Citrate ions tie up the calcium ions in the plasma and prevent premature clotting.

Partial Thromboplastin Time (PTT)

In this test, a plasma sample is mixed with chemicals that mimic the effects of activated platelets. Calcium ions are then introduced, and the clotting time is recorded. Clotting normally occurs in 35–50 seconds if the enzymes and clotting factors of the intrinsic pathway are present in normal concentrations. The PTT is prolonged by heparin therapy.

Plasma Prothrombin Time

This test checks the performance of the extrinsic pathway. The procedure is similar to that used in the PTT test, but the clotting process is triggered by exposure to a combination of tissue thromboplastin and calcium

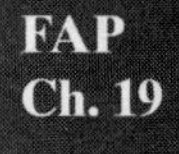

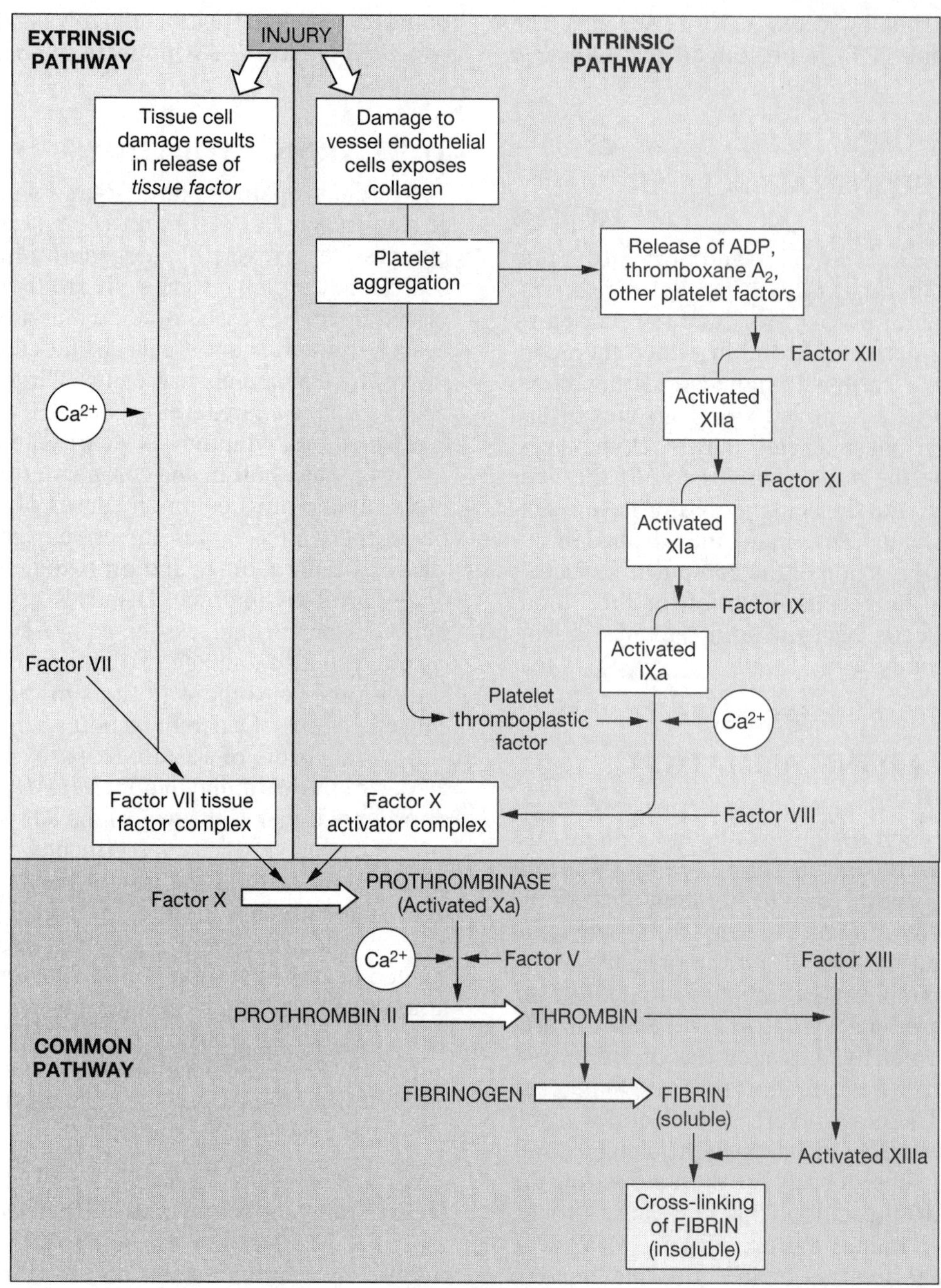

(a)

The Clotting Cascade

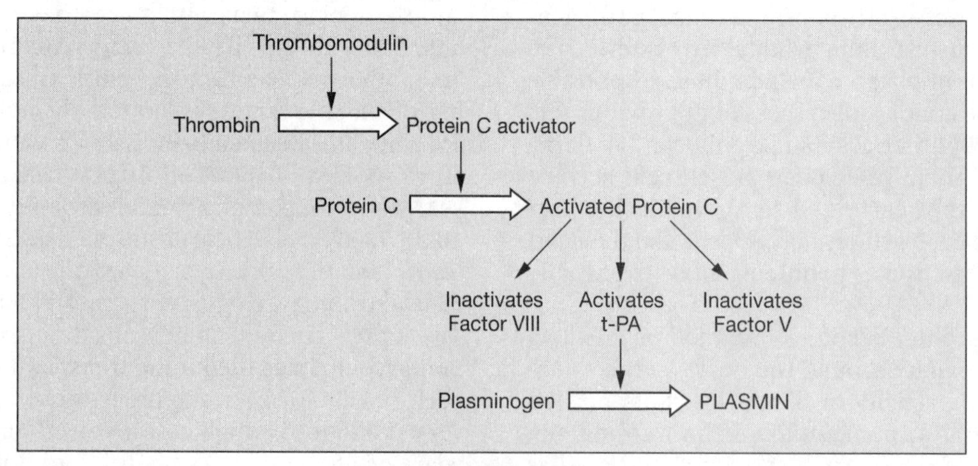

(b)

The Fibrinolytic System

•**FIGURE A-38** **The Clotting System**

ions. Clotting normally occurs in 13–17 seconds. The **prothrombin time** (PT) is prolonged by *coumadin* therapy.

DISSEMINATED INTRAVASCULAR COAGULATION FAP *p. 649*

The clotting process is complex and normally is precisely regulated. In *disseminated intravascular coagulation* (DIC), bacterial toxins or factors released by damaged tissues activate thrombin, which then converts fibrinogen to fibrin within the circulating blood. Much of the fibrin is removed by phagocytes or dissolved by plasmin, but small clots may block small vessels and damage the dependent tissues. If the liver cannot keep pace with the demand for fibrinogen, clotting abilities gradually decline and uncontrolled bleeding may occur. DIC is one of the complicating factors of *septicemia*, a dangerous infection of the bloodstream that spreads bacteria and bacterial toxins throughout the body.

INFECTION AND INFLAMMATION OF THE HEART FAP *p. 657*

Many different microorganisms may infect heart tissue, leading to serious cardiac abnormalities. **Carditis** (kar-DĪ-tis) is a general term for inflammation of the heart. Clinical conditions resulting from cardiac infection are usually identified by the primary site of infection. For example, those infections that affect the endocardium produce symptoms of **endocarditis.** Endocarditis primarily damages the chordae tendineae and heart valves; the mortality rate may reach 21–35 percent. The most-severe complications result from the formation of blood clots on the damaged surfaces. These clots subsequently break free, entering the bloodstream as drifting emboli that may cause strokes, heart attacks, or kidney failure (FAP *p. 649*). The destruction of heart valves by infection may lead to valve leakage, heart failure, and death.

Bacteria, viruses, protozoans, and fungal pathogens that attack the myocardium produce **myocarditis.** The microorganisms implicated include those responsible for many of the conditions discussed in previous sections, such as diphtheria, syphilis, polio, and malaria. The membranes of infected heart muscle cells become facilitated, and the heart rate rises dramatically. Over time, abnormal contractions may appear and the heart muscle weakens; these problems may eventually prove fatal.

If the pericardium becomes inflamed or infected, fluid may accumulate around the heart *(cardiac tamponade)* or the elasticity of the pericardium may be reduced *(constrictive pericarditis)*. In both conditions, the expansion of the heart is restricted and cardiac output is reduced. Treatment involves draining the excess fluid or cutting a window in the pericardial sac.

FAP Ch. 20

THE CARDIOMYOPATHIES FAP *p. 663*

The **cardiomyopathies** (kar-dē-ō-mī-OP-a-thēz) include an assortment of diseases with a common symptom: the progressive, irreversible degeneration of the myocardium. Cardiac muscle cells are damaged and replaced by fibrous tissue, and the muscular walls of the heart become thin and weak. As muscle tone declines, the ventricular chambers greatly enlarge. When the remaining cells cannot develop enough force to maintain cardiac output, symptoms of heart failure develop.

Chronic alcoholism and coronary artery disease are probably the most common causes of cardiomyopathy in the United States. Infectious agents, including viruses, bacteria, fungi, and protozoans, can also produce cardiomyopathies. Diseases affecting neuromuscular performance, such as muscular dystrophy (p. 66), can also damage cardiac muscle cells, as can starvation or chronic variations in the extracellular concentrations of calcium or potassium ions.

Several forms of cardiomyopathy are inherited. **Hypertrophic cardiomyopathy** *(HCM)* is an inherited disorder that makes the wall of the left ventricle thicken to the point at which it has difficulty pumping blood. Most people with HCM do not become aware of it until relatively late in life. However, HCM can also cause a fatal arrhythmia; it has been implicated in the sudden deaths of several young athletes. The implantation of an electronic cardiac pacemaker has proved to be beneficial in controlling these arrhythmias.

Finally, there are a significant number of cases of *idiopathic cardiomyopathy,* a term used when the primary cause cannot be determined.

Heart Transplants and Assist Devices

Individuals with severe cardiomyopathy may be considered as candidates for heart transplants. This surgery involves the removal of the weakened heart and the replacement with a heart taken from a suitable donor. To survive the surgery, the recipient must be in otherwise satisfactory health. Because the number of suitable donors is limited, the available hearts are generally assigned to individuals younger than age 50 years. Out of the 8000–10,000 U.S. patients each year who have potentially fatal cardiomyopathies, only about 1000 receive heart transplants. After successful transplantation, the 1-year survival rate is 80–85 percent and the 5-year survival rate is 50–70 percent. These rates are quite good, considering that these patients would have died if the transplant had not been performed. However, the procedure remains controversial due to the high cost involved and the limited years of life gained. As health-care dollars become

managed more closely, society will need to decide whether the probable life extension is worth the expense. Needless to say, this decision will not be an easy one to make.

Many individuals with cardiomyopathy who are initially selected for heart transplant surgery succumb to the disease before a suitable donor becomes available. For this reason, there continues to be considerable interest in the development of an artificial heart. One model, the *Jarvik-7,* had limited clinical use in the 1980s. Attempts to implant it on a permanent basis were unsuccessful, primarily because blood clots formed on the mechanical valves. When these clots broke free, they became drifting emboli that plugged peripheral vessels, producing strokes, kidney failure, and other complications. Modified versions of this unit and others now under development may still be used to maintain transplant candidates who are awaiting the arrival of a donor organ. These units are called *left ventricular assist devices* (LVAD). As the name implies, these devices assist, rather than replace, the damaged heart. In one study, the survival rate was 75 percent for patients relying on such a device for 3.5 months. Eventually these devices may be used to reduce the workload on damaged hearts, giving them time to heal rather than acting solely as a bridge to heart transplantation.

Another interesting approach, called *dynamic cardiomyoplasty,* involves the use of skeletal muscle tissue to apply permanent patches to injured hearts or to build small accessory pumps. For example, one procedure creates a biological LVAD by freeing a portion of the latissimus dorsi muscle from the side (FAP *p. 339*) and placing this flap, with its circulation intact, into the thoracic cavity. There it is folded to form a sling around the heart, and an electronic pacemaker is used to stimulate its contraction. Each time it contracts, the sling squeezes the heart and helps push blood into the major arteries. These methods are less stressful than heart transplants, because (1) they leave the damaged heart in place and (2) the transplanted tissue is taken from the same individual, so it will not be attacked by the immune system. Phase III clinical trials are now under way. However, until the data are analyzed, the use of skeletal muscle patches will remain an experimental concept rather than a recognized treatment for cardiomyopathy.

In 1996, attention focused on a Brazilian surgeon who developed a surgical procedure to improve cardiac function in patients with cardiomyopathy. The surgeon removes a portion of the weakened left ventricle. The ventricular muscle that remains is sewn together, forming a smaller chamber. In part because the muscle cells in the dilated heart were excessively stretched, the smaller, remodeled ventricle pumps blood more efficiently. Heart surgeons in the United States have confirmed the effectiveness of this therapy, which may reduce the demand for heart transplants in the years to come.

RHD AND VALVULAR STENOSIS FAP *p. 664*

Rheumatic (roo-MA-tik) **fever** is an inflammatory condition that can develop after untreated infection by streptococcal bacteria ("strep throat"). Rheumatic fever typically affects children of age 5–15 years; symptoms include high fever, joint pain and stiffness, and a distinctive full-body rash. Obvious symptoms generally persist for less than 6 weeks, although severe cases may linger for 6 months or more. The longer the duration of the inflammation, the more likely it is that carditis will develop. The carditis that does develop in 50–60 percent of individuals typically escapes detection, and scar tissue gradually forms in the myocardium and the heart valves. Valve condition deteriorates over time, and valve problems serious enough to affect cardiac function may not appear until 10–20 years after the initial infection.

During the interim, the affected valves become thickened and may also calcify to some degree. This thickening narrows the opening guarded by the valves, producing a condition called **valvular stenosis** (ste-NŌ-sis; *stenos,* narrow). The resulting clinical disorder is known as **rheumatic heart disease (RHD).** The thickened cusps stiffen in a partially closed position, but the valves do not completely block the blood flow, because the edges of the cusps are rough and irregular. Regurgitation may occur, and much of the blood pumped out of the heart may flow back in. The abnormal valves are also much more susceptible to bacterial infection, a type of endocarditis (p. 118). Fortunately, with the detection and prompt antibiotic treatment of strep infections, the number of cases of rheumatic heart disease has declined dramatically since the 1940s.

Mitral stenosis and **aortic stenosis** are the most common forms of rheumatic heart disease. About 40 percent of patients with RHD develop mitral stenosis, and two-thirds of them are women. The reason for the correlation between gender and mitral stenosis is unknown. In mitral stenosis, blood enters the left ventricle at a slower than normal rate; when the ventricle contracts, blood flows back into the left atrium as well as into the aortic trunk. As a result, the left ventricle has to work much harder to maintain adequate systemic circulation. The right and left ventricles discharge identical amounts of blood with each beat, so as the output of the left ventricle declines, blood "backs up" in the pulmonary circuit. Venous pressures then rise in the pulmonary circuit, and the right ventricle must develop greater pressures to force blood into the pulmonary trunk. In severe cases of mitral stenosis, the ventricular musculature is not up to the task. The heart weakens, and peripheral tissues begin

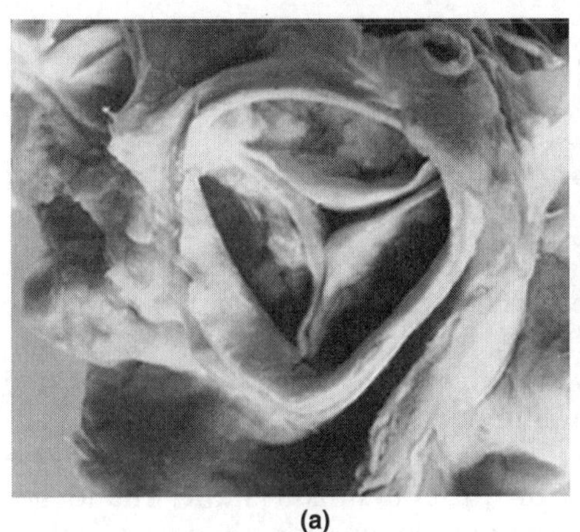

(a)

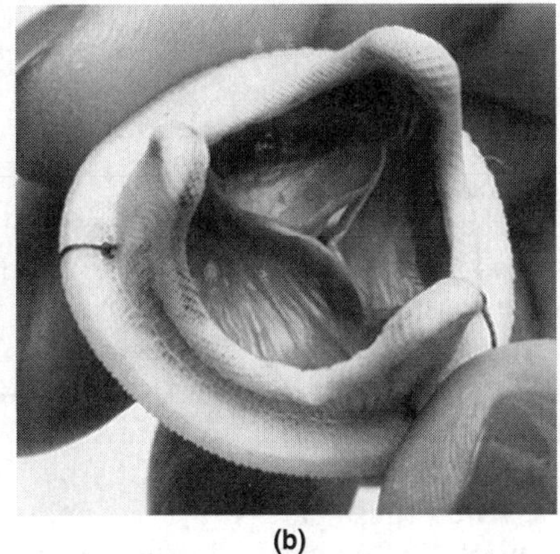

(b)

(c)

•**FIGURE A-39 Artificial Heart Valves. (a)** A stenotic semilunar valve; note the irregular, stiff cusps. **(b)** Intact Bioprosthetic™ heart valve, which uses the valve from a pig's heart. **(c)** Medtronic Hall™ prosthetic heart valve.

to suffer from oxygen and nutrient deprivation. (We discuss this condition, called heart failure, in more detail in a later section.)

Symptoms of aortic stenosis develop in roughly 25 percent of patients with RHD; 80 percent of these individuals are males. Symptoms of aortic stenosis are initially less severe than those of mitral stenosis. Although the left ventricle enlarges and works harder, normal circulatory function can typically be maintained for years. Clinical problems develop only when the opening narrows enough to prevent adequate blood flow. Symptoms then resemble those of mitral stenosis.

FAP Ch. 20

One reasonably successful treatment for severe stenosis involves the replacement of the damaged valve with a prosthetic (artificial) valve. Figure A-39a• shows a stenotic heart valve; two possible replacements are a valve from a pig (Figure A-39b•) and a synthetic valve (Figure A-39c•), one of a number of designs that have been employed. Pig valves do not require anticoagulant therapy but may wear out and begin leaking after roughly 10 years in service. The plastic or stainless steel components of the artificial valve are more durable but activate the clotting system of the recipient, leading to inflammation, clot formation, and other potential complications. Synthetic valve recipients must take anticoagulant drugs to prevent strokes and other disorders caused by embolus formation. Valve replacement operations are quite successful, with about 95 percent of the surgical patients surviving for 3 years or more and 70 percent surviving more than 5 years.

PROBLEMS WITH PACEMAKER FUNCTION FAP *p. 672*

Symptoms of severe bradycardia (below 50 beats per minute) include weakness, fatigue, fainting, and confusion. Drug therapies are seldom helpful, but artificial pacemakers can be used with considerable success. Wires run to the atria, the ventricles, or both, depending on the nature of the problem, and the unit delivers small electrical pulses to stimulate the myocardium. Internal pacemakers are surgically implanted, batteries and all. These units last 7–8 years or more before another operation is required to change the battery. External pacemakers are used for temporary emergencies, such as immediately after cardiac surgery. Only the wires are implanted, and an external control box is worn on a belt.

More than 50,000 artificial pacemakers are in use at present (Figure A-40•). The simplest provide constant stimulation to the ventricles at rates of 70–80 per minute. More-sophisticated pacemakers vary their rates to adjust to changing circulatory demands, as during exercise. Others are able to monitor cardiac activity and respond whenever the heart begins to function abnormally.

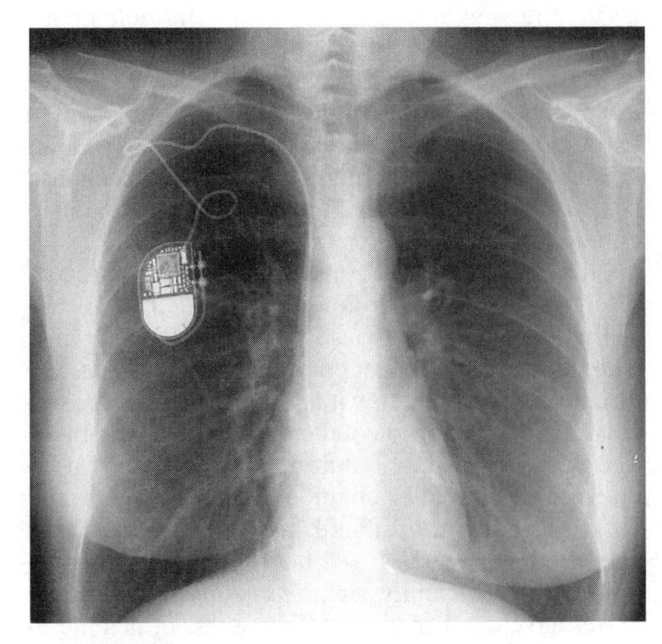

•**FIGURE A-40 An Artificial Pacemaker**

Tachycardia, usually defined as a heart rate of over 100 beats per minute, increases the work load on the heart. Cardiac performance suffers at very high heart rates, because the ventricles do not have enough time to refill with blood before the next contraction occurs. Chronic or acute incidents of tachycardia may be controlled by drugs that affect the permeability of pacemaker membranes or block the effects of sympathetic stimulation.

DIAGNOSING ABNORMAL HEARTBEATS

FAP pp. 672, 674

Damage to the conduction pathways caused by mechanical distortion, ischemia, infection, or inflammation can affect the normal rhythm of the heart. The resulting condition is called a **heart block,** or **conduction deficit.** Figure A-41a• shows the electrocardiograph of a normal heart; heart blocks of varying severity are represented in Figure A-41b•. In a **first-degree heart block** (Figure A-41b•), the AV node and proximal portion of the AV bundle slow the passage of impulses that are heading for the ventricular myocardium. As a result, a pause appears between the atrial and ventricular contractions. Although a delay exists, the regular rhythm of the heart continues, and each atrial beat is followed by a ventricular contraction.

If the delay lasts long enough, the nodal cells will still be repolarizing from the previous beat when the next impulse arrives from the SA node. The arriving impulse will then be ignored; the ventricles will not be stimulated; and the normal "atria–ventricles, atria–ventricles" pattern will disappear. This condition is a **second-degree heart block** (Figure A-41c•). A mild second-degree block may produce only an occasional skipped beat, but with more-substantial delays the ventricles will follow every second atrial beat. The resulting pattern of "atria, atria–ventricles, atria, atria–ventricles" is known as a **two-to-one (2:1) block.** Three-to-one or even four-to-one blocks are also encountered.

In a **third-degree heart block,** or **complete heart block,** the conducting pathway stops functioning (Figure A-41d•). The atria and ventricles continue to beat, but their activities are no longer synchronized. The atria follow the pace set by the SA node, beating 70–80 times per minute, and the ventricles follow the commands of the AV node, beating at a rate of 40–60 beats per minute. A temporary third-degree block can be induced by stimulating the vagus nerve. In addition to slowing the rate of impulse generation by the SA node, such stimulation inhibits the AV nodal cells to the point at which they cannot respond to normal stimulation. Comments such as "my heart stopped" or "my heart skipped a beat" generally refer to this phenomenon. The pause typically lasts just a few seconds. Longer delays end when a conducting cell, normally one of the Purkinje fibers, depolarizes to threshold. This phenomenon is called **ventricular escape,** because the ventricles are escaping from the control of the SA node. Ventricular escape can be a lifesaving event if the conduction system is damaged. Even without instructions from the SA or AV nodes, the ventricles will continue to pump blood at a slow but steady rate.

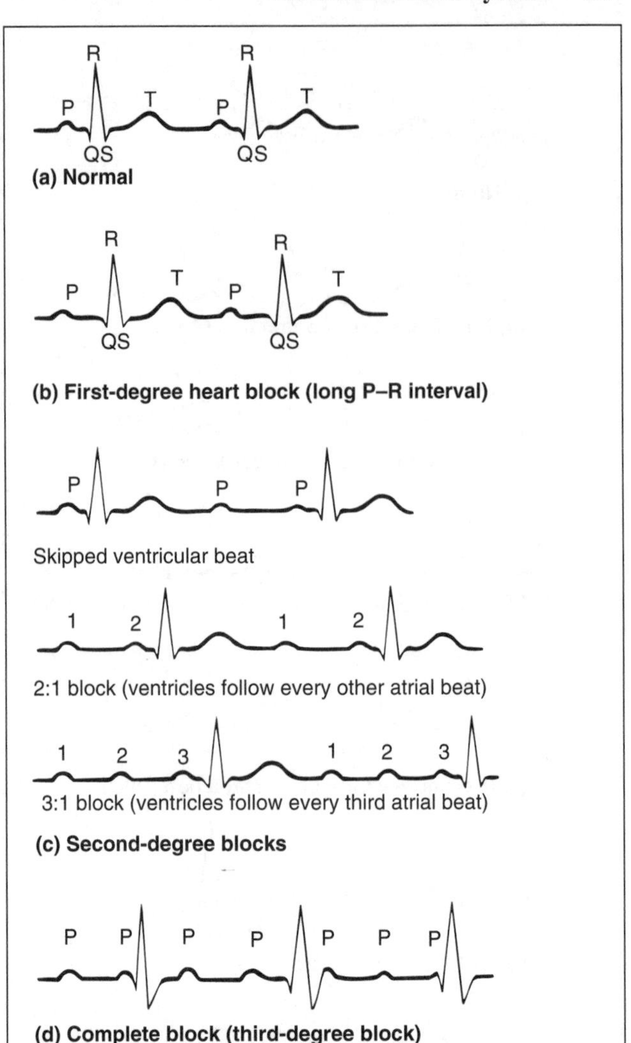

•**FIGURE A-41** Heart Blocks

Tachycardia and Fibrillation

Additional important examples of arrhythmias are shown in Figure A-42•. Figure A-42a• shows a normal heart rhythm. **Premature atrial contractions (PACs),** indicated in Figure A-42b•, often occur in healthy individuals. In a PAC, the normal atrial rhythm is momentarily interrupted by a "surprise" atrial contraction. Stress, caffeine, and various drugs may increase the frequency of PAC incidence, presumably by increasing the permeabilities of the SA pacemakers. The impulse spreads along the conduction pathway, and a normal ventricular contraction follows the atrial beat.

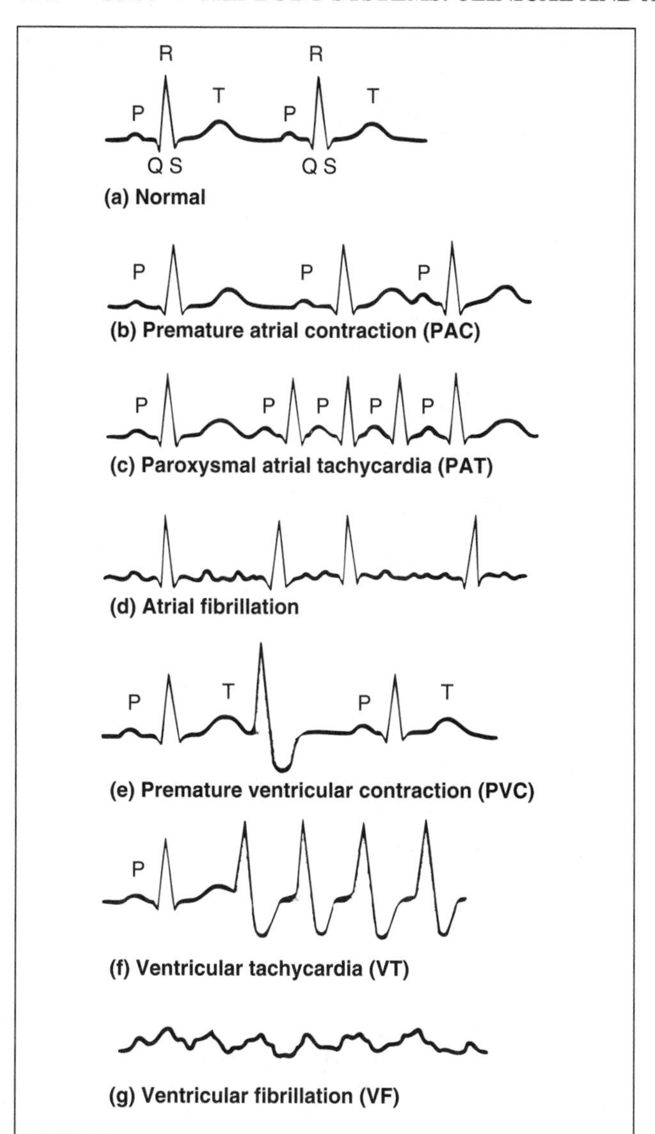

•**FIGURE A-42** Cardiac Arrhythmias

In **paroxysmal atrial tachycardia** (par-ok-SIZ-mal), or **PAT**, a premature atrial contraction triggers a flurry of atrial activity (Figure A-42c•). The ventricles are still able to keep pace, and the heart rate jumps to about 180 beats per minute. In **atrial flutter,** the atria contract in a coordinated manner, but the contractions occur very frequently. During a bout of **atrial fibrillation** (fi-bri-LĀ-shun), the impulses move over the atrial surface at rates of perhaps 500 beats per minute (Figure A-42d•). The atrial wall quivers instead of producing an organized contraction. The ventricular rate in atrial flutter or atrial fibrillation cannot follow the atrial rate and may remain within normal limits. Despite the fact that the atria are now essentially nonfunctional, the condition may go unnoticed, especially in older individuals who lead sedentary lives. In chronic atrial fibrillation, blood clots may form by the atrial walls. Clotting promotes the formation of emboli and increases the risks of stroke. As a result, most people diagnosed with this condition are placed on anticoagulant therapy. PACs, PAT, atrial flutter, and even atrial fibrillation are not considered very dangerous unless they are prolonged or associated with some more serious indications of cardiac damage, such as coronary artery disease or valve problems.

In contrast, ventricular arrhythmias can be fatal. Because the conduction system functions in one direction only, a ventricular arrhythmia is not linked to atrial activities. **Premature ventricular contractions (PVCs)** occur when a Purkinje cell or ventricular myocardial cell depolarizes to threshold and triggers a premature contraction (Figure A-42e•). Single PVCs are common and not dangerous. The cell responsible is called an *ectopic pacemaker.* The frequency of PVCs can be increased by exposure to epinephrine or other stimulatory drugs or to ionic changes that depolarize cardiac muscle cell membranes. Similar factors may be responsible for periods of **ventricular tachycardia,** also known as **VT** or *V-tach* (Figure A-42f•).

Multiple PVCs and VT often precede the most serious arrhythmia, **ventricular fibrillation (VF)** (Figure A-42g•). The resulting condition, also known as **cardiac arrest,** is rapidly fatal, because the heart stops pumping blood. During ventricular fibrillation, the cardiac muscle cells are overly sensitive to stimulation, and the impulses are traveling from cell to cell around and around the ventricular walls. A normal rhythm cannot become established, because the ventricular muscle cells are stimulating one another at such a rapid rate. The problem is exaggerated by a sustained rise in free intracellular calcium ion concentrations, due to a massive stimulation of alpha and beta receptors following sympathetic activation.

FAP Ch. 20

A **defibrillator** is a device that attempts to eliminate ventricular fibrillation and restore normal cardiac rhythm. Two electrodes are placed in contact with the chest, and a powerful electrical shock is administered. The electrical stimulus depolarizes the entire myocardium simultaneously. With luck, after repolarization the SA node will be the first area of the heart to reach threshold. Thus the primary goal of defibrillation is not just to stop the fibrillation but to give the ventricles a chance to respond to normal SA commands.

Several approaches are used to treat arrhythmias. Medications can be used to slow down rapid heart rates, or the abnormal portions of the conducting system can be destroyed. Pacemakers are used to accelerate slow heart rates. Implantable pacemakers that are able to sense ventricular fibrillation and deliver an immediate defibrillating shock have been successful in preventing sudden death in patients with previous episodes of ventricular tachycardia and ventricular fibrillation.

MONITORING THE HEART FAP *p. 674*

Many techniques can be used to examine the structure and performance of the heart. No single diagnostic procedure provides the complete picture, so the tests used will vary with the suspected nature of the

problem. A standard chest X-ray will show the basic size, shape, and orientation of the heart. Additional details require more-specialized procedures to enhance the clarity of the images.

Coronary arteriography (ar-tē-rē-OG-ra-fē) is often used to look for abnormalities in the coronary circulation. In this procedure, a catheter is inserted into a major artery in the arm or leg and then maneuvered back along the arterial passageways until its tip reaches the heart. A radiopaque dye can then be released at the openings of the coronary arteries, and its distribution can be followed in a series of high-speed X-rays. The images obtained are called **coronary angiograms** (Figure A-43a•). For direct analyses of cardiac performance and the collection of blood samples, a catheter may be introduced into the heart itself. The instrument can enter via the aorta or from the venous system by way of the inferior vena cava.

Because the heart is constantly moving, ordinary ultrasound, computerized tomography (CT), and magnetic resonance imaging (MRI) scans create blurred images. Special instruments and computers, however, that can generate images at high speed can be used with these techniques to develop three-dimensional still or moving pictures of the heart as it beats (Figure A-43b–d•). These procedures create dramatic images, but the cost and complexity of the equipment have so far limited their use to major research institutions. Although PET scans can be used to diagnose disorders of coronary circulation, cost factors also limit the clinical use of this technology.

Ultrasound analysis, called **echocardiography** (ek-ō-kar-dē-OG-ra-fē) (Figure A-43b•), provides images that lack the clarity of CT or MRI scans, but the equipment is relatively inexpensive and portable. Recent advances in data processing have made the images suitable for following details of cardiac contractions, including valve function and blood flow dynamics. Echocardiography is now an important diagnostic tool.

HEART FAILURE FAP pp. 677, 722

A condition of heart failure exists when the cardiac output is insufficient to meet the circulatory demand. The initial symptoms of heart failure depend on whether the problem is restricted to the left ventricle, the right ventricle, or both. However, over time these differences are eliminated; for example, the

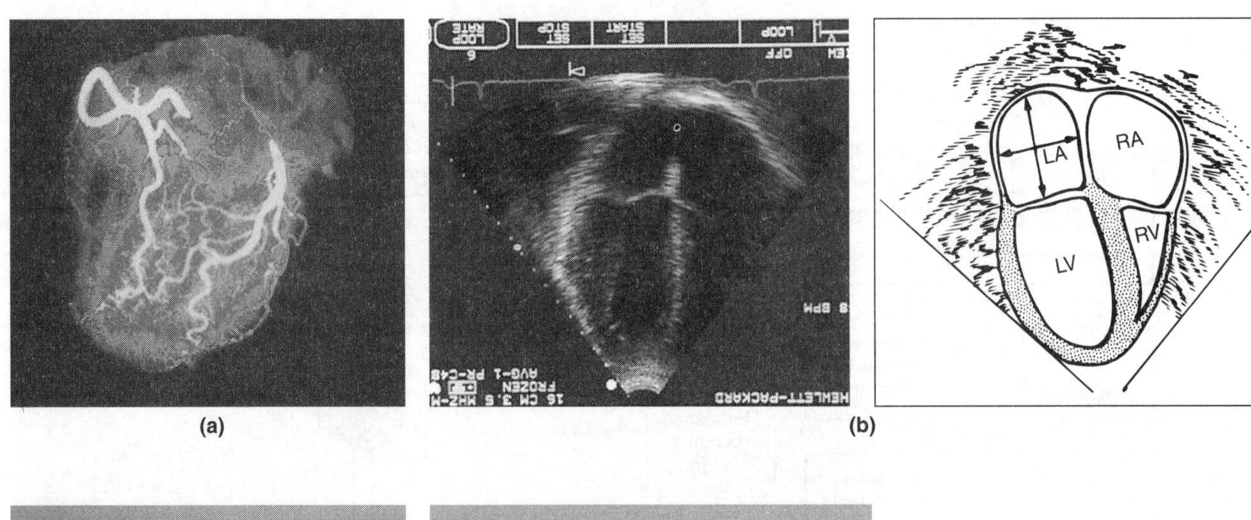

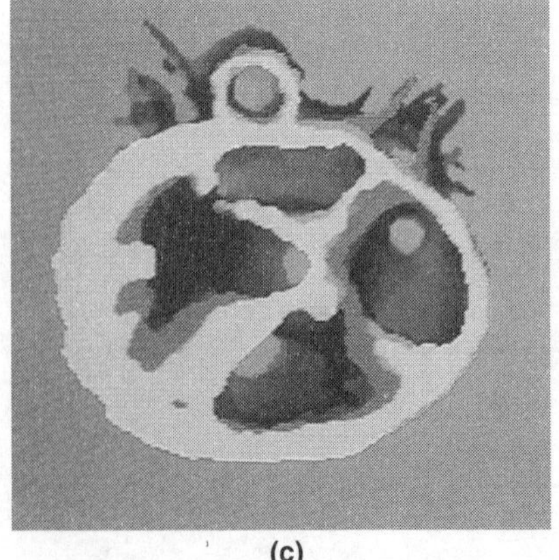

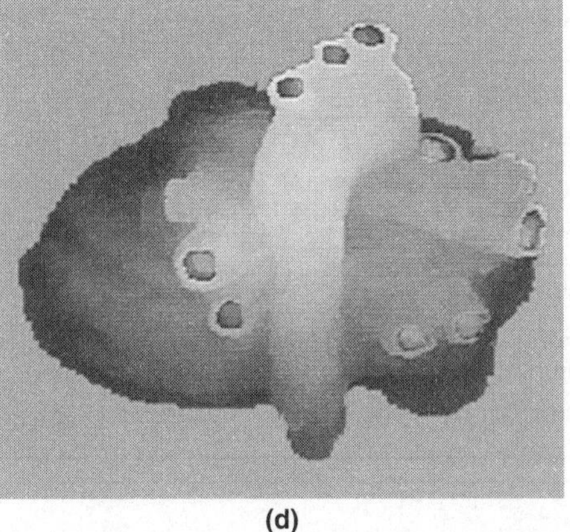

•**FIGURE A-43 Monitoring the Heart. (a)** A coronary angiogram. **(b)** An echocardiogram (left) with interpretive drawing (right). **(c)** A three-dimensional CT scan of a frontal section and **(d)** of a posterior/superior view of the heart.

major cause of right ventricular failure is left ventricular failure. Figure A-44• is a flow chart that shows the basic causes and effects of heart failure and indicates potential therapies.

Suppose that the left ventricle cannot maintain normal cardiac output due to damage to the ventricular muscle (see the discussion of myocardial infarctions on *p. 679* of the text) or high arterial pressures (hypertension, FAP *p. 706*). In effect, the left ventricle can no longer keep pace with the right ventricle, and blood backs up into the pulmonary circuit. This venous congestion is responsible for the term **congestive heart failure (CHF).** The right ventricle now works harder, elevating pulmonary arterial pressures and forcing blood through the lungs and into the weakened left ventricle.

At the capillaries of the lungs, arterial and venous pressures are now elevated. This elevated pressure pushes additional fluid out of the capillaries and into the interstitial fluids, most notably at the lungs. The fluid buildup and compression of the airways reduce the effectiveness of gas exchange, leading to shortness of breath, typically the first obvious sign of congestive heart failure. This fluid buildup begins at a pulmonary postcapillary pressure of about 25 mm Hg. At a capillary pressure of about 30 mm Hg, fluid not only enters the tissues of the lungs but crosses the alveolar walls and begins to fill the air spaces. This condition is called **pulmonary edema.** Abnormal sounds (*moist rales*) can be heard at the base of each lung on auscultation.

Over time, the less-muscular right ventricle may become unable to generate enough pressure to force

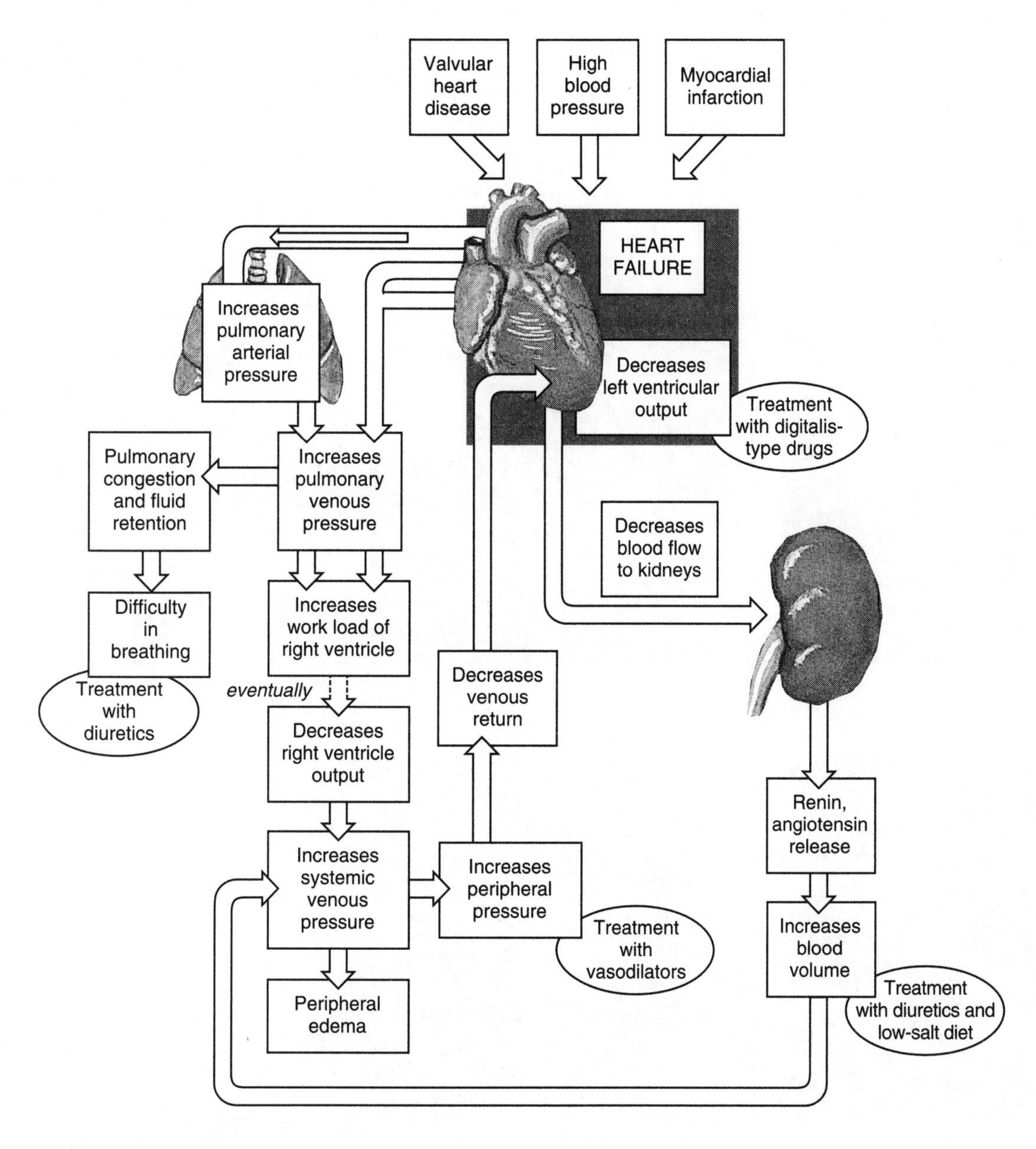

•**FIGURE A-44 Heart Failure**

FAP Ch. 20

blood through the pulmonary circuit. Venous congestion now occurs in the systemic circuit; the ankles become distended with fluid ("ankle edema"); and cardiac output declines further. When the reduction in systemic pressures lowers blood flow to the kidneys, renin and erythropoietin are released. This release in turn elevates blood volume as salt and water retention increase at the kidneys and RBC production accelerates. This rise in blood volume actually complicates the situation, as it tends to increase venous congestion and cause widespread edema.

The increased volume of blood in the venous system leads to a distension of the veins, making superficial veins more prominent. When the heart contracts, the rise in pressure at the right atrium produces a pressure pulse in the large veins. This venous pulse can be seen most easily over the right external jugular vein.

The treatment of congestive heart failure commonly includes the following:

- *The restriction of salt intake.* The expression "water follows salt" applies: When sodium ions and chloride ions are absorbed at the digestive tract, water follows by osmosis.
- *The administration of drugs to promote fluid loss.* These drugs, called **diuretics** (dī-ū-RET-iks; *diouretikos,* promoting urine), increase salt and water losses at the kidneys. (The mechanism is described in Chapter 27 of the main text.) The outmoded practice of "bloodletting," also called "bleeding," which was popular in the seventeenth and eighteenth centuries, did cause a temporary reduction in blood volume and thereby reduced symptoms of CHF. However, the relief was only temporary, and the practice had many disadvantages, such as anemia, in addition to the risk of infection.
- *Extended bed rest, to enhance venous return to the heart, coupled with physical therapy to maintain good venous circulation.*
- *The administration of drugs that enhance cardiac output.* These drugs may target the heart, the peripheral circulation, or both. When the heart has been weakened, drugs related to digitalis, an extract from the leaves of the foxglove plant, are commonly selected. *Digitoxin, digoxin,* and *ouabain* are examples. These compounds increase the force of cardiac muscle cell contractions. When high blood pressure is a factor, some type of vasodilator is given to lower peripheral resistance.
- *The administration of drugs that reduce peripheral vascular resistance,* such as *hydralazine* or ACE inhibitors (p. 127). The drop in peripheral resistance reduces the work load of the left ventricle and improves cardiac output.

ANEURYSMS FAP *p. 695*

An **aneurysm** (AN-ū-rizm) is a bulge in the weakened wall of a blood vessel, generally an artery. This bulge resembles a bubble in the wall of a tire. Like a bad tire, the affected artery may suffer a catastrophic blowout. The most dangerous aneurysms are those involving arteries of the brain, where they cause strokes, and of the aorta, where a blowout will cause fatal bleeding in a matter of minutes.

Aneurysms are normally caused by chronic high blood pressure, although any trauma or infection, such as syphilis, that weakens vessel walls can lead to an aneurysm. In addition, at least some aortic aneurysms have been linked to inherited disorders, such as *Marfan's syndrome,* that have weakened connective tissues in vessel walls (FAP *p. 695*). It is not known whether other genetic factors are involved in the development of other types of aneurysms.

Most aneurysms form gradually as vessel walls become less elastic. When a weak point develops, the arterial pressures distort the wall, creating an aneurysm. Unfortunately, because many aneurysms are painless, they are likely to go undetected.

When aneurysms are detected by ultrasound or other scanning procedures, the risk of rupture can sometimes be estimated on the basis of their size. For example, an aortic aneurysm larger than 6 cm has a 50 percent chance of rupturing within 10 years. Treatment commonly begins with the reduction of blood pressure by means of either vasodilators or beta-blockers (p. 127). An aneurysm in an accessible area, such as the abdomen, may be surgically removed and the vessel repaired. Figure A-45• shows a large aortic aneurysm and the steps involved in its surgical repair with a synthetic patch.

PROBLEMS WITH VENOUS VALVE FUNCTION FAP *p. 700*

Chapter 4 of the text notes that one of the consequences of aging is an increase in the fragility of connective tissues throughout the body (FAP *p. 136*). Blood vessels are no exception; with age, the walls of veins begin to sag. This change generally affects the superficial veins of the legs first, because at these locations gravity opposes blood flow. The situation is aggravated by a lack of exercise and by an occupation requiring long hours of standing or sitting. Because there is no muscular activity to help keep the blood moving, venous blood pools on the proximal (heart) side of each valve. As the venous walls are distorted, the valves become less effective, and gravity can then pull blood back toward the capillaries. This pulling further impedes normal blood flow, and the veins become grossly distended. The sagging, swollen vessels are called **varicose** (VAR-i-kōs) **veins.** Varicose veins are relatively harmless but unsightly. Surgical procedures are sometimes used to remove or constrict the distended vessels.

Varicose veins are not limited to the extremities. Another common site involves a network of veins in

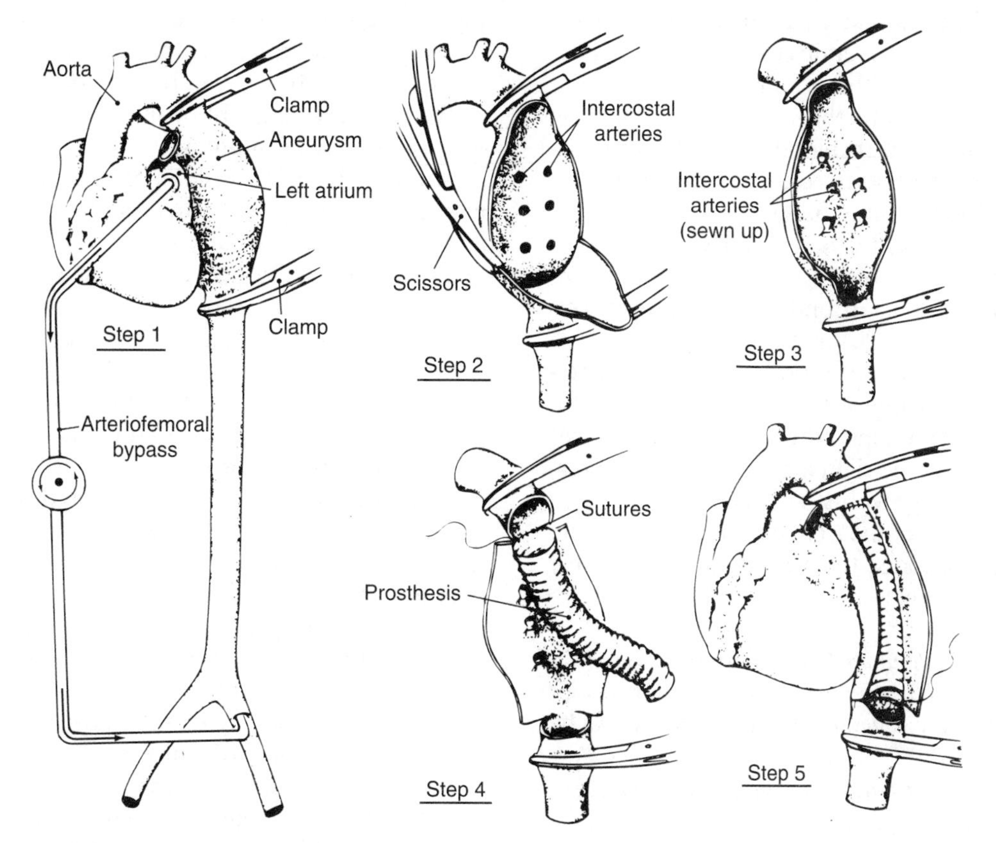

•**FIGURE A-45** **Repair of an Aneurysm**

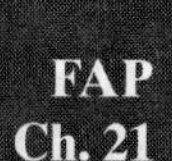
FAP Ch. 21

the walls of the anus. Pressures within the abdominopelvic cavity rise dramatically when the abdominal muscles are tensed. Straining to force defecation can force blood into these veins, and repeated incidents leave them permanently distended. These distended veins, known as **hemorrhoids** (HEM-o-roydz), can be uncomfortable and in severe cases extremely painful. Hemorrhoids and varicose veins are often associated with pregnancy, as a result of changes in blood flow and abdominal pressures. Minor cases can be treated by the topical application of drugs that promote the contraction of smooth muscles within the venous walls. More-severe cases may require the surgical removal or destruction of the distended veins.

HYPERTENSION AND HYPOTENSION FAP *p. 706*

Elevated blood pressure is called **primary hypertension,** or *essential hypertension,* if no obvious cause can be determined. Known risk factors include a hereditary history of hypertension, gender (males are at higher risk), high plasma cholesterol, obesity, chronic stresses, and cigarette smoking. **Secondary hypertension** appears as the result of abnormal hormonal production outside the cardiovascular system. For example, a condition resulting in excessive production of antidiuretic hormone (ADH), renin, aldosterone, or epinephrine typically produces hypertension, and many forms of kidney disease lead to hypertension caused by fluid retention or excessive renin production.

Hypertension significantly increases the work load on the heart, and the left ventricle gradually enlarges. The increased muscle mass requires a greater oxygen demand. When the coronary circulation cannot keep pace, symptoms of coronary ischemia appear.

Increased arterial pressures also place a physical stress on the walls of blood vessels throughout the body. This stress promotes or accelerates the development of arteriosclerosis and increases the risks of aneurysms, heart attacks, and strokes. Vessels supplying the retinas of the eyes are typically affected, and hemorrhages and associated circulatory changes can produce disturbances in vision. Because a routine physical exam includes the examination of these vessels, retinal changes may provide the first evidence that hypertension is affecting peripheral circulation.

One of the most difficult aspects of hypertension is that in most cases no obvious symptoms are present. As a result, clinical problems do not appear until the condition has reached the crisis stage. There is therefore considerable interest in the early detection and prompt treatment of hypertension.

Treatment consists of a combination of lifestyle changes and physiological therapies. Quitting smoking, getting regular exercise, and restricting dietary in-

take of salt, fats, and calories will improve peripheral circulation, prevent increases in blood volume and total body weight, and reduce plasma cholesterol levels. These strategies may be sufficient to control hypertension if it has been detected before significant cardiovascular damage has occurred. Most therapies involve antihypertensive drugs, such as calcium channel blockers, beta-blockers, diuretics, and vasodilators, singly or in combination. Beta-blockers reduce the effects of sympathetic stimulation on the heart, and the unopposed parasympathetic system lowers the resting heart rate and blood pressure. Diuretics promote the loss of water and sodium ions at the kidneys, lowering blood volume, and vasodilators further reduce blood pressure. A new class of antihypertensive drugs lowers blood pressure by preventing the conversion of angiotensin I to angiotensin II. These **angiotensin-converting enzyme (ACE) inhibitors,** such as *captopril,* are being used to treat chronic hypertension and congestive heart failure.

In hypotension, blood pressure declines and peripheral systems begin to suffer from oxygen and nutrient deprivation. One clinically important form of hypotension can develop after antihypertensive drugs have been administered. Problems may appear when the individual changes position from lying down to sitting or from sitting to standing. Each time you sit up or stand, blood pressure in your carotid sinus drops because your heart must suddenly counteract gravity to push blood up to your brain. The fall in pressure triggers the carotid reflex, and blood pressure returns to normal. But if the carotid response is prevented by beta-blockers or other drugs, blood pressure at the brain may fall so low that the individual becomes weak, dizzy, disoriented, or unconscious. This condition is known as **orthostatic hypotension** (or-thō-STAT-ik; *orthos,* straight + *statikos,* causing to stand), or simply **orthostasis** (or-thō-STĀ-sis). You may have experienced a brief episode of orthostasis when you stood suddenly after reclining for an extended period. The carotid reflex typically slows with age, so older people must sit and stand more carefully than they used to in order to avoid orthostatic hypotension.

OTHER TYPES OF SHOCK FAP *p. 720*

Although the text focuses on circulatory shock caused by low blood volume, shock can develop when the blood volume is normal. **Cardiogenic** (kar-dē-ō-JEN-ik) **shock** occurs when the heart becomes unable to maintain a normal cardiac output. The most common cause is failure of the left ventricle as a result of a myocardial infarction. Between 5 and 10 percent of patients surviving a heart attack must be treated for cardiogenic shock. The use of thrombolytic drugs, such as t-PA, can be very effective in restoring coronary circulation and ventricular function, thereby relieving the peripheral symptoms. Cardiogenic shock can also be the result of valvular heart disease, advanced coronary artery disease, cardiomyopathy, or ventricular arrhythmias.

In **obstructive shock,** ventricular output is reduced because tissues or fluids are restricting the expansion and contraction of the heart. For example, fluid buildup in the pericardial cavity (cardiac tamponade; see "Infection and Inflammation of the Heart," p. 118) can compress the heart and limit ventricular filling.

Distributive shock is the result of a widespread, uncontrolled vasodilation. The vasodilation produces a dramatic fall in blood pressure that leads to a reduction in blood flow and the onset of shock. Three important examples are neurogenic shock, septic shock, and anaphylactic shock:

1. **Neurogenic** (noo-rō-JEN-ik) **shock** can be caused by general or spinal anesthesia and by trauma or inflammation of the brain stem or spinal cord. The underlying problem is damage to the vasomotor center or to the sympathetic tracts or nerves, leading to a loss of vasomotor tone.
2. **Septic shock** results from the massive release of endotoxins, poisons derived from the cell walls of bacteria during a systemic infection. These compounds cause a vasodilation of precapillary sphincters throughout the body, resulting in drops in peripheral resistance and blood pressure. Symptoms of septic shock generally resemble those of other types of shock, but the skin is flushed, and the individual has a high fever. For this reason septic shock is also known as "warm shock."

   One form of septic shock, called **toxic shock syndrome (TSS),** results from an infection by the bacterium *Staphylococcus aureus.* Symptoms include high fever, sore throat, vomiting and diarrhea, and a generalized rash. As the condition progresses, shock, respiratory distress, and kidney or liver failure can develop; 10–15 percent of cases prove fatal. These symptoms result from the entry of bacteria or bacterial toxins into the bloodstream. Toxic shock syndrome was unrecognized before 1978, when it appeared in a group of children. Roughly 1250 cases are reported in the United States each year, 95 percent of them affecting women age 15–44. The mortality rate is 5 percent. Although other sources of infection are possible, infection most often appears to occur during menstruation, and the chances of infection are increased with the use of superabsorbent tampons. The brands involved were taken off the market, and the incidence has declined steadily since 1980. However, TSS continues to occur at a low but significant rate (6.2 persons per 100,000 menstruating women each year) in women who use ordinary tampons and in men or women after abrasion, burn injuries, or nasal surgery. The treatment for TSS includes fluid administration, the removal of the focus of infection (such as the removal of a tampon or the cleansing of a wound), and antibiotic therapy.

3. Extensive peripheral vasodilation also occurs in **anaphylactic** (an-a-fi-LAK-tik) **shock**, a dangerous allergic reaction. This type of shock, which can be life-threatening, is discussed in Chapter 22 of the text (see FAP *p. 789*).

THE CAUSES AND TREATMENT OF CEREBROVASCULAR DISEASE FAP *p. 722*

Cerebrovascular disease was introduced in Chapter 14 of the text (FAP *p. 443*); here we consider additional details related to the concepts in that chapter. Most symptoms of cerebrovascular disease appear when atherosclerosis reduces the circulatory supply to the brain. If the blood flow to a portion of the brain is completely shut off, a *cerebrovascular accident (CVA),* or *stroke,* occurs. The most common causes of strokes include **cerebral thrombosis** (clot formation at a plaque), **cerebral embolism** (drifting blood clots, fatty masses, or air bubbles), and **cerebral hemorrhages** (rupture of a blood vessel, commonly following the formation of an aneurysm). The observed symptoms and their severity vary with the vessel involved and the location of the blockage.

If the circulatory blockage disappears in a matter of minutes, the effects are temporary and the condition is called a **transient ischemic attack (TIA).** A TIA typically indicates that cerebrovascular disease exists, so preventive measures can be taken to forestall more-serious incidents. For example, taking aspirin each day slows blood clot formation in patients who experience TIAs and thereby reduces the risks of cerebral thrombosis and cerebral embolism.

FAP Ch. 21

If the blockage persists for a longer period, neurons die and the area degenerates. Stroke symptoms are initially exaggerated by the swelling and distortion of the injured neural tissues; if the individual survives, in many cases, brain function gradually improves. The management and treatment of strokes remain controversial. The surgical removal of the offending clot or blood mass may be attempted, but the results vary. Recent progress in the emergency treatment of cerebral thromboses and cerebral embolisms has involved the administration of clot-dissolving enzymes such as *tissue plasminogen activator* (t-PA; now sold as *Alteplase®*). The best results are obtained if the enzymes are administered within an hour after the stroke, although they may still be of use up to 24 hours after. Subsequent treatment involves anticoagulant therapy, typically with heparin (for 1 to 2 weeks) followed by coumadin (for up to 1 year) to prevent further clot formation. (These fibrinolytic and anticoagulant drugs were introduced in Chapter 19 of the text; see FAP *p. 649*.) A more complicated surgical procedure involves the insertion of a transplanted piece of a blood vessel that routes blood around the damaged area. None of these treatments is as successful as preventive surgery, in which plaques are removed before a stroke. The very best solution is to prevent or restrict plaque formation by controlling the risk factors involved.

[EXPLORE]

Additional resources, such as interactive tutorials, critical-thinking questions, clinical problems, and case studies, are available through the CD-ROM and Website for your textbook. To begin your exploration, launch the *Martini Interactive* CD-ROM or visit the Companion Website (http://www.prenhall.com/martini/fap5) and select Chapter 19, 20, or 21.

The Lymphatic System and Immunity

The lymphatic system consists of the fluid *lymph,* a network of *lymphatic vessels,* specialized cells called *lymphocytes,* and an array of *lymphoid tissues* and *lymphoid organs* throughout the body. This system has three major functions: (1) to protect the body through the *immune response*; (2) to transport fluid from interstitial fluid to the bloodstream; and (3) to help distribute hormones, nutrients, and wastes.

The immune response produced by activated lymphocytes is responsible for the detection and destruction of foreign or toxic substances that may disrupt homeostasis. For example, viruses, bacteria, and tumor cells are usually recognized and eliminated by cells of the lymphatic system. Immunity is the specific resistance to disease, and all the cells and tissues involved with the production of immunity are considered to be part of an *immune system.* Whereas the lymphatic system is an anatomically distinct system, the immune system is a physiological system that includes the lymphatic system as well as components of the integumentary, cardiovascular, respiratory, digestive, and other systems.

The role of the lymphatic system in the recirculation of extracellular fluid was detailed in Chapter 21 of the text (FAP *pp. 706, 709*).

THE PHYSICAL EXAMINATION AND THE LYMPHATIC SYSTEM

Individuals with lymphatic system disorders experience a variety of symptoms. The pattern observed depends on whether the problem affects the immune functions or the circulatory functions of the lymphatic system (Figure A-46•). Important symptoms and signs include the following:

- Infections are typically characterized by *enlarged lymph nodes.* Enlarged lymph nodes also develop in cancers of the lymphatic system, such as *lymphoma* (p. 133), or when primary tumors in other tissues have metastasized to regional lymph nodes. The status of regional lymph nodes is therefore important in the diagnosis and treatment of many cancers. The onset and duration of swelling; the size, texture, and mobility of

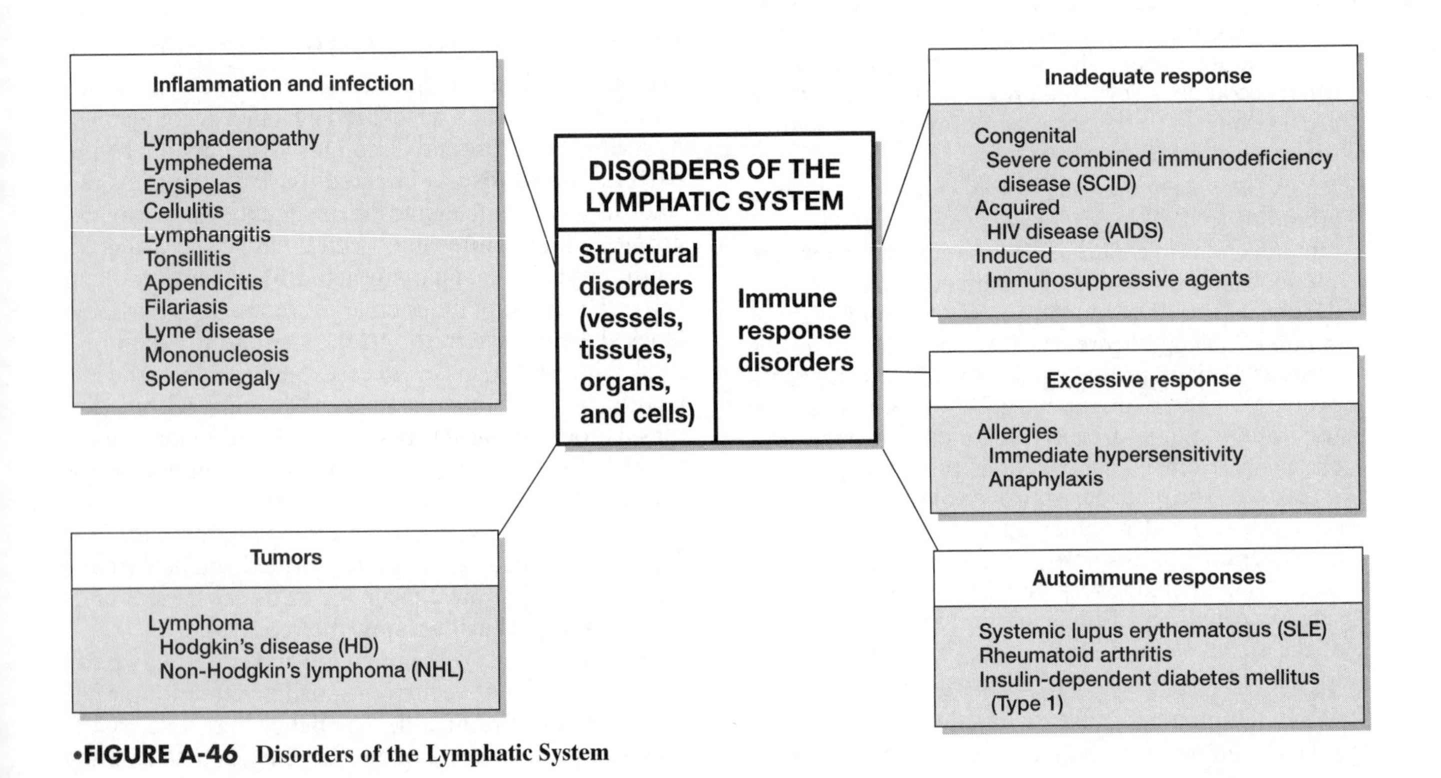

•FIGURE A-46 Disorders of the Lymphatic System

the nodes; the number of affected nodes; and the degree of tenderness are all important in diagnosis. For example, nodes containing cancer cells tend to be large, hard, locked in place, and nontender. On palpation, these nodes feel like dense masses rather than individual lymph nodes. In contrast, infected lymph nodes tend to be large, freely mobile, and very tender.

- *Lymphangitis* consists of *erythematous* (red) streaks on the skin that may develop with an inflammation of superficial lymph vessels. Lymphangitis commonly occurs in the lower limbs; the reddened streaks originate at an infection site. Before the linkage to the lymphatic system was known, this sign was called "blood poisoning."
- *Splenomegaly* (p. 134) is an enlargement of the spleen that can result from acute infections such as *endocarditis* (p. 118) and *mononucleosis* (p. 134) or chronic infections such as *malaria* (p. 113) or *leukemia* (p. 115). The spleen can be examined through palpation or percussion (p. 7) to detect splenic enlargement. In percussion, an enlarged spleen produces a distinctive dull sound. The patient history may also reveal important clues. For example, an individual with an enlarged spleen may report a feeling of fullness after eating a small meal, probably because the enlarged spleen limits gastric expansion.
- *Weakness* and *fatigue* typically accompany immunodeficiency disorders (p. 130), *Hodgkin's disease* and other *lymphomas* (p. 133), and *mononucleosis* (p. 134).
- *Skin lesions* such as hives, urticaria, or contact dermatitis (pp. 42, 43, 143), can develop during allergic reactions. Immune responses to a variety of allergens, including animal hair, pollen, dust, medications, and some foods, may cause such lesions.
- *Respiratory problems,* including rhinitis and wheezing, may accompany the allergic response to allergens such as pollen, hay, dust, and mildew. *Bronchospasms* are smooth-muscle contractions that constrict the airways and make breathing difficult. Bronchospasms, which often accompany severe allergic or asthmatic attacks, are a response to the appearance of antigens within the respiratory passageways.
- *Recurrent infections* occur for a variety of reasons. *Tonsillitis* (p. 130) and *adenoiditis* are common recurrent infections in children. More-serious infections are common among persons with immunodeficiency disorders such as AIDS (p. 138) or *severe combined immunodeficiency disease* (SCID) (p. 35). When the immune response is inadequate, the individual cannot overcome even a minor infection. Infections of the respiratory system are very common, and recurring, chronic gastrointestinal infections can produce chronic diarrhea. The pathogens involved may not affect persons with a normal immune response. Infections are also a problem for individuals who take medications that suppress the immune response. Examples of immunosuppressive drugs include anti-inflammatories such as the *corticosteroids (Prednisone®)* as well as more-specialized drugs such as *methotrexate* and *cyclosporine.*

FAP Ch. 22

When lymphatic circulatory functions are impaired, the most common sign is *lymphedema,* a tissue swelling caused by the buildup of interstitial fluid. Lymphedema can result from trauma and scarring of lymphatic vessels or from a lymphatic blockage due to a tumor or infections, including parasitic infection such as *filariasis* (FAP *p. 756*), or due to congenital malformations.

Diagnostic procedures and laboratory tests used to detect disorders of the lymphatic system are detailed in Tables A-29 and A-30 (p. 132).

DISORDERS OF THE LYMPHATIC SYSTEM

Disorders of the lymphatic system that affect the immune response can be sorted into three general categories, as diagrammed in Figure A-46•:

1. *Disorders resulting from an insufficient immune response.* This category includes immunodeficiency disorders, such as AIDS (p. 138) and SCID (p. 35). Individuals with depressed immune defenses can develop life-threatening diseases caused by microorganisms that are harmless to other individuals.
2. *Disorders resulting from an excessive immune response.* Conditions such as *allergies* (FAP *p. 738*) and *immune complex disorders* (p. 143) can result from an immune response that is out of proportion with the size of the stimulus.
3. *Disorders resulting from an inappropriate immune response. Autoimmune disorders* result when normal tissues are mistakenly attacked by T cells or the antibodies produced by activated B cells (FAP *p. 787*). For instance, in *thrombocytopenic purpura,* the body forms antibodies against its own platelets. We will consider representative disorders from each of these categories.

INFECTED LYMPHOID NODULES FAP *p. 759*

Lymphoid nodules can be overwhelmed by a pathogenic invasion. The result is a localized infection accompanied by regional swelling and discomfort. An individual with bacterial **tonsillitis** has infected tonsils. Symptoms include a sore throat, high fever, and leukocytosis (an abnormally high white blood cell count). The affected tonsil (normally, the pharyngeal tonsil) becomes swollen and inflamed, sometimes enlarging enough to partially block the entrance to the trachea. Breathing then becomes difficult or, in severe cases, impossible. If the infection proceeds, abscesses may develop within the tonsillar or peritonsillar tissues. The bacteria may enter the bloodstream by passing through the lymphatic capillaries and vessels to the venous system.

In the early stages, antibiotics may control the infection, but once abscesses have formed, the best treatment involves surgical drainage of the abscesses and tonsillectomy. **Tonsillectomy,** the removal of the tonsil, was once highly recommended and commonly performed to prevent recurring tonsillar infections. The procedure does reduce the incidence and severity of subsequent infections, but questions have arisen about

TABLE A-29 Representative Laboratory Tests for Disorders of the Lymphatic System and Immunity

Laboratory Test	Significance of Normal Values	Abnormal Values
Complete blood count	See laboratory tests for blood disorders, Table A-27c.	
WBC count	Adults: 5000–10,000/mm^3 (High-risk values: >30,000/mm^3 or <2500/mm^3)	Increased in chronic and acute infections, tissue death (MI, burns), leukemia, parasitic diseases, and stress; decreased in aplastic and pernicious anemias, systemic lupus erythematosus (SLE), and use of some medications
Differential WBC count	Neutrophils: 50–70%	Increased in acute infection, myelocytic leukemia, and stress; decreased in aplastic and pernicious anemias, viral infections, radiation treatment, and use of some medications
	Lymphocytes: 20–30%	Increased in chronic infections, lymphocytic leukemia, infectious mononucleosis, and viral infections; decreased in radiation treatment, AIDS, and corticosteroid therapy
	Monocytes: 2–8%	Increased in chronic inflammation, viral infections, and tuberculosis; decreased in aplastic anemia and corticosteroid therapy
	Eosinophils: 2–4%	Increased in allergies, parasitic infections, and some autoimmune disorders; decreased in steroid therapy
	Basophils: 0.5–1%	Increased in inflammatory processes and during healing; decreased in hypersensitivity reactions and corticosteroid therapy
Immunoglobulin electrophoresis (IgA, IgG, IgM)	Adults: IgA: 85–330 mg/dl IgG: 565–1765 mg/dl IgM: 55–145 mg/dl IgD and IgE: values should be minimal	Increased levels of IgG occur with infections; IgA levels increase with chronic infections and autoimmune disorders; IgE levels increase with allergic reactions and skin sensitivities; IgM levels are highest early in the immune response.
Antinuclear antibody	No antinuclear antibodies detected	Positive test occurs in up to 95% of persons diagnosed with SLE; false positive can occur with rheumatoid arthritis and other autoimmune disorders.
Anti-DNA antibody test	Low levels or none	Positive test with high levels of antibodies occurs in 40–80% of persons with SLE.
Total complement assay	Total complement: 41–90 hemolytic units C_3: 55–120 mg/dl C_4: 20–50 mg/dl	Total complement and C_3 are decreased in SLE and glomerulonephritis and increased in rheumatic fever, rheumatoid arthritis, and certain types of malignancies.
Rheumatoid factor test	Negative	Positive test in 80% of cases indicates rheumatoid arthritis, but results may also be positive in SLE and myositis, other inflammatory conditions.
AIDS serology		
Enzyme-linked immunosorbent assay (ELISA)	Negative	Positive test indicates detection of antibodies against HIV. Tests given in the early stages of infection yield a negative result; positive results may not develop for 6 months. HIV-positive status is assigned after two different tests are positive for the antibodies and the Western blot is positive.
Western blot	Negative	Positive test indicates detection of antibodies to specific viral proteins.

TABLE A-30 Representative Diagnostic Tests for Disorders of the Lymphatic System and Immunity

Diagnostic Procedure	Method and Results	Representative Uses
Skin tests		
Hypersensitivity response	Antigens are extracted and given in a sterile, diluted form; examples include animal dander, pollen, certain foods, medications that cause hypersensitivities (especially penicillin), and insect venom.	Detects specific antigens that may cause hypersensitivity reaction (allergy)
Prick test	A small amount of antigen is applied as a prick to the skin. Positive test: Erythema, hardening, and swelling appear around puncture area; usually the area affected is measured and must be of minimal size to qualify for a positive test.	As above
Intradermal test	Antigen is injected into the skin to form a 1–2 mm "bleb." Positive test: Within 15 minutes, a reddened wheal is produced that is larger than 5 mm in diameter and larger than those seen in the control group.	As above
Patch test	A patch is impregnated with the antigen and applied to the skin surface. Positive test: The patch provokes an allergic response, usually over several hours to several days.	As above
Exposure test	Positive test: A wheal of a certain diameter appears around the injection site within a specified period of time.	Diagnostic skin tests can be performed to identify prior exposure to fungal and other parasitic organisms; for example, the test for the organism that causes trichinosis can be performed and read in 15 minutes.
Tuberculin skin test	Injection of tuberculin protein into the skin. Positive test: Red, hardened area >10 mm wide appears around injection site 48–72 hours later.	Indicates presence of antibodies to the organism that causes tuberculosis (infection may be active or dormant).
Nuclear scan and CT scans of spleen	Scan images provided through energy from radiation emitted from radionuclides or through X-ray waves to reveal position, shape, and size of spleen	Detects abscesses, tumors of the liver or spleen, and infarcts of liver or splenic tissue
Biopsy of lymphoid tissue	Surgical excision of suspicious lymphoid tissue for pathological examination in the laboratory	Determines potential malignancy and staging of cancers in progress
Lymphangiography	Dye is injected into a lymphatic vessel in the distal portion of a limb and travels to nodes; X-rays are taken as the dye accumulates in the lymph nodes.	Identifies and determines the staging of Hodgkin's disease and other lymphomas and detects the cause of lymphedema (See Scan 17, p. 298.)

FAP Ch. 22

the overall cost to the individual. The tonsils are a first line of defense against bacterial invasion of the pharyngeal walls.

Appendicitis generally follows an erosion of the epithelial lining of the vermiform appendix. Several factors may be responsible for the initial ulceration—notably bacterial or viral pathogens. Bacteria that normally inhabit the lumen of the large intestine then cross the epithelium and enter the underlying tissues. Inflammation occurs, and the opening between the vermiform appendix and the rest of the intestinal tract may become constricted. Mucus secretion and pus formation accelerate, and the organ becomes increasingly distended. Eventually the swollen and inflamed appendix may rupture, or *perforate.* If it does, bacteria will be released into the warm, dark, moist confines of the abdominopelvic cavity, where they can cause a life-threatening peritonitis. The most effective treatment for appendicitis is the surgical removal of the organ, a procedure known as an **appendectomy.**

LYMPHOMAS

FAP p. 760

Lymphomas are malignant tumors consisting of cancerous lymphocytes or lymphocytic stem cells. Roughly 80,000 cases of lymphoma are diagnosed in the United States each year, and that number is steadily increasing. There are many types of lymphoma. One form, called **Hodgkin's disease (HD),** accounts for roughly 40 percent of all lymphoma cases. Hodgkin's disease most commonly strikes individuals at ages 15–35 years or those over age 50. The reason for this pattern of incidence is unknown. Although the cause of the disease is uncertain, an infectious agent (probably a virus) is suspected.

Other types of lymphoma are usually grouped together under the heading of **non-Hodgkin's lymphoma (NHL).** They are extremely diverse. More than 85 percent of NHL cases are associated with chromosomal abnormalities. They typically involve *translocations,* in which sections of chromosomes have been swapped from one chromosome to another. The shifting of genes from one chromosome to another interferes with the normal regulatory mechanisms, and the cells become cancerous. The nature of the cancer depends on which of the many types of lymphocyte are affected. A combination of inherited and environmental factors may be responsible for specific translocations. For example, one form, called **Burkitt's lymphoma,** develops only after genes from chromosome 8 have been translocated to chromosome 14. (There are at least three variations.) Burkitt's lymphoma normally affects male children in Africa and New Guinea who have been infected with the *Epstein–Barr virus (EBV).* This highly variable virus is also responsible for infectious mononucleosis (described below), and it has been suggested as a possible cause of chronic fatigue syndrome.

The EBV infects B cells, but under normal circumstances the infected cells are destroyed by the immune system. This virus is widespread in the environment, and childhood exposure generally produces lasting immunity. Children who develop Burkitt's lymphoma may have a genetic susceptibility to EBV infection; in addition, the presence of another illness, such as malaria, can weaken the immune system enough that the lymphoma cells are ignored and therefore are not destroyed.

The first symptom usually associated with any lymphoma is a painless enlargement of lymph nodes. The involved nodes have a firm, rubbery texture. Because the nodes are pain-free, the condition is typically overlooked until it has progressed far enough for secondary symptoms to appear. For example, patients seeking help for recurrent fevers, night sweats, gastrointestinal or respiratory problems, or weight loss may be unaware of any underlying lymph node changes. In the late stages of the disease, symptoms can include liver or spleen enlargement, central nervous system dysfunction, pneumonia, a variety of skin conditions, and anemia.

In planning treatment, clinicians consider the histological structure of the nodes and the stage of the disease. In a biopsy, the node is described as *nodular* or *diffuse.* A nodular node retains a semblance of normal structure, with follicles and germinal centers. In a diffuse node, the interior of the node has changed, and follicle structure has broken down. In general, the nodular lymphomas progress more slowly than do the diffuse forms, which tend to be more aggressive. Conversely, the nodular lymphomas are more resistant to treatment and are more likely to recur even after remission has been achieved.

The most important factor influencing treatment selection is the stage of the disease. Table A-31 presents a simplified staging classification of lymphomas. When the condition is diagnosed early (stage I or II), localized therapies may be effective. For example, the cancerous node(s) may be surgically removed and the

TABLE A-31 Cancer Staging in Lymphoma

Stage I. Involvement of a single node or region (or of a single extranodal site).

Typical treatment: surgical removal and/or localized irradiation; in slowly progressing forms of non-Hodgkin's lymphoma, treatment may be postponed indefinitely.

Stage II. Involvement of nodes in two or more regions (or of an extranodal site and nodes in one or more regions) on the same side of the diaphragm.

Typical treatment: surgical removal and localized irradiation that includes an extended area around the cancer site (the *extended field*).

Stage III. Involvement of lymph node regions on both sides of the diaphragm. This is a large category that is subdivided on the basis of the organs or regions involved. For example, in stage III, the spleen contains cancer cells.

Typical treatment: combination chemotherapy, with or without radiation; radiation treatment may involve irradiating all thoracic and abdominal nodes plus the spleen (*total axial nodal irradiation,* or *TANI*).

Stage IV. Widespread involvement of extranodal tissues above and below the diaphragm; involvement of bone marrow.

Treatment: highly variable, depending on the circumstances. Combination chemotherapy is always used; it may be combined with whole-body irradiation. The "last resort" treatment involves massive chemotherapy followed by a bone marrow transplant.

region(s) irradiated to kill residual cancer cells. Success rates are very high when a lymphoma is detected in an early stage. For Hodgkin's disease, localized radiation can produce remission that lasts 10 years or more in more than 90 percent of patients. The treatment of localized NHL is somewhat less effective. The 5-year remission rates average 60–80 percent for all types; success rates are higher for nodular forms than for diffuse forms.

Although these results are encouraging, few lymphomas are diagnosed in the early stages. For example, only 10–15 percent of NHL patients are diagnosed at stage I or II. For lymphomas at stages III and IV, most treatments involve chemotherapy. Combination chemotherapy, in which two or more drugs are administered simultaneously, is the most effective treatment. For Hodgkin's disease, a four-drug combination with the acronym **MOPP** (nitrogen **M**ustard, **O**ncovin® *[vincristine]*, **P**rednisone, and **P**rocarbazine) produces lasting remission in 80 percent of patients.

A **bone marrow transplant** is a treatment option for acute, late-stage lymphoma. When suitable donor marrow is available, the patient receives whole-body irradiation, chemotherapy, or some combination of the two sufficient to kill tumor cells throughout the body. This treatment also destroys normal bone marrow cells. Donor bone marrow is then infused. Within 2 weeks, the donor cells colonize the bone marrow and begin producing red blood cells, granulocytes, monocytes, and lymphocytes.

Potential complications of this treatment include the risk of infection and bleeding while the donor marrow is becoming established. The immune cells of the donor marrow may also attack the tissues of the recipient, a response called **graft-versus-host disease (GVH).** For a person with stage I or II lymphoma, without bone marrow involvement, bone marrow can be removed and stored (frozen) for over 10 years. If other treatment options fail or the person comes out of remission at a later date, an autologous marrow transplant can be performed. This procedure eliminates both the need for finding a matched donor and the risk of GVH disease.

FAP Ch. 22

DISORDERS OF THE SPLEEN FAP p. 762

The spleen responds like a lymph node to infection, inflammation, or invasion by cancer cells. The enlargement that follows is called **splenomegaly** (splen-ō-MEG-a-lē; *megas,* large); the spleen can rupture under these conditions. One relatively common condition that causes temporary splenomegaly is **mononucleosis.** This condition, also known as the "kissing disease," results from infection by the Epstein–Barr virus. In addition to enlargement of the spleen, symptoms of mononucleosis include fever, sore throat, widespread swelling of lymph nodes, increased numbers of atypical lymphocytes in the blood, and the presence of circulating antibodies to the virus. The condition typically affects young adults (ages 15–25 years) in the spring or fall. Treatment is symptomatic, as no drugs are effective against this virus. The most dangerous aspect of the disease is the risk of rupturing the enlarged spleen, which becomes fragile. Patients are therefore cautioned against heavy exercise and other activities that increase abdominal pressures. If the spleen does rupture, severe hemorrhaging can occur. Death will follow unless transfusion and an immediate splenectomy are performed.

An individual whose spleen is missing or nonfunctional has **hyposplenism** (hī-pō-SPLĒN-izm). Hyposplenism usually does not pose a serious problem. Hyposplenic individuals, however, are more prone to some bacterial infections, including infection by *Streptococcus pneumoniae,* than are individuals with normal spleens, so immunization against *S. pneumoniae* is recommended. In **hypersplenism,** the spleen becomes overactive; the increased phagocytic activities lead to anemia (low RBC count), leukopenia (low WBC count), and thrombocytopenia (low platelet count). Splenectomy is the only known cure for hypersplenism.

A CLOSER LOOK: THE COMPLEMENT SYSTEM FAP p. 767

The complement system contains 11 proteins (Figure A-47•). The proteins C1q, C1r, and C1s circulate as a single complex designated as C1. The proteins C2–C9 are individual proteins.

COMPLICATIONS OF INFLAMMATION FAP p. 770

When bacteria invade the dermis of the skin, the production of cellular debris and toxins reinforces and exaggerates the inflammation process. **Pus** is an accumulation of debris, fluid, dead and dying cells, and necrotic tissue components. Pus commonly forms at an infection site in the dermis; an **abscess** is an accumulation of pus in an enclosed tissue space. In the skin, an abscess can form as pus builds up inside the fibrin clot that surrounds the injury site. If the cellular defenses succeed in destroying the invaders, the pus will be either absorbed or surrounded by a fibrous capsule; this capsule is one type of **cyst.** (Cysts can also form in the absence of infection.)

Erysipelas (er-i-SIP-e-las; *erythros,* red + *pella,* skin) is a widespread inflammation of the dermis caused by bacterial infection. If the inflammation spreads into the subcutaneous layer and deeper tissues, the condition is called **cellulitis** (sel-ū-LĪ-tis). Erysipelas and cellulitis develop when bacterial invaders break through the fibrin wall. The bacteria involved produce large quantities of the enzymes *hyaluronidase,* which liquifies the

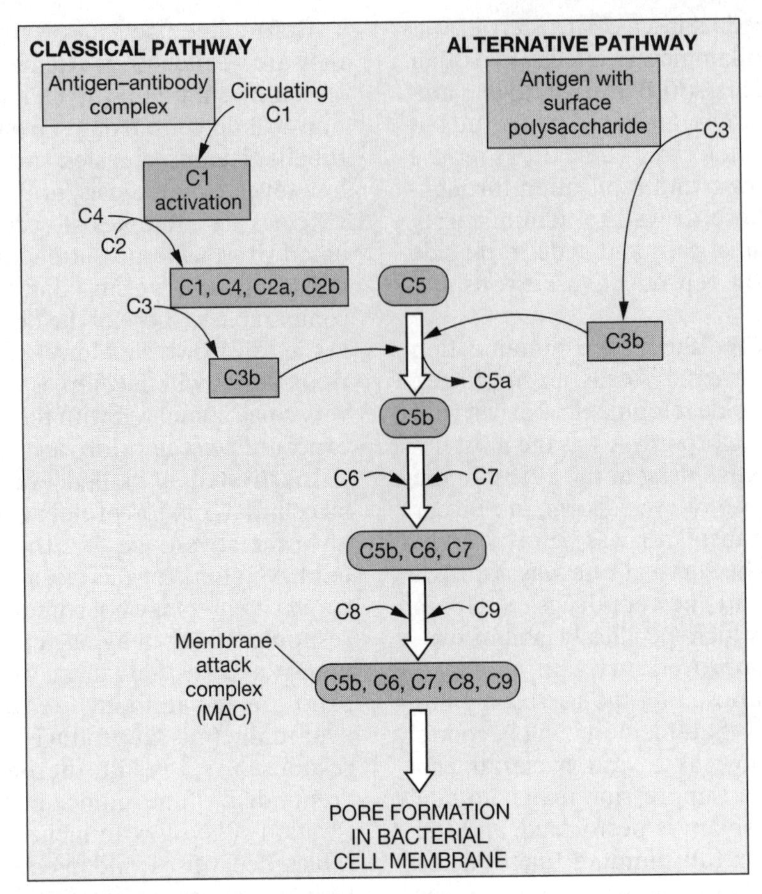

•**FIGURE A-47** The Complement Pathways

ground substance, and *fibrinolysin,* which breaks down fibrin and prevents clot formation. These are serious conditions that require prompt antibiotic therapy.

TRANSPLANTS AND IMMUNOSUPPRESSIVE DRUGS FAP p. 777

Organ transplantation is an important treatment option for patients with severe disorders of the kidneys, liver, heart, lungs, or pancreas. In early 2000, roughly 68,000 persons were awaiting organ transplants in the United States. However, in 1998, the most recent year for which data are available, U.S. surgeons transplanted approximately 12,000 kidneys, 4500 livers, 2400 hearts, 900 lungs, 50 heart–lung combinations, 1000 kidney–pancreas combinations, and 250 pancreases. Clearly there is a serious disparity between the demand for organ transplantation and the number of surgeries performed. The problem is that suitable organs are in short supply due to a lack of organ donors. During 1998, more than 30,000 patients who could have benefited from a kidney transplant did not receive one.

Graft rejection will not occur if the major histocompatibility complex (MHC) proteins of two individuals are identical; for this reason, most grafts made between identical twins are successful. The greater the difference in MHC structure between the donor and recipient, the greater the likelihood that the graft will be rejected. The process of tissue typing assesses the degree of similarity between the MHC complexes of two individuals. For this process, lymphocytes are collected and examined; they can easily be obtained from the blood and bear both Class I and Class II MHC molecules. There are many different MHC proteins, and finding a perfect match among nonidentical siblings, other relatives, or nonrelated persons can be difficult. Full siblings have only a 25 percent chance of being complete matches. For patients with a diverse ethnic background, finding a matched unrelated donor is almost impossible. With more people being tested and put on the international bone marrow donor registry, there is more hope for these patients. For example, in 1996 an adopted Korean-American soldier found a suitable donor in Korea, and a Chinese-American child from Hawaii found a donor in Taiwan. Unlike organ donors, bone marrow donors do not lose an irreplaceable organ. The bone marrow stem cells left in the donor soon replace the donated bone marrow, and the donor's health is unaffected.

Until recently, the drugs used to produce immunosuppression did not selectively target the immune

FAP Ch. 22

system. For example, Prednisone, a corticosteroid, was used for its broad anti-inflammatory effects. Two other drugs, **cyclophosphamide** (sī-klō-FOS-fa-mīd) and **azathioprine** (a-za-THĪ-ō-prēn), are more powerful, but they have greater associated risks. These drugs reduce the rates of cellular growth and replication throughout the body. When these drugs are administered, hematopoiesis slows dramatically and undesirable side effects may develop in the reproductive, nervous, and integumentary systems.

An understanding of the chemical communication among helper T and suppressor T cells, macrophages, and B cells has led to the development of drugs with more-selective effects. Cyclosporin A was the most important immunosuppressive drug in the 1980s. In the early 1980s, before the use of cyclosporin, the 5-year survival rate for liver transplants was below 20 percent. In the mid-1990s, the survival rate was approximately 80 percent, and newer drugs, such as **methotrexate** (*FK-506*), which specifically inhibits lymphocytes, have further improved survival.

An obvious problem posed by the use of any immunosuppressive drug is that the individual becomes more susceptible than normal to viral, bacterial, and fungal infections. Immunosuppression may continue indefinitely after a transplant is performed, and the patient may not recover full immune function for months after treatment has been discontinued. Over this period, the individual must be monitored for infection and, when necessary, treated with antibiotics.

A more subtle risk of immunosuppression is the reduction in immune surveillance by natural killer (NK) cells. Transplant patients are 100 times more likely to develop cancer than are others in their age group. Lymphoma-type cancers are the most common; these cancers appear to be linked to post-transplant infection with the Epstein–Barr virus. David, the SCID boy raised in a sterile bubble, died of a B-cell lymphoma after receiving a bone marrow transplant from his sister, who had an asymptomatic case of mononucleosis.

FAP Ch. 22

IMMUNIZATION FAP *p. 782*

Immunization is the manipulation of the immune system by the administration of antigens under controlled conditions or by the administration of antibodies that can combat an existing infection. In **active immunization,** a primary response to a particular pathogen is intentionally stimulated before an individual encounters the pathogen in the environment. The result is lasting immunity against that pathogen. Immunization is accomplished by the administration of a **vaccine,** a preparation of antigens derived from a specific pathogen. A vaccine is given orally or by intramuscular or subcutaneous injection. Most vaccines consist of the pathogen in whole or in part—living or dead. In some cases, a vaccine contains one of the metabolic products of the pathogen.

Before live bacteria or viruses are administered, they are weakened, or **attenuated** (a-TEN-ū-ā-ted), to lessen or eliminate the chance that a serious infection will develop from exposure to the vaccine. The rubella, mumps, measles, smallpox, chicken pox, yellow fever, tuberculosis, oral typhoid, and oral polio vaccines are examples of vaccines that use live attenuated viruses. Despite attenuation, the administration of live microorganisms may produce mild symptoms comparable to those of the disease itself, such as a low-grade fever or rash. However, the likelihood that serious illness will develop as a result of vaccination is very small compared with the risks posed by pathogen exposure *without* prior vaccination.

Inactivated, or "killed," vaccines consist of bacterial cell walls or viral protein coats only. These vaccines have the advantage that they cannot produce even mild symptoms of the disease. Unfortunately, inactivated vaccines may not stimulate as strong an immune response and so may not confer as long-lasting immunity as do live-organism vaccines. In the years after exposure, the antibody titer declines and the system eventually fails to produce an adequate secondary response. As a result, the immune system must be "reminded" of the antigen periodically by the administration of *boosters.* Influenza, typhoid, typhus, plague, rabies, hepatitis A, and injected polio vaccines use inactivated viruses or bacteria. In some cases, fragments of the bacterial or viral walls, or their toxic products, can be used to produce a vaccine. The tetanus, diphtheria, pneumonia, and hepatitis B vaccines are examples. Data about attenuated and inactivated vaccines are presented in Table A-32.

Gene-splicing techniques can now be used to incorporate antigenic compounds from pathogens into the cell walls of harmless bacteria. When exposed to these bacteria, the immune system responds by producing antibodies and memory B cells that are equally effective against the engineered bacterium and the pathogen.

Passive immunization is normally used if the individual has already been exposed to a dangerous pathogen or toxin and there is not enough time for active immunization to take effect. In passive immunization, the patient receives a dose of antibodies that will attack the pathogen and overcome the infection, even without the help of the host's own immune system. Passive immunization provides only short-term resistance to infection, because the antibodies are gradually removed from circulation and are not replaced.

The antibodies provided during passive immunization have traditionally been acquired by collecting and combining antibodies from the sera of many other individuals. This *pooled sera* is used to obtain large quantities of antibodies, but the procedure is very expensive. In addition, improper treatment of the sera carries the risk of accidental transmission of an infectious agent, such as a hepatitis virus or the virus responsible for AIDS. Antibodies can also be obtained from the blood

TABLE A-32 Immunizations Currently Available

Immunization Target	Type of Immunity Provided	Vaccine Type	Remarks
Viruses			
Poliovirus	Active	Live, attenuated	Oral
	Active	Inactivated	Boosters every few years
Chickenpox	Active	Live, attenuated	
Rubella	Active	Live, attenuated	
	Passive	Human antibodies (pooled)	
Mumps	Active	Live, attenuated	
Measles (rubeola)	Active	Live, attenuated	May need second booster
	Passive	Human antibodies (pooled)	
Hepatitis A	Active	Inactivated	May need second booster
	Passive	Human antibodies (pooled)	
Hepatitis B	Active	Inactivated	May need periodic boosters
	Passive	Human antibodies (pooled)	
Smallpox	Active	Live, attenuated (related) virus	Boosters every 10 years (no longer required, as disease has been eradicated)
Yellow fever	Active	Live, attenuated	Boosters every 10 years
Herpes zoster	Passive	Human antibodies (pooled)	
Haemophilus influenza B (HIB)	Active	Inactivated	
Rabies	Passive	Human antibodies (pooled)	
	Passive	Horse antibodies	
	Active	Inactivated	Boosters required
Bacteria			
Typhoid	Active	Inactivated or live, attenuated	Boosters every 2, 3, or 5 years, depending on the vaccine type
Tuberculosis	Active	Live, attenuated	
Plague	Active	Inactivated or live, attenuated	Boosters every 1–2 years
Tetanus	Active	Toxins only	Boosters every 5–10 years
	Passive	Human antibodies (pooled)	
Diphtheria	Active	Toxins only	Boosters every 10 years
	Passive	Horse antibodies	
Pertussis	Active	Antigens only	Not used over age 6 years
Streptococcal pneumonia	Active	Bacteria and cell wall components	
Botulism	Passive	Horse antibodies	
Rickettsia: typhus	Active	Inactivated	Boosters yearly if high risk of exposure
Lyme disease	Active	Inactivated	
Other			
Snake bite	Passive	Horse antibodies	
Spider bite	Passive	Horse antibodies	
Venomous fish spine	Passive	Horse antibodies	

of a domesticated animal (typically a horse) exposed to the same antigen. Unfortunately, recipients may suffer allergic reactions to horse serum proteins.

At present, antibody preparations are available to treat hepatitis A, hepatitis B, herpes zoster, diphtheria, tetanus, rabies, measles, rubella, botulism, and the venoms of certain fish, snakes, and spiders. Gene-splicing technology can also be used to reproduce pure antibody preparations free of antigenic or viral contaminants; this technology should eventually eliminate the need for pooled or foreign plasma.

Notice that passive immunity occurs naturally during fetal development. At that time, maternal IgG antibodies can cross the placental barriers and enter the

fetal circulation, where they persist for 3–6 months after birth.

LYME DISEASE

FAP *p. 784*

In November 1975, the town of Lyme, Connecticut, experienced an epidemic of adult arthritis and juvenile arthritis. Between June and September, 59 cases were reported, 100 times the statistical average for a town of its size. Symptoms were unusually severe; in addition to joint degeneration, victims experienced chronic fever and a prominent rash that began as a red bull's-eye centered around what appeared to be an insect bite (Figure A-48•). It took almost 2 years to track down the cause of this condition, now called **Lyme disease.**

Lyme disease is caused by the bacterium *Borrelia burgdorferi,* which normally lives in white-footed mice. The disease is transmitted to humans and other mammals by the bite of a tick that harbors that bacterium. Deer, which can carry infected adult ticks without becoming ill, have helped spread infected ticks through populated areas. The high rate of infection among children reflects the fact that they play outdoors during the summer in fields where deer may also be found. Children are thus more likely to encounter—and be bitten by—infected ticks. After 1975, the Lyme disease problem became regional and then national in scope. The incidence in the United States now averages 12,500 cases per year, and Lyme disease has also been reported in Europe.

Although some of the joint destruction results from the deposition of immune complexes by a mechanism comparable to that for rheumatoid arthritis, many of the symptoms (fever, pain, skin rash) develop in response to the release of interleukin-1 (IL-1) by activated macrophages. The cell walls of *B. burgdorferi* contain lipid–carbohydrate complexes that stimulate the secretion of IL-1 in large quantities. By stimulating the body's specific and nonspecific defense mechanisms, IL-1 exaggerates the inflammation, rash, fever, pain, and joint degeneration associated with the primary infection. Treatment for Lyme disease consists of the administration of antibiotics and anti-inflammatory drugs. A vaccine is now available.

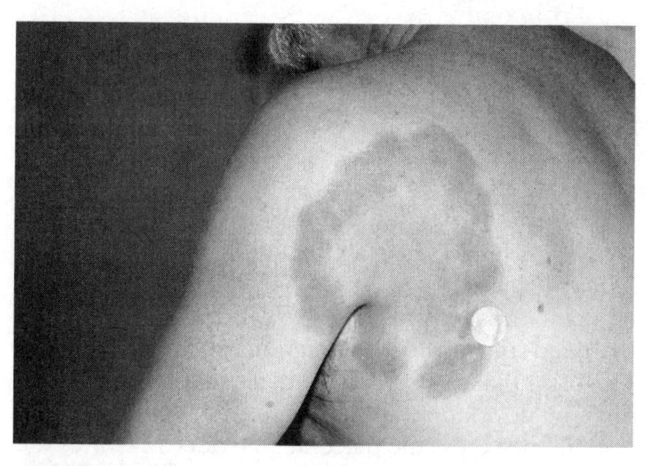

•**FIGURE A-48** **Bull's-eye Rash of Lyme Disease**

FAP Ch. 22

AIDS

FAP *pp. 784, 790*

Acquired immune deficiency syndrome (AIDS), or *late-stage HIV disease,* develops after infection by the *human immunodeficiency virus (HIV).* There are at least three types of HIV, designated *HIV-1, HIV-2,* and *HIV-3.* Most people with HIV in the United States are infected with HIV-1; HIV-2 infections are most common in West Africa. Because most of those infected with HIV-1 eventually develop AIDS but not all individuals infected with HIV-2 do so, HIV-2 may be a less dangerous virus. The distribution and significance of HIV-3 infection remain to be determined.

The discussion that follows, based on information pertaining to HIV-1 infection, expands on the discussion on *p. 790* of the main text.

Symptoms of HIV Disease

The initial infection may produce a flu-like illness with fever and swollen lymph nodes a few weeks after exposure to the virus. This exposure generally triggers the production of antibodies against the virus. These antibodies appear in the blood within 2–6 months of exposure; as we shall soon see, antibody tests can be used to detect infection. Further symptoms may not appear for 5–10 years or more. During this period, the virus content of the blood varies, but the viruses are at work within lymphoid tissues, especially in the lymph nodes. There is a steady decline in the numbers of CD4 T cells and, for reasons as yet unknown, of dendritic cells in lymphoid tissues.

The Centers for Disease Control and Prevention (CDC) in Atlanta, which monitors infectious diseases, now recognizes three categories of HIV disease on the basis of CD4 T cell counts:

1. In *Category 1,* the CD4 T cell count is above 500/μl, and these cells account for at least 29 percent of the total circulating lymphocytes.
2. In *Category 2,* the CD4 T cell count is between 200 and 499/μl, and these cells account for 14–28 percent of the total circulating lymphocytes.
3. In *Category 3,* the CD4 T cell count is below 200/μl, and these cells account for 14 percent or less of the total circulating lymphocytes. This category, **late-stage HIV disease,** includes all patients with the condition commonly known as AIDS.

Symptoms commonly do not appear until the CD4 T cell concentration falls below 500/μl (Category 2). The symptoms that first appear are typically mild, consisting of lymphadenopathy, chronic nonfatal infections, diarrhea, and weight loss. A person with AIDS (Category 3) develops a variety of life-threatening infections and disorders, and the average life expectan-

cy after diagnosis (without treatment) is 2 years. The life expectancy is short because HIV-1 selectively infects helper T cells. The reduction in helper T cell activity impairs the immune response; the effect is magnified, because suppressor T cells are relatively unaffected by the virus. Over time, circulating antibody levels decline, cellular immunity is reduced, and the body is left without defenses against a wide variety of bacterial, viral, fungal, and protozoan invaders.

This vulnerability is what makes AIDS dangerous. The effects of HIV on the immune system are not by themselves life threatening, but the infections that result when the immune system is weakened certainly are. With the depression of immune function, ordinarily harmless pathogens can initiate lethal infections known as *opportunistic infections.* In fact, the most common and dangerous pathogens for an individual with AIDS are microorganisms that seldom cause illnesses in individuals with normal immune systems. Persons with AIDS are especially prone to lung infections and pneumonia, commonly caused by infection with the fungus *Pneumocystis carinii.* These patients are also subject to a variety of other fungal infections, such as *cryptococcal meningitis,* and an equally broad array of bacterial and viral infections. Because infected persons are virtually defenseless, the symptoms and time course of these infections are very different from those in noninfected individuals.

In addition to pathogenic invasions, immune surveillance is depressed and the risk of cancer increases—as in transplant patients, who have to take immunosuppressive drugs (p. 130). One of the most common cancers in persons with AIDS is *Kaposi's sarcoma,* a condition that is extremely rare in uninfected individuals. Kaposi's sarcoma typically begins with rapid cell divisions in endothelial cells of cutaneous blood vessels. Associated lesions, blue or a deep brown-purple, generally appear first in the hands or feet and later occur closer to the trunk. In a small number of infected persons, the lesions develop in the epithelium of the digestive or respiratory tract. In normal individuals, the tumor usually does not metastasize; in individuals with AIDS, whose immune systems are relatively ineffective, the tumor typically converts to an aggressive, invasive cancer.

If a person with AIDS survives all these assaults, infection of the CNS by HIV eventually produces neurological disorders and a progressive dementia. So far, AIDS has been almost invariably fatal, and most of those who carry HIV may eventually die of AIDS. A few long-term survivors test as HIV-positive but have no signs of abnormal immune function. Their relative immunity has been linked to the presence of two mutated copies of a gene that directs the synthesis of an integral membrane protein. The protein, *CC-CKR-5,* is involved in the attachment and penetration of HIV. The most common strains of HIV cannot infect the immune cells of an individual with two mutant copies of this gene. If the individual has one normal gene and one mutant gene, infection occurs but the disease progresses more slowly than it does in individuals with two normal CC-CKR-5 genes.

Incidence

During 2000, an estimated 48,000 cases of AIDS will be diagnosed in the United States; the number of U.S. deaths attributed to this disorder since its discovery in 1982 exceeds 350,000. AIDS is now the leading cause of death in people aged 25–44 years. The CDC estimates that the number of AIDS cases in the United States will continue to increase (Figure A-49•). As of early 2000, there were approximately 746,000 people with AIDS in the U.S. alone.

Because HIV can remain in the body for years without producing clinical symptoms, the number of individuals infected and at risk is certain to be far higher than the number of reported cases. An estimated 1–2 million Americans are infected with HIV, and almost all will eventually develop AIDS, probably within the next decade. The numbers worldwide are even more frightening. The World Health Organization (WHO) estimates that as many as 60 million people have been infected with HIV over the last 20 years, and an estimated 18 million have died of AIDS. The number of persons with AIDS worldwide is expected to continue to climb rapidly. Several African nations are already on the verge of social and economic collapse due to devastation by AIDS. For instance, one-third of the population of Malawi is infected, and 60 percent of pregnant women there carry the virus; Botswana has similar statistics.

Modes of Infection

Infection with HIV occurs through intimate contact with the body fluids of an infected individual. Although all body fluids, including saliva, tears, and breast milk, carry the virus, the major routes of infection involve contact with blood, semen, or vaginal secretions. The transmission pattern has been analyzed for adults and adolescents with AIDS in the United States. Four major transmission routes have been identified: (1) sexual transmission, (2) intravenous drug use, (3) receipt of blood or tissue products, and (4) prenatal exposure.

Sexual Transmission In approximately 62 percent of all cases, exposure occurred through sexual contact with an infected individual. Male homosexual contact was involved in 57 percent of all cases (compared with 5 percent for heterosexual contact), and the ratio of males to females with AIDS is approximately 9 to 1. Canada, Europe, South America, Australia, and New Zealand exhibit a comparable transmission pattern. In the Caribbean and Africa, in contrast, AIDS began in the heterosexual community. It affects heterosexual men and women in roughly equal numbers, and the gender ratio for AIDS infection in those countries is approximately 1 to 1.

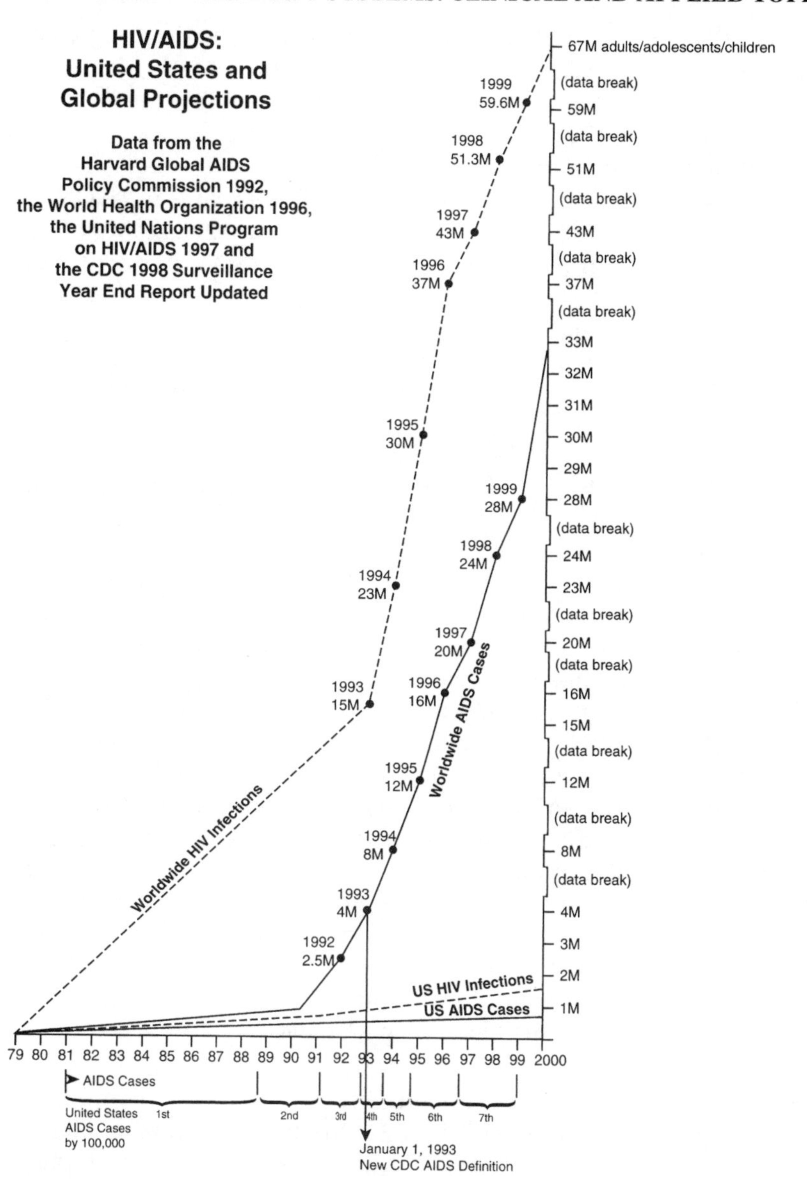

•**FIGURE A-49 United States and Global Projections for Total Number of HIV and AIDS Cases.** [From Gerald J. Stine (2000). *AIDS Update 2000.* © Prentice Hall, Upper Saddle River, NJ.]

Genital ulcers from other sexually transmitted diseases, such as syphilis, chancroid, or herpes, also increases the risk of HIV transmission.

Because homosexual transmission is predominant in the United States, many people still consider AIDS to be a disease of homosexuals. It is not. Over time, the number of cases in the heterosexual population has been steadily increasing; since 1987, the percentage of homosexual or bisexual AIDS patients has dropped by roughly 25 percent and the number of cases transmitted by heterosexual contact has steadily increased. It can be anticipated that the gender ratio will continue to shift toward the 1:1 male-to-female ratio typical of Africa and the Caribbean.

Intravenous Drug Use Roughly 19 percent of persons with AIDS contracted the disease by the shared use of needles. (Another 6 percent were homosexual males who shared needles with other drug users, making it unclear which risk factor was responsible for transmission.) Although only small quantities of blood are inadvertently transferred when needles are shared, this practice injects HIV directly into the bloodstream. It is thus a very effective way to transmit the infection.

Receipt of Blood or Tissue Products About 3 percent of persons with AIDS became infected with HIV after they received a transfusion of contaminated whole blood or plasma; an infusion of blood products, such as platelets or extracts of pooled sera for the treatment of hemophilia; or an organ transplant from an infected individual. With careful screening of blood and blood products, the rate of new transmission by this route is now essentially zero in the United States.

Two factors may account for the pattern of transmission observed in the United States. First, the disease appears to have spread through the homosexual community before it entered the heterosexual population via bisexual males or via intravenous drug use. Second, statistical evidence indicates that, on a per-exposure basis, the risk of male-to-male or male-to-female sexual transmission is many times greater than the risk of female-to-male transmission. As a result, it may take longer to spread the virus through the heterosexual population (male A to female A to male B) than through the homosexual population (male A to male B to male C). Whether homosexual or heterosexual contact is involved, a sex partner whose epithelial defenses are weakened is at increased risk of infection. This factor accounts for the relatively higher rate of transmission by anal intercourse, which tends to damage the delicate lining of the anorectal canal.

Prenatal Exposure In the United States, approximately 2000 infants are born each year already infected with HIV. Although this number is small compared with the total number of HIV-infected people, it is increasing rapidly. An untreated pregnant woman infected with HIV has a 20–33 percent chance of infecting her baby across the placenta. Treatment can reduce the rate of maternal–fetal transmission to 1 percent or less. As HIV spreads through the heterosexual population, more pregnant women will become infected, and maternal–fetal transmission will become more common. These unfortunate infants will place social and financial stresses on our society for the foreseeable future.

Prevention of AIDS

The best defense against AIDS is to avoid exposure to HIV. The most obvious precaution is to avoid sexual contact with infected individuals. All forms of sexual intercourse carry the potential risk of virus transmission. The use of synthetic (latex) condoms has been recommended when you do not know the sexual history of a partner. (Condoms that are not made of synthetic materials are effective in preventing pregnancy but do not block the passage of viruses.) Although condom use does not provide absolute protection, it drastically reduces the risk of infection. Contrary to popular belief, significant risks are associated with oral sex as well as other forms of sexual contact.

Attempts are under way to ensure that blood and blood products are adequately screened for the presence of HIV-1. A simple blood test exists for the detection of HIV-1 antibodies; a positive reaction indicates previous exposure to the virus. The assay, an example of an **ELISA** (enzyme-linked immunosorbent assay) **test,** is now used to screen blood donors, reducing the risk of infection by transfusion or the use of blood products from pooled sera. Pooled sera can also be heat-treated by exposure to temperatures sufficient to kill the virus but too low to denature blood proteins permanently.

Most public health facilities will perform the ELISA test on request for individuals who fear that they may have been exposed to HIV. Unfortunately, the test is not 100 percent reliable; false positive reactions occur at a rate of about 0.4 percent. In addition, the ELISA test does not detect HIV-2 or HIV-3. In the event of a positive test result, a retest should be performed by using the more sensitive **Western blot** procedure. Because the incubation period is variable, a positive test for HIV infection does not mean that the individual has AIDS. It does mean that the individual is likely to develop AIDS some time in the future and is a carrier capable of infecting others. By the time an individual develops AIDS, he or she is obviously sick and usually has little interest in sexual activity. In terms of the spread of this disease, the most-dangerous individuals are those who seem perfectly healthy and have no idea that they are carrying the virus.

Despite intensive efforts, a vaccine has yet to be developed that will provide immunity from HIV infection, although 40 clinical trials are under way. Current research programs are attempting to stimulate antibody production in response to (1) killed but intact viruses; (2) fragments of the viral envelope; (3) HIV proteins on the surfaces of other, less dangerous viruses; or (4) T cell proteins that are targeted by HIV. (The last approach is based on the assumption that the antibodies produced will cover the binding sites, preventing viral attachment and penetration.)

Treatment

There is no cure for AIDS. However, the length of survival for a person with AIDS has been steadily increasing, because (1) new drugs are available that slow the progression of the disease and (2) improved antibiotic therapies have helped overcome infections that would otherwise prove fatal. This combination is extending lives as the search for more-effective treatment continues. However, overcoming a bacterial infection in an individual with AIDS might require antibiotic doses up to 10 times greater than those used to fight infections in HIV-free individuals. Moreover, once the infection has been overcome, the patient may have to continue taking that drug for the rest of his or her life. As a result, some AIDS patients take 50–100 pills a day just to prevent recurrent infections.

The antiviral drug *azidothymidine,* or *AZT* (sold as *zidovudine* or *Retrovir®*), can slow the progression of AIDS. Low doses of AZT can be effective in delaying the transition of HIV disease to AIDS. Higher doses are used to treat AIDS, but they can lead to a variety of unpleasant side effects, including anemia and even bone marrow failure. However, AZT is effective in reducing the neurological symptoms of AIDS, because it can cross the blood–brain barrier to reach infected tissues of the central nervous system. Side effects are reduced when AZT treatment is alternated with other antiviral drugs, including *ddC (dideoxycytidine), ddI (dideoxyinosine), d4T,* and *3TC.* AZT and these four drugs inhibit the action of *reverse transcriptase,* the enzyme used to insert viral genes into the cell's DNA strands. (See the discussion of viral replication on p. 25.)

Protease inhibitors, such as *ritonavir, indinavir,* and *saquinavir mesylate,* prevent the assembly and dispersal of new viruses. Alone or in combination with AZT or other drugs, protease inhibitors can dramatically reduce the viral content of the blood, although viruses remain within infected immune cells. Treatment with protease inhibitors has dramatically reduced symptoms and extended the lives of individuals with AIDS. However, because some virus-infected cells are not destroyed, protease inhibitors so far have been unable to cure the disease.

Other viral inhibitors, such as *efavirenz* and *nevirapine,* represent a third category of anti-HIV drugs. Combined treatment known as *HAART* (*h*ighly *a*ctive *a*nti-*r*etroviral *t*herapy) involves two, three, or more antiviral drugs with different specific actions. HAART can reduce HIV viral counts and increase CD4 T cell counts to levels that prevent opportunistic infections for years. With more and different drugs available, various combinations may eventually provide prolonged control, if not an actual cure, of HIV infection.

AIDS and the Cost to Society

Treatment of AIDS is complex and expensive. The average cost is $70,000 per patient per year; the total projected annual cost is thus roughly $52 billion. As more individuals become infected and more of the

FAP Ch. 22

1–2 million HIV-infected people develop AIDS, the costs will continue to escalate rapidly.

There are other, less obvious costs to society. Individuals with depressed immune systems get sick more often and remain sick longer and, as a result, are more likely to spread the infecting pathogen to healthy individuals. Some diseases, such as tuberculosis, that had been present at very low levels in the U.S. population are now occurring with increased frequency. In part, the reason for this increase is that AIDS patients are succumbing to infection, but it also reflects the spread of these diseases through transmission from persons with AIDS into the general population. Long-term use of multiple antibiotics in AIDS patients can also promote the development of more-resistant disease organisms (which is always a risk of chronic antibiotic use; see p. 151).

TECHNOLOGY, IMMUNITY, AND DISEASE

FAP *p. 784*

Our understanding of disease mechanisms has been profoundly influenced by recent advances in genetic engineering. The information and technical capabilities developed during the 1980s have already affected clinical procedures. This trend is sure to continue, and several lines of research have the potential for broad application. These projects involve a mixture of genetic engineering, computer analysis, and protein biochemistry.

Researchers can identify the individual B cells responsible for producing a given antibody. That B cell can then be isolated in a petri dish and fused with a cultured cancer cell. This technique produces a **hybridoma** (hī-bri-DŌ-ma), a cancer cell that produces large quantities of a single antibody. Because hybridomas, like other cancer cells, undergo rapid mitotic divisions, culturing the original hybridoma cell soon produces an entire population, or **clone,** of genetically identical cells. This particular clone will produce large quantities of a **monoclonal** (mo-nō-KLŌ-nal; *mono,* one) **antibody.** Monoclonal antibodies are free from impurities. They are identical molecules, because they are produced from genetically identical cells.

FAP Ch. 22

One important use for this technology has been the development of antibody tests for the clinical analysis of body fluids. Labeled antibodies can be used to detect small quantities of specific antigens in a sample of plasma or other body fluids. For example, a popular home pregnancy test relies on monoclonal antibodies that detect small amounts of a placental hormone, *human chorionic gonadotropin (hCG),* in the urine of a pregnant woman. Other monoclonal antibodies are used in standard blood screening tests for sexually transmitted diseases and urine tests for ovulation. Monoclonal antibodies can also be used to provide passive immunity to disease. Passive immunizations that include monoclonal antibodies cause fewer of the unpleasant side effects associated with antibodies from pooled sera, because the product does not contain plasma proteins, viruses, or other contaminants. The antibodies can be made to order by exposing a population of B cells to a particular antigen, then isolating any B cells that produce the desired antibodies.

Genetic engineering can be used to promote immunity in other ways as well. One interesting approach involves gene-splicing techniques. The genes coding for an antigenic protein of a viral or bacterial pathogen are identified, isolated, and inserted into a harmless bacterium that can be cultured in the laboratory. The clone that eventually develops will produce large quantities of pure antigen that can then be used to stimulate a primary immune response. Vaccines against malaria and hepatitis were developed in this manner, and a similar strategy may be used to design an AIDS vaccine.

A more controversial experimental technique involves taking a pathogen and adding or removing genes to make it harmless. The modified pathogen can then be used to produce active immunity without the risk of severe illness. Fears that the engineered organism could mutate or regain its pathogenic properties have so far limited the use of this approach, even in animal trials.

Hybridomas that manufacture other products of the immune system can also be produced. Interferons are not effective against all viruses, but interferon nasal sprays appear to provide resistance against the viruses responsible for the common cold. (Unfortunately, these sprays can cause nasal bleeding and other unpleasant side effects, so they have not been approved for sale to the public.) Interferons can also control certain forms of virus-induced cancers and hepatitis. Interleukins may prove useful in increasing the intensity of the immune response.

FETAL INFECTIONS

FAP *p. 786*

Fetal infections are rare, because the developing fetus acquires passive immunity from IgG antibodies produced by the mother. These defenses break down, however, if the maternal antibodies are unable to cope with a bacterial or viral infection. The fetus may then begin producing IgM antibodies. Blood drawn from a newborn infant or taken from the umbilical cord of a developing fetus can be tested for the presence of IgM antibodies. This procedure provides concrete evidence of congenital infection, because IgM antibodies cannot cross the placenta. For example, a newborn infant with congenital syphilis will have IgM antibodies that target the pathogenic bacterium involved *(Treponema pallidum).* Fetal or neonatal (newborn) blood can also be tested for antibodies against the rubella (German measles) virus or other pathogens.

In the case of congenital syphilis, antibiotic treatment of the mother can prevent fetal damage. In the absence of antibiotic treatment, fetal syphilis can cause liver and bone damage, hemorrhaging, and a susceptibility to secondary infections in the newborn infant. There is no satisfactory treatment for congenital rubella infection, which can cause severe developmental ab-

normalities. For that reason, rubella immunization has been recommended for young children (to slow the spread of the disease in the population) and for women of childbearing age. The vaccination, which contains live attenuated viruses, must be administered before pregnancy to prevent maternal and fetal infection during pregnancy and subsequent fetal damage.

DELAYED HYPERSENSITIVITY AND SKIN TESTS

FAP *pp. 787, 788*

Delayed hypersensitivity begins with the sensitization of cytotoxic T cells. At the initial exposure, macrophages called *antigen-presenting cells (APCs)* present antigens to T cells. On subsequent exposures, the T cells respond by releasing cytokines, which stimulate macrophage activity and produce a massive inflammatory response in the immediate area (Figure A-50•). Examples of delayed hypersensitivity include the many types of contact dermatitis, such as poison ivy (FAP *p. 152*).

Skin tests can be used to check for delayed hypersensitivity. The antigen is administered by shallow injection, and the site is inspected minutes to days later. Most skin tests inject antigens taken from bacteria, fungi, viruses, or parasites. If the individual has previously been exposed to the antigen and has developed antibodies to it, in 2–4 days the injection site will become red, swollen, and invaded by macrophages, T cells, and neutrophils. These signs are considered a positive test, which indicates previous exposure but does not necessarily indicate the presence of the disease. The skin test for tuberculosis, the *ppd,* is the most commonly used test of this sort.

IMMUNE COMPLEX DISORDERS

FAP *pp. 787, 788*

Under normal circumstances, immune complexes are promptly eliminated by phagocytosis. But when an antigen appears suddenly in high concentrations, the local phagocytic population may not be able to cope with the situation. The immune complex may then enlarge further, eventually forming insoluble granules that are deposited in the affected area. The presence of these complexes triggers the extensive activation of complement, leading to inflammation and tissue damage at that site (Figure A-51•). This condition is known as an **immune complex disorder.** The process of immune complex formation is further enhanced by neutrophils, which release enzymes that attack the inflamed cells and tissues. The most serious immune complex disorders involve deposits within blood vessels and in the filtration membranes of the kidneys.

SYSTEMIC LUPUS ERYTHEMATOSUS

FAP *p. 787*

Systemic lupus erythematosus (LOO-pus e-rith-ē-ma-TŌ-sis), or **SLE,** appears to result from a generalized breakdown in the antigen recognition mechanism. An individual with SLE manufactures *autoantibodies* against the body's own nucleic acids, ribosomes, clotting factors, blood cells, platelets, and lymphocytes. The immune complexes form deposits in peripheral tissues, producing anemia, kidney damage, arthritis, and vascular inflammation. CNS function deteriorates if the blood flow through damaged cranial or spinal vessels slows or stops.

The most obvious sign of this condition is the presence of a butterfly-shaped rash centered over the bridge of the nose (Figure A-52•). SLE affects women nine times as often as it affects men, and the incidence in the United States averages 2–3 cases per 100,000 population. It has no known cure, but almost 80 percent of persons with SLE survive 5 years or more after diagnosis. Treatment consists of controlling the symptoms and depressing the immune response through the administration of specialized drugs or corticosteroids.

AUTOIMMUNE DISORDERS

FAP *p. 787*

Autoimmune disorders result when plasma cells begin producing antibodies that will bind to normal "self" antigens throughout the body. Table A-33 lists representative autoimmune disorders that are discussed in the text and elsewhere in the *Applications Manual.*

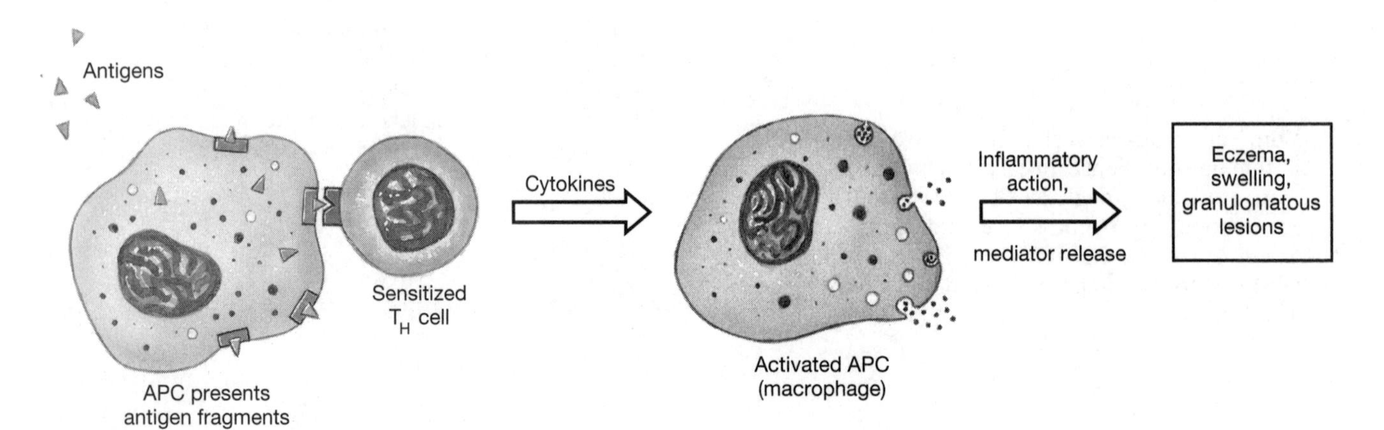

•**FIGURE A-50** Delayed Hypersensitivity

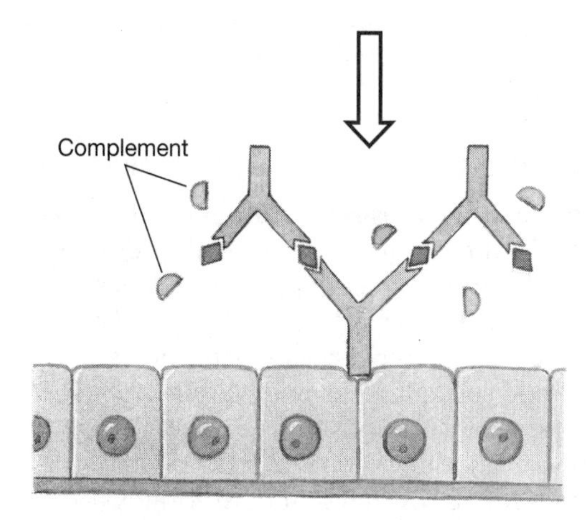
IgG
antibody
Antigens
Antigen–antibody immune complex is formed
Complement
Immune complex is deposited in tissue, activating complement

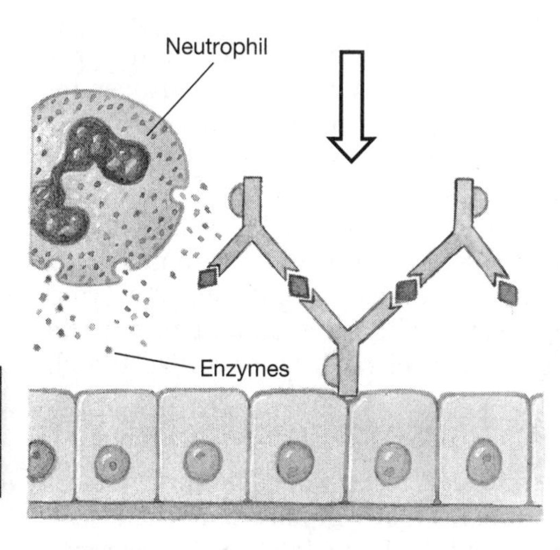
Neutrophil
Enzymes
Reactions of complement with immune complex attract neutrophils that release lysosomal enzymes, causing inflammation

•**FIGURE A-51** **Immune Complex Hypersensitivity**

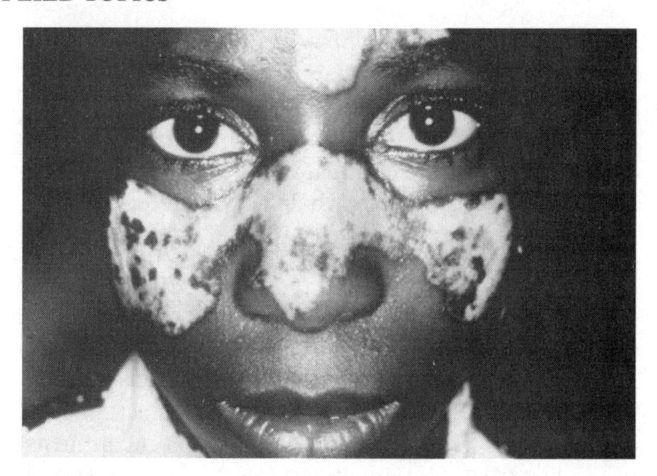

•**FIGURE A-52** **Butterfly Rash of Systemic Lupus Erythematosus**

TABLE A-33 Autoimmune Disorders

Disorder	Antibody Target	Discussion
Psoriasis	Epidermis of skin	p. 44 and FAP *p. 158*
Vitiligo	Melanocytes of skin	FAP *p. 149*
Rheumatoid arthritis	Connective tissues at joints	p. 60 and FAP *p. 265*
Myasthenia gravis	Synaptic ACh receptors	p. 67 and FAP *p. 281*
Multiple sclerosis	Myelin sheaths of axons	p. 74 and FAP *p. 370*
Addison's disease	Adrenal cortex	p. 102 and FAP *p. 602*
Graves' disease	Thyroid follicles	p. 99 and FAP *p. 597*
Hypopara-thyroidism	Chief cells of parathyroid	p. 99 and FAP *p. 598*
Thyroiditis	Thyroid-binding globulin	p. 97
Type I diabetes	Beta cells of pancreatic islets	p. 103 and FAP *p.* 607
Rheumatic fever	Myocardium and heart valves	p. 119 and FAP *p. 664*
Systemic lupus erythema-tosus	DNA, cytoskeletal proteins, other tissue components	p. 143
Thrombo-cytopenic purpura	Platelets	p.130
Pernicious anemia	Parietal cells of stomach	p. 115
Chronic hepatitis	Hepatocytes of liver	p. 166 and FAP *p. 878*

[EXPLORE]

Additional resources, such as interactive tutorials, critical-thinking questions, clinical problems, and case studies, are available through the CD-ROM and Website for your textbook. To begin your exploration, launch the *Martini Interactive* CD-ROM or visit the Companion Website (http://www.prenhall.com/martini/fap5) and select Chapter 22.

FAP Ch. 22

The Respiratory System

The anatomical components of the respiratory system can be divided into two parts: an *upper respiratory system,* which includes the nose, nasal cavity, paranasal sinuses, and pharynx; and a *lower respiratory system,* composed of the larynx, trachea, bronchi, and lungs. The *respiratory tract* consists of the airways that carry air to and from the exchange surfaces of the lungs. The respiratory tract can be divided into a *conducting portion* and a *respiratory portion.* The conducting portion begins at the entrance to the nasal cavity and extends through the pharynx and larynx and along the trachea and bronchi to the terminal bronchioles. The respiratory portion of the tract includes the respiratory bronchioles and the alveoli, which are part of the respiratory membrane, where gas exchange occurs.

SYMPTOMS OF RESPIRATORY DISORDERS

When they seek medical attention, individuals with lower respiratory disorders generally do so as a result of one or two major symptoms, specifically *chest pain* and *dyspnea:*

1. The chest pain associated with a respiratory disorder usually worsens when the person takes a deep breath or coughs. This pain with breathing is distinct from the chest pain experienced by individuals with angina (pain appears during exertion) or a myocardial infarction (pain is continuous, even at rest). Several disorders, such as those affecting the pleural membranes, cause chest pain that is localized to specific regions of the thorax. A person with such a condition will usually press against the sensitive area and avoid coughing or deep breathing in an attempt to reduce the pain.
2. *Dyspnea,* or difficulty in breathing, can be a symptom of pulmonary disorders, cardiovascular disorders, metabolic disorders, or environmental factors such as hypoxia at high altitudes. It may be a chronic problem, or it may develop only during exertion or when the person is lying down.

Dyspnea due to respiratory problems generally indicates one of the following classes of disorders:

- *Obstructive disorders* result from increased resistance to airflow along the respiratory passageways. The individual usually struggles to breathe, even at rest, and exhalation is more difficult than inhalation. Examples of obstructive disorders include *emphysema* (p. 153) and *asthma* (p. 151).
- *Restrictive disorders* include (1) arthritis; (2) paralysis or weakness of respiratory muscles, which result from trauma and scarring, muscular dystrophy, myasthenia gravis, multiple sclerosis, polio, or other factors; (3) physical trauma or congenital structural disorders, such as scoliosis, that limit lung expansion; and (4) pulmonary fibrosis, in which abnormal fibrous tissue in the alveolar walls slows oxygen diffusion into the bloodstream. Individuals with restrictive disorders usually experience dyspnea during exertion, because pulmonary ventilation cannot increase enough to meet the respiratory demand.

Cardiovascular disorders that produce dyspnea include *heart disease, congestive heart failure,* and *pulmonary embolism.* In *paroxysmal nocturnal dyspnea,* a person awakens at night, gasping for air. In most cases, the underlying cause is a reduced cardiac output due to advanced heart disease or heart failure. *Cheyne–Stokes respiration* consists of cycles of rapid, deep breathing separated by periods of respiratory arrest. This breathing pattern is most commonly seen in persons with CNS disorders or congestive heart failure.

Dyspnea may also be related to metabolic problems, such as the acute acidosis associated with *diabetes mellitus* (p. 103) and with *uremia* (p. 180). The fall in blood pH can trigger *Kussmaul breathing,* which consists of rapid, deep respiratory cycles.

THE PHYSICAL EXAMINATION AND THE RESPIRATORY SYSTEM

Several components of the physical examination will detect signs of respiratory disorders:

1. *Inspection* can reveal abnormal dimensions, such as the "barrel chest" that develops in emphysema or other obstructive disorders (p. 153), or *clubbing* of the fingers (p. 42). Clubbing is typically a late sign of disorders such as emphysema or congestive heart failure. *Cyanosis,* a blue color of the skin and mucous membranes,

generally indicates hypoxia (low tissue oxygen content). Laboratory testing of arterial blood gases will assist in determining the cause and extent of the hypoxia.

2. *Palpation* of the bones and muscles of the thoracic cage can detect structural problems or asymmetry. For example, the asymmetrical contraction of respiratory muscles during breathing may indicate a restrictive disorder.
3. *Percussion* on the surface of the thoracic cage over the lungs should yield sharp, resonant sounds. Dull or flat sounds can indicate structural changes in the lungs, such as those that accompany pneumonia, or the collapse of part of a lung *(atelectasis).* Increased resonance can result from obstructive disorders, such as emphysema, due to hyperinflation of the lungs as the individual attempts to improve alveolar ventilation.
4. *Auscultation* of the lungs with a stethoscope yields the distinctive sounds of inhalation and exhalation. These sounds vary in intensity, pitch, and duration. Abnormal breath sounds accompany several pulmonary disorders:
   - *Rales* (rahlz) are hissing, whistling, scraping, or rattling sounds associated with increased airway resistance. The sounds are created by turbulent airflow past accumulated pus or mucus or through airways narrowed by inflammation. Descriptions and interpretations of these sounds are very subjective, but in general, *moist rales* are gurgling sounds produced as air flows over fluids within the respiratory tract. They are heard in conditions such as *bronchitis* (p. 153), *tuberculosis* (p. 151), and *pneumonia* (FAP *p. 813*). *Dry rales* are produced as air flows over thick masses of mucus, through inflamed airways, or into fluid-filled alveoli. Dry rales are characteristic of *asthma* (p. 151), congestive heart failure (p. 124), and *pulmonary edema* (FAP *p. 722*). *Rhonchi* are loud dry rales produced by mucus buildup in the air passages.
   - *Stridor* is a very loud, high-pitched sound that can be heard without a stethoscope. Stridor generally indicates acute airway obstruction, such as the partial blockage of the glottis by a foreign object.
   - *Wheezing* is a whistling sound that can occur with inhalation or exhalation. It generally indicates airway obstruction due to mucus buildup or bronchospasms.
   - *Coughing* is a familiar sign of several respiratory disorders. Although primarily a reflex mechanism that clears the airway, coughing may also indicate irritation of the lining of the respiratory passageways. The duration, pitch, causative factors, and productivity may be important clues in the diagnosis of a respiratory disorder. (A *productive cough* ejects *sputum,* a mixture of mucus, cell debris, and pus; a *nonproductive cough,* or *dry cough,* does not.) If the cough is productive, the sputum ejected can be analyzed. This analysis will provide information about the presence of epithelial cells, macrophages, blood cells, or pathogens. If pathogens are present, the sputum can be cultured so that the specific microorganism involved can be identified.
   - A *friction rub* is a distinctive crackling sound produced by abrasion between abnormal serous membranes. A *pleural rub* accompanies respiratory movements and indicates problems with the pleural membranes, such as *pleurisy* (FAP *p. 814*). A *pericardial rub* accompanies the heartbeat and indicates inflammation of the pericardium, as in *pericarditis* (p. 106).
5. During the assessment of vital signs, the respiratory rate (number of breaths per minute) is recorded, along with notations about the general rhythm and depth of respiration. *Tachypnea* is a respiratory rate faster than 20 breaths per minute in an adult; *bradypnea* is an adult respiratory rate below 12 breaths per minute.

FAP Ch. 23

Table A-34 introduces important procedures and laboratory tests useful in diagnosing respiratory disorders.

DISORDERS OF THE RESPIRATORY SYSTEM

The respiratory system provides a route for air movement into and out of the lungs and supplies a large, warm, moist surface area for the exchange of oxygen and carbon dioxide between the air and circulating blood. Disorders affecting the respiratory system may therefore involve the following three mechanisms:

1. *Interfering with the movement of air along the respiratory passageways.* Internal or external factors may be involved. Within the respiratory tract, the constriction of small airways, as in *asthma* (p. 151), can reduce airflow to the lungs. The blockage of major airways, as in choking or a swollen epiglottis, can completely shut off the air supply. External factors that interfere with air movement include (1) the introduction of air *(pneumothorax)* or blood *(hemothorax)* into the pleural cavity, with subsequent lung collapse; (2) the buildup of fluid within the pleural cavities (a *pleural effusion*), which compresses and collapses the lungs; and (3) arthritis, muscular paralysis, or other conditions that prevent the normal skeletal or muscular activities responsible for moving air into and out of the respiratory tract.
2. *Damaging or otherwise impeding the diffusion of gases at the respiratory membrane.* The walls of the alveoli are part of the respiratory membrane, where gas exchange occurs. Any disease that affects the alveolar walls will reduce the efficiency of gas exchange. In *emphysema* (p. 153) and *lung cancer* (p. 155), alveoli are destroyed. Respiratory exchange is disrupted by the buildup of fluid or mucus within the alveoli. This

TABLE A-34 Representative Diagnostic and Laboratory Tests for Respiratory Disorders

Diagnostic Procedure	Method and Result	Representative Uses
Pulmonary function studies	A spirometer is used to determine lung volumes and capacities, including V_T, IC, ERV, IRV, FRC, vital capacity, and total lung capacity on exertion. (See Figure 23-18 in FAP, *p. 821*.) Forced vital capacity (FVC) is the amount of air forcibly exhaled after maximal inhalation.	Differentiate obstructive from restrictive lung diseases; determine extent of pulmonary disease. Increased functional residual capacity (FRC) occurs in obstructive diseases, such as emphysema and chronic bronchitis; FRC is normal in restrictive diseases, such as pulmonary fibrosis.
Peak expiratory flow (PEF)	Determined after forceful exhalation into a PEF meter	Useful in evaluating asthma; permits self-monitoring and appropriate medication of airway obstruction
Bronchoscopy	Fiber-optic tubing is inserted into oral cavity and further into trachea, larynx, and bronchus for internal viewing.	Detects abnormalities such as inflammation and tumors; used to remove aspirated foreign objects or mucus from bronchi and to obtain samples of secretions or of a small specimen of lung tissue for biopsy
Mediastinoscopy	A lighted instrument is inserted into an incision made at the jugular notch superior to the manubrium of the sternum.	Detects abnormalities of mediastinal lymph nodes; permits biopsy specimen removal; detects pulmonary disorders and metastasis of lung cancer
Lung biopsy	Lung tissue is removed for pathological analysis via bronchoscopy or during exploratory surgery of the thoracic area.	Differentiates pulmonary pathologies and determines the presence of malignancy
Chest X-ray study	Standard X-ray produces film sheet with radiodense tissues shown in white on a negative image.	Detects abnormalities of lungs, such as tumors, inflammation of the lungs, rib or sternal fractures, or pneumothorax; detects fluid accumulation (pulmonary edema), pneumonia, and atelectasis; determines heart size
Thoracentesis	A needle is inserted into the intrapleural space for removal of fluid for analysis or for relieving pressure due to fluid accumulation.	See Pleural fluid analysis on the next page.
Pulmonary angiography	A catheter is inserted in the femoral vein and threaded through the right ventricle and into the pulmonary arteries. Contrast dye is intermittently injected as X-ray films are taken.	Detects pulmonary embolism
Lung scan	Radionuclide is injected intravenously; radiation that is emitted is captured to create an image of the lungs.	Determines areas with decreased blood flow due to pulmonary embolism, tumor, or other pulmonary disease
Computerized tomography (CT) scan	Standard CT; contrast media are usually used.	Detects tumors, cysts, or other structural abnormalities

disruption can occur as the result of inflammation or infection of the lungs, as in the various types of pneumonia.

3. *Blocking or reducing the normal circulation of blood through the alveolar capillaries.* Blood flow to portions of the lungs may be prevented by a *pulmonary embolism,* a circulatory blockage (FAP *p. 744, 813*). Not only does a pulmonary embolism prevent normal gas exchange in the affected regions of a lung, but also it results in tissue damage and, if the blockage persists for several hours, permanent alveolar collapse. Pulmonary blood pressure may then rise (a condition called *pulmonary hypertension*), leading to pulmonary edema and a reduction in alveolar function in other portions of the lungs.

These problems can result from trauma, congenital or degenerative problems, the formation of tumors,

TABLE A-34 Representative Diagnostic and Laboratory Tests for Respiratory Disorders *(continued)*

Laboratory Test	Normal Values in Blood Serum or Plasma	Significance of Abnormal Values
Arterial blood gases and pH		
pH	7.35–7.45	<7.35 indicates acidosis; >7.45 indicates alkalosis. See text (Table 27-4, FAP *p. 1009*).
P_{CO_2}	35–45 mm Hg	>45 mm Hg with pH <7.35 indicates respiratory acidosis present in pulmonary disorders such as emphysema, in chronic obstructive pulmonary disease, and in CNS depression leading to irregular breathing; <35 mm Hg with a pH >7.45 indicates respiratory alkalosis that occurs during prolonged hyperventilation; seen in emphysema. See text (Table 27-4, FAP *p. 1009*).
P_{O_2}	75–100 mm Hg	<75 mm Hg may occur in pneumonia, asthma, emphysema, and COPD.
HCO_3^-	22–28 mEq/l	>28 mEq/l (with elevated P_{CO_2} and decreased pH) indicates renal compensation for respiratory acidosis; <22 mEq/l (with decreased P_{CO_2} and elevated pH) is characteristic of renal compensation for respiratory alkalosis.
Sputum studies		
Cytology	No malignant cells are present.	Sloughed malignant cells indicate cancerous process in lungs.
Culture and sensitivity (C&S)	Sputum sample is placed on growth medium.	Identifies causative pathogenic organism and the organism's susceptibility to antibiotics
Acid-fast bacilli	Staining technique reveals no acid-fast bacilli.	Presence of stained rod-shaped microbes may indicate tuberculosis.
Alpha$_1$-antitrypsin determination	Adults: >213 mg/dl	Decreased value (<50 mg/dl) indicates a possible genetic predisposition for emphysema.
Tuberculin skin test	Skin wheal, produced by intradermal application of tuberculin, at 48–72 hours should be <10 mm (area that is hardened as opposed to the reddened area).	A skin papule at injection site that measures > 10 mm after 48–72 hours is a positive test for tuberculosis.
Pleural fluid analysis		
Fluid color and clarity	No pus; fluid is clear, with WBC <1000 mm^3	WBC >1000 mm^3 indicates potential infectious or inflammatory process; detects lung or pleural malignancy.
Culture	No bacterial growth (fluid sample is placed on growth medium)	Presence of bacteria indicates infection; fungi can be present in immunocompromised person.

FAP Ch. 23

inflammation, or infection of the lungs (Figure A-53•). Illnesses caused by infections of the upper respiratory tract include some of the most common diseases. Many respiratory infections are transmitted by droplets in the air, typically emitted in a sneeze or cough. Infections of the lower respiratory tract include two of the most deadly diseases in human history: pneumonia and tuberculosis. Table A-35 (p. 150) summarizes information about some of the most important infectious diseases of the respiratory system, along with the causative organisms.

Respiratory system disorders also occur secondarily, as a consequence of dysfunctions of other body systems. For instance, asthma is the result of a prob-

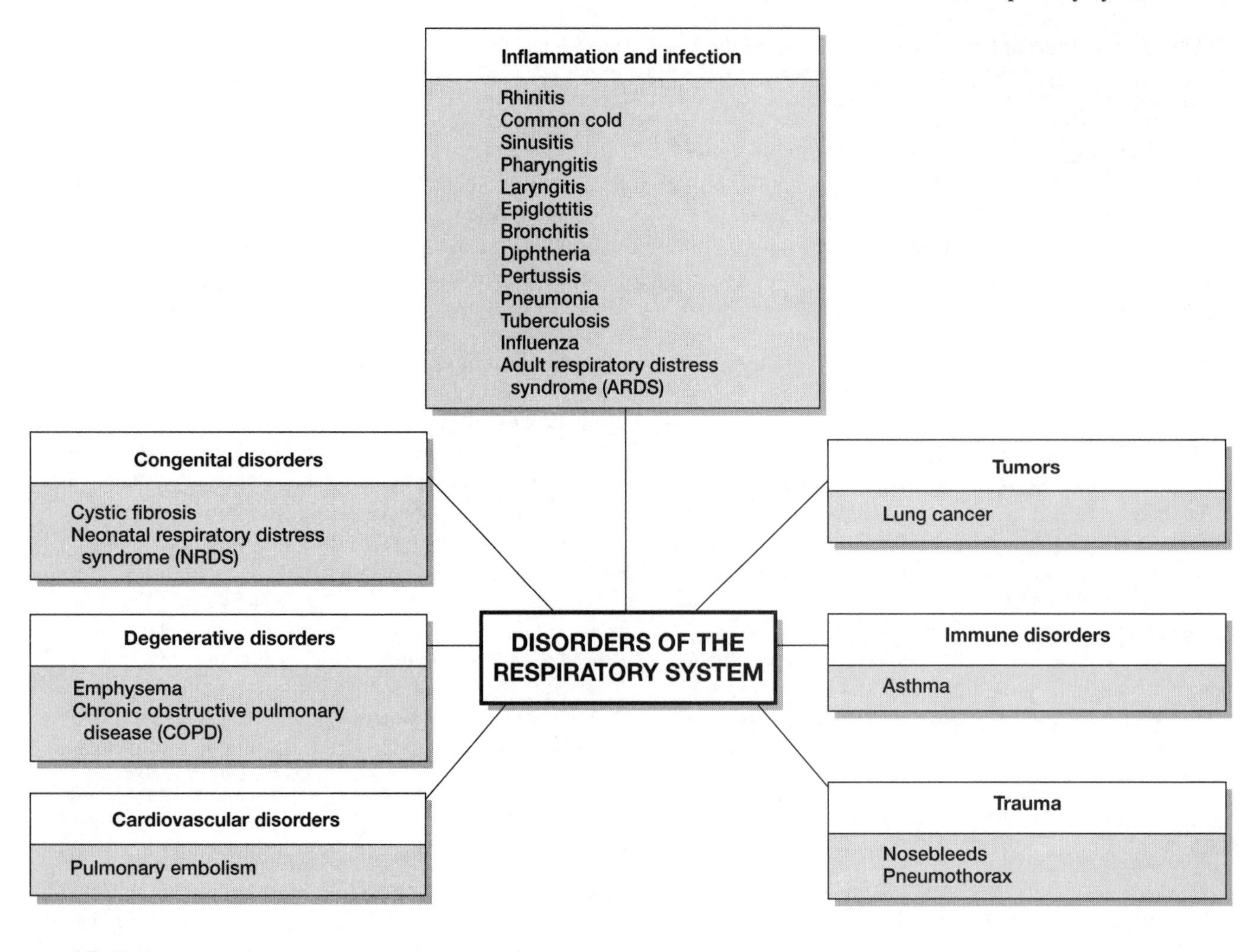

•**FIGURE A-53** **Disorders of the Respiratory System**

lem with immune function, and pulmonary emboli result from cardiovascular problems affecting lung perfusion.

OVERLOADING THE RESPIRATORY DEFENSES FAP *p. 800*

Large quantities of airborne particles can overload the respiratory defenses and produce a variety of illnesses. Chemical or physical irritants that reach the lamina propria or underlying tissues promote the formation of scar tissue *(fibrosis),* reducing the elasticity of the lung, and can restrict airflow along the passageways. Irritants or foreign particles may also enter the lymphatic vessels of the lung, producing inflammation of the regional lymph nodes. Chronic irritation and the stimulation of the epithelium and its defenses cause changes in the epithelium that increase the likelihood of lung cancer.

Severe symptoms of such disorders develop slowly; they may take 20 years or more to appear. **Silicosis** (sil-i-KŌ-sis, produced by the inhalation of silica dust), **asbestosis** (as-bes-TŌ-sis, from the inhalation of asbestos fibers), and **anthracosis** (an-thra-KŌ-sis, the "black lung disease" of coal miners, caused by the inhalation of coal dust) are conditions caused by the overloading of the respiratory defenses (Figure A-54•).

FAP Ch. 23

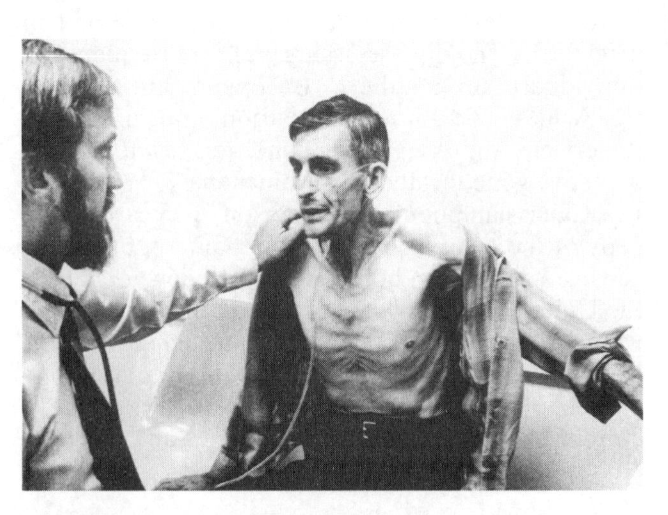

•**FIGURE A-54** **A Person with Anthracosis (Black Lung Disease)**

TABLE A-35 Examples of Infectious Diseases of the Respiratory System

Disease	Organism(s)	Description
Bacterial diseases		
Sinusitis	*Streptococcus pneumoniae* or *Haemophilus influenzae*	Inflammation of the paranasal cavities; headaches, pain, and pressure in facial bones
Pharyngitis	*Streptococcus pyogenes*	Inflammation of the throat; strep throat; sore throat, fever, no cough or sputum
Laryngitis	*Streptococcus pneumoniae* or *Haemophilus influenzae*	Inflammation of the larynx; dry, sore throat; hoarse voice or loss of voice
Epiglottitis	*Haemophilus influenzae*	Inflammation of the epiglottis; most common in children; fever, noisy breathing, drooling; swelling can block trachea and lead to suffocation
Bronchitis	*Streptococcus pneumoniae* or *Mycoplasma pneumoniae*	Inflammation of the bronchial passageways; productive cough with sputum; may enter alveoli and cause pneumonia
Diphtheria	*Corynebacterium diphtheriae*	Inflammation of the pharynx; pseudomembrane in pharynx; bacterial toxin affects heart and other tissues
Pertussis (whooping cough)	*Bordetella pertussis*	Highly contagious disease of children; mucus production; severe coughing ends in a "whoop" sound during inhalation
Pneumonia	*Streptococcus pneumoniae, Staphylococcus aureus,* or *Klebsiella pneumoniae*	Inflammation of the lungs; alveoli fill with fluids; chills, high fever; cough with sputum of mucus, pus, and blood
Tuberculosis	*Mycobacterium tuberculosis*	Highly contagious infection of the lungs; can spread to other tissues; abscesses, or tubercles, form in lungs; bacteria multiply in WBCs and alveolar macrophages and break down alveoli
Viral diseases		
Common cold (coryza)	Rhinoviruses or coronaviruses	Nasal obstruction, nasal discharge, sneezing, and headache
Influenza	Orthomyxoviruses (influenza virus A, B, C)	Frequent variations in Type A and B viruses can produce new epidemics; fever, headache, sore throat, nasal discharge, muscle weakness, fatigue, chest pain, and cough.

FAP Ch. 23

CYSTIC FIBROSIS

FAP *p. 800*

Cystic fibrosis (CF) is the most common lethal inherited disease that affects Caucasians of Northern European descent; it occurs at a frequency of 1 in 2500 births. The disease occurs with less frequency in individuals of Southern European ancestry, in the Ashkenazic Jewish population, and in African Americans and Asians. The condition results from a defective gene located on chromosome 7. Within the U.S. Caucasian population, 1 person in 25 carries one copy of the gene for this disorder, and an infant receiving a copy from both parents will develop CF. In the United States, 2000 babies are born with CF each year and roughly 30,000 persons are living with this condition. Individuals with classic CF seldom survive past age 30; death is generally the result of chronic, recurrent bacterial infection of the lungs and associated heart failure.

The causative gene carries instructions for a transmembrane protein responsible for the active transport of chloride ions. This membrane protein is abundant in exocrine cells that produce watery secretions. In persons with CF, the protein does not function normally. The secretory cells cannot transport salts and water effectively, and the secretions produced are thick and gooey. Mucous glands of the respiratory tract and secretory cells of the pancreas, salivary glands, and digestive tract are affected.

The most-serious symptoms appear because the respiratory defense system cannot transport such dense mucus. The mucus escalator (FAP *p. 799*) stops working, and mucus plugs block the smaller respiratory passageways. This blockage reduces the diameter of the airways, and the inactivation of the normal respiratory defenses leads to frequent bacterial infections. The most-dangerous infections involve the bacterium

Pseudomonas aeruginosa, which colonizes the stagnant mucus and further stimulates mucus production by epithelial cells.

Treatment has primarily been limited to supportive care and antibiotic therapy to control infections. In a few instances, lung transplants have provided relief, but the technical and logistical problems involved with this approach are formidable. The structures of the abnormal gene that causes CF and of its normal counterpart have been determined. The current goal is to correct the defect by inserting normal genes into the cells in critical areas of the body. In the meantime, one of the factors contributing to the thickness of the mucus has been discovered to be the presence of DNA released from degenerating cells in areas of inflammation. Inhaling an aerosol spray containing an enzyme that breaks down DNA has proved to be remarkably effective in improving respiratory performance.

TUBERCULOSIS

FAP *p. 808*

Tuberculosis (TB) is a major health problem throughout the world. With roughly 3 million deaths each year, it is the leading cause of death from infectious diseases. An estimated 2 *billion* people are infected, and 8 million cases are diagnosed each year. Unlike other deadly diseases, such as AIDS, TB is transmitted through casual contact. Anyone who breathes is at risk of contracting this disease; all it takes is exposure to the causative bacterium. Coughing, sneezing, or speaking by an infected individual spreads the pathogen through the air in the form of tiny droplets that can be inhaled by other people.

If untreated, the disease progresses in stages. At the site of infection, macrophages and fibroblasts wall off the area, forming an abscess. If the scar-tissue barricade fails, the bacteria move into the surrounding tissues, and the process repeats itself. The resulting masses of fibrous tissue distort the conducting passageways, increasing resistance and decreasing airflow. In the alveoli, the attacked surfaces are destroyed. The combination severely reduces the area available for gas exchange.

Treatment for TB is complex, because (1) the bacteria can spread to many different tissues and (2) they can develop a resistance to standard antibiotics relatively quickly. As a result, several drugs are used in combination over a period of 6–9 months. The most-effective drugs now available include *isoniazid,* which interferes with bacterial replication, and *rifampin,* which blocks bacterial protein synthesis.

The TB problem is much less severe in developed nations, such as the United States, than in developing nations. However, tuberculosis was extremely common in the United States early in the twentieth century. An estimated 80 percent of Americans born around 1900 became infected with tuberculosis during their lives. Although many were able to meet the bacterial challenge, it was the leading cause of death in 1906. These statistics have been drastically changed with the advent of antibiotics and techniques for early detection of infection. Between 1906 and 1984, the death rate fell from 200 deaths per 100,000 population to 1.5 deaths per 100,000 population. From 1984 to 1994, the TB incidence and death rates were on the rise; roughly 24,000 TB-related deaths occurred in 1994. Although the death rates in 1995 were slightly lower (22,000), as of 2000 an estimated 10–15 million people in the United States are infected with tuberculosis. Most individuals with TB are not infectious unless the disease reactivates, causing symptoms of coughing, fever, and weight loss.

Today, tuberculosis is unevenly distributed through the U.S. population, with several groups at relatively high risk of infection. For example, Hispanics, African Americans, prison inmates, individuals with immune disorders (such as persons with AIDS), and hospital employees who work near infected patients are more likely to be infected than are other members of the population. At present 2–5 percent of young American adults have been infected. The percentage of these cases that are caused by antibiotic-resistant strains is growing. An estimated 14 percent of TB patients diagnosed in the United States each year have infections that are resistant to both isoniazid and rifampin. The statistics for some parts of the country are worse; in New York City, 33 percent of new cases are drug-resistant. The most frightening part about this surge in resistant TB is that the fatality rate for infections resistant to two or more antibiotics is 50 percent. Unless efforts are made to combat the spread of TB, over the next decade the United States may face a public health crisis as deadly as HIV infection and even more difficult to control.

ASTHMA

FAP *pp. 809, 822*

FAP Ch. 23

Asthma (AZ-muh) affects an estimated 3–6 percent of the U.S. population. There are several forms of asthma, but each is characterized by unusually sensitive and irritable air-conducting passageways. In many cases, the trigger appears to be an immediate hypersensitivity reaction to an allergen in the inhaled air. Drug reactions, air pollution, chronic respiratory infections, exercise, and emotional stress can also induce an asthma attack in sensitive individuals.

The most obvious and potentially dangerous symptoms include (1) the constriction of smooth muscles all along the bronchial tree, (2) edema and swelling of the mucosa of the respiratory passageways, and (3) the accelerated production of mucus. The combination makes breathing very difficult. Exhalation is affected

more than inhalation; the narrowed passageways often collapse before exhalation is completed. Although mucus production increases, mucus transport slows, so fluids accumulate along the passageways. Coughing and wheezing then develop. The bronchoconstriction and mucus production occur in a few minutes, in response to the release of histamine and prostaglandins by mast cells. The activated mast cells also release interleukins, leukotrienes, and platelet-activating factors. As a result, over a period of hours, neutrophils and eosinophils migrate into the area. The area then becomes inflamed, further reducing airflow and damaging respiratory tissues. Because the inflammation compounds the problem, antihistamines alone cannot control a severe asthma attack.

When a severe attack occurs, it reduces the functional capabilities of the respiratory system. Peripheral tissues gradually become oxygen starved; this condition can prove fatal. Asthma fatalities have been increasing in recent years. The annual death rate from asthma in the United States is approximately 4 deaths per million population (for ages 5–34 years). Mortality among asthmatic African Americans is twice that among Caucasian Americans.

The treatment of asthma involves the dilation of the respiratory passageways by administering **bronchodilators** (brong-kō-DĪ-lā-torz), drugs that relax bronchial smooth muscle, and by reducing inflammation and swelling of the respiratory mucosa through anti-inflammatory medication. Important bronchodilators include *theophylline, epinephrine, albuterol,* and other beta-adrenergic drugs. Although the strongest beta-adrenergic drugs are very useful in a crisis, they are effective only for relatively brief periods, and over-use of them can lead to reduced efficiency. Asthmatic individuals must be closely monitored due to the drugs' potential effects on cardiovascular function. Anti-inflammatory medication with less-acute effects, such as inhaled or ingested steroids, is becoming increasingly important.

FAP Ch. 23

RESPIRATORY DISTRESS SYNDROME (RDS) FAP *p. 812*

Septal cells begin producing surfactant at the end of the sixth fetal month. By the eighth month, surfactant production has risen to the level required for normal respiratory function. **Neonatal respiratory distress syndrome (NRDS),** also known as *hyaline membrane disease (HMD),* develops when surfactant production fails to reach normal levels. Although some forms of HMD are inherited, the condition most commonly accompanies premature delivery.

In the absence of surfactant, the alveoli tend to collapse during exhalation. Although the conducting passageways remain open, the newborn infant must then inhale with extra force to reopen the alveoli on the next breath. In effect, every breath must approach the power of the first, so the infant rapidly becomes exhausted. Respiratory movements become progressively weaker; eventually the alveoli fail to expand and gas exchange ceases.

One method of treatment involves assisting the infant by administering air under pressure so that the alveoli are held open. This procedure, known as **positive end-expiratory pressure (PEEP),** can keep the newborn alive until surfactant production increases to normal levels. Surfactant from other sources can also be provided; suitable surfactants can be extracted from cow lungs *(Survanta®),* obtained from the liquid (amniotic fluid) that surrounds full-term infants, or synthesized by gene-splicing techniques *(Exosurf®).* These preparations are usually administered in the form of a fine mist of surfactant droplets. Clinical trials are under way to deliver oxygen to the respiratory membrane by means of a perfluorocarbon solution *(Liquivent®).* The lungs are filled with the solution, which readily absorbs oxygen. Because the alveoli are filled with fluid rather than with air, they remain open despite the lack of surfactant. This procedure avoids the necessity of administering air under pressure, which can have undesirable side effects.

Surfactant abnormalities can also develop in adults as the result of severe respiratory infections or other sources of pulmonary injury. Alveolar collapse follows, producing a condition known as **adult respiratory distress syndrome (ARDS).** The PEEP procedure is typically used to maintain life until the underlying problem can be corrected, but at least 50–60 percent of ARDS cases result in fatalities.

BOYLE'S LAW AND AIR OVEREXPANSION SYNDROME FAP *p. 815*

Swimmers descending 1 m or more beneath the surface experience significant increases in pressure, due to the weight of the overlying water. Air spaces throughout the body decrease in volume. The pressure increase normally produces mild discomfort in the middle ear, but some people experience acute pain and disorientation. The water pressure first collapses the auditory tubes (see Figure 17-22 of the text, *p. 558*). As the volume of air in the middle ears decreases, the tympanic membranes are forced inward. This uncomfortable situation can be remedied by closing the mouth, pinching the nose and exhaling gently; elevating the pressure in the nasopharynx forces air through the auditory tubes and into the middle ear spaces. As the volume of air in each middle ear increases, the tympanic membrane returns to its normal position. When the swimmer returns to the surface, the pressure drops and the air in the middle ear expands. This expansion usually goes unnoticed, because the air simply forces its way out along the auditory tube and into the nasopharynx.

Scuba divers breathe air under pressure, and that air is delivered at the same pressure as that of their

surroundings. A descent to a depth of 10 m doubles the pressure on a diver due to the weight of the overlying water. Consider what happens if that diver then takes a full breath of air and heads for the surface. As the pressure declines, the volume increases (Boyle's law, FAP *p. 815*). Thus, at the surface, the volume of air in the lungs will have doubled. Such a drastic increase cannot be tolerated. The usual result is a tear in the wall of the lung. The symptoms and severity of this *air overexpansion syndrome* depend on where the air ends up. If the air flows into the pleural cavity, the lung may collapse; if it enters the mediastinum, it may compress the pericardium and produce symptoms of cardiac tamponade. Worst of all, the air may rupture blood vessels and enter the bloodstream. The air bubbles then form emboli that can block blood vessels in the heart or brain, producing a heart attack or stroke. These are all serious conditions, so divers are trained to avoid holding their breath and to exhale when swimming toward the surface.

CPR

FAP *p. 819*

Cardiopulmonary resuscitation, or **CPR,** restores blood flow and ventilation to an individual whose heart has stopped beating. Compression applied to the rib cage over the sternum reduces the volume of the thoracic cavity, squeezing the heart and propelling blood into the aorta and pulmonary trunk. When the pressure is removed, the thorax expands and blood moves into the great veins. Cycles of compression are interspersed with cycles of mouth-to-mouth breathing that maintain pulmonary ventilation.

This practice has been credited with saving many thousands of lives. Basic CPR techniques can be mastered in about 8 hours of intensive training, using special equipment. Yearly recertification courses must be taken, because the skills fade with time and because CPR techniques cannot be practiced on a living person without causing severe injuries, such as broken ribs. Training is available at minimal cost through charitable organizations such as the Red Cross and the American Heart Association.

DECOMPRESSION SICKNESS

FAP *p. 824*

Decompression sickness can develop when an individual experiences a sudden change in pressure. Nitrogen is the gas responsible for this condition. Nitrogen, which accounts for 78.6 percent of the atmospheric gas mixture, has a relatively low solubility in body fluids. Under normal atmospheric pressures, blood contains few nitrogen molecules, but at higher than normal pressures, additional nitrogen molecules diffuse across the alveolar surfaces and into the bloodstream.

As more nitrogen enters the blood, the gas is distributed throughout the body. Over time, nitrogen diffuses into peripheral tissues and into body fluids such as the cerebrospinal fluid, aqueous humor, and synovial fluids. If the atmospheric pressure decreases, the change must occur slowly enough that the excess nitrogen can diffuse out of the tissues, into the blood, and across the alveolar surfaces. If the pressure falls suddenly, this gradual movement of nitrogen from the periphery to the lungs cannot occur. Instead, the nitrogen leaves solution and forms bubbles of nitrogen gas in the blood, tissues, and body fluids.

A few bubbles in peripheral connective tissues may not be particularly dangerous, at least initially. However, these bubbles can fuse together, forming larger bubbles that distort tissues, causing pain. Bubbles typically develop in joint capsules first. These bubbles cause severe pain, and the afflicted individual tends to bend over or curl up. This symptom accounts for the popular name of this condition, "the bends." Bubbles in the systemic or pulmonary circulation can cause infarcts, and those in the cerebrospinal circulation can cause strokes, leading to sensory losses, paralysis, or respiratory arrest.

Treatment consists of recompression, exposing the individual to pressures that force the nitrogen back into solution and alleviate the symptoms. Pressures are then reduced gradually over a period of 1 day or more. Breathing air with more oxygen and less nitrogen than are in atmospheric air accelerates the removal of excess nitrogen from the blood.

Today most cases of the bends involve scuba divers who have gone too deep or stayed at depth too long. The condition is not restricted to divers, however; the first reported cases involved construction crews who worked in pressurized surroundings. Although such accidents are exceedingly rare, the sudden loss of cabin pressure in a commercial airplane can also produce symptoms of decompression sickness.

BRONCHITIS, EMPHYSEMA, AND COPD

FAP *pp. 824, 839*

Bronchitis (brong-KĪ-tis) is an inflammation of the bronchial lining. The most characteristic symptom is the overproduction of mucus, which leads to frequent coughing. An estimated 20 percent of adult males have *chronic bronchitis.* This condition is most commonly related to cigarette smoking, but it also results from other environmental irritants, such as chemical vapors. Over time, the increased mucus production can block smaller airways and reduce respiratory efficiency. This condition is called **chronic airways obstruction.**

Emphysema (em-fi-SĒ-muh) is a chronic, progressive condition characterized by shortness of breath and an inability to tolerate physical exertion. The underlying problem is the destruction of respiratory exchange surfaces. In essence, respiratory bronchioles and alveoli are functionally eliminated. The alveoli gradually expand, capillaries deteriorate, and gas exchange in the affected region comes to a halt.

Emphysema has been linked to the inhalation of air that contains fine particulate matter or toxic vapors, such as those in cigarette smoke. Early in the disease, local regulation shunts blood away from the damaged areas, and the individual may not notice problems, even with strenuous activity. As the condition progresses, the reduction in exchange surface limits the ability to provide adequate oxygen. However, obvious clinical symptoms typically fail to appear until the damage is extensive.

Alpha$_1$-antitrypsin, an enzyme that is normally present in the lungs, helps prevent degenerative changes in lung tissue. Most people requiring treatment for emphysema are adult smokers; this group includes individuals with alpha$_1$-antitrypsin deficiency and those with normal tissue enzymes. In the United States, 1 person in 1000 carries two copies of a gene that codes for an abnormal, inactive form of this enzyme. A single change in the amino acid sequence appears responsible for this inactivation. At least 80 percent of nonsmokers with abnormal alpha$_1$-antitrypsin will develop emphysema, generally at ages 45–50 years. *All* smokers will develop at least some emphysema, typically by ages 35–40 years.

Unfortunately, the loss of alveoli and bronchioles in emphysema is permanent and irreversible. Further progression can be limited by the cessation of smoking. The only effective treatment for severe cases is the administration of oxygen, but lung transplants have helped some patients, as has the surgical removal of nonfunctional lung tissue. For persons with alpha$_1$-antitrypsin deficiency who are diagnosed early, attempts are under way to provide enzyme supplements by daily infusion or periodic injection.

Two patterns of symptoms may appear in individuals with advanced emphysema. In one pattern, other aspects of pulmonary structure and function are relatively normal. The respiratory rate in these individuals increases dramatically. The lungs are fully inflated at each breath, and expirations are forced. These individuals maintain near-normal arterial P_{O_2}. Their respiratory muscles are working hard, and they use a lot of energy just breathing. As a result, these people tend to be thin. Because blood oxygenation is near normal, skin color in Caucasians in this group is pink. The combination of heavy breathing and pink coloration has led to the descriptive term *pink puffers* for individuals with this condition. Because they expend extra energy in breathing, pink puffers tend to be underweight.

FAP Ch. 23

In the second group, emphysema has been complicated by chronic bronchitis and chronic airways obstruction. This combination, known as **chronic obstructive pulmonary disease** (COPD), is particularly dangerous. Individuals with COPD commonly expand their chests permanently in an effort to enlarge their lung capacities and make the best use of the remaining functional alveoli. This adaptation gives them a distinctive "barrel-chested" appearance. Persons with COPD also have symptoms of heart failure, including widespread edema. Blood oxygenation is low, and the skin has a bluish color. The combination of widespread edema and bluish coloration has led to the descriptive term *blue bloaters* for individuals with this condition.

MOUNTAIN SICKNESS FAP p. 829

Mountain sickness commonly develops after a rapid ascent to altitudes of at least 2500 m. Symptoms, which appear within a day after arrival, include severe headache, insomnia, tachypnea, and nausea. The underlying cause is thought to be a combination of mild cerebral edema and respiratory alkalosis from hyperventilation induced by hypoxia. The symptoms, which may persist for a week before subsiding, can be prevented or reduced by the administration of acetazolamide *(Diamox®)*. This drug opposes the alkalosis by inhibiting carbonic anhydrase activity and promoting bicarbonate loss at the kidneys. (Airplanes are pressurized to 2000 m [7000 ft], so mountain sickness is not a problem for air travelers.)

Acute mountain sickness is a life-threatening condition that affects (1) individuals performing strenuous physical activities shortly after arriving at high altitudes (typically well above 2500 m) or (2) individuals adapted to high altitudes who descend to sea level for 1–2 weeks and then return. Fatalities result from the combination of cerebral edema and pulmonary edema. Treatment includes rest, breathing pure oxygen, and immediate transport to lower altitudes.

SHALLOW WATER BLACKOUT FAP p. 834

In preparing to dive under water, misguided swimmers may attempt to outwit their chemoreceptor reflexes. The usual method involves taking some extra-deep breaths before submerging. These individuals are intentionally hyperventilating, usually with the stated goal of "taking up extra oxygen."

From the hemoglobin saturation curve, it should be obvious that this explanation is incorrect. At a normal alveolar P_{O_2}, hemoglobin is 97.5 percent saturated; no matter how many breaths the swimmer takes, the alveolar P_{O_2} cannot rise significantly. But the P_{CO_2} will be affected, because the increased ventilation rate lowers the carbon dioxide concentrations of the alveoli and blood. This lowering produces the desired effect by temporarily shutting off the chemoreceptors that monitor the P_{CO_2}. As long as CO_2 levels remain depressed, the swimmer does not feel the need to breathe, despite the fact that P_{O_2} continues to fall.

Under normal circumstances, breath holding causes a decline in P_{O_2} and a rise in P_{CO_2}. As indicated in Figure A-55•, by the time the P_{O_2} has fallen to 60 percent of normal levels, carbon dioxide levels will have risen enough to make breathing an unavoidable ne-

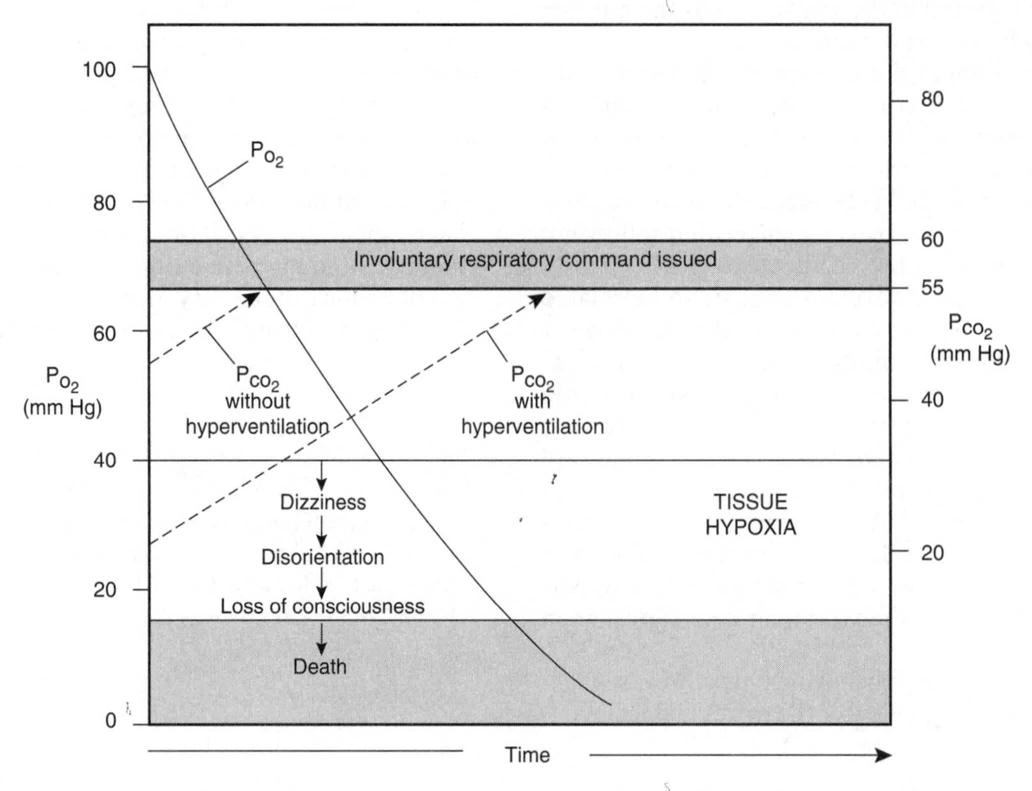

•**FIGURE A-55 Carbon Dioxide and Respiratory Demand.** The stimulation of respiration by rising P_{CO_2} normally prevents a dangerous decline in oxygen levels. After hyperventilation, the P_{CO_2} may be so low that respiratory stimulation will not occur before low P_{O_2} impairs neural function.

cessity. But after hyperventilation, oxygen levels can fall so low that the swimmer becomes unconscious (generally at a P_{O_2} of 15–20 mm Hg) before the P_{CO_2} rises enough to stimulate breathing. This *shallow water blackout* has fatal consequences in most cases. Hence swimmers and divers should not hyperventilate.

CHEMORECEPTOR ACCOMMODATION AND OPPOSITION FAP *p. 834*

Carbon dioxide receptors show accommodation to sustained P_{CO_2} levels above or below normal. Although these receptors register the initial change quite strongly, over a period of days they adapt to the new level as "normal." As a result, after several days of elevated P_{CO_2} levels, the effects on the respiratory centers begin to decline. Fortunately, the response to a low arterial P_{O_2} remains intact, and the response to arterial oxygen concentrations becomes increasingly important.

Some individuals with severe chronic lung disease cannot maintain normal P_{O_2} and P_{CO_2} levels in the blood. The arterial P_{CO_2} rises to 50–55 mm Hg, and the P_{O_2} falls to 45–50 mm Hg. At these levels, the carbon dioxide receptors accommodate, and most of the respiratory drive comes from the arterial oxygen receptors. If these individuals are given too much oxygen, they may simply stop breathing. Vigorous stimulation to encourage breathing or a mechanical respirator may be required.

LUNG CANCER FAP *p. 839*

Lung cancers now account for about 30 percent of all cancer deaths, making this condition the primary cause of cancer death in the U.S. population. Despite advances in the treatment of other forms of cancer, the survival statistics for lung cancer have not changed significantly. Even with early detection, the 5-year survival rates are only 30 percent for men and 50 percent for women, and most lung cancer patients die within a year of diagnosis.

Detailed statistical and experimental evidence has shown that *85–90 percent of all lung cancers are the direct result of cigarette smoking.* Claims to the contrary are simply unjustified and insupportable. The data are far too extensive to detail here, but the incidence of lung cancer for nonsmokers is 3.4 per 100,000 population, whereas the incidence for smokers ranges from 59.3 per 100,000 for those who smoke between a half-pack and a pack per day to 217.3 per 100,000 for those who smoke one to two packs per day. Before about 1970, this disease primarily affected middle-aged men, but as the number of women smokers has increased

(a trend that started in the 1940s), so has the number of women who die from lung cancer.

Smoking changes the quality of the inspired air, making it drier and contaminated with several carcinogenic compounds and particulate matter. The combination overloads the respiratory defenses and damages the epithelial cells throughout the respiratory system. The histological changes that follow are described in Chapter 4 of the text (FAP *p. 137*). Whether lung cancer develops appears to be related to the total cumulative exposure to the carcinogens. The more cigarettes smoked, the greater the risk, whether those cigarettes are smoked over a period of weeks or years. Up to the point at which tumors form, the histological changes induced by smoking are reversible; a normal epithelium will return if the carcinogens are removed. At the same time, the statistical risks decline to significantly lower levels. Ten years after quitting, a former smoker stands only a 10 percent greater chance of developing lung cancer than does a nonsmoker.

The fact that cigarette smoking typically causes cancer is not surprising in view of the toxic chemicals contained in the smoke. What is surprising is that more smokers do not develop lung cancer. Evidence suggests that some smokers have a genetic predisposition to developing one form of lung cancer. Dietary factors may also play a role in preventing lung cancer, although the details are controversial. In terms of their influence on the risk of lung cancer, there is a general agreement that (1) vitamin A has no effect; (2) vegetables containing beta-carotene reduce the risk, but pills of beta-carotene may increase the risk; and (3) a high-cholesterol, high-fat diet increases the risk.

[EXPLORE]

Additional resources, such as interactive tutorials, critical-thinking questions, clinical problems, and case studies, are available through the CD-ROM and Website for your textbook. To begin your exploration, launch the *Martini Interactive* CD-ROM or visit the Companion Website (http://www.prenhall.com/martini/fap5) and select Chapter 23.

FAP
Ch. 23

The Digestive System

The digestive system consists of the *digestive tract* and *accessory digestive organs.* The digestive tract is divided into the oral cavity, pharynx, esophagus, stomach, small intestine, and large intestine. Most of the absorptive functions of the digestive system occur in the small intestine, with lesser amounts in the stomach and large intestine. The accessory digestive organs provide acids, enzymes, and buffers that assist in the chemical breakdown of food. The accessory organs include the salivary glands, the liver, the gallbladder, and the pancreas. The salivary glands produce saliva, a lubricant that contains enzymes that aid in the digestion of carbohydrates. The liver produces bile, which is concentrated in the gallbladder and released into the small intestine for fat emulsification. The pancreas secretes enzymes and buffers important to the digestion of proteins, carbohydrates, and lipids.

The activities of the digestive system are controlled through a combination of local reflexes, autonomic innervation, and the release of gastrointestinal hormones such as *gastrin, secretin,* and *cholecystokinin.*

SYMPTOMS AND SIGNS OF DIGESTIVE SYSTEM DISORDERS

The functions of the digestive organs are varied, and the symptoms and signs of digestive system disorders are equally diverse. Common symptoms of digestive disorders include the following:

- Pain is a common symptom of digestive disorders. Widespread pain in the oral cavity results from (1) trauma; (2) infection of the oral mucosa by bacteria, fungi, or viruses; or (3) a deficiency in vitamin C (*scurvy,* p. 6) or one or more of the B vitamins. Focal pain in the oral cavity accompanies (1) the infection or blockage of salivary gland ducts; (2) tooth disorders such as tooth fractures, *dental caries, pulpitis,* abscess formation, and *gingivitis* (FAP *p. 856*); and (3) oral lesions, such as those produced by the *herpes simplex* virus.

  Abdominal pain is characteristic of a variety of digestive disorders. In most cases the pain is perceived as distressing but tolerable; if the pain is acute and severe, a surgical emergency may exist. Abdominal pain can cause rigidity in the abdominal muscles in the painful area. This rigidity is easily felt on palpation. The muscle contractions *(guarding)* may be voluntary, in an attempt to protect a painful area, or an involuntary spasm resulting from irritation of the peritoneal lining, as in *peritonitis* (FAP *p. 851*). Persons with peritoneal inflammation also experience *rebound tenderness,* in which pain appears when fingertip pressure on the abdominal wall is suddenly removed.

  Abdominal pain can result from disorders of the digestive, circulatory, urinary, or reproductive system. Digestive tract disorders producing abdominal pain include *appendicitis* (p. 132), *peptic ulcers* (p. 162), *pancreatitis* (FAP *p. 874*), *cholecystitis* (p. 167), *hepatitis* (p. 165), *intestinal obstruction* (p. 158), *diverticulitis* (FAP *p. 884*), *peritonitis,* and certain cancers.
- *Dyspepsia,* or indigestion, is pain or discomfort in the epigastric region. Digestive tract disorders associated with dyspepsia include *esophagitis* (p. 162), *gastritis* (FAP *p. 865*), *peptic ulcers, gastroesophageal reflux,* and *cholecystitis* (p. 167).
- *Nausea* is a sensation that usually precedes or accompanies vomiting. Nausea results from digestive disorders or from disturbances of central nervous system function.
- *Anorexia* is a decrease in appetite that, if prolonged, is accompanied by weight loss. Digestive disorders that cause anorexia include *stomach cancer* (FAP *p. 868*), pancreatitis, hepatitis, and several forms of *diarrhea* (p. 168). Anorexia may also accompany disorders that involve other systems. *Anorexia nervosa,* an eating disorder with a psychological basis, is discussed in the text *(p. 932).*
- *Dysphagia* is difficulty in swallowing. This difficulty may result from trauma, infection, inflammation, or a blockage of the posterior oral cavity, pharynx, or esophagus. For example, the infections of *tonsillitis, pharyngitis,* and *laryngitis* may cause dysphagia.

THE PHYSICAL EXAMINATION AND THE DIGESTIVE SYSTEM

Physical assessment can provide information useful in the diagnosis of digestive system disorders. The abdominal region is particularly important, because most of the digestive system is located within the abdominopelvic cavity. We will now discuss four methods of physical assessment of the digestive system: inspection, palpation, percussion, and auscultation (p. 7):

1. *Inspection* can provide a variety of diagnostic clues:
   - Bleeding of the gums, as in gingivitis, and characteristic oral lesions can be seen on inspection of the oral cavity. Examples of distinctive lesions include those of oral herpes simplex infections and *thrush,* lesions produced by infection of the mouth by

Candida albicans. This fungus (also called a yeast) is a normal resident of the digestive tract. However, the fungus causes widespread oral infections in immunodeficient persons, such as individuals who have AIDS or who are undergoing immunosuppressive therapies. (Healthy infants can also get a mild oral fungal infection.)

- Peristalsis in the stomach and intestines may be seen as waves passing across the abdominal wall in persons who do not have a thick layer of abdominal fat. The waves become very prominent during the initial stages of intestinal obstruction.
- A general yellow discoloration of the skin, a sign called *jaundice,* results from excessive levels of bilirubin in body fluids. Jaundice is commonly seen in individuals with cholecystitis (p. 167) or liver diseases such as hepatitis (p. 165) and *cirrhosis* (p. 166).
- Abdominal distention is caused by (1) fluid accumulation in the peritoneal cavity, as in *ascites* (FAP *p. 851*); (2) air or gas (flatus) within the digestive tract or peritoneal cavity; (3) obesity; (4) abdominal masses, such as tumors, or enlargement of visceral organs; (5) pregnancy; (6) the presence of an abdominal *hernia* (p. 68); or (7) fecal impaction, as in severe and prolonged constipation.
- *Striae* are multiple scars, 1–6 cm in length, that are visible through the epidermis. Striae develop in damaged dermal tissues after stretching; they are typically seen in the abdominal region after a pregnancy or other rapid weight gain. Abnormal striae may develop after ascites or in cases of subcutaneous edema. Purple striae are signs of *Cushing's disease* (adrenocortical hypersecretion, p. 102).

2. *Palpation* of the abdomen may reveal specific details about the status of the digestive system, including:
   - The presence of abnormal masses, such as tumors, within the peritoneal cavity
   - Abdominal distention from (1) excess fluid within the digestive tract or peritoneal cavity or (2) gas within the digestive tract
   - Herniation of digestive organs through the inguinal canal or weak spots in the abdominal wall (p. 68)
   - Changes in the size, shape, or texture of visceral organs. For example, in several liver diseases, the liver becomes enlarged and firm, and these changes can be detected on palpation of the right upper quadrant.
   - Voluntary or involuntary abdominal muscle contractions (guarding)
   - Rebound tenderness
   - Specific areas of tenderness and pain. For example, someone with acute hepatitis (p. 166), a liver disease, generally experiences pain on palpation of the right upper quadrant. In contrast, a person with appendicitis (p. 132) generally experiences pain when the right lower quadrant is palpated.
3. *Percussion* of the abdomen is less instructive than percussion of the chest, because the visceral organs do not contain extensive air spaces that reflect the sounds conducted through surrounding tissues. However, the stomach usually contains a small air bubble, and percussion over this area produces a sharp, resonant sound. The sound becomes dull or disappears when the stomach fills, the spleen enlarges, or the peritoneal cavity contains abnormal quantities of peritoneal fluid, as in ascites (FAP *p. 851*). Percussion may also detect gas accumulation within the peritoneal cavity or within portions of the colon.
4. *Auscultation* can detect gurgling abdominal sounds, or *bowel sounds,* produced by peristaltic activity along the digestive tract. Increased bowel sounds occur in persons with acute diarrhea, and bowel sounds may disappear in persons with (1) advanced intestinal obstruction; (2) peritonitis, an infection of the peritoneum; or (3) spinal cord injuries that prevent normal innervation.

Diagnostic procedures, such as endoscopy, are commonly used to provide additional information. Information on representative diagnostic procedures and laboratory tests is given in Tables A-36 and A-37 (p. 160).

TABLE A-36 Examples of Diagnostic Tests Used in the Diagnosis of Gastrointestinal Disorders

Diagnostic Procedure	Method and Result	Representative Uses
ORAL CAVITY Sialography	X-ray film is taken after contrast medium is injected into salivary ducts while patient's salivary glands are stimulated.	Identifies calculi, inflammation, and tumors in salivary duct or gland
Periapical (PA) X-rays	Periapical film is taken of crown and root area of several teeth.	Detect tooth decay, tooth impactions, fractures, progression of bone loss with periodontal disease, inflammation of periodontal ligament, and periapical abscesses (at end of root)
Bitewing X-rays of teeth	Bitewing film is taken interproximally. X-ray cone is pointed perpendicularly toward the spaces between the crowns of adjacent posterior teeth.	Reveal tooth decay between teeth and early bone loss in periodontal disease

FAP Ch. 24

TABLE A-36 Examples of Diagnostic Tests Used in the Diagnosis of Gastrointestinal Disorders *(continued)*

Diagnostic Procedure	Method and Result	Representative Uses
UPPER GI (esophagus, stomach, and duodenum)		
Esophagogastroduodenoscopy, esophagoscopy, gastroscopy	Fiber-optic endoscope is inserted into oral cavity and further into esophagus, stomach, and duodenum to permit visualization of lining and lumen of one area.	Detects tumors, ulcerations, polyps, esophageal varices, inflammation, and obstructions; provides biopsy of tissues in upper GI tract
Barium swallow	Series of X-rays of esophagus is taken after barium sulfate is ingested to increase contrast of structures; normally done with upper GI series.	Determines abnormalities of pharynx and esophagus, especially to determine cause of dysphagia; identifies tumors, hiatal hernia, and diverticuli
Upper GI series	Series of X-rays of stomach and duodenum is taken after barium sulfate is ingested to increase contrast.	Determines cause of epigastric pain; detects ulcers, polyps, gastritis, tumors, and inflammation in upper GI tract
Esophageal function studies Acid reflux Acid clearing	Slender tube, through which dilute hydrochloric acid solution is passed, is inserted from oral cavity into stomach. A pH probe is placed in the esophagus. The pH at the level of the gastroesophageal sphincter is measured periodically to detect acid reflux. Hydrochloric acid solution is then given again, and patient swallows until the acid is cleared from esophagus.	Detects gastroesophageal reflux Acid clearing normally occurs in less than 10 swallows. If more are required, patient may have severe esophagitis from chronic gastric acid reflux (gastroesophageal reflux disease, or GERD).
Gastric analysis Basal gastric secretion	When patient is fasting, tube is inserted into nose and routed along esophagus to sample stomach contents. Acid levels are periodically monitored through this *nasogastric (NG) tube* to determine basal levels of gastric secretion.	Identifies cause of ulceration of gastric lining; may be increased acid levels. If no increased acidity with ulceration is present, a malignancy may exist.
Gastric acid stimulation	Chemical stimulus is administered and acid levels are monitored to determine results of gastric acid stimulation test.	
Arteriography (celiac and mesenteric)	Catheter is inserted into femoral artery and threaded to celiac trunk or a mesenteric artery, where contrast medium is injected; X-ray films are then taken.	Determines cause of bleeding in gastrointestinal tract or blockage of blood supply
LOWER GI (colon)		
Barium enema	Barium sulfate enema is administered to provide contrast in intestinal lumen, and X-ray film series is taken.	Determines cause of abdominal pain and bloody stools; detects tumors and obstructions of bowel; identifies polyps and diverticula
Sigmoidoscopy **Colonoscopy**	Fiber-optic tube is inserted into rectum for viewing of anus, rectum, and sigmoid colon; colonoscopy allows viewing of the entire large intestine to cecum; biopsy of intestinal tissue can be performed.	Detects tumors, polyps, and ulcerations of intestinal lining. Polyps are removed with scope.

TABLE A-36 Examples of Diagnostic Tests Used in the Diagnosis of Gastrointestinal Disorders *(continued)*

Diagnostic Procedure	Method and Result	Representative Uses
LIVER, PANCREAS, GALLBLADDER		
X-ray of abdomen	Radiodense tissues appear in white on negative film image.	Detects abdominal masses and obstructions
Computerized tomography of liver, pancreas, gallbladder	Cross-sectional radiation is applied to provide three-dimensional images of liver, gallbladder, and pancreas.	Identifies tumors, pancreatitis (acute, chronic), hepatic cysts and abscesses, and biliary calculi (stones); determines cause of jaundice
Ultrasonography of liver, pancreas, gallbladder	Liver, pancreas, and gallbladder are examined by means of sound waves emitted by transducer placed on abdomen. Sound waves deflect off dense structures and produce echoes, which are amplified and graphically recorded.	Detects polyps, tumors, abscesses, hepatic cysts, and gallstones; determines cause of jaundice
Cholangiography Intravenous or percutaneous	Intravenously administered dye is concentrated in liver and released into bile duct; X-ray films are taken. Percutaneous method involves insertion of needle with catheter into bile duct, using ultrasonography for guidance; dye is then administered through catheter.	Identifies biliary calculi obstructing ducts
Liver biopsy	Needle is inserted through small skin incision and into liver; liver tissue is removed by excision or aspiration.	Liver tissue is examined for evidence of cirrhosis, hepatitis, tumors, and granuloma.
Endoscopic retrograde cholangiopancreatography (ERCP)	Duodenoscope (fiber-optic tube) is inserted into oral cavity and threaded through stomach to the duodenum; through scope, catheter is inserted into duodenal ampulla for dye injection; X-rays are then taken to visualize the bile duct system.	Detects calculi, tumors, or cysts of pancreatic or bile ducts; determines presence of pancreatic tumor

TABLE A-37 Examples of Laboratory Tests Used in the Diagnosis of Gastrointestinal Disorders

Laboratory Test	Normal Values in Blood Plasma or Serum	Significance of Abnormal Values
UPPER AND LOWER GI		
Serum electrolytes		
Potassium	Adults: 3.5–5.0 mEq/l	Vomiting, diarrhea, and nasogastric intubation can cause electrolyte losses and create imbalances.
Sodium	135–145 mEq/l	
Magnesium	1.5–2.5 mEq/l	
Serum gastrin level	Adults: 45–200 pg/ml	Elevated levels occur with pernicious anemia and some gastric ulcers; very high levels in Zollinger–Ellison syndrome (pancreatic tumor that produces excessive gastrin).
Lactose tolerance test	Fasting patient receives 50 mg of lactose. The hydrogen gas content of exhaled air is measured, or blood is tested at periodic intervals up to 2 hours to show plasma glucose increasing to >20 mg/dl.	Unabsorbed lactose is metabolized by bacteria, producing hydrogen gas that is exhaled at the lungs. Blood glucose levels do not rise to normal levels after lactose ingestion if patient lacks lactase.

FAP Ch. 24

TABLE A-37 Examples of Laboratory Tests Used in the Diagnosis of Gastrointestinal Disorders *(continued)*

Laboratory Test	Normal Values in Blood Plasma or Serum	Significance of Abnormal Values
Carcinoembryonic antigen (CEA) (in plasma)	Adults: Nonsmokers <2.5 ng/ml Smokers <3.5 ng/ml	Colorectal cancer is a possibility with elevated levels. This antigen is not specific for colorectal cancer; elevated levels may occur in other types of carcinomas and in ulcerative colitis.
Stool culture		
Culture and sensitivity (C&S). Stool sample is cultured for bacterial content.	Only normal intestinal microbial inhabitants should be isolated.	Bacteria such as *Shigella, Campylobacter,* enterotoxic *Escherichia coli, Vibrio* cholera, and *Salmonella* can cause acute diarrhea.
Ova and parasites (O&P). A stool sample is microscopically examined for parasite eggs or adults.	None found	Typical parasites found in feces that cause intestinal disturbances are tapeworms, entamoeba, and some protozoans, such as *Giardia.*
Fecal analysis		
Occult blood	None found	Hidden (occult) blood in feces can occur when bleeding from GI tract is minimal due to inflammation, ulceration, or with a tumor that is causing small amounts of bleeding.
Fat content	Adults: <6 g/24 hours	Fat content >6 g/24 hours, a condition called steatorrhea, often indicates malabsorption syndrome; also occurs in cystic fibrosis and pancreatic diseases.
LIVER AND GALLBLADDER		
Serum bilirubin (total)	Total serum bilirubin: 0.1–1.2 mg/dl	Can be elevated with hepatitis (infectious or toxic) and biliary obstruction
	Indirect bilirubin: 0.2–1.0 mg/dl (unconjugated)	Excessive hemolysis as in transfusion reactions or erythroblastosis fetalis; elevation also occurs with hepatitis.
	Direct bilirubin: 0.1–0.3 mg/dl (conjugated)	Elevated levels occur with obstructions of bile duct system, hepatitis, cirrhosis of liver, and liver cancer.
	Infant (newborn) total serum bilirubin: 1–12 mg/dl	Bilirubin >15 mg/dl can result in serious neurological problems.
Urine bilirubin	None	Detectable bilirubin is present when direct bilirubin is elevated, as in obstruction of bile duct system, hepatitis, cirrhosis of liver, and liver cancer.
Liver enzyme tests		
Aspartate aminotransferase (AST or SGOT)	Adults: 0–35 U/l	Elevated levels occur with hepatitis, acute pancreatitis, and cirrhosis.
Alanine aminotransferase (ALT or SGPT)	Adults: 0–35 IU/l	Elevated levels occur with hepatitis and other liver diseases.
Alkaline phosphatase (isoenzyme ALP_1)	Adults: 30–120 mU/ml	Elevated levels occur in biliary obstructions, hepatitis, and liver cancer.
5'-nucleotidase	Adults: 0–1.6 units	Elevated levels are useful for early detection of liver disease.
Gamma-glutamyl transferase/transpeptidase (GGT or GGTP)	Adults: 8–38 U/l	Elevated levels occur with liver disease; levels also become elevated after ingestion of alcohol.
Serum protein	Adults: Albumin 3.5–5.5 g/dl (or 52–68% of the total protein) Adults: Globulin 2.0–3.0 g/dl (or 32–48% of the total protein)	Decreased levels of albumin and increased levels of globulin occur in chronic liver disease.
Ammonia (plasma)	Adults: 15–45 µg/dl	Elevated level occurs with liver failure and hepatic encephalopathy.

TABLE A-37 Examples of Laboratory Tests Used in the Diagnosis of Gastrointestinal Disorders *(continued)*

Laboratory Test	Normal Values in Blood Plasma or Serum	Significance of Abnormal Values
Hepatitis virus studies	None	Antibodies can be detected against virus that causes hepatitis A (HAV); hepatitis B (HBV) can be detected by testing for both viral antigen and antibodies; hepatitis C (HCV) can be detected by testing for antibodies against the virus. Quantitative viral counts are done for hepatitis B and C.
Prothrombin time (PT) test	Adults: 11–15 seconds	Prolonged time occurs with liver disease, vitamin K deficiency, and warfarin (coumadin) administration.
PANCREAS		
Amylase		
Urine	Adults: 35–260 U/l	Elevated levels occur for 7–10 days after pancreatic disease begins.
Serum	Adults: 60–180 U/l	Elevated levels occur with pancreatic disease and with obstruction of pancreatic duct.
Lipase (serum)	Adults: 0–110 U/l	Elevated levels most commonly occur in acute pancreatitis.
Sweat electrolytes test		
Sodium Chloride	Children: <70 mEq/l Children: <50 mEq/l	Elevated levels of both sodium and chloride occur in cystic fibrosis; does not reflect severity of disorder.

DISORDERS OF THE DIGESTIVE SYSTEM

Due to the number and diversity of digestive organs, there are many types of digestive system disorders. One common method of categorizing these disorders uses a combination of four anatomical and functional characteristics (Figure A-56a•). The largest category includes inflammation and infection of the digestive organs. This group is usually broken down regionally (by oral cavity, esophagus, stomach, and so forth). Table A-38 (p. 164) summarizes information about infectious diseases of the digestive system. Digestive system cancers are also relatively common and diverse. *Malabsorption disorders* are characterized by problems with the absorption of one or more nutrients. *Congenital disorders* result from developmental problems affecting the structure of the digestive tract or accessory organs. *Nutritional and metabolic disorders* (Figure A-56b•), which affect the entire body, are often considered with disorders of the digestive system, because they typically accompany digestive system disorders. However, nutritional and metabolic disorders may also reflect (1) disorders involving the endocrine or nervous system, (2) congenital metabolic problems, or (3) dietary abnormalities.

FAP Ch. 24

ACHALASIA AND ESOPHAGITIS FAP *p. 861*

In the condition known as **achalasia** (ak-a-LĀ-zē-uh), a swallowed bolus descends along the esophagus relatively slowly, as a result of abnormally weak peristaltic waves; its arrival does not trigger the opening of the lower esophageal sphincter. Materials then accumulate at the base of the esophagus like cars at a stoplight. Secondary peristaltic waves may occur repeatedly, adding to the individual's discomfort. The most successful treatment involves cutting the circular muscle layer at the base of the esophagus or expanding a balloon in the lower esophagus until the muscle layer tears.

A weakened or permanently relaxed sphincter can cause **esophagitis** (ē-sof-a-JĪ-tis), or inflammation of the esophagus, as powerful stomach acids enter the inferior portion of the esophagus. The esophageal epithelium has few defenses against attack by acids and enzymes; inflammation, epithelial erosion, and intense discomfort result. Frequent episodes of backflow results in *gastroesophageal reflux disease* (GERD), which causes the symptoms commonly known as *heartburn.* This condition supports a multimillion-dollar industry devoted to producing and promoting antacids.

PEPTIC ULCERS FAP *p. 865*

Peptic ulcers are lesions of the gastric or duodenal mucosa caused by bacterial action or erosion by acids and pepsin in gastric juice. Regardless of the primary cause of a gastric ulcer, once gastric juices have destroyed the epithelial layers, the virtually defenseless lamina

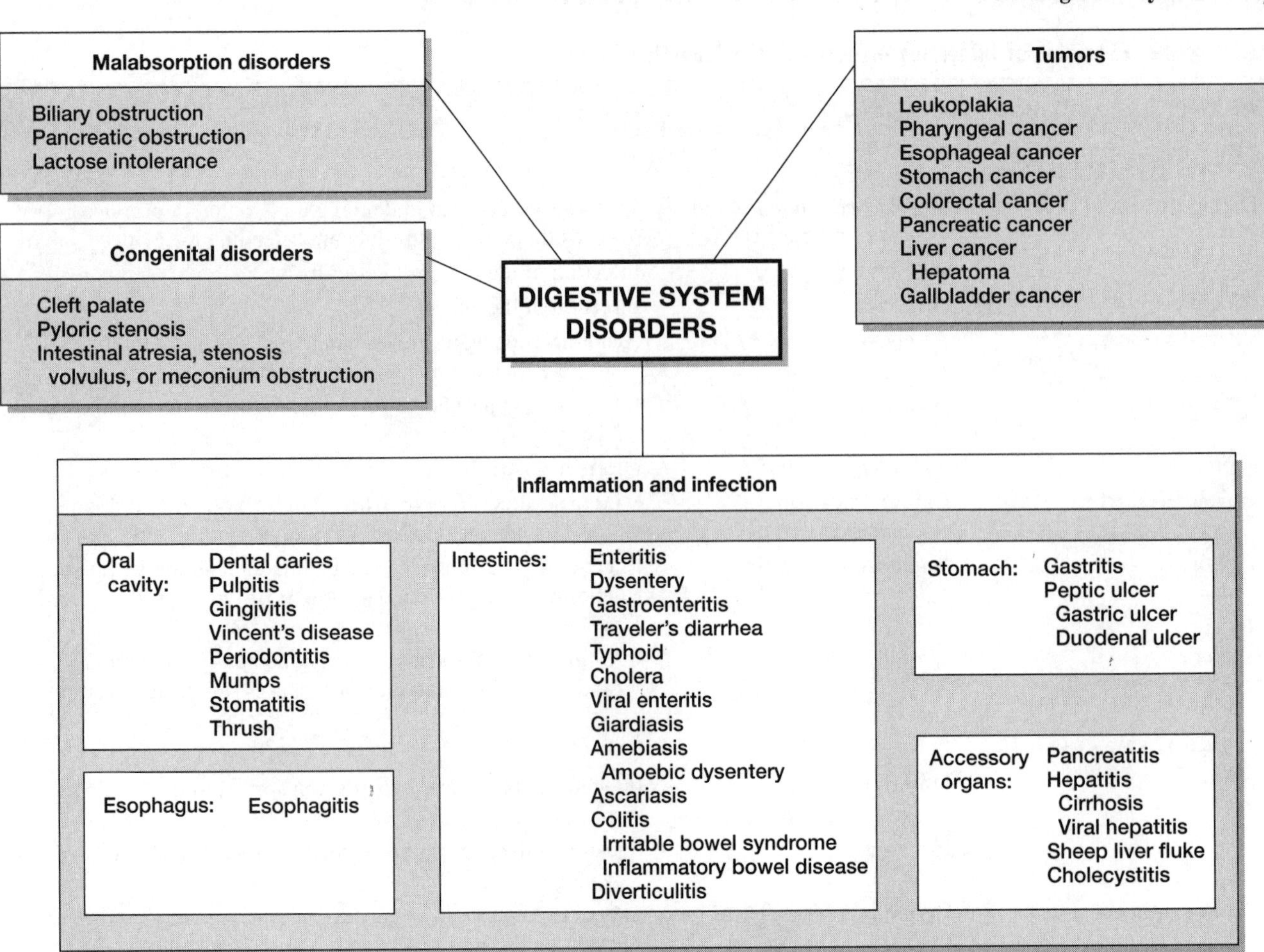

(a)

Eating disorders

- Inadequate food intake
  - Anorexia
  - Anorexia nervosa
- Excessive food intake
  - Bulimia
  - Obesity

NUTRITIONAL AND METABOLIC DISORDERS

Thermoregulatory disorders

- Elevated body temperature
  - Fever
  - Heat exhaustion
  - Heat stroke
  - Hyperthermia
- Lowered body temperature
  - Accidental hypothermia
  - Induced hypothermia

Metabolic disorders

- Catabolic problems
  - Ketosis
- Congenital disorders
  - Phenylketonuria (PKU)
- Protein deficiency diseases
  - Marasmus
  - Kwashiorkor
- Mineral disorders
  - Deficiencies
  - Excesses
- Vitamin disorders
  - Hypervitaminosis
  - Avitaminosis
- Water balance disorders
  - Dehydration
  - Overhydration
    - Water intoxication

(b)

•FIGURE A-56 Disorders of the Digestive System. (a) Major categories of digestive system disorders. **(b)** Nutritional and metabolic disorders.

TABLE A-38 Examples of Infectious Diseases of the Digestive System

Disease	Organism(s)	Description
BACTERIAL DISEASES		
Dental caries	*Streptococcus mutans* and other oral bacteria	Tooth decay; bacteria in dental plaque on teeth produce acids that dissolve tooth enamel, leading to cavities
Pulpitis	As above	Infection of the pulp of the tooth
Gingivitis	As above	Infection of the gums
Vincent's disease	As above	Acute necrotizing ulcerative gingivitis, or trenchmouth; bacterial infection and ulcer formation
Periodontitis	As above	Infection of gums and bone; results in loosening and loss of teeth
Peptic ulcers	*Helicobacter pylori*	Ulcers in gastric lining
Traveler's diarrhea	Enterotoxic form of *Escherichia coli*	Mild to severe watery diarrhea, nausea, vomiting, abdominal pain, and general lack of energy
Typhoid	*Salmonella typhi*	Infection of the intestines and gallbladder; abdominal pain, abdominal distention, pain, low WBC count, and enlarged spleen
Cholera	*Vibrio cholerae*	Intestinal infection; symptoms include nausea, vomiting, abdominal pain, and diarrhea; causes severe dehydration
VIRAL DISEASES		
Mumps	Mumps virus (paramyxovirus)	Infection of the salivary glands; can spread to the meninges or gonads
Viral enteritis	Rotaviruses	Intestinal infection; causes watery diarrhea, especially in young children
Viral hepatitis	Hepatitis A virus (HAV)	Infectious hepatitis; transmitted by fecal-contaminated water, milk, shellfish, or other food
	Hepatitis B virus (HBV)	Serum hepatitis; transmitted by exchange of body fluid through blood transfusions, wounds, shared needles (intravenous drug use), or sexual contact; pregnant carrier of the virus can pass it on to her baby
	Hepatitis C virus (HCV)	Formerly non-A, non-B hepatitis; transmitted through blood, intravenous drug use, and sexual contact
	Hepatitis D virus	Occurs only in persons infected with HBV; transmission as for HBV
	Hepatitis E virus	Transmission as for HAV
	Hepatitis G virus	Little is known
FUNGAL DISEASES		
Candidiasis (thrush)	*Candida albicans*	Yeast infection of the oral mucosa; forms white, milky patches in the mouth
PARASITIC DISEASES		
Protozoa		
Giardiasis	*Giardia lamblia*	Intestinal infection; symptoms include diarrhea, dehydration, and weight loss; more common in children than adults
Amebiasis, or amoebic dysentery	*Entamoeba histolytica*	Infection of the large intestine; can produce ulcers and peritonitis; diarrhea contains blood
Helminths		
Ascariasis	*Ascaris lumbricoides*	Roundworm infestation of the intestines; larval movement to pharynx causes damage to intestinal wall; adult worms eat contents of the intestine
Sheep liver fluke	*Fasciola hepatica*	Flatworm infestation of the bile-conducting passageways in the liver and gallbladder; causes inflammation of the liver; flukes consume blood

propria will be exposed to digestive attack. Sharp abdominal pain results. In severe cases, the damage to the mucosa of the stomach can cause significant bleeding, and the acids can even erode through the wall of the stomach and into the peritoneal cavity. This condition, called a **perforated ulcer,** requires immediate surgical correction.

The administration of antacids can typically control gastric or duodenal ulcers by neutralizing the acids and allowing time for the mucosa to regenerate. The drug *cimetidine* (sī-MET-i-dēn), or *Tagamet®*, inhibits the secretion of acid by the parietal cells. Dietary restrictions limit the intake of foods that promote acid production (caffeine and pepper) or that damage unprotected mucosal cells (alcohol and aspirin).

The treatment of peptic ulcers has changed radically since the identification of *Helicobacter pylori* as a likely causative agent. These bacteria are able to resist gastric acids long enough to penetrate the mucous coating of the epithelium. Once within the protective layer of mucus, they bind to the epithelial surfaces, where they are safe from the action of gastric acids and enzymes. Over time, the bacteria release toxins that damage the epithelial lining. These toxins ultimately result in the erosion of the epithelium and the destruction of the lamina propria by gastric juices. Individuals whose stomachs harbor *H. pylori* are also at higher than normal risk of gastric cancer, although the reason is not known.

Current treatment for *H. pylori*–related ulcers consists of combinations of up to three antibiotics (*tetracycline, amoxicillin,* and *metronidazole*), Pepto-Bismol™, and treatment with drugs that suppress acid production. Because strains of *H. pylori* resistant to at least one of these antibiotics have already appeared, research is under way to find alternative methods of controlling these infections.

DRASTIC WEIGHT-LOSS TECHNIQUES FAP *p. 871*

At any moment, an estimated 20 percent of the U.S. population is dieting to promote weight loss. In addition to the appearance of "fat farms" and exercise clubs, the use of surgery to promote weight loss has been on the rise. Many of the techniques involve the surgical remodeling of the gastrointestinal tract. **Gastric stapling** attempts to correct an overeating problem by reducing the size of the stomach. A large portion of the gastric lumen is stapled shut, leaving only a small pouch in contact with the esophagus and duodenum. After this surgery, the individual can eat only a small amount before the stretch receptors in the gastric wall become stimulated and a feeling of fullness results. Gastric stapling is a major surgical procedure with many potential complications. In addition, the smooth muscle in the wall of the functional portion of the stomach gradually becomes increasingly tolerant of distension, and the operation may have to be repeated.

In a *jejunal–ileal bypass,* the proximal jejunum is surgically attached to the ileum. This procedure bypasses most of the absorptive area of the small intestine and causes rapid weight loss. However, much of the weight loss is due to chronic diarrhea, and metabolic problems occur. A *gastric bypass* is a surgical procedure that connects the proximal small intestine to a small pouch formed by a superior portion of the stomach. This procedure seems more effective than gastric stapling and is less likely to cause diarrhea or metabolic problems than is a jejunal–ileal bypass.

A more drastic approach involves the surgical removal or bypass of a large portion of the jejunum. This procedure reduces the effective absorptive area, producing a marked weight loss. After the operation, the individual must take dietary supplements to ensure that all the essential nutrients and vitamins can be absorbed before the chyme enters the large intestine. Chronic diarrhea and serious liver disease are potential complications.

LIVER DISEASE FAP *pp. 877, 878*

The liver is the largest and most important visceral organ, and liver disorders affect almost every other vital system in the body. A variety of clinical tests are used to check the functional and physical state of the liver (see Tables A-36 and A-37):

- **Liver function tests** can assess specific functional capabilities.
- A **serum bilirubin assay** indicates how efficiently the liver has been able to extract and excrete bilirubin.
- **Serum and plasma protein assays** can detect changes in the liver's rate of plasma protein synthesis, and **serum enzyme tests** can reveal liver damage by detecting intracellular enzymes in the circulating blood.
- **Liver scans** involve the injection of radioisotope-labeled compounds into the bloodstream. Compounds are chosen that will be selectively absorbed by Kupffer cells, liver cells, or abnormal liver tissues.
- CT scans of the abdominal region are commonly used to provide information about cysts, abscesses, tumors, or hemorrhages in the liver.
- A **liver biopsy** can also be taken by laparoscopy, in which a long needle, commonly guided by CT scans to avoid large blood vessels, is inserted through the abdominal wall. Laparoscopic examination can also reveal gross structural changes in the liver or gallbladder.

The term *liver disease* includes a variety of disorders. We will focus here on two major liver diseases, hepatitis and cirrhosis.

Hepatitis

Hepatitis (hep-a-TĪ-tis) is inflammation of the liver. Viruses that target the liver are responsible for most cases of hepatitis, although some environmental toxins can cause similar symptoms. Six forms of viral hepatitis, A, B, C, D, E, and G, have been identified:

1. **Hepatitis A,** or *infectious hepatitis,* typically results from the ingestion of water, milk, shellfish, or other food contaminated by infected fecal wastes. It has a relatively short incubation period of 2–6 weeks. The disease generally runs its course in a matter of months, and fatalities are rare among individuals under age 40. There is no ongoing chronic infection.
2. **Hepatitis B,** or *serum hepatitis,* is transmitted by the exchange of body fluids during intimate contact. For example, infection can occur through the transfusion of blood products, through a break in the skin or mucosa, or by sexual contact. The incubation period ranges from 1 to 6 months. If a pregnant woman is infected, the newborn baby may become infected at birth. Most people with hepatitis B eliminate the infection. About 5 percent, however, are chronic carriers; these individuals are infectious and experience cumulative liver damage.
3. **Hepatitis C,** originally designated *non-A, non-B hepatitis,* is most commonly transferred from individual to individual through the collection and transfusion of contaminated blood. Since 1990, blood screening procedures have been used to lower the incidence of transfusion-related hepatitis C. Hepatitis C can also be transmitted among intravenous drug users through shared needles, and evidence suggests that it is rarely sexually transmitted. The chronic infectious carrier state of hepatitis C infection produces significant liver damage in roughly half the individuals infected with the virus. Interferon treatment early in the disease can slow the progression of the disease in some patients.
4. **Hepatitis D** is caused by a virus that produces symptoms only in persons already infected with hepatitis B. The transmission of hepatitis D resembles that of hepatitis B. In the United States, the disease is most common among intravenous drug users. The combination of hepatitis B and hepatitis D causes progressive and severe liver disease.
5. **Hepatitis E** resembles hepatitis A in that it is transmitted by the ingestion of contaminated food or water. Hepatitis E is the most common form of hepatitis worldwide, but cases seldom occur in the United States. Hepatitis E infections are most acute and potentially lethal for pregnant women.
6. **Hepatitis G,** the most recently described form, appears in the same populations as does hepatitis C. Little else is known about this type of hepatitis.

FAP Ch. 24

The hepatitis viruses disrupt liver function by attacking and destroying liver cells. An infected individual may develop a high fever, and the liver may become inflamed and tender. As the disease progresses, several hematological parameters change markedly. For example, enzymes normally confined to the cytoplasm of functional liver cells appear in the circulating blood. Normal metabolic regulatory activities become less effective, and blood glucose levels decline. Plasma protein synthesis slows, and the clotting time becomes unusually long. The injured hepatocytes stop removing bilirubin from the circulating blood, and symptoms of jaundice appear.

Hepatitis is either acute or chronic. *Acute hepatitis* is characteristic of hepatitis forms A and E. Almost everyone who contracts hepatitis A or hepatitis E (except in pregnancy) eventually recovers, although full recovery can take several months. Once recovered, infected individuals cannot transmit the disease. Symptoms of acute hepatitis include severe fatigue and jaundice. In *chronic hepatitis,* fatigue is less pronounced and jaundice is rare. *Chronic active hepatitis* is a progressive disorder that can lead to severe medical problems as liver function deteriorates and cirrhosis develops. Common complications include the following:

- The formation of *esophageal varices* (FAP *p. 860*) due to portal hypertension
- Ascites (FAP *p. 851*) due to increased peritoneal fluid production at the elevated venous pressures
- *Bacterial peritonitis,* which may recur for unknown reasons
- *Hepatic encephalopathy,* which is characterized by disorientation and confusion. These symptoms are the result of the disruption of central nervous system function due to some combination of factors in the blood. High ammonia levels and the presence of abnormal concentrations of fatty acids, amino acids, and waste products have been implicated.

Hepatitis B, C, D, and E infections can produce either acute or chronic hepatitis. Individuals with chronic forms are potentially infectious; they may eventually experience liver failure and death as a result of the infection. Roughly 10 percent of hepatitis B patients develop potentially dangerous complications; the percentage is higher for hepatitis C. The prognosis is significantly worse for those infected with both hepatitis B and hepatitis D. Chronic active hepatitis infections (especially B or C) may also lead to liver cancer. The long-term effects of hepatitis G infection remain to be determined.

Passive immunization with pooled immunoglobulins is available for both the hepatitis A and B viruses. (Active immunization for hepatitis A and B is also available.) No vaccines are available to stimulate immunity to hepatitis forms C, D, E, or G.

Cirrhosis

The underlying problem in **cirrhosis** (sir-Ō-sis) appears to be the widespread destruction of hepatocytes by exposure to drugs (especially alcohol), viral infection, ischemia, or a blockage of the hepatic ducts. Two processes are involved in producing the symptoms. Initially, the damage to hepatocytes leads to the formation of extensive areas of scar tissue that branch throughout the liver. The surviving hepatocytes then undergo repeated cell divisions, but the fibrous tissue prevents the new hepatocytes from achieving a normal arrangement of lobules. As a result, the liver

gradually converts from an organized assemblage of lobules to a fibrous aggregation of poorly functioning cell clusters. Jaundice, ascites, and other symptoms may appear as the condition progresses.

CHOLECYSTITIS FAP *p. 880*

An estimated 16–20 million people in the United States have gallstones that go unnoticed; small stones are commonly flushed down the bile duct and eliminated. If gallstones enter and jam the cystic duct or bile duct, the painful symptoms of **cholecystitis** (kō-lē-sis-TĪ-tis) appear. Approximately 1 million people develop acute symptoms each year. The gallbladder becomes swollen and inflamed, infections can develop, and symptoms of *obstructive jaundice* develop.

A blockage that does not work its way down the bile duct to the duodenum must be removed or destroyed. Small gallstones can be chemically dissolved. In most cases, surgery is required to remove large gallstones. (The gallbladder is also removed to prevent recurrence.) Many surgeons are now using a laparoscope, inserted through three or four small abdominal incisions, to perform this surgery.

COLON INSPECTION AND CANCER FAP *p. 884*

Each year in the United States, roughly 133,600 new cases of *colorectal cancer* are diagnosed and 56,800 deaths result from this condition. This deadly form of cancer is normally diagnosed in persons over 50 years of age. Primary risk factors for colorectal cancer include (1) a diet rich in animal fats and low in fiber, (2) *inflammatory bowel disease,* and (3) a number of inherited disorders that promote epithelial tumor formation along the intestines.

Successful treatment of colorectal cancer depends on the early identification of the condition. It is believed that most colorectal cancers begin as small, localized mucosal tumors, or **polyps** (POL-ips), that grow from the intestinal wall. The prognosis improves dramatically if cancerous polyps are removed before metastasis has occurred. If the tumor is restricted to the mucosa and submucosa, the 5-year survival rate is higher than 90 percent. If it extends into the serosa, the survival rate drops to 70–85 percent; after metastasis to other organs, the rate drops to about 5 percent.

One of the early signs of polyp formation is the appearance of blood in the feces. Unfortunately, many people ignore small amounts of blood in fecal materials, because they attribute the bleeding to "harmless" hemorrhoids. This offhand diagnosis should always be professionally verified.

When blood is detected in the feces, X-ray techniques are commonly used as a first step in diagnosis. In the usual procedure, a large quantity of a liquid barium solution is introduced by enema. Because this solution is radiopaque, the X-rays will reveal any intestinal masses, such as a large tumor, blockages, or structural abnormalities.

Typically, the most precise surveys can be obtained with the aid of a flexible **colonoscope** (ko-LON-o-skōp). This instrument permits direct visual inspection of the lining of the large intestine. The colonoscope is also used to biopsy the mucosal lining and to remove polyps, thereby avoiding the potential complications of traditional surgery.

INFLAMMATORY BOWEL DISEASE FAP *p. 884*

The general term **colitis** (ko-LĪ-tis) is used to indicate a condition characterized by an inflammation of the colon. **Irritable bowel syndrome** is characterized by diarrhea, constipation, or an alternation between the two. When constipation is the primary problem, this condition is called a *spastic colon,* or *spastic colitis.* Irritable bowel syndrome may have a partly psychological basis; the mucosa is usually normal in appearance but the motility can be abnormal. Bulking agents such as fiber or psyllium *(Metamucil®)* and drugs that affect the enteric nervous system can help control symptoms. **Inflammatory bowel disease,** such as *ulcerative colitis,* involves chronic inflammation of the digestive tract, most commonly affecting the colon. The mucosa becomes inflamed and ulcerated; extensive areas of scar tissue develop; and colonic function deteriorates. Acute bloody diarrhea and cramps are common symptoms. Fever and weight loss are also typical complaints. The treatment of inflammatory bowel disease normally involves anti-inflammatory drugs and corticosteroids that reduce inflammation. In severe cases, oral or intravenous fluid replacement is required. In cases that do not respond to other therapies, immunosuppressive drugs, such as *cyclosporine,* may be used to good effect.

The treatment of severe inflammatory bowel disease may also involve a **colectomy** (ko-LEK-to-mē), the removal of all or a portion of the colon. If a large part or even all of the colon must be removed, normal connection with the anus cannot be maintained. Instead, the end of the intact digestive tube is sutured to the abdominal wall, and wastes then accumulate in a plastic pouch or sac attached to the opening. If the attachment involves the colon, the procedure is a **colostomy** (ko-LOS-to-mē); if the ileum is involved, it is an **ileostomy** (il-ē-OS-to-mē).

DIARRHEA FAP *p. 887*

Diarrhea is characterized as the production of copious, watery stools. Many conditions result in diarrhea, and we will consider only a representative sampling. Most

infectious diarrhea involves organisms that are shed in the stool and then spread from person to person (or from person to food to person). Proper hand washing after defecation, clean drinking water, and good sewage disposal are the best preventive measures.

Gastroenteritis

An irritation of the small intestine can lead to a series of powerful peristaltic contractions that eject the contents of the small intestine into the large intestine. An extremely powerful irritating stimulus produces a "clean sweep" of the absorptive areas of the digestive tract. Vomiting clears the stomach, duodenum, and proximal jejunum, and peristaltic contractions evacuate the distal jejunum and ileum. Bacterial toxins, viral infections, and various poisons may produce these extensive gastrointestinal responses. Conditions affecting primarily the small intestine are usually referred to as a form of **enteritis** (en-ter-Ī-tis). If both vomiting and diarrhea are present, the term **gastroenteritis** (gas-trō-en-ter-Ī-tis) may be used instead.

Traveler's Diarrhea

Traveler's diarrhea, a form of infectious diarrhea generally caused by a bacterial or viral infection, develops because the irritated or damaged mucosal cells are unable to maintain normal absorption levels. The irritation stimulates the production of mucus, and the damaged cells and mucous secretions add to the volume of feces produced. Despite the inconvenience, this type of diarrhea is usually temporary, and mild diarrhea is probably a reasonably effective method of rapidly removing an intestinal irritant. Drugs, such as *Lomotil®*, that prevent peristaltic contractions in the colon slow the diarrhea but leave the irritant intact, and the symptoms may return with a vengeance when the drug effects fade. A 5-day course of antibiotics may be effective in controlling diarrhea due to bacterial infection.

Giardiasis

Giardiasis is an infection caused by the protozoans *Giardia intestinalis* and *G. lamblia* (Figure A-57•). These pathogens can colonize the duodenum and jejunum and interfere with the normal absorption of lipids and carbohydrates. Many people do not develop acute symptoms but act as carriers who can spread the disease. Acute symptoms usually appear within 3 weeks of initial exposure. Diarrhea, abdominal pains, cramps, nausea, and vomiting are the primary complaints. These symptoms persist for 5–7 days; some patients are subject to relapses and chronic bloating, diarrhea, and weight loss. Treatment typically consists of the oral administration of drugs, such as *quinacrine* or *metronidazole,* that can kill the protozoan.

The transmission of giardiasis requires that food or water be contaminated with feces that contain *cysts,* resting stages of the protozoan that are produced during passage through the large intestine. Rates of infection are highest (1) in developing countries with poor sanitation, (2) among campers drinking surface water, (3) among individuals with impaired immune systems (as in AIDS), and (4) among toddlers and young children. The cysts can survive in the environment for months, and they are not killed during the chlorine treatment used to kill bacteria in drinking water. Travelers are advised to boil or ultrafilter water and to heat food properly before eating it, as these preventive measures will destroy the cysts.

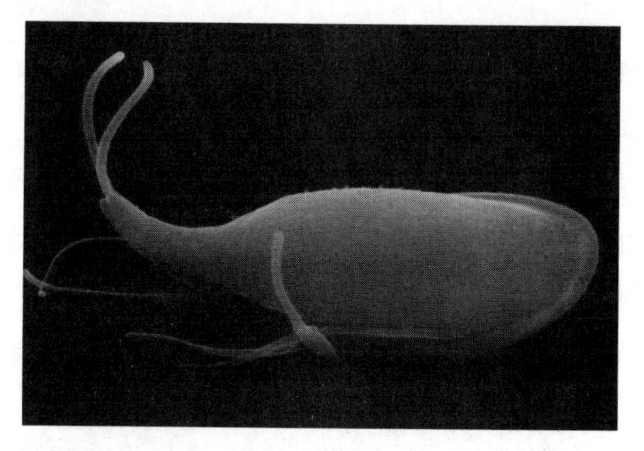

•**FIGURE A-57** ***Giardia.*** *Giardia* is a flagellated protozoan that infects the small intestine.

FAP Ch. 24

Cholera Epidemics

Cholera epidemics are most common in areas where sanitation is poor, where drinking water is contaminated by fecal wastes, and where eating raw fish or shellfish is popular. After an incubation period of 1–2 days, the symptoms of nausea, vomiting, and diarrhea persist for 2–7 days. Fluid loss during the worst stage of the disease can approach 1 liter per hour. This dramatic loss causes a rapid drop in blood volume, leading to acute hypovolemic shock and damage to the kidneys and other organs.

Treatment consists of oral or intravenous fluid replacement while the disease runs its course. Antibiotic therapy may also prove beneficial. A vaccine is available, but its low success rate (40–60 percent) and short duration (4–6 months of protection) make it relatively ineffective in preventing or controlling cholera outbreaks. More than 500,000 cases of cholera were reported during an epidemic that began in Peru in 1991. The death rate was 0.5 percent—a remarkably low rate, compared with death rates in other outbreaks in the twentieth century, which were as high as 60 percent.

LACTOSE INTOLERANCE FAP *p. 890*

Lactose intolerance is a malabsorption syndrome, induced by the ingestion of dairy foods, that results from the lack of the enzyme *lactase* at the brush border of the intestinal epithelium. This condition poses more of a problem than would be expected if the only out-

come were the inability to use the disaccharide lactose. Undigested lactose provides a particularly stimulating energy source for the bacterial inhabitants of the colon. The result is increased intestinal gas generation, cramps, and diarrhea. These problems can develop if the individual drinks more than 8 oz. of milk or eats similar amounts of other dairy products.

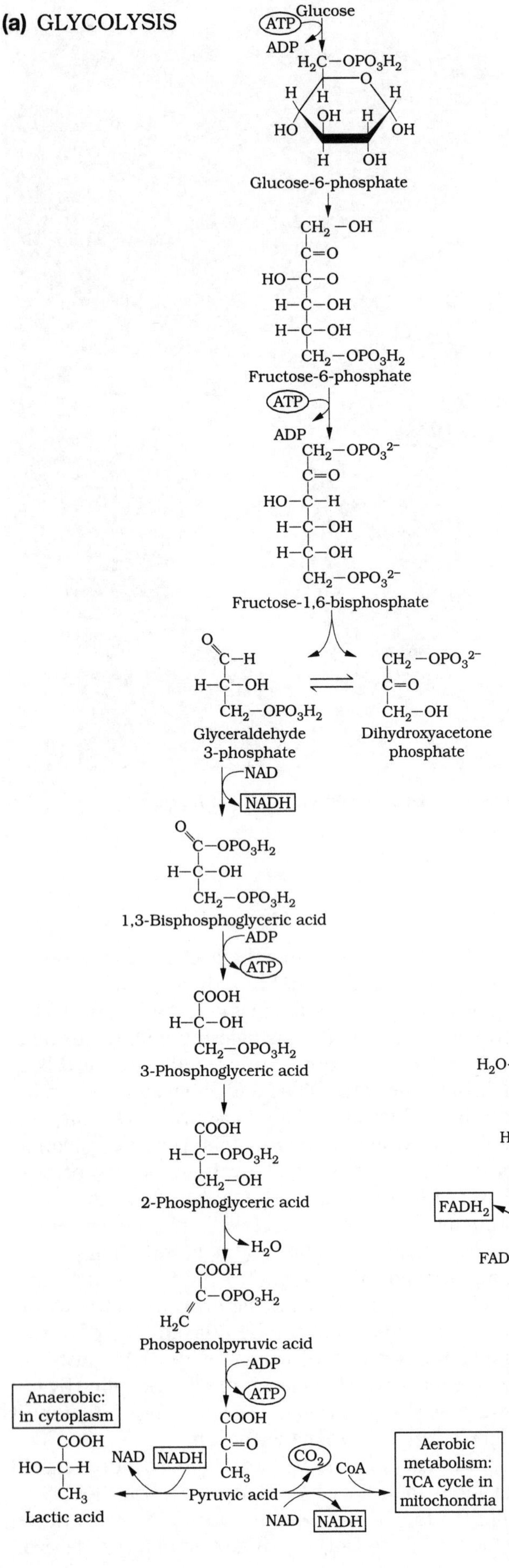

Lactose intolerance appears to have a genetic basis. Infants produce lactase to digest milk, but older children and adults may stop producing this enzyme. In some populations, lactase production continues throughout adulthood. Only about 15 percent of Caucasians develop lactose intolerance, whereas estimates ranging from 80 to 90 percent have been suggested for the adult African and Asian populations. These differences affect dietary preferences in these groups. Food relief efforts must take such preferences into account; for example, shipping powdered milk to starvation areas in Africa can make matters worse if supplies are distributed to adults rather than to children.

A CLOSER LOOK: AEROBIC METABOLISM

FAP p. 906

Mitochondria use organic substrates, such as the pyruvic acid produced by glycolysis (Figure A-58a•), from the surrounding cytoplasm. Aerobic metabolism includes the

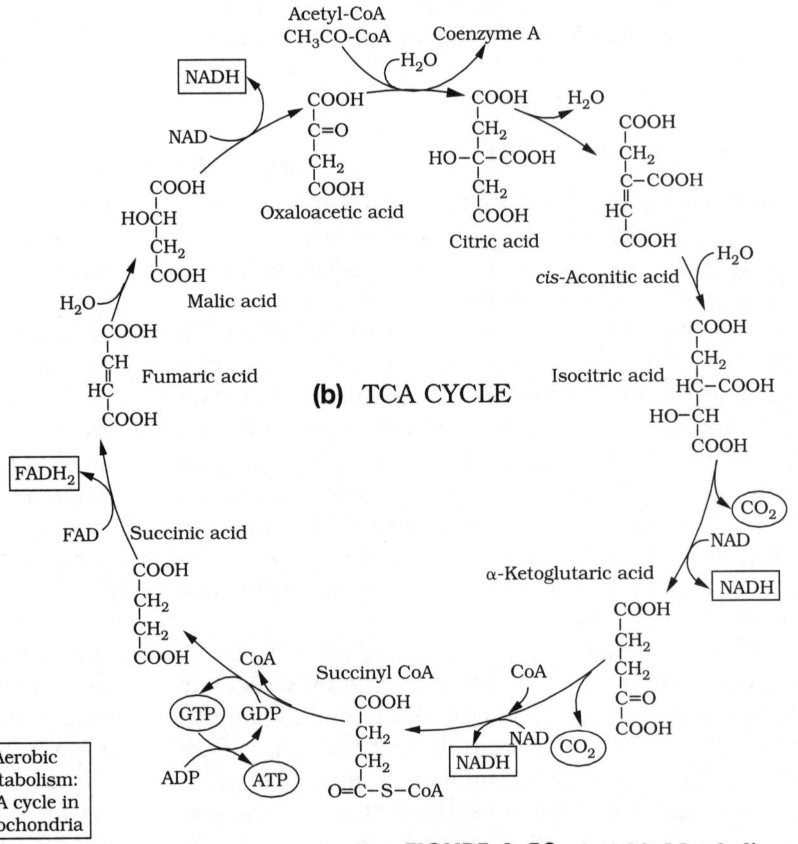

•**FIGURE A-58 Aerobic Metabolism.** **(a)** Glycolysis. **(b)** The TCA cycle.

FAP Ch. 25

tricarboxylic acid, or TCA, cycle (Figure A-58b•) and the oxidation–reduction reactions of the electron transport system, or ETS (FAP *p. 908*).

PROTEIN DEFICIENCY DISEASES FAP *p. 919*

Regardless of the energy content of the diet, if it is deficient in essential amino acids, the individual will be malnourished to some degree. In a **protein deficiency disease,** protein synthesis decreases throughout the body. As protein synthesis in the liver fails to keep pace with the breakdown of plasma proteins, plasma osmolarity falls. This reduced osmolarity results in a fluid shift as more water moves out of the capillaries and into interstitial spaces, the peritoneal cavity, or both. The longer the individual remains in this state, the more severe the ascites and edema that result.

This clinical scenario is relatively common in developing countries, where dietary protein is often scarce or prohibitively expensive. Growing infants suffer from **marasmus** (ma-RAZ-mus) when deprived of adequate proteins and calories. **Kwashiorkor** (kwash-ē-OR-kor) occurs in children whose protein intake is inadequate, even if the caloric intake is acceptable (Figure A-59•). In each case, additional complications include damage to the developing brain. It is estimated that more than 100 million children worldwide suffer from protein deficiency diseases such as these. War and civil unrest that disrupt local food production and distribution have been more instrumental than a shortage of food in itself in producing recent famines.

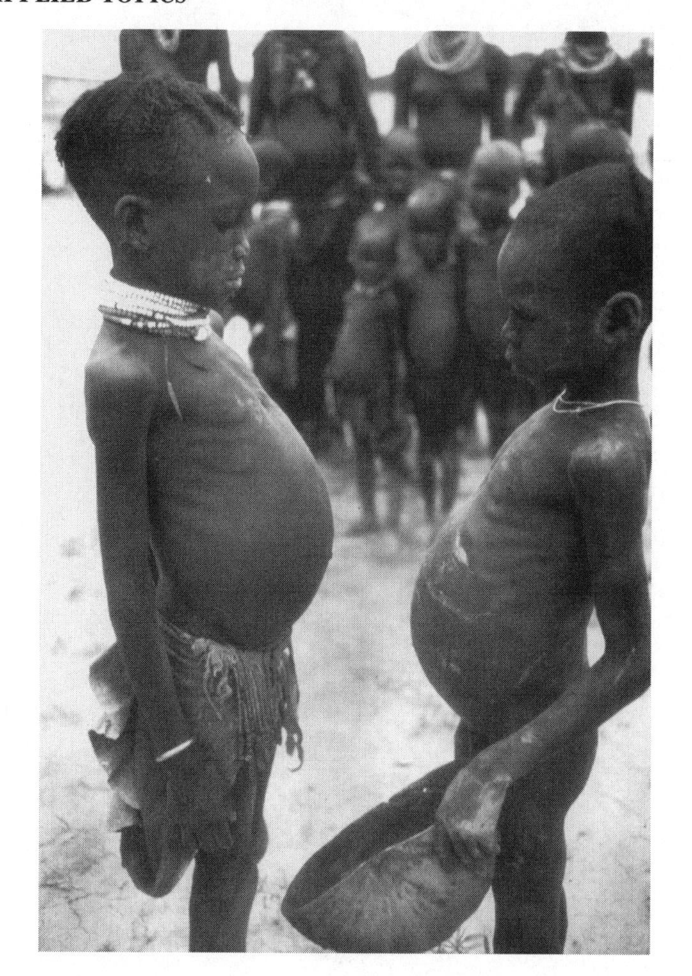

•FIGURE A-59 Children with Kwashiorkor

PHENYLKETONURIA FAP *p. 919*

Phenylketonuria (fen-il-kē-tō-NOO-rē-uh), or **PKU,** is one of about 130 disorders that have been traced to the lack of a specific enzyme. Individuals with PKU are deficient in a key enzyme, *phenylalanine hydroxylase,* responsible for the conversion of the amino acid phenylalanine to tyrosine. This reaction is a necessary step in the synthesis of tyrosine, an important component of many proteins and the structural basis for a pigment (melanin), two hormones (epinephrine and norepinephrine), and two neurotransmitters (dopamine and norepinephrine). In addition, this conversion must occur before the carbon chain of a phenylalanine molecule can be recycled or broken down in the TCA cycle.

FAP Ch. 25

If PKU is undetected and untreated at birth, plasma concentrations of phenylalanine gradually escalate from normal (about 3 mg/dl) to levels above 20 mg/dl. High plasma concentrations of phenylalanine affect overall metabolism, and a number of unusual byproducts are excreted in the urine. The synthesis and degradation of proteins and other amino acid derivatives are affected. Developing neural tissue is most strongly influenced by these metabolic changes, and severe brain damage and mental retardation result.

Fortunately, this condition is detectable shortly after birth, when the infant has digested milk or formula, because this digestion produces elevated levels of phenylalanine in the blood and phenylketone, a metabolic byproduct, in the blood and urine. (During pregnancy, the normal mother metabolizes phenylalanine for the PKU fetus, so fetal levels are normal prior to delivery.) Treatment consists of controlling the amount of phenylalanine in the diet while plasma concentrations are monitored. This treatment is most important in infancy and childhood, when the nervous system is developing. Once the child has grown, dietary restriction of phenylalanine can be relaxed, except during pregnancy. A pregnant woman with PKU must protect the fetus from high levels of phenylalanine by following a strict phenylalanine-restricted diet that must begin before the pregnancy occurs.

Although the dietary restrictions are more relaxed for adults than for children, persons with PKU must still monitor the ingredients used in the preparation of their meals. Because tyrosine cannot be synthesized

from dietary phenylalanine, the diet of these persons must also contain adequate amounts of tyrosine. One popular artificial sweetener, *Nutrasweet®*, consists of phenylalanine and aspartic acid. The consumption of food or beverages that contain this sweetener can therefore cause problems for PKU sufferers.

In its most severe form, PKU affects approximately 1 infant in 20,000. Individuals who carry only a single gene for PKU will produce the affected enzyme but in lesser amounts. These individuals are asymptomatic but have slightly elevated phenylalanine levels in their blood. Statistical analysis of the incidence of fully developed PKU indicates that as many as 1 person in 70 may carry a gene for this condition.

GOUT

FAP *p. 919*

At concentrations above 7.4 mg/dl, body fluids are supersaturated with uric acid. Although symptoms may not appear at once, uric acid crystals begin to precipitate in body fluids. The condition that results is called **gout.** The severity of the symptoms depends on the amount and location of the crystal deposits.

Initially, the joints of the extremities, especially the metatarsal/phalangeal joint of the great toe, are likely to be affected. This intensely painful condition, called *gouty arthritis,* may persist for several days and then disappear for a period of days to years. Recurrences often involve other joints and may produce generalized fevers. Precipitates may also form within cartilages, in synovial fluids, in tendons or other connective tissues, or in the kidneys and urine. At serum concentrations of over 12–13 mg/dl, half the patients will develop kidney stones, and kidney function can be affected to the point of kidney failure.

The incidence of gout, ranging from 0.13 to 0.37 percent of the population, is much lower than that of hyperuricemia. Only about 5 percent of persons with gout are women, and most affected males are over 50. Foods high in purines, such as meats or fats, may aggravate or initiate the onset of gout. These foods tend to cost more than carbohydrates, so "rich foods" have often been associated with this condition.

OBESITY

FAP *p. 923*

Regulatory obesity results from a failure to regulate food intake so that appetite, diet, and activity are in balance. Most instances of obesity fall within this category. In most instances, there is no obvious organic cause, although in rare cases the problem arises because some disorder, such as a tumor, affects the hypothalamic centers that deal with appetite and satiation. Typically, chronic overeating is thought to result either from inactivity or from psychological or sociological factors, such as stress, neurosis, long-term habits, and family or ethnic traditions. Genetic factors may also be involved, but because the psychological and social environment plays such an important role in human behavior, the exact connections have been difficult to assess. In short, individuals with regulatory obesity overeat for some reason and thereby extend the duration and magnitude of the absorptive state.

In **metabolic obesity,** the condition is secondary to some underlying organic malfunction that affects cell and tissue metabolism. For example, some cases of obesity have been linked to reduced insulin sensitivity due to a reduction in the number of insulin receptors in adipose tissue and in skeletal muscle. Metabolic obesity is commonly associated with chronic hypersecretion or hyposecretion of metabolically active hormones, such as insulin, glucocorticoids, or thyroxine.

Categorizing an obesity problem is less important in a clinical setting than is determining the degree of obesity and the number and severity of the related complications. The affected individuals are at a high risk of developing diabetes, hypertension, and coronary artery disease as well as gallstones, thrombi or emboli, hernias, arthritis, varicose veins, and some forms of cancer. A variety of treatments may be considered, ranging from behavior modification, nutritional counseling, psychotherapy, and exercise programs to gastric stapling, a jejunal–ileal bypass, or a partial gastric bypass.

Liposuction

One much-publicized method of battling obesity is the process of liposuction. **Liposuction** is a surgical procedure for the removal of unwanted adipose tissue. Adipose tissue is flexible but not as elastic as areolar tissue, and it tears relatively easily. In liposuction, a small incision is made through the skin and a tube is inserted into the underlying adipose tissue. Suction is then applied. Because adipose tissue tears easily, chunks of tissue containing adipocytes, other cells, fibers, and ground substance can be vacuumed away. Estimates of the number of liposuctions, among the most common cosmetic surgeries performed today, range from 175,000–300,000 in 2000.

This practice has received a lot of news coverage, and many advertisements praise the technique as easy, safe, and effective. In fact, it is not always easy, and it can be dangerous and have limited effectiveness. The density of adipose tissue varies from place to place in the body and from individual to individual, and it is not always easy to suck through a tube. An anesthetic must be used to control pain, and anesthesia always poses risks; blood vessels are stretched and torn, and extensive bleeding can occur. Heart attacks, pulmonary embolism, and fluid balance problems can develop, with fatal results. The death rate for this procedure is 1 in 5000. Finally, adipose tissue can repair itself, and adipocyte populations recover over time. The only way to ensure that fat lost through liposuction will not return is to adopt a lifestyle that includes a proper diet and adequate exercise. Over time, such

a lifestyle can produce the same weight loss, *without liposuction,* eliminating the surgical expense and risk.

ADAPTATIONS TO STARVATION FAP p. 926

Figure A-60• shows changes in the metabolic stores of a 70-kg individual during prolonged starvation. Carbohydrate utilization declines almost immediately as the stores are depleted. As blood glucose levels decline, gluconeogenesis accelerates, using glycerol, amino acids, and lactic acid. The glycerol is provided by adipocytes; the amino acids and lactic acid are provided primarily by skeletal muscle. At this point, the kidneys begin to assist the liver by deaminating amino acids and generating additional glucose molecules.

Gluconeogenesis is accompanied by an increase in circulating ketone bodies—some derived from ketogenic amino acids, others from the catabolism of fatty acids. As the starvation stress continues, peripheral tissues further restrict their glucose utilization. The ketone bodies generated by fatty acid catabolism become the primary energy source.

The fasting individual gradually becomes weak and lethargic as peripheral systems are weakened by protein catabolism and stressed by pH changes. Buffer systems are challenged by the circulating amino acids, lactic acid, and ketone bodies, and ketoacidosis becomes a potential problem. Under these circumstances, most tissues begin catabolizing ketone bodies almost exclusively, and in extreme starvation more than 90 percent of the daily energy demands are met by the oxidation of ketone bodies. At this stage, even neural tissue relies on ketone bodies to supplement declining glucose supplies.

Structural proteins are the last to be mobilized, with other forms, such as the contractile proteins of skeletal muscle, more readily available. When peripheral tissues catabolize proteins, the amino acids are exported to the liver, where they can be safely deaminated. The carbon fragments are then catabolized to provide ATP, or they are used to manufacture glucose molecules or ketone bodies that can be broken down by peripheral tissues.

When lipid reserves are exhausted, crises soon follow. On a gram-for-gram basis, cells must catabolize almost twice as much protein as lipid to obtain the same energy benefits. Making matters worse, by this time most of the easily mobilized proteins have already been broken down. As structural proteins are disassembled, a variety of dangerous effects may appear. Accelerated protein catabolism causes problems with fluid balance, because the nitrogenous and acidic wastes must be eliminated in urine. These waste products are excreted in solution, and the more waste products eliminated, the greater the associated water loss. An increase in urinary water losses can lead to dehydration, and the combination of dehydration and acidosis can cause kidney damage.

When glucose concentrations can no longer be sustained above 40–50 mg/dl, the individual becomes disoriented and confused. The eventual cause of death is kidney failure, ketoacidosis, protein deficiency, or hypoglycemia.

How long does it take to reach this critical state? That essentially depends on the size of the person's lipid reserves. Prolonged starvation for most people would last about 8 weeks, but the truly obese can hold out far longer. With adequate water and vitamin supplements, an 8-month fast has been used as a weight-loss technique. (This technique was an emergency treatment rather than a diet plan, because prolonged fasting can result in kidney damage or severe ketoacidosis.)

Metabolic adjustments during starvation are coordinated primarily by the endocrine system. The glucocorticoids produced by the adrenal cortex are the

FAP Ch. 25

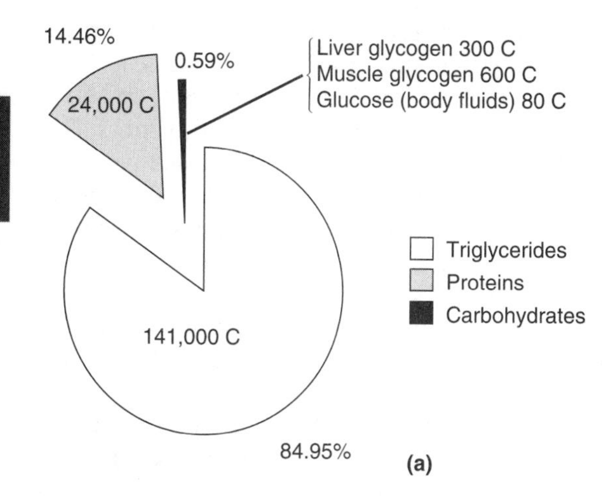

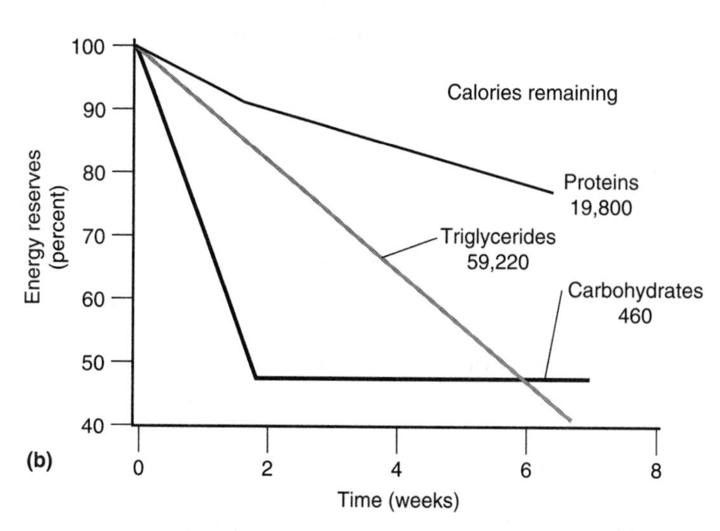

•**FIGURE A-60 Metabolic Reserves and the Effects of Starvation. (a)** Estimated metabolic reserves of a 70-kg individual. **(b)** Projected effects of prolonged starvation on the metabolic reserves of the same individual.

most important hormones, aided by growth hormone from the pituitary gland. The effects of these hormones on peripheral tissues are included in Table 25-1 of the main text *(p. 921)*.

THERMOREGULATORY DISORDERS

ACCIDENTAL HYPOTHERMIA **FAP *p. 936***

If body temperature drops significantly below normal levels, the thermoregulatory system begins to lose sensitivity and effectiveness. Cardiac output and respiratory rate decrease, and if the core temperature falls below 28°C (82°F), cardiac arrest is likely. The individual then has no heartbeat, no respiratory rate, and no response to external stimuli, even painful ones. Body temperature continues to decline, and the skin turns blue or pale and cold.

At this point, we would probably assume that the individual has died. But because metabolic activities have decreased systemwide, the victim may still be saved, even after several hours have elapsed. Treatment consists of cardiopulmonary support and gradual rewarming, both external and internal. The skin can be warmed up to 45°C (110°F) without damage; warm baths or blankets can be used. One effective method of raising internal temperatures involves the introduction of warm saline solution into the peritoneal cavity.

Hypothermia is a significant risk for those engaged in water sports, and it may complicate treatment of a drowning victim. Water absorbs heat roughly 27 times as fast as air does, and the body's heat-gain mechanisms are unable to keep pace over long periods or when faced with a large temperature gradient. But hypothermia in cold water does have a positive side. On several occasions, small children who have drowned in cold water have been successfully revived after periods of up to 4 hours. Children lose body heat quickly, and their systems stop functioning very quickly as body temperature declines. This rapid drop in temperature prevents the oxygen starvation and tissue damage that would otherwise occur when breathing stops.

Resuscitation is not attempted if the individual has actually frozen. Water expands roughly 7 percent during ordinary freezing, and cell membranes throughout the body are destroyed in the process. Very small organisms can be frozen and subsequently thawed without ill effects, because their surface-to-volume ratio is enormous and the freezing process occurs so rapidly that ice crystals never form.

FEVERS **FAP *p. 937***

When interleukin-1 increases the "thermostat setting" of the preoptic center of the hypothalamus, the heat-gain center is activated. The individual feels cold and may curl up in a blanket. Shivering may begin and continue until the temperature at the preoptic area corresponds to the new setting. The fever passes when the thermostat is reset to normal. The **crisis phase** then ensues as the heat-loss center is stimulated. The individual feels unbearably warm and discards the blanket; the skin is flushed, and the sweat glands work furiously to bring the temperature down. Repeated cycles of this type constitute the "chills and fever" pattern of many febrile illnesses.

Fevers are classified as *chronic* or *acute*. Chronic fevers may persist for weeks or months as the result of infections, cancers, or thermoregulatory disorders. In some cases, a discrete cause cannot be determined, leading to a classification as a *fever of unknown origin (FUO)*. Acute hyperthermia, as seen during heat stroke, in certain diseases, or in some marathon runners, is life-threatening. Immediate treatment may involve cooling the individual in an ice bath (to increase conduction) or giving alcohol rubs (to increase evaporation) in combination with the administration of **antipyretic drugs,** such as aspirin or acetaminophen.

[EXPLORE]

Additional resources, such as interactive tutorials, critical-thinking questions, clinical problems, and case studies, are available through the CD-ROM and Website for your textbook. To begin your exploration, launch the *Martini Interactive* CD-ROM or visit the Companion Website (http://www.prenhall.com/martini/fap5) and select Chapter 24 or 25.

The Urinary System

The urinary system consists of the kidneys, where urine production occurs, and the conducting system, which transports and stores urine prior to its elimination from the body. The conducting system includes the ureters, the urinary bladder, and the urethra. Although the kidneys perform all the vital functions of the urinary system, problems with the conducting system can have direct and immediate effects on renal function.

THE PHYSICAL EXAMINATION AND THE URINARY SYSTEM

The primary symptoms of urinary system disorders are pain and changes in the frequency of urination. The nature and location of the pain can provide clues to the source of the problem (see Figure 15-4, FAP *p. 485*). For example,

- Pain in the superior pubic region may be associated with urinary bladder disorders.
- Pain in the superior lumbar region or the flank that radiates to the right upper quadrant or left upper quadrant can be caused by kidney infections such as *pyelonephritis* (FAP *p. 976*).
- *Dysuria* (painful or difficult urination) can occur with *cystitis* or *urethritis* (FAP *p. 976*) or with *urinary obstructions* (FAP *p. 974*). In males, enlargement of the prostate gland can lead to compression of the urethra and dysuria.

Individuals with urinary system disorders may urinate more or less frequently than usual and may produce normal or abnormal amounts of urine:

- An irritation of the lining of the ureters or urinary bladder can lead to the desire to urinate with increased frequency, although the total amount of urine produced each day remains normal. When these problems exist, the individual feels the urge to urinate when the urinary bladder volume is very small. The irritation may result from trauma, urinary bladder infection (cystitis) or tumors, or increased acidity of the urine.
- *Incontinence,* an inability to control urination voluntarily, may involve periodic involuntary urination—a continual, slow trickle of urine from the urethra. Incontinence results from urinary bladder or urethral problems, damage or weakening of the muscles of the pelvic floor, or interference with normal sensory or motor innervation in the region. Renal function and daily urinary volume are normal.
- In *urinary retention,* renal function is normal, at least initially, but urination does not occur. Urinary retention in males commonly results from enlargement of the prostate and compression of the prostatic urethra. In both genders, urinary retention can result from the obstruction of the outlet of the urinary bladder or from central nervous system damage, such as a stroke or damage to the spinal cord.
- Changes in the volume of urine produced by a normally hydrated person indicate problems either at the kidneys or with the control of renal function. **Polyuria,** the production of excessive amounts of urine, results from hormonal or metabolic problems, such as those associated with *diabetes* (p. 103), or from damage to the glomeruli, as in *glomerulonephritis* (FAP *p. 952*). **Oliguria** (a urine volume of 50–500 ml/day) and **anuria** (0–50 ml/day) are conditions that indicate serious kidney problems and potential renal failure. Renal failure can occur with *heart failure* (p. 124), renal ischemia, *circulatory shock* (FAP *p. 720*), burns (FAP *p. 162*), pyelonephritis (FAP *p. 976*), hypovolemia, and a variety of other disorders.

Important clinical signs of urinary system disorders include the following:

- *Hematuria,* the presence of red blood cells in urine, indicates bleeding at either the kidneys or the conducting system. Hematuria producing dark red or tea-colored urine typically indicates bleeding in the kidney, and hematuria producing bright red urine indicates bleeding in the inferior portion of the urinary tract (urinary bladder or urethra). Hematuria most commonly occurs with trauma to the kidneys, calculi (kidney stones), tumors, or urinary tract infections.
- *Hemoglobinuria* is the presence of hemoglobin in urine. Hemoglobinuria indicates increased hemolysis of red blood cells in the bloodstream due to cardiovascular or metabolic problems. Conditions that result in hemoglobinuria include the *thalassemias* (p. 109), *sickle cell anemia* (p. 112), *hypersplenism* (p. 134), and some autoimmune disorders.

- Changes in urine color accompany some renal disorders. For example, urine becomes (1) cloudy due to the presence of bacteria, lipids, crystals, or epithelial cells; (2) red or brown from hemoglobin or myoglobin; (3) blue-green from bilirubin; or (4) brown-black from excessive concentration. Not all color changes are abnormal, however. Some foods and several prescription drugs can cause changes in urine color. A serving of beets can give urine a reddish color, whereas eating rhubarb can give urine an orange tint, and B vitamins turn it a vivid yellow.
- Renal disorders with pronounced *proteinuria* (plasma proteins in urine) typically lead to a generalized edema in peripheral tissues. Facial swelling, especially around the eyes, is common.
- A fever commonly develops when the urinary system is infected by pathogens. Urinary bladder infections (cystitis) typically result in a low-grade fever; kidney infections, such as pyelonephritis, can produce very high fevers.

During the physical assessment, percussion or palpation can be used to check the status of the kidneys and urinary bladder. The kidneys lie in the *costovertebral area,* the region bounded by the lumbar spine and the twelfth rib on either side. To detect tenderness due to kidney inflammation, the examiner gently thumps a fist over each flank. This usually does not cause pain unless the underlying kidney is inflamed.

The urinary bladder can be palpated just superior to the pubic symphysis. However, on the basis of palpation alone, urinary bladder enlargement due to urine retention can be difficult to distinguish from the presence of an abdominal mass.

Many procedures and laboratory tests are used in the diagnosis of urinary system disorders. The functional anatomy of the urinary system can be examined by using a variety of sophisticated procedures. For example, an X-ray of the kidneys, ureter, and urinary bladder (Figure 26-17, FAP *p. 972*) is taken after the administration of a radiopaque compound that will enter the urine, creating an **intravenous pyelogram** (PĪ-el-ō-gram), or **IVP.** This procedure, sometimes called an *EU (excretory urogram),* permits the detection of unusual kidney, ureter, or urinary bladder structures and masses. Computerized tomography (CT) scans or ultrasound scans may also provide useful information about localized abnormalities; representative scans of the kidneys are shown in the Scanning Atlas, Scans 12 and 13. Other diagnostic procedures and important laboratory tests are detailed in Table A-39. Figure A-61• (p. 177) outlines the major classes of disorders of the urinary system.

TABLE A-39 Examples of Tests Used in the Diagnosis of Urinary System Disorders

Diagnostic Procedure	Method and Result	Representative Uses
Cystoscopy	A small tube (cystoscope) is inserted along urethra into urinary bladder to permit visualization of lining of the urethra, urinary bladder, and ureteral openings within the bladder.	Used to obtain a biopsy specimen or to remove stones (calculi) and small tumors; provides direct visualization of urethra, urinary bladder and ureteral openings
Retrograde pyelography	Radiopaque dye is injected into ureters through a catheter in cystoscope inserted into urinary bladder; X-ray films are then taken.	Detects obstructions of ureter caused by tumors, calculi, or strictures; visualizes renal pelvis and ureters without relying on renal filtration (useful if renal function is impaired)
Renal biopsy	Using ultrasound as a guide, biopsy needle is inserted through back and into kidney. Specimen is then removed for analysis.	Determines cause of renal disease; detects rejection of transplanted kidney; used to perform tumor biopsy
Intravenous pyelography (IVP)	Dye injected intravenously is filtered at kidney and excreted into urinary tract; X-rays are then taken to view kidneys, ureters, and urinary bladder.	Determines presence of kidney disease, obstructions such as calculi or tumors, or anatomical abnormalities; relies on renal filtration of contrast medium
Cystography	Dye is inserted through catheter placed in urethra and threaded into urinary bladder. X-rays are then taken.	Identifies tumors of urinary bladder and rupture of bladder by trauma; if X-rays are taken as patient voids, determines reflux of urine from urinary bladder to ureter

FAP Ch. 26

TABLE A-39 Examples of Tests Used in the Diagnosis of Urinary System Disorders *(continued)*

Laboratory Test	Normal Values	Significance of Abnormal Values
URINALYSIS		
pH	4.6–8.0	Alkaline or acidic urine may indicate increased or decreased blood pH.
Color	Pale yellow-amber	Color may change with certain drugs and foods; dark red-brown urine may indicate bleeding from kidney; bright red blood comes from lower urinary tract; some bacterial infections cause green tint.
Appearance	Clear	Clouded urine may result from bacterial infection or from certain foods.
Odor	Aromatic	Acetone odor occurs in diabetic ketoacidosis; asparagus in the diet gives a distinctive odor.
Specific gravity	1.005–1.030	Increased in dehydration, increased ADH production, heart failure, glycosuria, or proteinuria; decreased in diabetes insipidus or renal failure
Protein	<100 mg/24 h	Increased protein loss occurs in kidney infections or inflammation and after strenuous exercise.
Glucose	None	Glucose appears in urine in cases of diabetes mellitus or Cushing's disease and after corticosteroid therapy.
Ketones	None	Appear in ketoacidosis or poorly controlled cases of diabetes mellitus, during dehydration, and after several hours of fasting
URINE ELECTROLYTES		
Sodium	40–220 mEq/day	Increased levels occur with dehydration; decreased levels occur with renal, liver, or congestive heart failure.
Potassium	40–80 mEq/l over 24 hours	Elevated levels occur with diuretics, dehydration, and starvation.
BLOOD TESTS		
Electrolytes (serum)		
Sodium	Adults: 135–145 mEq/l	Increased levels occur with severe dehydration; decreased levels occur with SIADH, renal failure, and diuretic use.
Potassium	Adults: 3.5–5.5 mEq/l	Increased levels occur with acute renal failure, acidosis; decreased levels can occur in some renal diseases that affect tubules and after use of diuretics.
Renin (plasma)	Adults: 0.1–4.3 ng/ml per hour	Increased levels occur in hypertension and Addison's disease; decreased levels occur with ADH therapy.
Acid phosphatase (plasma)	Adults: 0.11–5.5 U/l	Elevated levels occur with prostate cancer and bone cancer.
Antistreptolysin O titer (ASO titer)	<160 Todd units/ml	Increased levels occur in glomerulonephritis, rheumatic fever, bacterial endocarditis, and scarlet fever.
Bicarbonate (serum)	Adults: 24–28 mEq/l	Elevation or reduction of bicarbonate levels is important in diagnosis of acid–base disorders.
Urea (urea nitrogen) (serum)	5–25 mg/dl	Estimates GFR; values increase in renal disease, dehydration, gastrointestinal bleeding, liver disease, and gout.
Uric acid (serum)	Adults: 3.0–8.5 mg/dl	Elevated levels occur with renal failure, gout, increased metabolism of nucleotides, leukemias, and lymphomas. Decreased levels occur with drug treatment for gout.

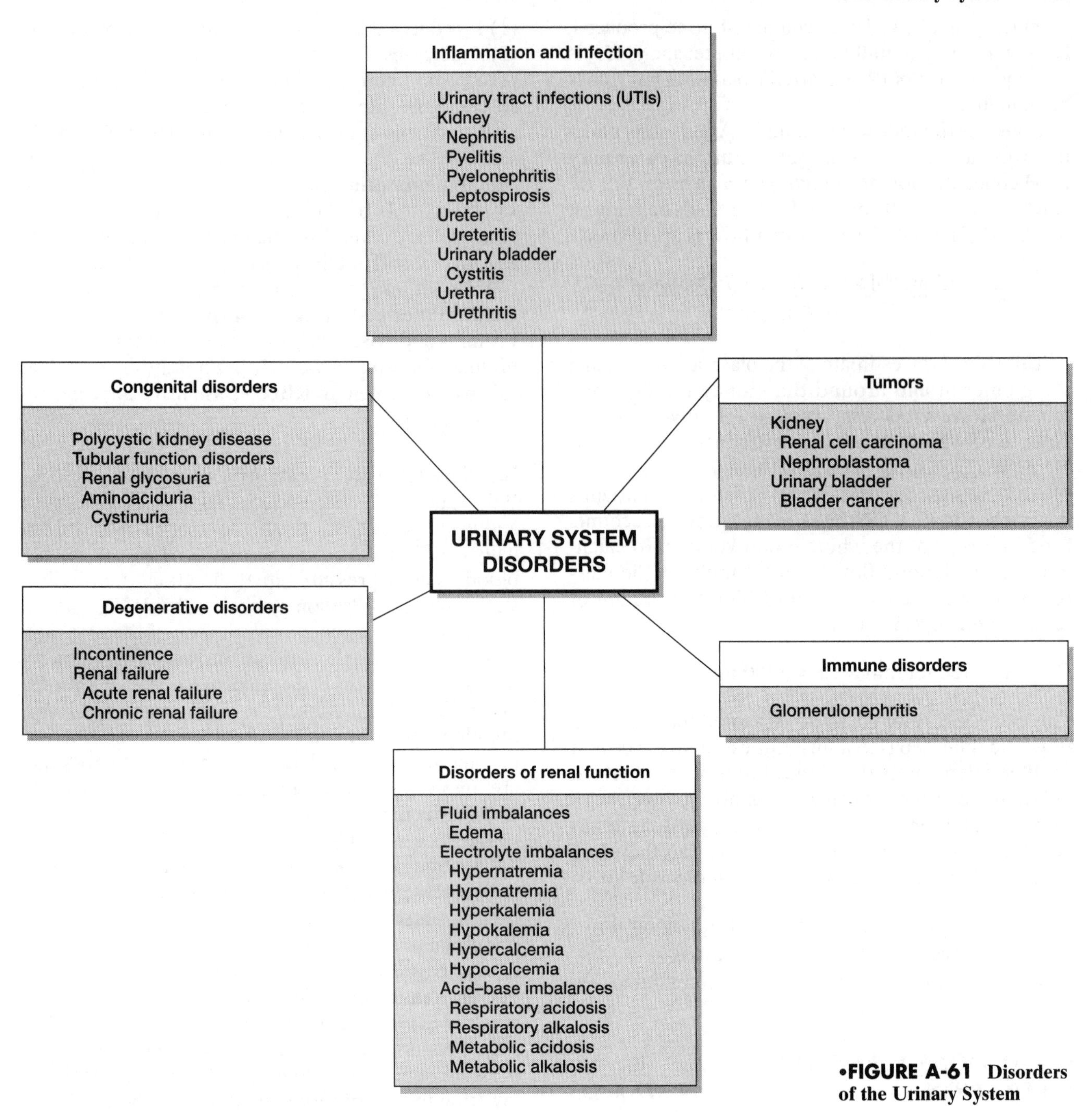

•**FIGURE A-61** Disorders of the Urinary System

DISORDERS OF THE URINARY SYSTEM

PAH AND THE CALCULATION OF RENAL BLOOD FLOW FAP *p. 946*

Although seldom used in clinical practice, *para-aminohippuric acid,* or *PAH,* can be administered to determine the rate of blood flow through the kidneys. PAH enters the filtrate through filtration at the glomerulus. As blood flows through the peritubular capillaries, any remaining PAH diffuses into the peritubular fluid, and the tubular cells actively secrete it into the filtrate. By the time blood leaves the kidney, virtually all the PAH has been removed from the bloodstream and filtered or secreted into the urine. You can therefore calculate renal blood flow if you know the PAH concentrations of the arterial plasma and urine. The calculation proceeds in a series of steps. The first step is to determine the plasma flow through the kidney by using the formula

$$P_f = \frac{PAH_u \times V_u}{PAH_p}$$

where P_f is the plasma flow, also known as the *effective renal plasma flow,* or *ERPF;* PAH_u is the concentration of PAH in the urine, usually expressed in milligrams per

FAP Ch. 26

milliliter (mg/ml); V_u is the volume of urine produced, usually in terms of milliliters per minute; and PAH_p is the concentration of PAH in arterial plasma in milligrams per milliliter.

Consider the following example: A person producing urine at a rate of 1 ml per minute has a urinary PAH concentration of 15 mg/ml with an arterial PAH concentration of 0.02 mg/ml. That person's hematocrit is normal (Hct = 45). The plasma flow is as follows:

$$P_f = \frac{15 \text{ mg/ml} \times 1 \text{ ml/min} = 750 \text{ ml/min}}{0.02 \text{ mg/ml}}$$

This value is an estimate of the plasma flow through the glomeruli and around the kidney tubules each minute. However, plasma accounts for only part of the volume of whole blood; the rest consists of formed elements. To have a plasma flow of 750 ml/min, the blood flow must be considerably greater. The patient's hematocrit is 45, which means that plasma accounts for 55 percent of the whole blood volume. To calculate the renal blood flow, we must multiply the plasma flow by 1.8 (each 100 ml of blood has 55 ml of plasma, and 100/55 = 1.8):

$$750 \text{ ml/min} \times 1.8 = 1350 \text{ ml/min}$$

This value, 1350 ml/min, is the estimated tubular blood flow. The final step is to adjust this estimate to account for blood that enters the kidney but flows to the renal pelvis, the capsule, or other areas not involved with urine production. This value is usually estimated as 10 percent of the total blood flow. We can therefore complete the calculation for this example as follows:

$$\begin{aligned} 1350 \text{ ml/min} &= 90 \text{ percent of total blood flow} \\ 10 \text{ percent} &= 1350/9 = 150 \text{ ml/min} \\ \text{Total blood flow} &= 1350 + 150 = 1500 \text{ ml/min} \end{aligned}$$

CONDITIONS AFFECTING FILTRATION

FAP *p. 955*

Changes in net filtration pressure *(NFP)* can result in significant alterations in kidney function. Factors that can disrupt normal filtration rates include physical damage to the filtration apparatus and interference with normal filtrate or urine flow.

Physical Damage to the Filtration Apparatus

FAP Ch. 26

The lamina densa and podocytes can be injured by mechanical trauma, such as a blow to the kidneys, bacterial infection, circulating immune complexes, or exposure to metabolic poisons, such as mercury. The usual result is a sudden increase in the permeability of the glomerulus. When damage is severe, plasma proteins and even blood cells enter the capsular spaces. The loss of plasma proteins has two immediate effects: (1) It reduces the osmotic pressure of the blood, and (2) it increases the osmotic pressure of the filtrate. The result is an increase in both the net filtration pressure and the rate of filtrate production.

Blood cells entering the filtrate will not be reabsorbed. The presence of blood cells in the urine is called **hematuria** (hēm-a-TOOR-ē-uh). Although small amounts of protein can be reabsorbed, when glomeruli are severely damaged, the nephrons are unable to reabsorb all the plasma proteins that enter the filtrate. Plasma proteins then appear in the urine, a condition termed **proteinuria** (prō-tēn-OOR-ē-uh). Proteins and RBCs can form masses within the renal tubules; abnormal masses or precipitates in urine are called *casts*. Protein or RBC casts in urine indicate kidney damage.

Interference with Filtrate or Urine Flow

If the tubule, collecting duct, or ureter becomes blocked and urine flow cannot occur, capsular pressures gradually rise. When the capsular hydrostatic pressure and blood osmotic pressure equal the glomerular hydrostatic pressure, filtration stops completely. The severity of the problem depends on the site of the blockage. If it involves a single nephron, only a single glomerulus will be affected. If the blockage occurs within the ureter, filtration in that kidney will come to a halt. If the blockage occurs in the urethra, both kidneys will become nonfunctional. (For examples of factors that are involved in urinary blockage, see "Problems with the Conducting System" in FAP, *p. 974.*)

Elevated capsular pressures can also result from inflammation of the kidneys, a condition called **nephritis** (nef-RĪ-tis). A generalized nephritis may result from bacterial infections, exposure to toxic or irritating drugs, or autoimmune disorders. One of the major problems in nephritis is that the inflammation causes swelling, but the renal capsule prevents the kidney from increasing in size. The result is an increase in the hydrostatic pressures in the peritubular fluid and filtrate. This pressure opposes the glomerular hydrostatic pressure, lowering the net filtration pressure and the GFR.

INHERITED PROBLEMS WITH TUBULAR FUNCTION

FAP *p. 962*

The tubular absorption of specific ions or compounds involves many different carrier proteins. Some individuals have an inherited inability to manufacture one or more of these carrier proteins, so they experience impaired tubular function. For example, in **renal glycosuria** (glī-kō-SOO-rē-uh) a defective carrier protein makes it impossible for the proximal convoluted tubule (PCT) to reabsorb glucose from the filtrate. Although renal glucose levels are abnormally high, blood glucose is normal, which distinguishes this condition from diabetes mellitus. Affected individuals gen-

erally do not have any clinical problems except when demand for glucose is high, as in starvation, acute stress, or pregnancy.

There are several types of **aminoaciduria** (a-mē-nō-as-i-DŪ-rē-uh), differing according to the identity of the missing carrier protein. Some of these disorders affect the reabsorption of an entire class of amino acids; others involve individual amino acids, such as lysine or histidine. **Cystinuria** is the most common disorder of amino acid transport, occurring in approximately 1 person in 12,500. Persons with this condition have difficulty reabsorbing cystine and amino acids with similar carbon structures, such as lysine, arginine, and ornithine. The most obvious and painful symptom is the formation of kidney and bladder stones that contain crystals of these amino acids. In addition to removal of these stones, treatment for cystinuria involves the maintenance of a high rate of urinary flow, so that amino acid concentrations do not rise high enough to promote stone formation, and a reduction of urinary acidity, because stone formation is enhanced by acidic conditions.

Any of these problems with tubular absorption and secretion will have a direct effect on urinary volume. Urine cannot be concentrated past 1200 mOsm/l. Hence the greater the number of solutes in the tubular fluid that enters the collecting ducts, the less water can be extracted by osmosis.

DIURETICS

FAP *p. 968*

Diuretics (dī-ū-RET-iks) are drugs that promote the loss of water in the urine. Diuretics have many different mechanisms of action, but each affects transport activities or water reabsorption along the nephron and collecting system. Some affect both. Important diuretics in use today include the following:

- *Osmotic diuretics.* Osmotic diuretics are metabolically harmless substances that are filtered at the glomerulus and ignored by the tubular epithelium. Their presence in the urine increases its osmolarity and limits the amount of water reabsorption possible. **Mannitol** (MAN-i-tol) is the most frequently administered osmotic diuretic. It is used to accelerate fluid loss and speed the removal of toxins from the blood and to elevate the GFR after severe trauma or other conditions have impaired renal function.
- *Drugs that block sodium and chloride transport.* A class of drugs called **thiazides** (THĪ-a-zīdz) reduces sodium and chloride transport in the proximal and distal tubules. Thiazides such as *hydrochlorothiazide* are often used to accelerate fluid losses in the treatment of hypertension and peripheral edema.
- *High-ceiling, or loop, diuretics.* The **high-ceiling diuretics,** such as *furosemide* and *bumetanide,* inhibit transport along the loop of Henle, reducing the osmotic gradient and the ability to concentrate the urine. They are called high-ceiling diuretics because they produce a much higher degree of diuresis than do other drugs. They are fast-acting and are commonly used in a clinical crisis—for example, in treating acute pulmonary edema or renal failure. In both the thiazide and the furosemide diuretics, water, Na^+, and K^+ are lost in the urine.
- *Aldosterone-blocking agents.* Blocking the action of aldosterone prevents the reabsorption of sodium along the distal convoluted tubule (DCT) and collecting tubule and so accelerates fluid losses. The drug *spironolactone* is this type of diuretic. It is often used in conjunction with other diuretics, because blocking the aldosterone-activated exchange pumps helps reduce the potassium ion loss. These drugs are also known as *potassium-sparing diuretics.* Atrial natriuretic peptide may be used as a diuretic, because it counteracts the effects of both aldosterone and ADH at the kidneys.
- *ACE inhibitors.* **ACE** *(angiotensin-converting enzyme inhibitors)* **inhibitors** prevent the conversion of angiotensin I to angiotensin II by converting enzyme. In turn, that prevents the stimulation of aldosterone production and promotes water loss.
- *Drugs with diuretic side effects.* Many drugs prescribed for other conditions promote diuresis as a side effect. For example, drugs that block carbonic anhydrase activity, such as acetazolamide *(Diamox®),* have an indirect effect on sodium transport (see Figures 26-11 and 26-14, FAP *pp. 961, 967*). Although they cause diuresis, these drugs are seldom prescribed with that effect in mind. (Because carbonic anhydrase is also involved in aqueous humor secretion, Diamox is used to reduce intraocular pressure in glaucoma patients.) Two more-familiar drugs, caffeine and alcohol, have pronounced diuretic effects. Caffeine produces diuresis directly by reducing sodium reabsorption along the tubules. Alcohol works indirectly by suppressing the release of ADH at the posterior pituitary gland.

URINALYSIS

FAP *p. 972*

Table 26-7 (FAP *p. 973*) indicates representative values for the most important components of normal urine. There are also several basic screening tests that can be performed by recording changes in the color of test strips that are dipped in the sample. Urine pH and urinary concentrations of glucose, ketones, bilirubin, urobilinogen, plasma proteins, and hemoglobin can be monitored by this technique. In addition, the density or *specific gravity* of the urine is typically determined by using a simple device known as a **urinometer** (ū-ri-NOM-e-ter), or **densitometer** (den-si-TOM-e-ter). The sample may also be spun in a centrifuge, and any sediment is examined under the microscope. Mineral crystals, bacteria, red or white blood cells, and deposits, known collectively as **casts,** can be detected in this way. Figure A-62• provides an

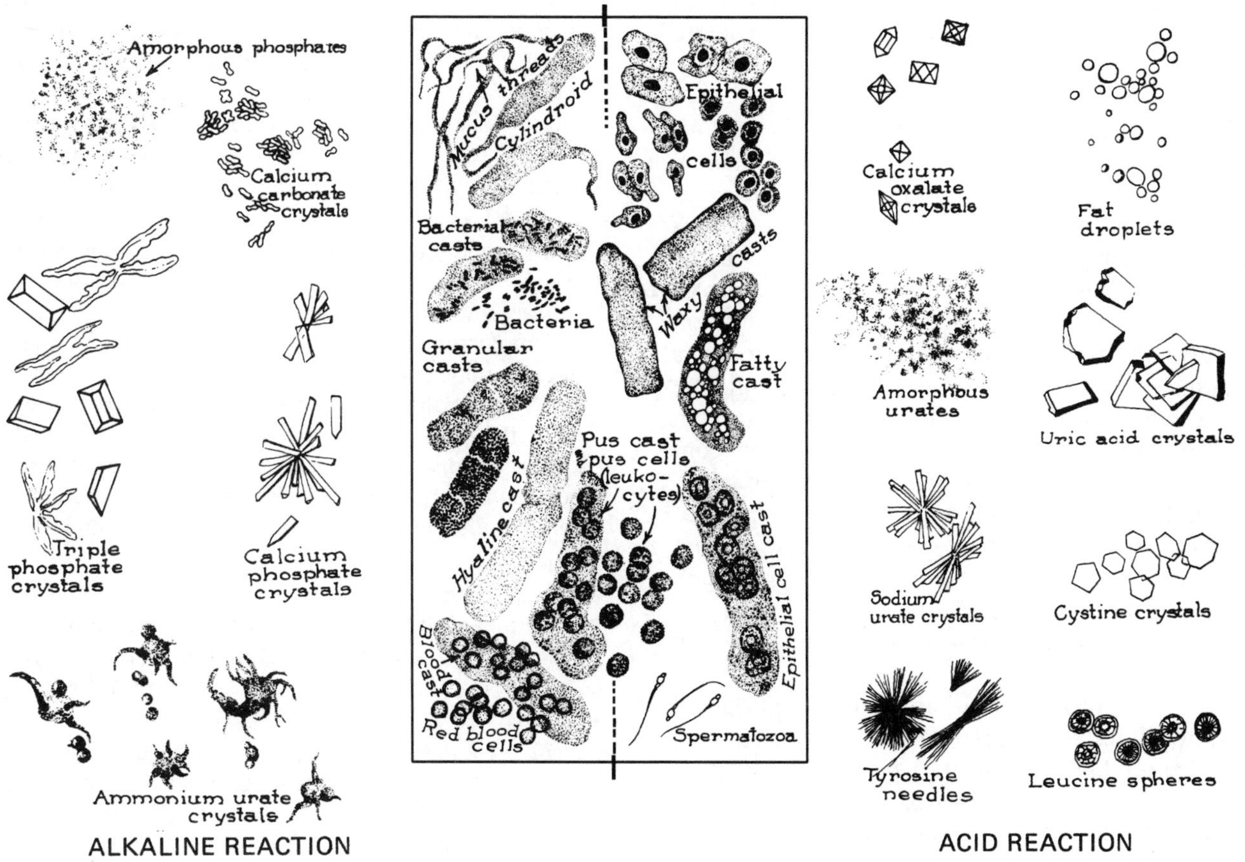

•**FIGURE A-62 Microscopic Examination of Urine Sediment.** [Redrawn after Todd and Sanford.]

overview of the major categories of urinary casts. During a urinary tract infection, bacteria may be cultured to determine their specific identities. New test strips can detect WBCs and measure specific gravity.

More-comprehensive analyses can determine the total osmolarity of the urine and the concentration of individual electrolytes and minor metabolites, metabolic wastes, vitamins, and hormones. A test for one hormone in the urine, *human chorionic gonadotropin* (hCG), provides an early and reliable proof of pregnancy.

The information provided by urinalysis can be especially useful when correlated with the data obtained from blood tests. The term **azotemia** (a-zō-TĒ-mē-uh) refers to the presence of excess metabolic wastes in the blood. This condition may result from the overproduction of urea or other nitrogenous wastes by the liver ("pre-renal syndrome"). Conversely, in **uremia** (ū-RĒ-mē-uh), all normal kidney functions are adversely affected. (The symptoms of uremia, which are those of kidney failure, are discussed in the clinical discussion on renal failure, FAP *p. 970*.)

FAP Ch. 27

The total volume of urine produced in a 24-hour period may also be of interest. Polyuria (pol-ē-Ū-rē-uh) refers to excessive production of urine—well over 2 liters per day. Polyuria most commonly results from endocrine disorders, such as the various forms of diabetes, metabolic disorders, or damage to the filtration apparatus, as in glomerulonephritis. **Oliguria** (o-li-GŪ-rē-uh) refers to inadequate urine production (50–500 ml/day). In **anuria** (a-NŪ-rē-uh), a negligible amount of urine is produced (0–50 ml/day)—a potentially fatal problem.

WATER AND WEIGHT LOSS FAP *p. 990*

The safest way to lose weight is to reduce the intake of food while ensuring that all dietary essentials are available in adequate quantities. Water must be included on the list of essentials along with the amino acids, fatty acids, vitamins, and minerals. Because nearly half of our normal water intake comes from food, a person who eats less becomes more dependent on drinking fluids and on whatever water is generated metabolically. At the start of a diet, the body conserves water and catabolizes lipids. That is why the first week of dieting may seem rather unproductive. Over that week, the level of circulating ketone bodies gradually increases. During subsequent weeks, the rate of water loss at the kidneys increases in order to excrete waste products, such as the hydrogen ions released by ketone bodies and the urea and ammonia gener-

ated during protein catabolism. The rate of fluid intake must also increase, or else the individual risks dehydration. Dehydration under these conditions is a major problem, because as water is lost, the solute concentration in the extracellular fluid (ECF) climbs, further increasing the concentration of waste products and acids. This effect can be opposed by fluid movement from the intracellular fluid (ICF) into the ECF, but only up to a point.

DIAGNOSTIC CLASSIFICATION OF ACID–BASE DISORDERS

FAP *p. 1009*

THE ANION GAP

Under normal circumstances, sodium ions are the primary cations in the ECF, and their positive charges are roughly balanced by the negative charges of the major anions (chloride and bicarbonate), minor anions (phosphate and sulfate), and plasma proteins. The concentrations of Na^+, Cl^-, and HCO_3^- are relatively easy to determine. The concentration of Na^+ is greater than the concentration of Cl^- plus that of HCO_3^-. The difference is called the anion gap:

$$\text{anion gap} = [Na^+] - ([HCO_3^-] + [Cl^-])$$

The anion gap in healthy individuals is 10–12 mEq/l. The calculation of the anion gap is useful in diagnosis, because it can be used to distinguish among different types of metabolic acidosis. The ECF of an individual in metabolic acidosis will have low pH, and the P_{CO_2} and HCO_3^- will be reduced as the carbonic acid–bicarbonate system attempts to buffer the excess H^+.

If the anion gap is normal, the problem results either from the generation or ingestion of HCl or from the loss of bicarbonate:

1. HCl is a strong acid, and it dissociates completely into H^+ and Cl^-. The chloride ion gained is balanced by the loss of a bicarbonate ion, which buffers the hydrogen ion. The net result is that the anion gap remains unchanged, although the HCO_3^- level drops.
2. In diarrhea, the body loses the HCO_3^- contained in the buffers secreted into the intestinal tract. As you may recall from Chapters 19 and 26 of the text, the movement of a bicarbonate ion across a cell membrane involves a countertransport mechanism that exchanges it for a chloride ion. Thus, for every bicarbonate ion secreted into the digestive tract and lost, a chloride ion is absorbed and retained. Again, the net result is that the anion gap remains unchanged although the HCO_3^- level drops.

If the individual in metabolic acidosis has an increased anion gap, the problem must be either the production or ingestion of organic acids or toxins or renal failure:

1. Conditions such as *ketoacidosis* and *lactic acidosis* are caused by organic acids that on dissociation release H^+ and anions that are not considered in the calculation of the anion gap. As a result, when the hydrogen ions are buffered by bicarbonate ions, HCO_3^- levels decline. Because this decline is not accompanied by an elevation in Cl^- concentrations, the anion gap increases.
2. In renal failure, acids that are normally excreted in the urine, such as sulfuric acid and phosphoric acid, are retained. The anions released by the dissociation of these acids are not considered in the calculation of the anion gap. Again, bicarbonate levels decline as they buffer the H^+, and the anion gap increases.

THE NOMOGRAM

When reviewing blood test results, clinicians may refer to a graphical representation of the data on bicarbonate, carbon dioxide, pH, and P_{CO_2} values. This graph, called a *nomogram,* is shown in Figure A-63•. A nomogram provides a visual summary of the information provided in Table 27-4 of the main text (FAP *p. 1009*).

The horizontal axis represents blood pH, and the vertical axis represents plasma HCO_3^- concentration. The curving lines indicate the relationship between pH and bicarbonate levels at a specific value of P_{CO_2}. For example, at a P_{CO_2} of 30 mm Hg (the curve that starts in the upper right-hand corner of the nomogram), the pH and bicarbonate values must lie somewhere along that line. The area at the center of the graph corresponds to the normal range of pH, bicarbonate levels, and P_{CO_2}. When acid–base disorders occur, these values change. The new values, when plotted on the nomogram, will typically fall in one of the shaded areas.

[EXPLORE]

Additional resources, such as interactive tutorials, critical-thinking questions, clinical problems, and case studies, are available through the CD-ROM and Website for your textbook. To begin your exploration, launch the *Martini Interactive* CD-ROM or visit the Companion Website (http://www.prenhall.com/martini/fap5) and select Chapter 26 or 27.

FAP Ch. 27

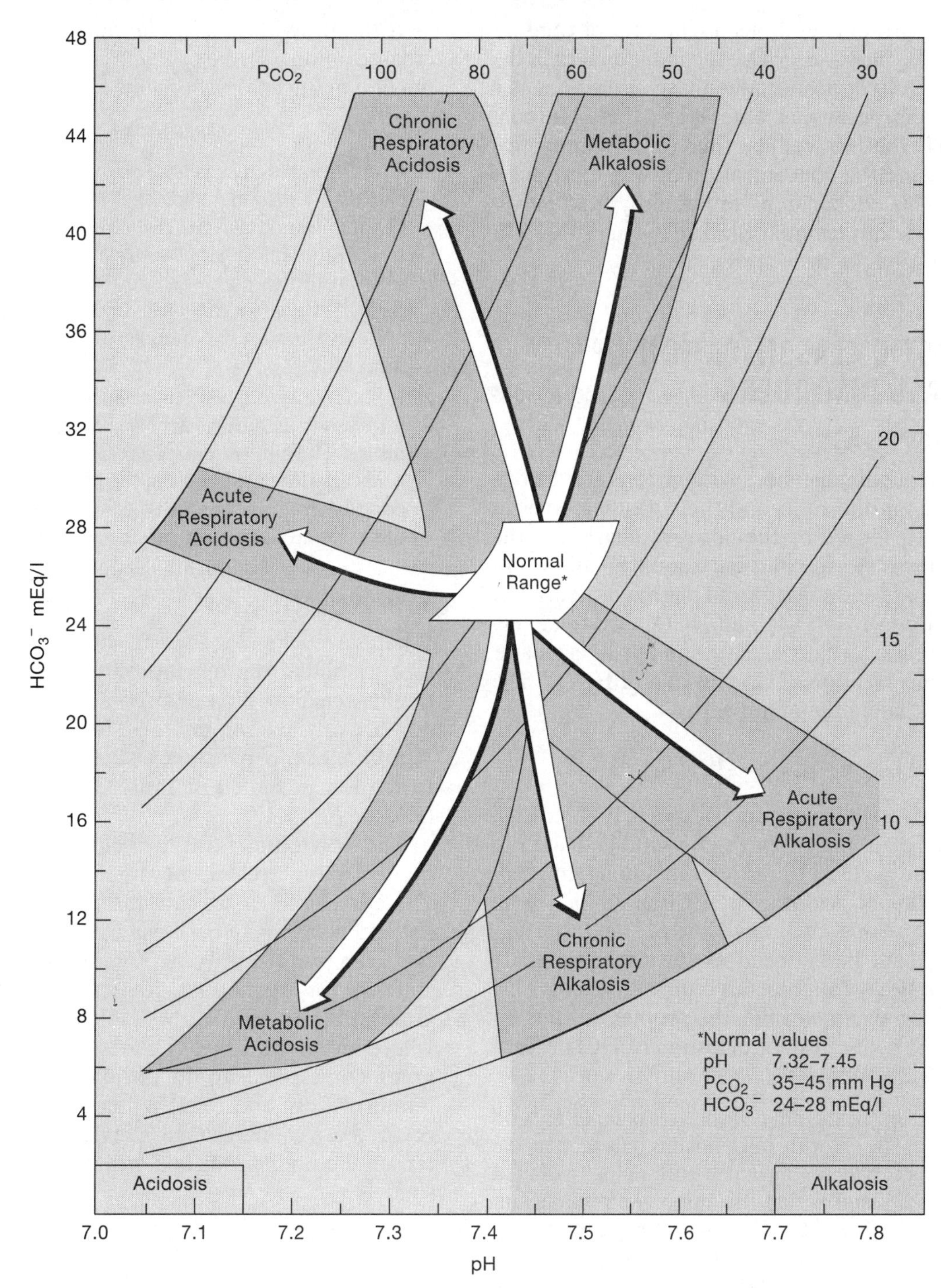

•**FIGURE A-63 An Acid–Base Nomogram.** A nomogram is a graphical summary of the information and patterns reported in Table 27-4, FAP *p. 1009.* The central box indicates the normal range of values for pH, P_{CO_2}, and HCO_3^-. The shaded areas indicate the values observed in representative acid–base disorders.

FAP Ch. 27

The Reproductive System and Development

In this section, we consider applied topics related to the continuation of the human species and the life histories of individuals. We will discuss aspects of the male and female reproductive systems, pregnancy, development, aging, and death.

THE PHYSICAL EXAMINATION AND THE REPRODUCTIVE SYSTEM

The male reproductive system consists of the gonads (testes), a series of specialized ducts (the epididymis, ductus deferens, ejaculatory duct, and urethra), accessory glands (the seminal vesicles, prostate gland, and bulbourethral glands), and the external genitalia (penis and scrotum). The female reproductive system consists of the gonads (ovaries), derivatives of an embryonic system of ducts (the uterine tubes, uterus, and vagina), accessory glands (the greater and lesser vestibular glands), the external genitalia (the clitoris, labia majora, and labia minora), and secondary sexual organs (the mammary glands of the breasts).

ASSESSMENT OF THE MALE REPRODUCTIVE SYSTEM

An assessment of the male reproductive system begins with a physical examination. Common signs and symptoms of male reproductive disorders include the following:

- *Testicular pain* results from various infections, including *gonorrhea* or other sexually transmitted diseases (STDs; p. 190), *mumps* (FAP *p. 855*), *typhoid, rheumatic fever* (p. 119), or *influenza.* Testicular pain also results from *testicular torsion* (twisting of the spermatic cord, with resulting ischemia), testicular cancer (FAP *p. 1025*), *cryptorchidism* (FAP *p. 1018*), or the presence of a *hernia* (p. 68). The pain may instead originate elsewhere along the reproductive tract, such as along the ductus deferens or within the prostate gland, or in other systems, as in *appendicitis* (p. 132) or a urinary obstruction, such as a ureteral kidney stone.
- *Urethral discharge* and *dysuria* are commonly associated with STDs. These symptoms also accompany disorders, such as *epididymitis* or *prostatitis* (FAP *p. 1029*), that may be infectious or noninfectious.
- *Impotence* (FAP *p. 1053*) can occur as a result of psychological factors, such as fear or anxiety, medications, or alcohol abuse. It can also develop secondarily to cardiovascular or hematological problems that affect blood pressure or blood flow to the penile arteries.

Inspection of the male reproductive system normally involves the examination of the external genitalia and palpation of the prostate gland. Inspection of the external genitalia entails the following observational steps:

1. *The inspection of the penis and scrotum for skin lesions, such as vesicles, chancres, warts, and* condylomas *(wart-like growths).* For example, painful vesicles often appear in clusters after infection with the herpes simplex virus. Distinctive skin lesions commonly indicate the presence of specific STDs; other, less apparent STDs may be present as well. A chancre is a painless ulceration associated with early-stage *syphilis* (p. 190). In the examination of uncircumcised males, the foreskin is retracted to observe the lining of the prepuce. **Phimosis,** an inability to retract the foreskin in an adult uncircumcised male, generally indicates an inflammation of the prepuce and adjacent tissues.
2. *The palpation of each testis, epididymis, and ductus deferens to detect the presence of abnormal masses, swelling, or tumors.* Possible abnormal findings include the following:
   - *Scrotal swelling* due to distortion of the scrotal cavity by blood (a *hematocele*), lymph (a *chylocele*), or serous fluid (a *hydrocele*).
   - *Testicular swelling* due to an enlargement of the testis or the presence of a nodular mass. **Orchitis** is a general term for inflammation of the testis. This inflammation can be the result of an infection, such as syphilis (p. 190), *mumps,* or *tuberculosis* (p.151). Testicular swelling may also accompany testicular cancer or testicular torsion.
   - *Epididymal swelling* due to cyst formation *(spermatocele),* tumor formation, or infection. **Epididymitis** is an acute inflammation of the epididymis that may indicate an infection of the reproductive or urinary tract. This condition may also develop from irritation caused by the backflow, or *reflux,* of urine into the ductus deferens.
   - *Swelling of the spermatic cord* may indicate (1) an inflammation of the ductus deferens *(deferentitis),*

(2) serous fluid accumulation in a pocket of serous membrane (a *hydrocele*), (3) bleeding within the spermatic cord, (4) testicular torsion, or (5) the formation of *varicose veins* (p. 125) within the pampiniform plexus, a condition known as a *varicocele.*

3. *A* digital rectal examination (DRE) *screens for prostatitis, tumors, or an inflammation of the seminal vesicles.* In this procedure, a gloved finger is inserted into the rectum and pressed against the anterior rectal wall to palpate the posterior walls of the prostate gland and seminal vesicles.

If urethral discharge is present or if discharge occurs in the course of any of these procedures, the fluid can be cultured to check for the presence of pathogens. Table A-40 summarizes information about pathogens that are responsible for infections of the reproductive system. Other potentially useful diagnostic procedures and laboratory tests are included in Table A-41.

TABLE A-40 Examples of Infectious Diseases of the Reproductive System

Disease	Organism(s)	Description
Bacterial diseases		
Bacterial vaginitis	Varied, but often *Gardnerella vaginalis*	Vaginitis caused by resident bacteria; symptoms include watery discharge.
Chancroid	*Haemophilus ducreyi*	A relatively rare STD; symptoms include soft chancres, which become ulcerated lesions, and enlarged lymph nodes in the groin.
Chlamydia	*Chlamydia trachomatis*	Chlamydial infections cause PID, nongonococcal urethritis, and LGV.
Gonorrhea	*Neisseria gonorrhoeae*	Infection of the epithelial cells of the male and female reproductive tracts; majority of females show no symptoms, but others may develop PID; majority of males develop painful urination (dysuria) and produce a viscous discharge.
Lymphogranuloma venereum (LGV)	*Chlamydia trachomatis*	One type of chlamydia STD; symptoms include enlarged lymph nodes, which may abscess and form ulcers
Pelvic inflammatory disease (PID)	*Neisseria gonorrhoeae* *Chlamydia trachomatis*	An infection of the uterine tubes (salpingitis); symptoms include fever, abdominal pain, and elevated WBC counts; can cause peritonitis in severe cases; sterility can result from formation of scar tissue in the uterine tubes.
Syphilis	*Treponema pallidum*	STD with a long period of chronic illness; symptoms of primary syphilis include chancres and enlarged lymph nodes; symptoms of secondary syphilis involve a reddish skin rash, fever, and headaches; tertiary syphilis affects CNS and cardiovascular system.
Toxic shock syndrome	*Staphylococcus aureus*	Vaginitis; symptoms include high fever, sore throat, vomiting, and diarrhea, which may lead to shock, respiratory distress, kidney or liver failure, and death.
Viral diseases		
Genital herpes	Herpes simplex viruses (HSV-1 and HSV-2)	Most cases caused by HSV-2; ulcers develop on external genitalia, heal, and recur.
Genital warts	Human papillomavirus	Warts appear on external genitalia, perineum, and anus, and on vagina and cervix of females; associated with cervical cancer.
Fungal diseases		
Candidiasis	*Candida albicans*	Yeast infection that causes vaginitis; symptoms include itching, burning, and lumpy white discharge.
Parasitic diseases		
Trichomoniasis	*Trichomonas vaginalis*	Flagellated protozoan parasite of both male and female urinary and reproductive tracts; infection produces white or greenish-gray discharge in both genders and intense vaginal itching in females; may be asymptomatic.

FAP Ch. 28

TABLE A-41 Examples of Tests Used in the Diagnosis of Reproductive Disorders

Diagnostic Procedure	Method and Result	Representative Uses
FEMALES		
Mammography	X-ray film is taken of breast.	Detects cysts or tumors of breast; effective in detecting early breast cancer
Thermography	Heat energy emitted from breast is detected by infrared camera and recorded.	Tumors, cysts, fibrocystic disease, and infection cause localized hot areas.
Laparoscopy	Fiber-optic tubing is inserted through incision in abdominal wall to view pelvic organs, remove tissue for biopsy, or perform surgical procedures.	Detects pelvic organ abnormalities such as cysts or adhesions; determines cause of pelvic pain; enables diagnosis of pelvic inflammatory disease and endometriosis
Papanicolaou (Pap) smear	Cells from cervix are removed for cytological analysis.	Detects cervical cancer; reported results: *Class I:* no abnormal cells; *Class II:* some abnormal cells, but none that suggest a malignancy (normally due to inflammation); *Class III:* some abnormal cells, possible malignancy; *Class IV:* some abnormal cells, probable malignancy; *Class V:* definite malignancy
Colposcopy	Special instrument is used to view cervical tissue microscopically in situ and to guide removal of tissue for biopsy.	Detects areas of dysplasia and malignancies of cervix; follow-up to abnormal PAP smear
Cervical biopsy	Tissue is removed from cervix for examination.	Detects dysplasia and malignancy
Transvaginal sonography	A small ultrasonic probe is inserted into the vagina.	Obtains high-definition echograms of the ovaries
MALES		
Transrectal ultrasonography	Ultrasound transducer is inserted rectally and scan is performed.	Detects prostatic tumor and nodules or abnormalities of seminal vesicles and surrounding structures; used to guide biopsy of nodules

ASSESSMENT OF THE FEMALE REPRODUCTIVE SYSTEM

Important signs and symptoms of female reproductive disorders include the following:

- *Acute pelvic pain,* a symptom that may accompany disorders such as *pelvic inflammatory disease* (PID) (FAP *p. 1040*), ruptured tubal pregnancy, a ruptured ovarian cyst, or inflammation of the uterine tubes *(salpingitis).*
- *Bleeding outside normal menses,* which can result from oral contraceptive use, tumors, hormonal fluctuation, pelvic inflammatory disease, or *endometriosis* (FAP *p. 1045*).
- *Amenorrhea* (FAP *p. 1045*), which may occur in women with *anorexia nervosa* (FAP *p. 932*), women who overexercise and are underweight, in extremely obese women, in postmenopausal women, and during pregnancy.
- *Abnormal vaginal discharge,* which may be the result of a bacterial, fungal, or protozoan infection, including some STDs.
- Dysuria, which may accompany an infection of the reproductive system due to the migration of the pathogen to the urethral entrance, even though the female reproductive and urinary tracts are distinct.
- *Infertility,* which may be related to hormonal disturbances, a variety of ovarian disorders (FAP *p. 1079*), or anatomical problems along the reproductive tract.

A physical examination generally includes the following steps:

1. *The inspection of the external genitalia for skin lesions, trauma, or related abnormalities.* Swelling of the labia majora results from (a) regional cellulitis with lymphedema, (b) a *labioinguinal hernia* (rare), (c) bleeding within the labia as the result of local trauma, or (d) *bartholinitis,* an abscess that develops after infection of one of the greater vestibular glands *(Bartholin's glands).*
2. *The inspection and/or palpation of the perineum, vaginal opening, labia, clitoris, urethral meatus, and vestibule to detect lesions, abnormal masses, or discharge from the vagina or urethra.* Samples of any discharge present can be tested to detect and identify any pathogens involved.

TABLE A-41 Examples of Tests Used in the Diagnosis of Reproductive Disorders *(continued)*

Laboratory Test	Normal Values in Blood Plasma or Serum	Significance of Abnormal Values
FEMALES		
Estrogen (serum)	Early uterine cycle: 60–400 pg/ml Middle: 100–600 pg/ml Late: 150–350 pg/ml Postmenopausal: <30 pg/ml	Detects hypofunctioning ovaries and helps determine timing of ovulation
Estradiol (serum)	Follicular phase: 20–150 pg/ml Ovulation: 100–500 pg/ml Luteal phase: 60–260 pg/ml	Decreased levels occur in ovarian dysfunction and in amenorrhea.
FSH (serum)	Before and after ovulation: 4–20 mIU/ml Midcycle: 10–40 mIU/ml	Helps determine the cause of infertility and menstrual dysfunction; increased levels occur in absence of estrogens (as during menopause); decreased levels occur with anorexia nervosa or hypopituitarism.
LH (serum)	Follicular phase: 3–30 mIU/ml Midcycle: 30–150 mIU/ml	Determines whether ovulation has occurred; increased levels occur in ovarian hypofunction.
Progesterone (serum)	Before ovulation: <70 ng/dl Midcycle: 250–2800 ng/dl	Determines timing of ovulation; levels are increased in early pregnancy.
Pregnanediol (urine)	Before ovulation: 0.5–1.5 mg over 24 hours Midcycle: 2–7 mg over 24 hours	Increased levels occur with pregnancy or an ovarian cyst; decreased levels occur with impending miscarriage, ovarian tumor, or preeclampsia.
Prolactin (serum)	Nonlactating females: 0–23 ng/ml	Values >100 ng/ml in a nonlactating female may indicate pituitary tumor.
MALES		
Semen analysis		
Volume	2–5.0 ml	Decreased sperm count causes infertility; infertility could also result if >40% of sperm are immotile or >30% of sperm are abnormal.
Sperm count	60–150 million/ml	
Motility	60–80% are motile	
Sperm morphology	70–90% normal structure	
Testosterone (serum)	Adults: 0.3–1.0 μg/dl	Decreased level could indicate testicular disorder, alcoholism, or pituitary hypofunction.
Acid phosphatase (ACP) (serum)	Adults: 0.0–0.8 U/l at 37° C	Increased levels occur with carcinoma of prostate gland, Paget's disease, and multiple myeloma of bone marrow.
Prostate-specific antigen (PSA)	<4 ng/ml	Increased levels occur with prostatic cancer, benign prostatic hypertrophy, and increasing age.
FEMALES AND MALES		
Serologic test for syphilis	Negative	Presence of antibodies indicates past or present infection with syphilis.
Gonorrhea culture test	Negative	Positive test indicates gonorrheal infection.
Herpes simplex virus	Negative	Positive test indicates presence of virus in culture.
***Chlamydia* smear**	Negative	Positive culture indicates presence of *Chlamydia.*

3. *The inspection of the vagina and cervix by using a* speculum, *an instrument that retracts the vaginal walls to permit direct visual inspection.* Changes in the color of the vaginal walls may be important diagnostic clues. For example:

   - Cyanosis of the vaginal and cervical mucosa normally occurs during pregnancy (see below), but it may also occur when a pelvic tumor exists or in persons with congestive heart failure.
   - Reddening of the vaginal walls occurs in *vaginitis* (p. 189), bacterial infections such as gonorrhea, protozoan infection by *Trichomonas vaginalis,* and yeast infections. It can also appear postmenopausally in some women (a condition known as *atrophic vaginitis*).

FAP Ch. 28

The cervix is inspected to detect lacerations, ulceration, polyps, or cervical discharge. A spatula and brush is then used to collect cells from the cervical

os and to transfer them to a glass slide. After the sample is fixed with a chemical spray, cytological examination is performed. This technique is the best-known example of a *Papanicolaou (Pap) test* (Table A-2, p. 10), and the sampling process is commonly called a *Pap smear.* This test screens for the presence of cervical cancer. Ratings of PAP smear results are given in Table A-41.

4. *A* bimanual examination *for the palpation of the uterus, uterine tubes, and ovaries.* The physician inserts two fingers vaginally and places the other hand against the lower abdomen to palpate the uterus and surrounding structures. The contour, shape, size, and location of the uterus can be determined, and any swellings or masses will be apparent. Abnormalities in other reproductive organs, such as ovarian cysts, endometrial growths, or tubal masses, can also be detected in this way.

NORMAL AND ABNORMAL SIGNS ASSOCIATED WITH PREGNANCY

Pregnancy imposes a number of stresses on maternal body systems. The major physiological changes are discussed in Chapter 29 (FAP *p. 1081*). Several clinical signs may be apparent in the course of a physical examination, including the following:

- *Chadwick's sign* is a normal cyanosis of the vaginal wall and cervix during pregnancy.
- The size of the uterus changes drastically during pregnancy; the uterus at full-term extends almost to the level of the xiphoid process.
- Significant uterine bleeding, causing vaginal discharge of blood, most commonly occurs in *placenta previa* (p. 194), in which the placenta forms near the cervix. Subsequent cervical stretching leads to tearing and bleeding of the vascular channels of the placenta. Vaginal bleeding may also occur prior to miscarriage.
- Nausea and vomiting tend to occur in pregnancy, especially during the first 3 months.
- Edema of the extremities, especially the legs, typically occurs because the increased total blood volume and the weight of the uterus compress the inferior vena cava and its tributaries. As venous pressures rise in the lower limbs and inferior trunk, varicose veins and *hemorrhoids* (p. 126) may develop.
- Back pain due to increased stress on muscles of the lower back is common. These muscles are strained as the weight of the uterus accentuates the lumbar curvature.
- A weight gain of 10–12.5 kg (22–27.5 lbs) is now considered desirable, although 20 years ago weight increases of 20–25 kg (44–55 lbs) were considered acceptable. Failure to gain adequate weight during a pregnancy can indicate serious problems.
- The combination of estrogen, progesterone, prolactin, human placental lactogen (hPL), and other hormones that are elevated during pregnancy appears to promote the development of *insulin resistance,* a decrease in target cell sensitivity to insulin. As a result, pregnant diabetic women are at an increased risk of ketoacidosis. Previously nondiabetic women unable to increase insulin levels sufficiently to compensate for the increased insulin resistance can develop *gestational diabetes.* Glucose levels must be monitored and stabilized to prevent the increased risk of fetal mortality and developmental defects. Gestational diabetes develops in 1–3 percent of pregnancies.
- In some cases, a dangerous combination of hypertension, proteinuria, edema, and seizures occurs. We will consider this condition, called *eclampsia,* in a later section (p. 194).

DISORDERS OF THE REPRODUCTIVE SYSTEM

Representative disorders of the reproductive system are diagrammed in Figure A-64•.

THE DIAGNOSIS AND TREATMENT OF OVARIAN CANCER FAP *p. 1039*

A woman in the United States has a lifetime risk of 1 chance in 70 of developing ovarian cancer. In 2000 in the United States, 23,100 ovarian cancers are expected to be diagnosed and 14,000 women are expected to die from this condition. Ovarian cancer is the third most common reproductive cancer among women. It is also the most dangerous, because it is seldom diagnosed in its early stages. The prognosis is relatively good for cancers that originate in the general ovarian tissues or from abnormal oocytes. These cancers respond well to some combination of chemotherapy, radiation, and surgery. However, 85 percent of ovarian cancers develop from epithelial cells, and sustained remission can be obtained in only about one-third of these cases. Early diagnosis greatly improves the chances of successful treatment, but as yet there is no standardized, effective screening procedure. *(Transvaginal sonography* can detect ovarian cancer at Stage I or Stage II, but there is a high incidence of false-positive results.)

The minimal treatment required at Stage I or Stage II involves a unilateral removal of an ovary and uterine tube (a *salpingo-oophorectomy*). For more-advanced cancer, a *bilateral salpingo-oophorectomy* (BSO) and *total hysterectomy* (removal of the uterus) are performed. The treatment of more-dangerous forms of early-stage ovarian cancer includes radiation and chemotherapy in addition to surgery.

Treatment of Stage III or Stage IV ovarian cancer commonly involves the removal of the omentum, in addition to a BSO and total hysterectomy and aggressive chemotherapy. A bone marrow transplant may be required, because stem cells in the bone marrow are destroyed by these chemicals. Some chemotherapy agents are introduced into the peritoneal cavity, where higher concentrations can be administered

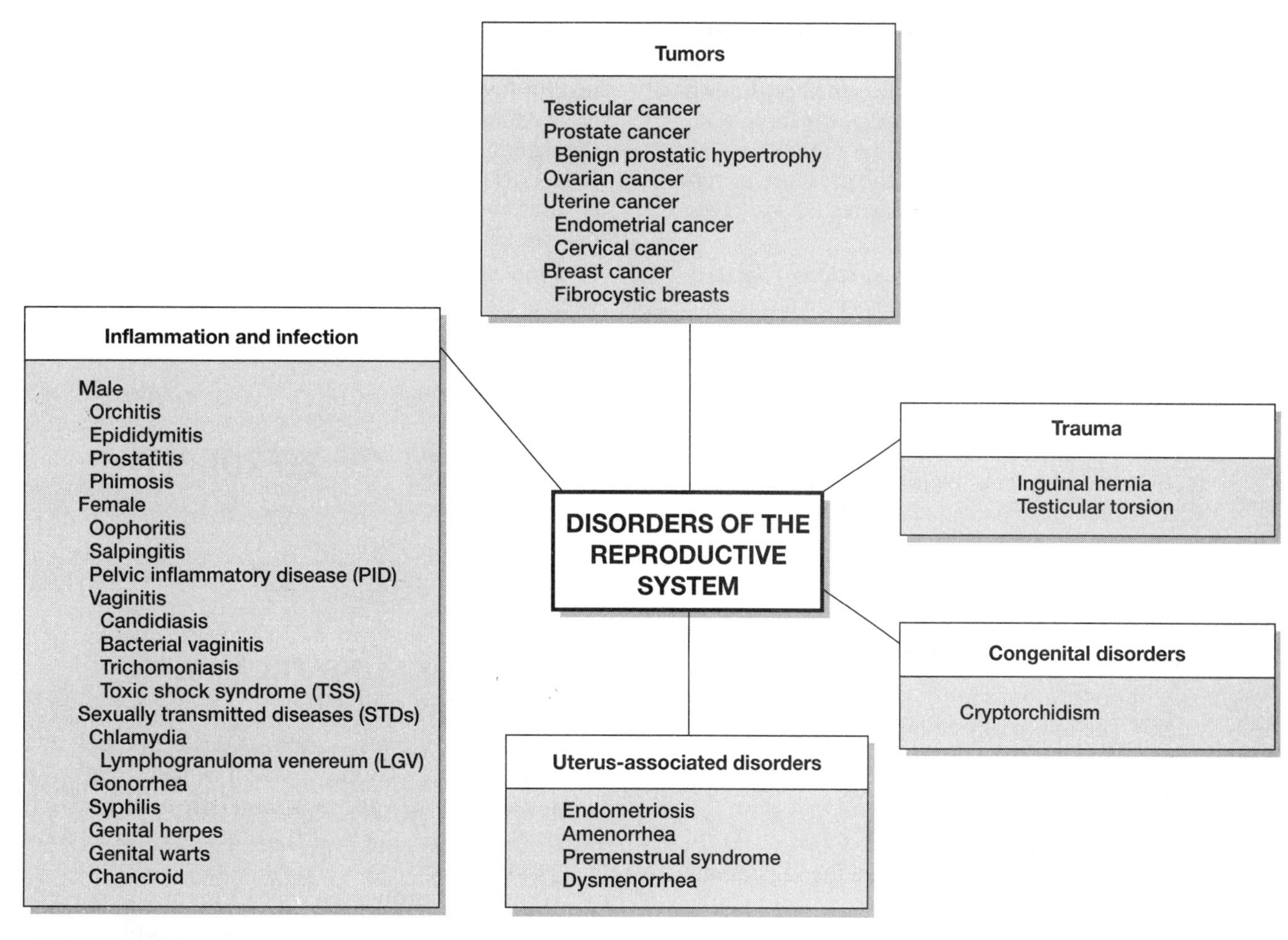

•**FIGURE A-64** **Disorders of the Reproductive System**

without the systemic effects that would acompany the infusion of these drugs into the bloodstream. This procedure is called *intraperitoneal therapy.*

UTERINE TUMORS AND CANCERS FAP p. 1042

Uterine tumors are the most common tumors in women. It has been estimated that 40 percent of women over age 50 have benign uterine tumors involving smooth muscle and connective tissue cells. If small, these *leiomyomas* (lē-ō-mī-Ō-maz), or *fibroids,* generally cause no problems. If stimulated by estrogens, they can grow quite large, reaching weights as great as 13.6 kg (30 lb). Occlusion of the uterine tubes, distortion of adjacent organs, and compression of blood vessels may then lead to complications. In symptomatic young women, observation or conservative treatment with drugs or restricted surgery may be utilized to preserve fertility. In older women, a decision may be made to remove the uterus.

FAP Ch. 28

Benign epithelial tumors in the uterine lining are called *endometrial polyps.* Roughly 10 percent of women probably have polyps, but because the polyps tend to be small and cause no symptoms, the condition passes unnoticed. If bleeding occurs, if the polyps become excessively enlarged, or if they protrude through the cervical os, they can be removed.

Uterine cancers are less common, affecting approximately 11.9 per 100,000 women. In 2000 in the United States, roughly 48,900 new cases are expected to be reported and approximately 11,100 women are expected to die from the disease. There are two types of uterine cancers: (1) *endometrial* and (2) *cervical.*

Endometrial cancer is an invasive cancer of the endometrium. The condition most commonly affects women age 50–70. Estrogen therapy, used to treat osteoporosis in postmenopausal women, increases the risk of endometrial cancer by 2–10 times. Adding progesterone therapy to the estrogen therapy seems to reduce this risk.

There is no satisfactory screening test for endometrial cancer. The most common symptom is irregular bleeding, and diagnosis typically involves the examination of a biopsy of the endometrium by suction or scraping. The prognosis varies with the degree of spread. The treatment of early-stage endometrial cancer involves a hysterectomy, perhaps followed by

localized radiation therapy. In advanced stages, more-aggressive radiation treatment is recommended. Chemotherapy has not proved to be very successful in treating endometrial cancers; only 30–40 percent of patients benefit from this approach.

Cervical cancer is the most common reproductive system cancer in women age 15–34. Most women with cervical cancer develop no symptoms until late in the disease. At that stage, vaginal bleeding, especially after intercourse, pelvic pain, and vaginal discharge may appear. Early detection is the key to reducing the mortality rate for cervical cancer. The standard screening test is the Pap smear, named for Dr. George Papanicolaou, an anatomist and cytologist. The cervical epithelium normally sheds its superficial cells, and a sample of cells scraped or brushed from the epithelial surface can be examined for abnormal or cancerous cells. The American Cancer Society recommends yearly Pap tests at ages 20 and 21, followed by smears at 1-year to 3-year intervals until age 65.

The primary risk factor of cervical cancer is a history of multiple sexual partners. It appears likely that these cancers develop after viral infection by one of several *human papillomaviruses* (HPVs), which are transmitted through sexual contact.

Early treatment of abnormal but not cancerous lesions detected by mildly abnormal Pap smears may prevent the progression to cancer formation. The treatment of localized, noninvasive cervical cancer involves the removal of the affected portion of the cervix. The treatment of more-advanced cancers typically involves a combination of radiation therapy, hysterectomy, lymph node removal, and chemotherapy.

VAGINITIS

FAP *p. 1046*

There are several forms of vaginitis, and minor cases are relatively common. **Candidiasis** (kan-di-DĪ-a-sis) results from a fungal (yeast) infection. The organism responsible appears to be a normal component of the vaginal environment in 30–80 percent of healthy women. Antibiotic administration, immunosuppression, stress, pregnancy, and other factors that change the local environment can stimulate the unrestricted growth of the fungus. Symptoms include itching and burning, and a lumpy white discharge may also be produced. Topical antifungal medications are used to treat this condition.

Bacterial (nonspecific) vaginitis results from the combined action of several bacteria. The bacteria involved are normally present in about 30 percent of adult women. In this form of vaginitis, the vaginal discharge contains epithelial cells and large numbers of bacteria. The discharge has a homogeneous, watery texture and a characteristic odor sometimes described as fishy or aminelike. Topical or oral antibiotics are effective in controlling this condition.

Trichomoniasis (trik-ō-mō-NĪ-a-sis) involves infection by the parasite *Trichomonas vaginalis,* introduced by sexual contact with a carrier. Because it is a sexually transmitted disease, both partners must be treated to prevent reinfection. A foamy, green discharge that is intensely itchy and watery is characteristic, but women can be asymptomatic carriers.

A vaginal infection by *Staphylococcus* bacteria is responsible for *toxic shock syndrome (TSS),* a form of septic shock that is discussed on p. 127.

EXPERIMENTAL CONTRACEPTIVE METHODS

FAP *p. 1055*

A number of experimental contraceptive methods are being investigated. For example, researchers are attempting to determine whether low doses of inhibin will suppress the release of gonadotropin-releasing hormone (GnRH) and thereby prevent ovulation. Another approach is to develop a method of blocking human chorionic gonadotropin (hCG) receptors at the corpus luteum. Produced by the placenta, hCG maintains the corpus luteum for the first 3 months of pregnancy. If the corpus luteum were unable to respond to hCG, normal menses would occur despite implantation of a blastocyst.

Several male contraceptives are also under development:

- *Gossypol,* a yellow pigment extracted from cottonseed oil, produces a dramatic decline in sperm count and sperm motility after 2 months. It can be administered topically, because it is readily absorbed through the skin. Fertility returns within a year after treatment is discontinued. Unfortunately, gossypol has not been approved as yet, because it has about a 10 percent risk of permanent sterility and may lead to hypokalemia.
- Weekly doses of testosterone suppress GnRH secretion over a period of 5 months. The result is a drastic reduction in the sperm count. The combination of a testosterone implant, comparable to that of the Norplant® system, with a GnRH antagonist, *cetrorelix,* effectively suppresses spermatogenesis. A new synthetic form of testosterone, *alpha-methyl-nortestosterone (MENT),* appears to be even more effective than testosterone in suppressing GnRH production.
- A drug used to control blood pressure appears to cause temporary, reversible sterility in males. This drug is now being evaluated to see if low dosages will affect fertility in normal males without affecting blood pressure.

If contraceptive methods fail, options exist to either prevent implantation or terminate the pregnancy. The "morning-after pills" contain estrogens or progestins. They must be taken within 72 hours of intercourse, and they appear to alter the transport of the zygote or to prevent its attachment to the uterine wall. The drug known as *RU-486 (Mifepristone®)* blocks the action of progesterone at the endometrial lining. The result is a normal menses and the degeneration of the endometrium whether or not a pregnancy has occurred.

SEXUALLY TRANSMITTED DISEASES

FAP p. 1056

Close physical contact can spread infectious diseases from person to person, and sexual contact is as close as two individuals can get. Some infections are spread almost exclusively by sexual contact. These infections are called **sexually transmitted diseases,** or **STDs**. A variety of bacterial, viral, and fungal infections are included in this category. At least two dozen STDs are currently recognized, and roughly 15 million people become infected each year in the United States. All STDs are unpleasant, and some are deadly. Here we will discuss six of the most common STDs: *chlamydia, gonorrhea, syphilis, herpes, genital warts,* and *chancroid.*

Chlamydia

As diagnostic procedures improve, infections by the bacterium *Chlamydia trachomatis* are proving to be the most frequent cause of STDs. Roughly 4 million people become infected each year in the United States. The incidence of asymptomatic infections among college students has been estimated at 5 percent; estimates of the incidence of infection in women age 15–24 in the general population range from 2.4–11.3 percent. Chlamydial infection can have a variety of clinical effects. It is responsible for the majority of cases of pelvic inflammatory disease (resulting in tubal blockage and infertility) as well as *nongonococcal urethritis,* another STD called **lymphogranuloma venereum (LGV),** and conjunctivitis in the newborn. In LGV, the lymph nodes in the groin become enlarged and inflamed. Abscesses and ulcers then develop over lymph nodes in the region. Chlamydial infections can be controlled with antibiotics (tetracycline and sulfa drugs), but all sexual partners must be treated to prevent reinfection.

Gonorrhea

The bacterium *Neisseria gonorrhoeae* is responsible for gonorrhea, one of the most common STDs in the United States. Nearly 2 million cases were reported in the early 1970s; roughly 400,000 cases are expected to be reported during 2000. These bacteria normally invade epithelial cells that line the male or female reproductive tract. In relatively rare cases, they will also colonize the pharyngeal or rectal epithelium.

The symptoms of genital infection differ according to the gender of the infected individual. It has been estimated that up to 80 percent of women infected with gonorrhea experience no symptoms or symptoms so minor that medical treatment is not sought. As a result, these women act as carriers, spreading the infection through their sexual contacts. An estimated 10–15 percent of women infected with gonorrhea experience more-acute symptoms because the bacteria invade the epithelia of the uterine tubes. This infection probably accounts for many of the cases of pelvic inflammatory disease (PID) in the U.S. population. As many as 80,000 women become infertile each year as the result of scar tissue formation along the uterine tubes after gonorrheal and/or chlamydial infections.

FAP Ch. 28

Seventy to eighty percent of infected males develop symptoms painful enough to make them seek antibiotic treatment. The asymptomatic 20–30 percent are male carriers who unknowingly spread the infection. The urethral invasion is accompanied by pain on urination (dysuria) and typically by a viscous urethral discharge. A sample of the discharge can be cultured to permit the positive identification of the organism involved.

Syphilis

Syphilis (SIF-i-lis) results from infection by the bacterium *Treponema pallidum.* The first reported syphilis epidemics occurred in Europe during the sixteenth century, possibly introduced by early explorers returning from the New World. The death rate from the "Great Pox" was appalling, far greater than it is today, even after we take into account the absence of antibiotic therapies at that time. It appears likely that the syphilis bacterium has mutated during the interim. These changes have reduced the immediate mortality rate but have prolonged the period of chronic illness and have increased the likelihood of successful transmission. Syphilis still remains a life-threatening disease. Untreated syphilis can cause serious cardiovascular and neurological illness years after infection, and it can be spread to a fetus during pregnancy, producing congenital malformations. The annual reported incidence of this disease has declined over the last decade to roughly 2 cases per 100,000 population, the lowest rate since 1960.

Primary syphilis begins as the bacteria cross the mucous epithelium and enter the lymphatic vessels and bloodstream. At the invasion site, the bacteria multiply. After an incubation period of 1.5 to 6 weeks, their activities produce a painless raised lesion, or **chancre** (SHANG-ker) (Figure A-65•). This lesion remains infectious and persists for several weeks before fading away, even without treatment. In heterosexual men, the chancre tends to appear on the penis; in women, it may develop on the labia, vagina, or cervix. Lymph nodes in the region often enlarge and remain swollen even after the chancre has disappeared.

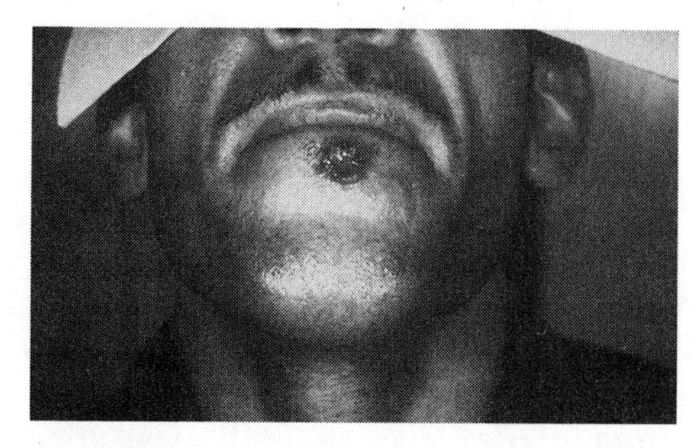

•**FIGURE A-65** A Syphilitic Chancre

Symptoms of *secondary syphilis* appear roughly 6 weeks later. Secondary syphilis is also infectious. It generally involves a diffuse, reddish skin rash. Like the chancre, the rash fades over a period of 2–6 weeks. These symptoms may be accompanied by fever, headaches, and malaise. The combination is so vague that the disease may easily be overlooked or misdiagnosed. In a few instances, more-serious complications such as *meningitis* (p. 76), *hepatitis* (p. 165), or *arthritis* (p. 60) develop.

The individual then enters the *latent phase*, which is noninfectious. The duration of the latent phase varies widely. Fifty to seventy percent of untreated individuals with latent syphilis fail to develop the symptoms of *tertiary syphilis,* or *late syphilis,* although the bacterial pathogens remain within their tissues. Those who develop tertiary syphilis may do so 10 or more years after infection.

The most-severe symptoms of tertiary syphilis involve the central nervous system and the cardiovascular system. **Neurosyphilis** may result from a bacterial infection of the meninges or the tissues of the brain or spinal cord. **Tabes dorsalis** (TĀ-bēz dor-SAL-is) results from the invasion and demyelination of the posterior columns of the spinal cord and the sensory ganglia and nerves. In the cardiovascular system, the disease affects the major vessels, leading to *aortic stenosis* (p. 119), *aneurysms* (p. 125), or *focal calcification* (FAP *p. 696*).

Equally disturbing are the effects of transmission from mother to fetus across the placenta. These cases of congenital syphilis are marked by infections of the developing bones and cartilages of the skeleton and progressive damage to the spleen, liver, bone marrow, and kidneys. The risk of fetal transmission may be as high as 95 percent, so maternal blood testing is recommended early in pregnancy. Blood donations are screened to prevent transfer through blood transfusion. The treatment of syphilis involves the administration of *penicillin* or other antibiotics.

Herpes

Genital herpes results from infection by herpes viruses. Two different viruses are involved. Eighty to ninety percent of genital herpes cases are caused by the virus known as HSV-2 (herpes simplex virus Type 2), which is usually associated with the external genitalia. The remaining cases are caused by HSV-1, the virus that is also responsible for cold sores on the mouth. Typically, within a week of the initial infection, the individual develops painful, ulcerated lesions on the external genitalia, with associated lymphadenopathy. In women, ulcerations may also appear on the cervix. These ulcerations gradually heal over the next 2–3 weeks. Recurring lesions are common, although subsequent incidents are less severe.

During delivery, infection of the newborn with herpes viruses present in the mother's vagina can lead to serious illness, because the infant has few immunological defenses. The antiviral agent *acyclovir* has helped in treating initial infections and in reducing recurrences.

Genital Warts

Genital warts, or *condyloma acuminata,* result from infection by one of a number of strains of *human papillomavirus* (HPV). Several of these strains are thought to be responsible for cases of cervical, anal, vaginal, and penile cancer. Roughly 1.2 million cases of genital warts are diagnosed each year in the United States. There is no satisfactory treatment for this problem. The traditional treatments have included cryosurgery, erosion by caustic chemicals, surgical removal, and laser surgery to remove the warts. These treatments remove the visible signs of infection, but the virus remains within the epidermis. Alpha-interferons have been tried with limited success.

Chancroid

Chancroid is an STD caused by the bacterium *Haemophilus ducreyi.* Chancroid cases were rarely seen in the United States before 1984, but since then the number of cases has risen dramatically, reaching 4000–5000 cases per year. The primary sign of this disease is the development of *soft chancres,* soft lesions otherwise resembling those of syphilis. The majority of chancroid patients also develop prominent inguinal lymphadenopathy (FAP *p. 760*).

DISORDERS OF DEVELOPMENT

Development is a complex process, and developmental disorders are extremely diverse. Figure A-66• surveys representative disorders of development.

TERATOGENS AND ABNORMAL DEVELOPMENT FAP *pp. 1068, 1092*

Teratogens (TER-a-tō-jenz) are stimuli that disrupt normal development by damaging cells, altering chromosomal structure, or that interfere with normal induction. **Teratology** (ter-a-TOL-o-jē)—literally, the "study of monsters"—deals with extensive departures from the pathways of normal development. Teratogens that affect the embryo in the first trimester will potentially disrupt cleavage, gastrulation, or neurulation. The embryonic survival rate will be low, and most survivors will have severe anatomical and physiological defects that affect all the major organ systems. Errors introduced into the developmental process during the second or third trimester will be likely to affect specific organs or organ systems, for the major organizational patterns are already established. Nevertheless, the alterations reduce the chances for long-term survival.

We encounter many powerful teratogens in everyday life. The location and severity of the resulting defects vary with the nature of the stimulus and the time

FAP Ch. 29

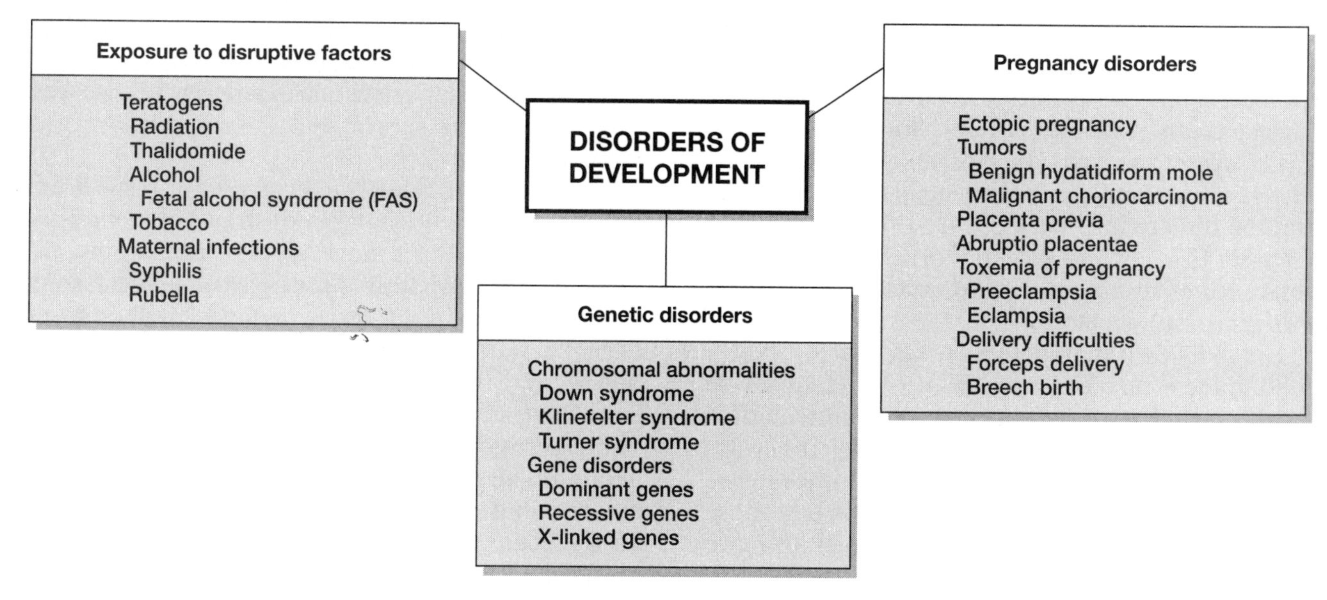

•**FIGURE A-66** **Disorders of Development**

of exposure. Radiation is a powerful teratogen that can affect all cells. Even the X-rays used in diagnostic procedures can break chromosomes and produce developmental errors; thus nonionizing procedures such as ultrasound are used to track embryonic and fetal development. Fetal exposure to the microorganisms responsible for syphilis (p.190) or *rubella* ("German measles") can also produce serious developmental abnormalities, including congenital heart defects, mental retardation, and deafness.

Some chemical agents are teratogenic only if present at a time when embryonic or fetal targets are vulnerable to their effects. Thousands of critical inductions are under way during the first trimester, initiating developmental sequences that will produce the major organs and organ systems of the body. In almost every case, the nature of the inducing agent remains unknown, and the effects of unusual compounds within the maternal circulation cannot be predicted. As a result, virtually any unusual chemical that reaches an embryo has the potential for producing developmental abnormalities. For example, during the 1960s, the European market was strong for **thalidomide,** a drug effective in promoting sleep and preventing nausea. Thalidomide was commonly prescribed for women in early pregnancy, with disastrous results. The drug crossed the placenta and entered the fetal circulation, where for a few days it interfered with the induction process responsible for limb development. Many such infants were born without limbs or with drastically deformed ones. Thalidomide had not been approved by the U.S. Food and Drug Administration (FDA), and it could not be sold legally in the United States. Although today the FDA is often criticized for the slow pace of its approval process, in this case the combination of rigorous testing standards and complex bureaucratic procedures protected the public.

However, even when extensive testing is performed with laboratory animals, uncertainties still remain because the chemical nature of the inducer responsible for a specific process may vary from one species to another. For example, thalidomide produces abnormalities in humans and monkeys, but developing mice, rats, and rabbits are unaffected by the drug.

More-powerful teratogens will have an effect regardless of the time of exposure. Pesticides, herbicides, and heavy metals are common around agricultural and industrial environments, and these substances can contaminate the drinking water in the area. A number of prescription drugs, including certain antibiotics, tranquilizers, sedatives, steroid hormones, diuretics, anesthetics, and analgesics, also have teratogenic effects. Pregnant women should read the "Caution" label before using any drug without the advice of a physician. Most "natural" herbs and substances have not been tested, and their chemical composition may vary from source to source, so their effects during pregnancy are unknown. (We do know that some plants produce teratogens and store them in their leaves, presumably as a defense against herbivorous animals.)

Fetal alcohol syndrome (FAS) occurs when maternal alcohol consumption produces developmental defects, such as skeletal deformation, cardiovascular defects, and neurological disorders. Mortality rates can be as high as 17 percent, and the survivors are plagued by problems in later development. The most severe cases involve mothers who consume the alcohol content of at least 7 ounces of hard liquor, 10 beers, or

FAP Ch. 29

several bottles of wine each day. But because the effects produced are directly related to the degree of exposure, there is probably no level of alcohol consumption that can be considered completely safe. Fetal alcohol syndrome is the number one cause of mental retardation in the United States today, affecting roughly 7500 infants each year.

Smoking presents another major risk to the developing fetus. In addition to introducing potentially harmful chemicals, such as nicotine, smoking lowers the P_{O_2} of maternal blood and reduces the amount of oxygen that reaches the placenta. A fetus carried by a smoking mother will not grow as rapidly as one carried by a nonsmoking mother, and smoking increases the risks of spontaneous abortion, prematurity, and fetal death. The rate of infant mortality after delivery is also higher when the mother smokes, and postnatal development can be adversely affected.

Induction and Sexual Differentiation

The physical (phenotypic) gender of a newborn infant depends on the hormonal cues received during development, not on the genetically determined gender of the individual. If something disrupts the normal inductive processes, the individual's genetic and anatomical genders may be different. Such a person is called a **pseudohermaphrodite** (soo-dō-her-MAF-ro-dīt). For example, if a female embryo becomes exposed to male hormones, it will develop the sexual characteristics of a male. Such situations are relatively rare. The most common cause is the hypertrophy of the fetal adrenal glands and their production of androgens in high concentrations; in some cases, this condition has been linked to genetic abnormalities. Maternal exposure to androgens, in the form of anabolic steroids or as a result of an endocrine tumor, can also produce a female pseudohermaphrodite.

Male pseudohermaphrodites may result from an inability to produce adequate concentrations of androgens due to some enzymatic defect. In **testicular feminization syndrome,** the infant appears to be a normal female at birth. Typical physical changes occur at puberty, and the individual develops the overt physical and behavioral characteristics of an adult woman. Menstrual cycles do not begin, however, because the vagina ends in a blind pocket, and there is no uterus. Biopsies performed on the gonads reveal normal testicular structure, and the interstitial cells are busily secreting testosterone. The problem apparently involves a defect in the cellular receptors sensitive to circulating androgens. Neither the embryo nor the adult tissues can respond to the testosterone produced by the gonads, so the person develops as, and remains physically, a female.

If detected in infancy, many cases of pseudohermaphroditism can be treated with hormones and surgery to produce males or females of normal appearance. Depending on the arrangement of the internal organs and gonads, normal reproductive function may be more difficult to achieve. Sex hormones also affect the brain's development, and sex assignment is behavioral as well as anatomical. In one instance, a genetic male who had been raised as a female after undergoing surgery on indeterminate external genitalia later adopted a male name, attire, and behavior after hormonal changes occurred at puberty.

Pseudohermaphroditism is one example of a developmental problem caused by hormonal miscues or by an inability to respond appropriately to hormonal instructions. Another example is provided by male infertility associated with maternal exposure to *diethylstilbestrol* (DES), a synthetic steroid prescribed in the 1950s to prevent miscarriages. An estimated 28 percent of male offspring produced abnormally small amounts of semen with marginal sperm counts at maturity. Daughters also have higher than normal infertility rates, due to uterine, vaginal, and uterine tube abnormalities, and they have an increased risk of developing vaginal cancer.

ECTOPIC PREGNANCIES FAP *p. 1069*

Implantation normally occurs at the endometrial surface that lines the uterine cavity. The precise location within the uterus varies, although in most cases implantation occurs in the uterine body. In an **ectopic pregnancy,** implantation occurs outside the uterus.

The incidence of ectopic pregnancies is approximately 0.6 percent of all pregnancies. Women who douche regularly have a 4.4 times higher risk of experiencing an ectopic pregnancy, presumably because the flushing action pushes the zygote away from the uterus. If the uterine tube has been scarred by a previous episode of pelvic inflammatory disease, the risk of an ectopic pregnancy increases. Although implantation may occur within the peritoneal cavity, in the ovarian wall, or in the cervix, 95 percent of ectopic pregnancies involve implantation within a uterine tube. Because it cannot expand enough to accommodate the developing embryo, the tube normally ruptures during the first trimester. At that time, the bleeding that occurs in the peritoneal cavity can be severe enough to pose a threat to the woman's life.

In a few instances, the ruptured uterine tube releases the embryo with an intact umbilical cord, so further development can occur. About 5 percent of these abdominal pregnancies actually complete full-term development; normal birth cannot occur, but the infant can be surgically removed from the abdominopelvic cavity. Because abdominal pregnancies are possible, it has been suggested that men as well as women could act as surrogate mothers if a zygote were surgically implanted into the peritoneal wall. It is not clear how

FAP Ch. 29

the endocrine, cardiovascular, nervous, and other systems of a man would respond to the stresses of pregnancy. However, the procedure has been tried successfully in mice, and experiments continue.

PROBLEMS WITH PLACENTATION

FAP *p. 1075*

In a **placenta previa** (PRĒ-vē-uh; "in the way"), implantation occurs in or near the cervix. This condition causes problems as the growing placenta approaches the internal os (internal cervical orifice). In a **total placenta previa,** the placenta extends across the internal os, whereas a partial placenta previa only partially blocks the os. The placenta is characterized by a rich fetal blood supply and the erosion of maternal blood vessels within the endometrium. Where the placenta passes across the internal os, the delicate complex hangs like an unsupported water balloon. As the pregnancy advances, even minor mechanical stresses can be enough to tear the placental tissues, leading to massive fetal and maternal bleeding.

Most cases can be diagnosed by ultrasound in the second trimester. As the uterus enlarges, the placenta may retract from covering the cervical os. If not, by the seventh month, the placenta reaches its full size, the cervical canal is dilated, and the uterine contents push against the placenta where it spans the internal cervical os without the support of the uterine wall. Minor, painless bleeding may occur. The treatment of total placenta previa involves bed rest for the mother until the fetus reaches a size at which cesarean delivery can be performed with a reasonable chance of neonatal (newborn) survival.

In an **abruptio placentae** (ab-RUP-shē-ō pla-SEN-tē), part or all of the placenta tears away from the uterine wall sometime after the fifth month of gestation. This is a serious and dangerous condition, and the bleeding and pain are usually sufficient to prompt an immediate visit to a physician. In severe cases, the bleeding leads to maternal anemia, shock, and kidney failure. Although maternal mortality is low, the fetal mortality rate from this condition ranges from 30 to 100 percent, depending on the severity of the fetal blood loss.

PROBLEMS WITH THE MAINTENANCE OF A PREGNANCY

FAP *p. 1083*

The rate of maternal complications during pregnancy is relatively high. Pregnancy stresses maternal systems, and the stresses can overwhelm homeostatic mechanisms. The term **toxemia** (tok-SĒ-mē-uh) **of pregnancy** refers to disorders that affect the maternal cardiovascular system. Chronic hypertension is the most characteristic symptom, but fluid imbalances, proteinuria, and central nervous system (CNS) disturbances, leading to coma or convulsions, can also occur. Some degree of toxemia occurs in 6–7 percent of third-trimester pregnancies. Severe cases account for 20 percent of maternal deaths and contribute to an estimated 25,000 neonatal deaths each year. Prenatal care involves monitoring the mother's vital signs and urine to detect early signs of toxemia so that treatment can prevent further progression.

Toxemia of pregnancy includes **preeclampsia** (prē-ē-KLAMP-sē-uh) and **eclampsia** (ē-KLAMP-sē-uh). Preeclampsia is most common during a woman's first pregnancy. The mother's systolic and diastolic pressures become elevated, reaching levels at or above 180/110. Other symptoms include fluid retention and edema, along with CNS disturbances and changes in kidney function. Roughly 4 percent of individuals with preeclampsia develop eclampsia.

Eclampsia is heralded by the onset of severe convulsions lasting 1–2 minutes, followed by a variable period of coma. Other symptoms resemble those of preeclampsia, with additional evidence of liver and kidney damage. The mortality rate from eclampsia is approximately 5 percent; the mother can be saved only if the fetus is delivered immediately. Once the fetus and placenta have been removed from the uterus, symptoms of eclampsia disappear over a period of hours to days.

COMPLEXITY AND PERFECTION

FAP *pp. 1086, 1094*

The expectation of prospective parents that every pregnancy will be idyllic and every baby will be perfect reflects deep-seated misconceptions about the nature of the developmental process. These misconceptions lead to the belief that when serious developmental errors occur, someone or something is at fault and that blame might be assigned to maternal habits (such as smoking, alcohol consumption, or improper diet), maternal exposure to toxins or prescription drugs, or the presence of other disruptive stimuli in the environment. The prosecution of women who give birth to severely impaired infants for "fetal abuse" (exposing a fetus to known or suspected risk factors) is an extreme example of this philosophy.

Although environmental stimuli can indeed lead to developmental problems, such factors are only one component of a complex system normally subject to considerable variation. Even if every pregnant woman were packed in cotton and confined to bed from conception to delivery, developmental accidents and errors would continue to occur with regularity.

Spontaneous mutations are the result of random errors in replication; such incidents are relatively common. At least 10 percent of fertilizations produce zygotes with abnormal chromosomes. Because most spontaneous mutations fail to produce visible defects, the actual number of mutations must be far larger. Most of the affected zygotes die before completing development, and only about 0.5 percent of newborn in-

FAP Ch. 29

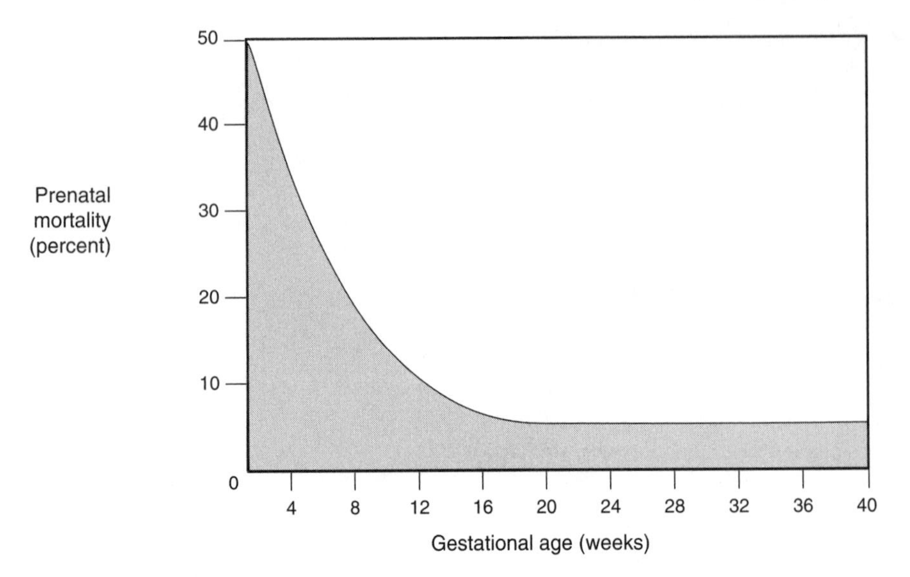

•**FIGURE A-67 Prenatal Mortality**

fants show chromosomal abnormalities that result from spontaneous mutations.

Due to the nature of the regulatory mechanisms, prenatal development does not follow precise, predetermined pathways. For example, much variation exists in the pathways of blood vessels and nerves, because it does not matter how blood or neural impulses get to their destinations, as long as they do get there. If the variations fall outside acceptable limits, however, the embryo or fetus fails to complete development. Very minor changes in heart structure can result in the death of a fetus, whereas large variations in venous distribution are common and relatively harmless. Virtually everyone can be considered abnormal to some degree, because no one has characteristics that are statistically average in every respect. An estimated 20 percent of your genes are subtly different from those found in the majority of the population, and minor defects such as extra nipples or birthmarks are quite common.

Current evidence suggests that as many as half of all conceptions produce zygotes that do not survive the cleavage stage. These zygotes disintegrate within the uterine tubes or uterine cavity; because implantation never occurs, there are no obvious signs of pregnancy. These instances of preimplantation mortality are commonly associated with chromosomal abnormalities. Of those embryos that implant, roughly 20 percent fail to complete 5 months of development, with an average survival time of 8 weeks. In most cases, severe problems affecting early embryogenesis or placenta formation are responsible. Figure A-67• graphically shows the relation of prenatal mortality to gestational age.

Prenatal mortality tends to eliminate the most severely affected fetuses. Those with less-extensive defects may survive, completing full-term gestation or arriving via premature delivery. **Congenital malformations** are structural abnormalities, present at birth, that affect major systems. Spina bifida, hydrocephalus, anencephaly, cleft lip, and Down syndrome are among the most common congenital malformations; we described those conditions in earlier chapters of the main text. The incidence of congenital malformations at birth averages about 6 percent, but only 2 percent are categorized as severe. Of these congenital problems, only 10 percent can be attributed to environmental factors in the absence of chromosomal abnormalities or genetic factors, including a family history of similar or related defects.

Medical technology continues to improve our abilities to understand and manipulate physiological processes. Genetic analysis of potential parents may now provide estimates of the likelihood of specific problems, although the problems themselves remain outside our control. But even with a better understanding of the genetic mechanisms involved, we will probably never be able to control every aspect of development and thereby prevent spontaneous abortions and congenital malformations. Too many complex, interdependent steps are involved in prenatal development, and malfunctions of some kind are statistically inevitable.

MONITORING POSTNATAL DEVELOPMENT FAP *p. 1087*

Each newborn infant is closely scrutinized after delivery. The maturity of the newborn may also be determined prior to delivery by means of ultrasound or amniocentesis (Table A-42). Immediately on delivery, the newborn is checked and assigned an **Apgar rating.** This rating evaluates the heart rate, respiratory rate, muscle tone, response to stimulation, and color at 1 and 5 minutes after birth. In each category, the infant receives a score ranging from 0 (poor) to 2 (excellent),

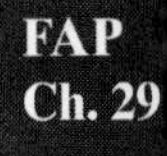

TABLE A-42 Tests Performed During Pregnancy and on the Neonate

Diagnostic Procedure	Method and Result	Representative Uses
Amniocentesis	A needle, inserted through abdominal wall into uterine cavity, collects amniotic fluid for analysis.	Detects chromosomal abnormalities and level of hemolysis in erythroblastosis fetalis; determines fetal lung maturity; detects birth defects such as spina bifida; evaluates fetal distress
Pelvic ultrasonography	Standard ultrasound	Detects multiple fetuses, fetal abnormalities, and placenta previa; estimates fetal age, growth, and gender
External fetal monitoring	Monitoring devices on external abdominal surface measure fetal heart rate and force of uterine contraction.	Detects irregular heart rate or fetal stress
Internal fetal monitoring	Electrode is attached to fetal scalp to monitor heart rate; catheter is placed in uterus to monitor uterine contractions.	As above
Chorionic villi biopsy	Test performed during weeks 8–10 of gestation; small pieces of villi are suctioned into a syringe.	Detects chromosomal abnormalities and biochemical disorders
Laboratory Test	**Normal Values**	**Significance of Abnormal Values**
Amniotic fluid analysis		
Karyotyping	Normal chromosomes	Detects chromosomal defects such as those in Down syndrome
Bilirubin	Traces only	Increased values may indicate amount of hemolysis of fetal RBCs by mother's Rh antibodies.
Meconium	Not present	Present in fetal distress
Lecithin/sphingomyelin ratio (L/S ratio)	≥2:1 ratio	Ratio below 2:1 indicates fetal immaturity.
Creatinine	≥2 mg/dl of amniotic fluid indicates week 36 of gestation.	Less than 2 mg/dl indicates fetal immaturity (not as accurate as L/S ratio).
Alpha-fetoprotein (AFP)	Week 16 of gestation: 5.7–31.5 ng/ml (lowers with increasing gestational age)	Increased values indicate possible neural tube defect such as spina bifida.
Blood tests		
Human chorionic gonadotropin (hCG) (maternal serum or urine)	Nonpregnant females (serum): <0.005 IU/ml Nonpregnant females (urine): negative Pregnant females (urine): <500,000 IU over 24 hours	Determines pregnancy; used in home pregnancy tests
TORCH Toxoplasmosis Other (syphilis, group B beta-hemolytic strep, and *Varicella*) Rubella Cytomegalovirus Herpes simplex Type II virus	Pregnant females: negative for IgM antibodies to these pathogens Neonates: negative for IgM antibodies to these pathogens	Pathogens causing these disorders can cross placenta and infect fetus; these fetal infections cause mild to severe problems, such as stillbirth; mother must be free of active herpes lesions to deliver vaginally.
Alpha-fetoprotein, AFP (serum)	Adults: <40 ng/ml	>500 ng/ml occurs in liver tumors; in pregnancy, levels peak at weeks 16–18; elevated levels occur with Down syndrome, anencephaly, and spina bifida.

TABLE A-42 Tests Performed During Pregnancy and on the Neonate *(continued)*

Laboratory Test	Normal Values	Significance of Abnormal Values
Blood type, Rh factor	Rh^+ or Rh^-	A sensitized Rh^- mother carrying an Rh^+ baby can result in erythroblastosis fetalis.
2-hour postprandial glucose test	Adults: 70–140 mg/dl (2 hours after glucose administration)	Increased level indicates possible diabetic state.
Human placental lactogen (hPL)	>4 mg/dl	<4 mg/dl can result from fetal distress; indicates possible miscarriage; useful in evaluating placental function
Vaginal culture for group B beta-hemolytic strep	None	Presence increases risk of prematurity and neonatal sepsis.
NEONATES		
Blood from umbilical cord	Blood typing titer	Detects maternal Rh antibody
Bilirubin	<12 mg/dl	Increased levels occur in jaundice due to immaturity of newborn's liver.
Phenylalanine (serum)	1–3 mg/dl	>4 mg/dl occurs in phenylketonuria.
Galactose-1-phosphate	18.5–28.5 U/g hemoglobin	Decreased value indicates galactosemia.

and the scores are then totaled. An infant's Apgar rating (0–10) has been shown to be an accurate predictor of newborn survival and of the presence of neurological damage. For example, newborn infants with *cerebral palsy* (FAP *p. 489*) tend to have a low Apgar rating.

In the course of this examination, the breath sounds, the depth and rate of respiration, and the heart rate are noted. Both the respiratory rate and the pulse rate are considerably higher in infants than in adults (see Table A-1, p. 8). Later, a more complete physical examination of the newborn focuses on the status of vital systems. Inspection of the infant normally includes the following:

- The head of a newborn infant may be misshapen after vaginal delivery, but it generally assumes its normal shape within the next few days. However, the size of the head must be checked to detect hydrocephalus (FAP *p. 444*), and the cranial vault is checked to ensure that *anencephaly* (FAP *p. 493*) does not exist.
- The eyes, nose, mouth, and ears are inspected for reflex responses and for obstruction.
- The abdomen is palpated to detect abnormalities of internal organs.
- The heart and lungs are auscultated to check for breath sounds and heart murmurs.
- The external genitalia are inspected. The scrotum of a male infant is checked to see if the testes have descended.
- Cyanosis of the hands and feet is normal in newborns, but the rest of the body should be pink. A generalized cyanosis may indicate congenital circulatory disorders, such as *erythroblastosis fetalis* (FAP *p. 638*) or patent foramen ovale, ductus arteriosus, or tetralogy of Fallot (FAP *p. 743*).

Measurements of body length, head circumference, and body weight are taken. A weight loss in the first 48 hours is normal, because fluid shifts occur as the infant adapts to the change from weightlessness (floating in amniotic fluid) to normal gravity. (Comparable fluid shifts occur in astronauts returning to Earth after extended periods in space.)

The nervous and muscular systems of newborns are assessed for normal reflexes and muscle tone. Reflexes commonly tested include the following:

- The *Moro reflex* is triggered when support for the head of a supine infant is suddenly removed. The reflex response consists of trunk extension and a rapid cycle of extension–abduction and flexion–adduction of the limbs. This reflex normally disappears at an age of about 3 months.
- The *stepping reflex* consists of walking movements triggered by holding the infant upright, with a forward slant, and placing the soles of the feet against the ground. This reflex normally disappears at an age of about 6 weeks.
- The *placing reflex* can be triggered by holding the infant upright and drawing the top of one foot across the bottom edge of a table. The reflex response is to flex and then extend the leg on that side. This reflex also disappears at an age of about 6 weeks.
- The *sucking reflex* is triggered by stroking the lips. The associated *rooting reflex* is initiated by stroking the cheek, and the response is to turn the mouth toward the site of stimulation. These reflexes persist until age 4–7 months.

- The *Babinski reflex* is positive, with fanning of the toes in response to stroking of the side of the sole of the foot (FAP *p. 430*). This reflex disappears at about age 3 years as descending motor pathways become established.

These procedures check for the presence of anatomical and physiological abnormalities. They also provide baseline information useful in assessing postnatal development. In addition, newborn infants are typically screened for genetic or metabolic disorders, such as phenylketonuria (PKU) (p. 170), congenital hypothyroidism (p. 99), galactosemia (p. 22), and sickle cell anemia (p. 112).

The excretory systems of the newborn infant are assessed by the examination of urine and feces. The first urination may be pink, owing to the presence of uric acid derivatives. The first bowel movement consists of a mixture of epithelial cells and mucus. This mixture, called *meconium,* is greenish-black.

Pediatrics is a medical specialty focusing on postnatal development from infancy through adolescence. Infants and young children cannot clearly describe the problems they are experiencing, so pediatricians and parents must be skilled observers. Standardized testing procedures are also used to assess developmental progress. In the **Denver Developmental Screening Test** (DDST), infants and children are checked repeatedly during their first 5 years. The test checks gross motor skills, such as sitting up or rolling over, language skills, fine motor coordination, and social interactions. The results are compared with normal values determined for individuals of similar age. These screening procedures assist in identifying children who may need special teaching and attention.

Too often parents tend to focus on a single ability or physical attribute, such as the age when the infant takes a first step or the growth rate versus standardized growth charts. This kind of one-track analysis has little practical value, and parents can become overly concerned with how their infant compares with the norm. *Normal values are statistical averages,* not absolute realities. For example, most infants begin walking at 11 to 14 months of age. But about 25 percent start before then, and another 10 percent have not started walking by the fourteenth month. Walking early does not indicate true genius, and walking late does not mean that the infant will need physical therapy. The questions on screening tests such as the DDST are intended to identify any *patterns* of developmental deficits. Such patterns appear only when a broad range of abilities and characteristics is considered.

DEATH AND DYING

FAP *p. 1090*

Despite exaggerated claims, few cases of individuals who have reached an age of 120 years have been substantiated. Estimates for the life span of individuals born in the United States during 1997 are 74 years for males and 80 years for females. Interestingly enough, the causes of death vary with age group. Consider the graphs shown in Figure A-68•, which indicate the mortality statistics for various age groups. The major cause of death in young people is accidents; in adults over age 40–45, it is cardiovascular disease. More-specific information about the major causes of death is given in Table A-43. Many of the characteristic differences in mortality values result from changes in the functional capabilities of the individuals linked to development or senescence. These values would differ significantly if tabulated for countries and cultures with different genetic and environmental pressures.

The differences in mortality values for males and females are related to differences in the accident rates among young people and in the rates of heart disease and cancer among older individuals. For instance, an upswing in cancer rates among females reflects a rising breast cancer incidence for those over age 34, whereas lung cancer is the primary cancer killer of older men. Among women, the incidence of lung cancers and related killers, including pulmonary disease,

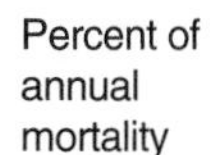

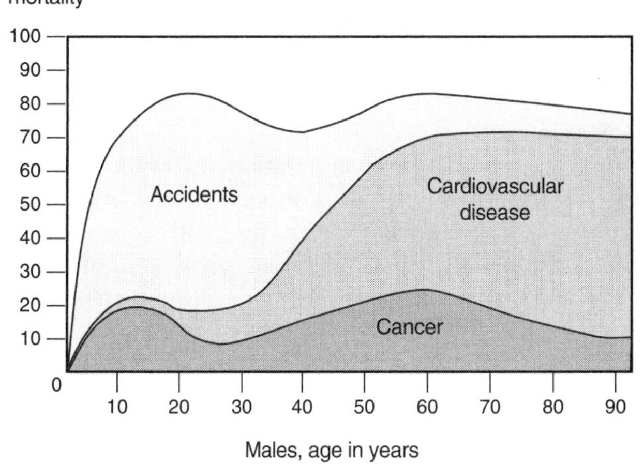

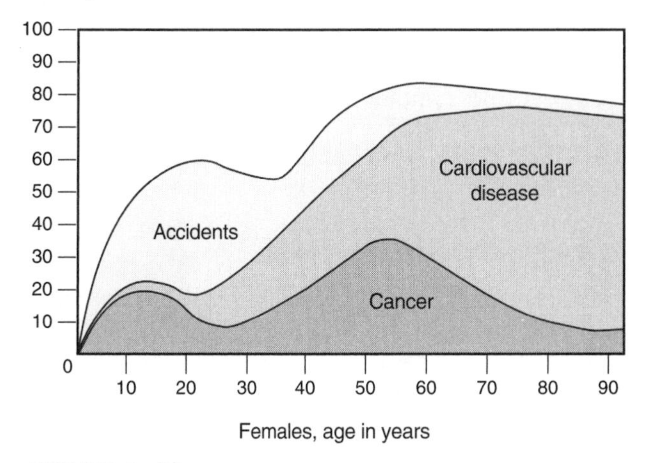

•**FIGURE A-68 Major Causes of Postnatal Mortality**

TABLE A-43 The Five Major Causes of Death in the U.S. Population

Rank	Age 1–14	Age 15–44	Age 45–64	Age 64+
1	Congenital anomalies	Accidents	Cancer	Heart disease
2	Accidents	Cancer	Heart disease	Cancer
3	Cancer	Heart disease	Accidents	Cerebrovascular disease
4	Homicide	Suicide	Cerebrovascular disease	COPD*
5	Pneumonia, influenza	Homicide	Pneumonia, influenza, COPD	Pneumonia, influenza

*COPD = chronic obstructive pulmonary disease.

heart disease, and pneumonia, has been steadily increasing as the number of women smokers has increased. This change has narrowed the difference between male and female life expectancies.

Experimental evidence and calculations suggest that the human life span has an upper limit of about 150 years. As medical advances continue, research must focus on two related issues: (1) extending the average life span toward that maximum and (2) improving the functional capabilities of long-lived individuals. The first objective may be the easiest from a technical standpoint. It is already possible to reduce the number of deaths attributed to specific causes. For example, new treatments promote remission in a variety of cancer cases, and anticoagulant therapies reduce the risks of death or permanent damage after a stroke or heart attack. Many defective organs can be replaced with functional transplants, and the use of controlled immunosuppressive drugs increases the success rates of these operations. Artificial hearts have been used, though with limited success thus far, and artificial kidneys and endocrine pancreases are under development.

The second objective poses more of a problem. Few people past their mid-nineties lead active, stimulating lives, and most would find the prospect of living another 50 years rather horrifying unless the quality of their lives could be significantly improved. Our abilities to prolong life now involve making stopgap corrections in systems on the brink of complete failure. Reversing the process of senescence would entail manipulating the biochemical operations and genetic programming of virtually every organ system. Although investigations continue, breakthroughs cannot be expected in the immediate future.

In the interim, we are left with some serious ethical and moral questions. If we could postpone the moment of death almost indefinitely with some combination of resuscitators and pharmacological support, how would we decide when it is appropriate to do so? How can medical and financial resources be fairly allocated? Who gets the limited number of organs available for transplant? Who should be selected for experimental therapies of potential significance? Should we take into account that care of an infant or child may add decades to a life span, whereas the costly insertion of an artificial heart in a 60-year-old may add only months to years? How shall we allocate the costs of sophisticated procedures that can reach hundreds of thousands of dollars per individual over the long run? Are these individual or family responsibilities? Will only the rich be able to survive into a second century of life? Should the government provide the funds? If yes, what will happen to tax rates as the baby boomers become elderly citizens? And what about the role of the individual involved? If you decline treatment, are you mentally and legally competent? Could your survivors bring suit if you were forced to survive or if you were allowed to die? These and other difficult questions will not go away. In the years to come, we will have to find answers we are content to live and die with.

[EXPLORE]

Additional resources, such as interactive tutorials, critical-thinking questions, clinical problems, and case studies, are available through the CD-ROM and Website for your textbook. To begin your exploration, launch the *Martini Interactive* CD-ROM or visit the Companion Website (http://www.prenhall.com/martini/fap5) and select Chapter 28 or 29.

FAP Ch. 29

COMPANION ATLAS

Embryology Summaries

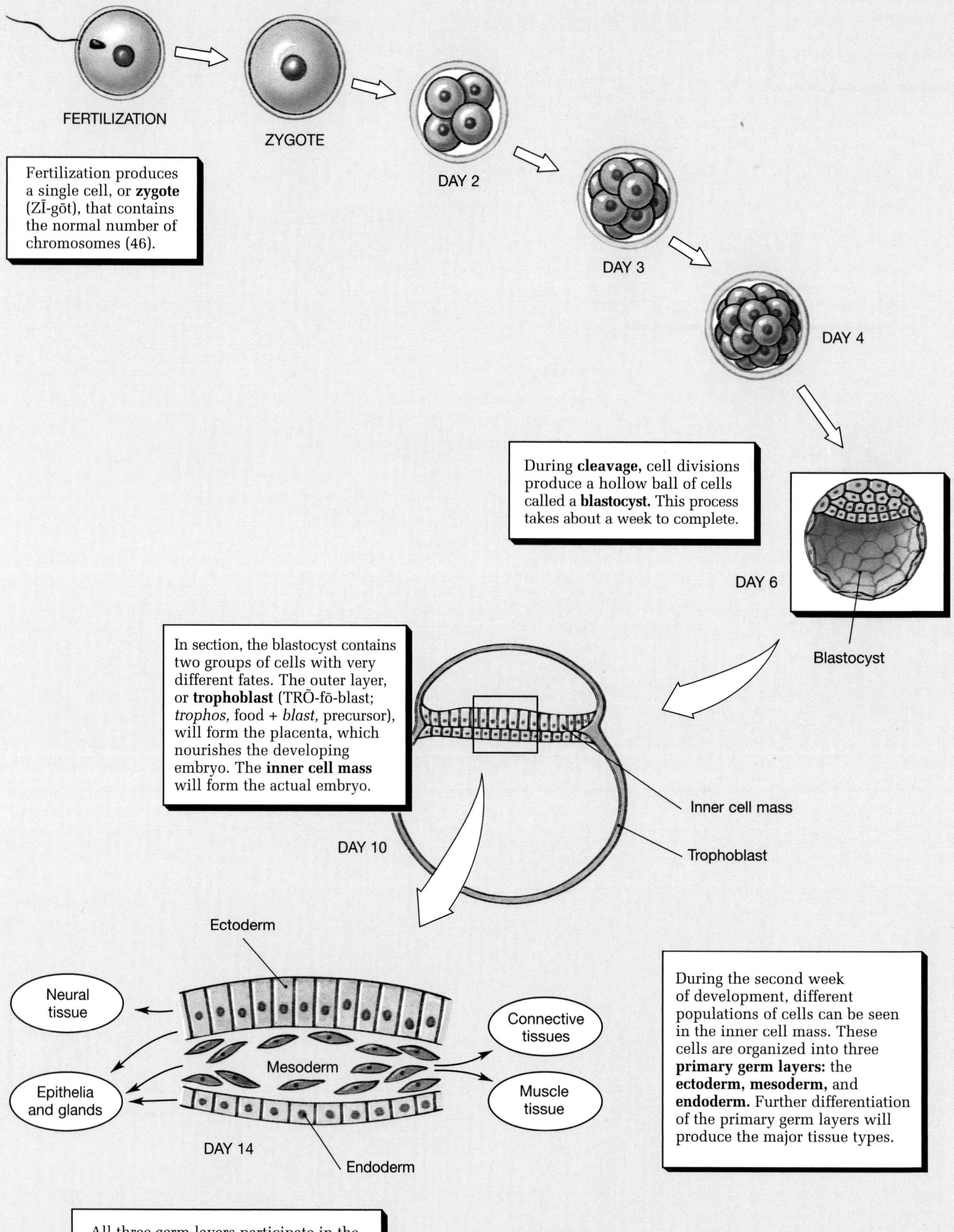

Fertilization produces a single cell, or **zygote** (ZĪ-gōt), that contains the normal number of chromosomes (46).

During **cleavage,** cell divisions produce a hollow ball of cells called a **blastocyst.** This process takes about a week to complete.

In section, the blastocyst contains two groups of cells with very different fates. The outer layer, or **trophoblast** (TRŌ-fō-blast; *trophos,* food + *blast,* precursor), will form the placenta, which nourishes the developing embryo. The **inner cell mass** will form the actual embryo.

During the second week of development, different populations of cells can be seen in the inner cell mass. These cells are organized into three **primary germ layers:** the **ectoderm, mesoderm,** and **endoderm.** Further differentiation of the primary germ layers will produce the major tissue types.

All three germ layers participate in the formation of functional organs and organ systems. Their interactions will be detailed in later Embryology Summaries dealing with specific systems.

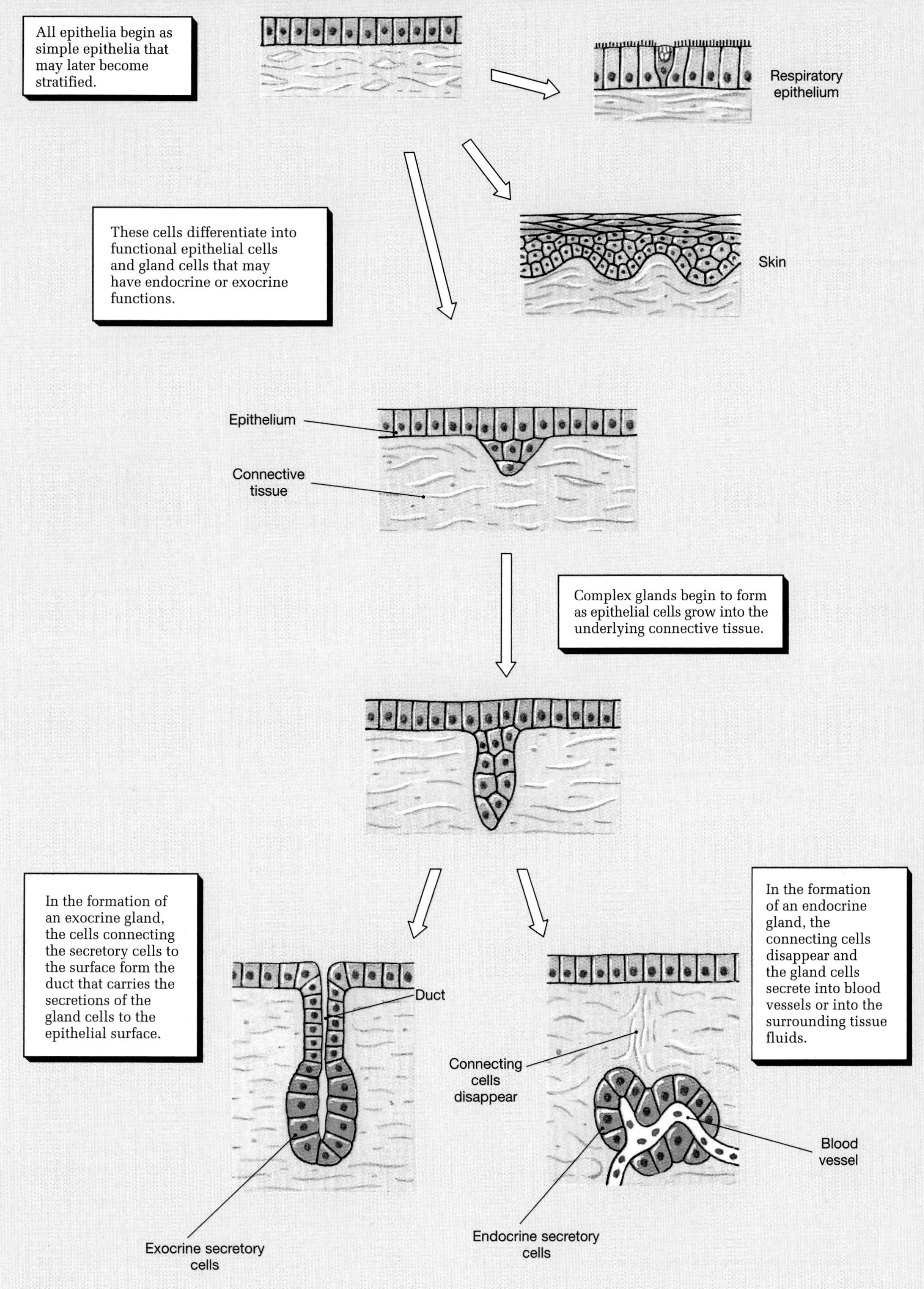
All epithelia begin as simple epithelia that may later become stratified.
Respiratory epithelium
These cells differentiate into functional epithelial cells and gland cells that may have endocrine or exocrine functions.
Skin
Epithelium
Connective tissue
Complex glands begin to form as epithelial cells grow into the underlying connective tissue.
In the formation of an exocrine gland, the cells connecting the secretory cells to the surface form the duct that carries the secretions of the gland cells to the epithelial surface.
Duct
Exocrine secretory cells
In the formation of an endocrine gland, the connecting cells disappear and the gland cells secrete into blood vessels or into the surrounding tissue fluids.
Connecting cells disappear
Blood vessel
Endocrine secretory cells

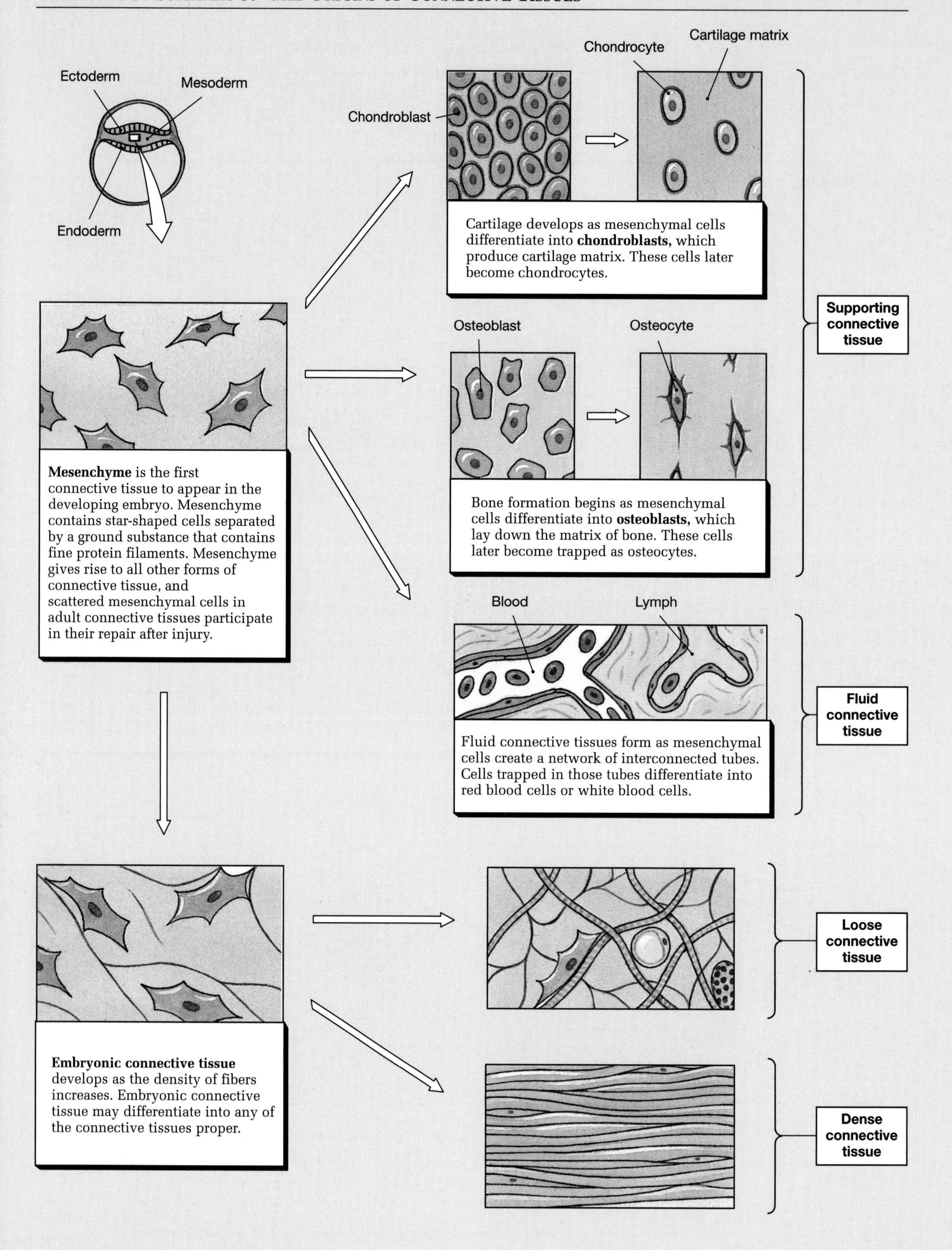
Ectoderm
Mesoderm
Endoderm
Chondroblast
Chondrocyte
Cartilage matrix
Cartilage develops as mesenchymal cells differentiate into **chondroblasts,** which produce cartilage matrix. These cells later become chondrocytes.
Osteoblast
Osteocyte
Bone formation begins as mesenchymal cells differentiate into **osteoblasts,** which lay down the matrix of bone. These cells later become trapped as osteocytes.
Supporting connective tissue
Mesenchyme is the first connective tissue to appear in the developing embryo. Mesenchyme contains star-shaped cells separated by a ground substance that contains fine protein filaments. Mesenchyme gives rise to all other forms of connective tissue, and scattered mesenchymal cells in adult connective tissues participate in their repair after injury.
Blood
Lymph
Fluid connective tissues form as mesenchymal cells create a network of interconnected tubes. Cells trapped in those tubes differentiate into red blood cells or white blood cells.
Fluid connective tissue
Embryonic connective tissue develops as the density of fibers increases. Embryonic connective tissue may differentiate into any of the connective tissues proper.
Loose connective tissue
Dense connective tissue

Many organ systems show similar patterns of organization. For example, the digestive, respiratory, urinary, and reproductive systems each include passageways lined by epithelia and surrounded by layers of smooth muscle. These patterns are the result of developmental processes under way in the first 2 months of embryonic life.

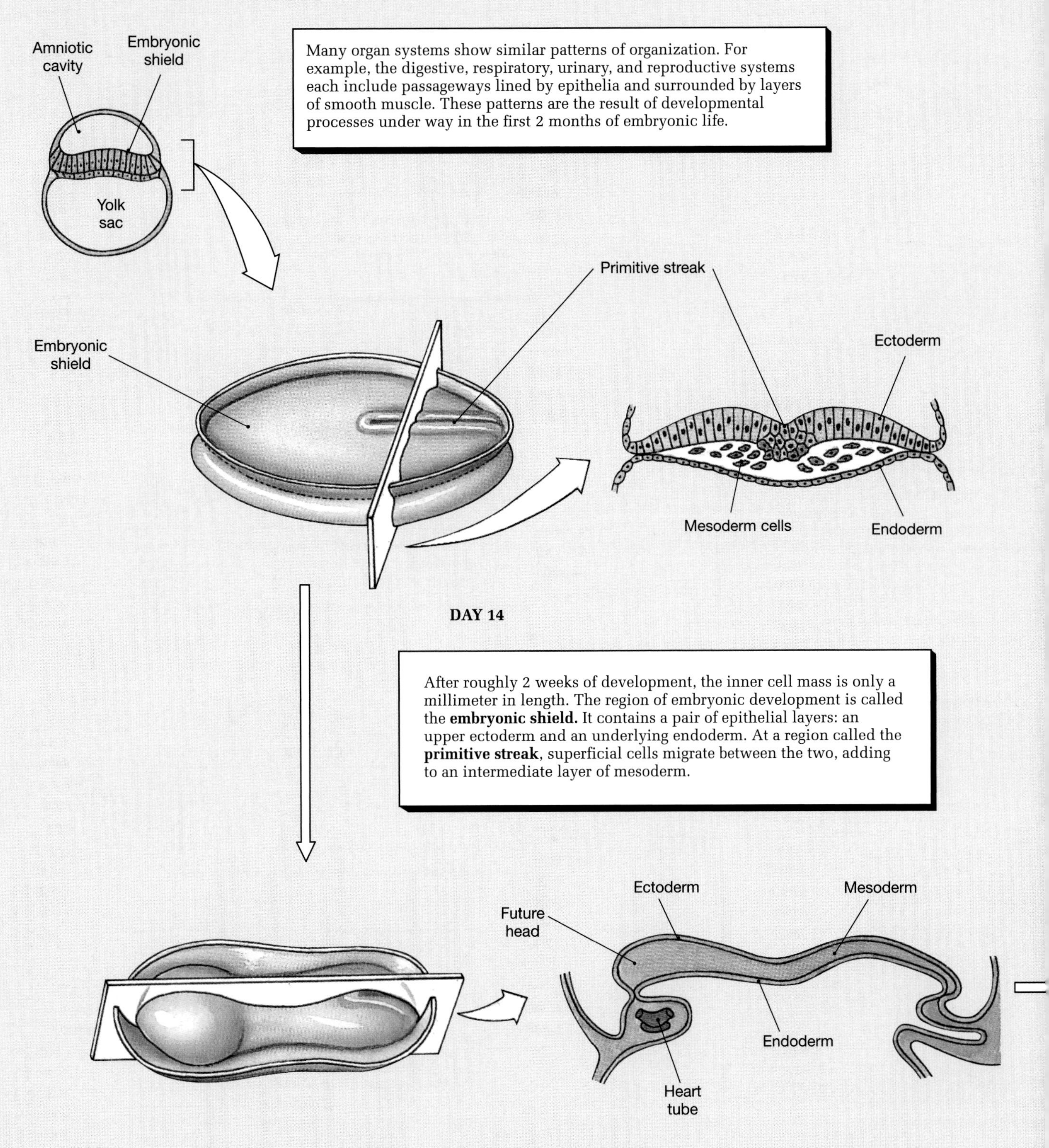

After roughly 2 weeks of development, the inner cell mass is only a millimeter in length. The region of embryonic development is called the **embryonic shield.** It contains a pair of epithelial layers: an upper ectoderm and an underlying endoderm. At a region called the **primitive streak**, superficial cells migrate between the two, adding to an intermediate layer of mesoderm.

By day 18, the embryo has begun to lift off the surface of the embryonic shield. The heart and many blood vessels have already formed, well ahead of the other organ systems. Unless otherwise noted, discussions of organ system development in later Embryology Summaries will begin at this stage.

DERIVATIVES OF PRIMARY GERM LAYERS	
Ectoderm forms:	Epidermis and epidermal derivatives of the integumentary system, including hair follicles, nails, and glands communicating with the skin surface (sweat, milk, and sebum) Lining of the mouth, salivary glands, nasal passageways, and anus Nervous system, including brain and spinal cord Portions of the endocrine system (pituitary gland and parts of adrenal glands) Portions of the skull, pharyngeal arches, and teeth
Mesoderm forms:	Lining of the body cavities (pleural, pericardial, peritoneal) Muscular, skeletal, cardiovascular, and lymphatic systems Kidneys and part of the urinary tract Gonads and most of the reproductive tracts Connective tissues supporting all organ systems Portions of the endocrine system (parts of adrenal glands and endocrine tissues of the reproductive tracts)
Endoderm forms:	Most of the digestive system: epithelium (except mouth and anus), exocrine glands (except salivary glands), liver, and pancreas Most of the respiratory system: epithelium (except nasal passageways) and mucous glands Portions of the urinary and reproductive systems (ducts and the stem cells that produce gametes) Portions of the endocrine system (thymus, thyroid gland, parathyroid glands, and pancreas)

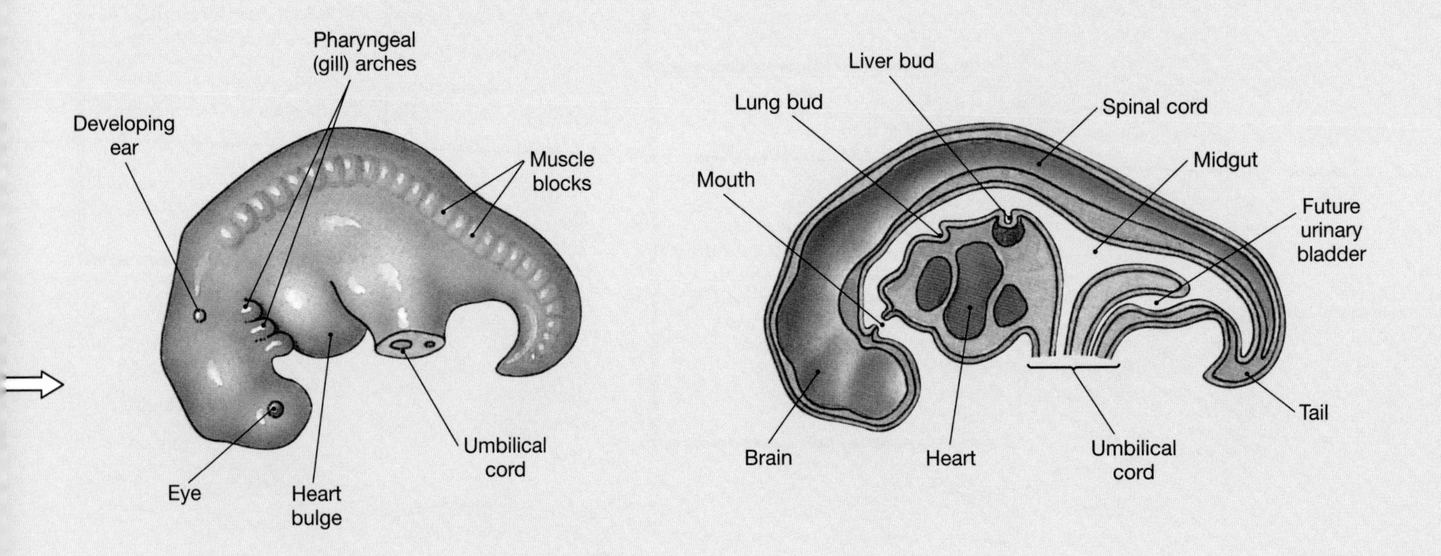

DAY 28

After 1 month, all major organ systems have begun to form. The role of each of the primary germ layers in the formation of organs is summarized in the accompanying table; details are given in later Embryology Summaries.

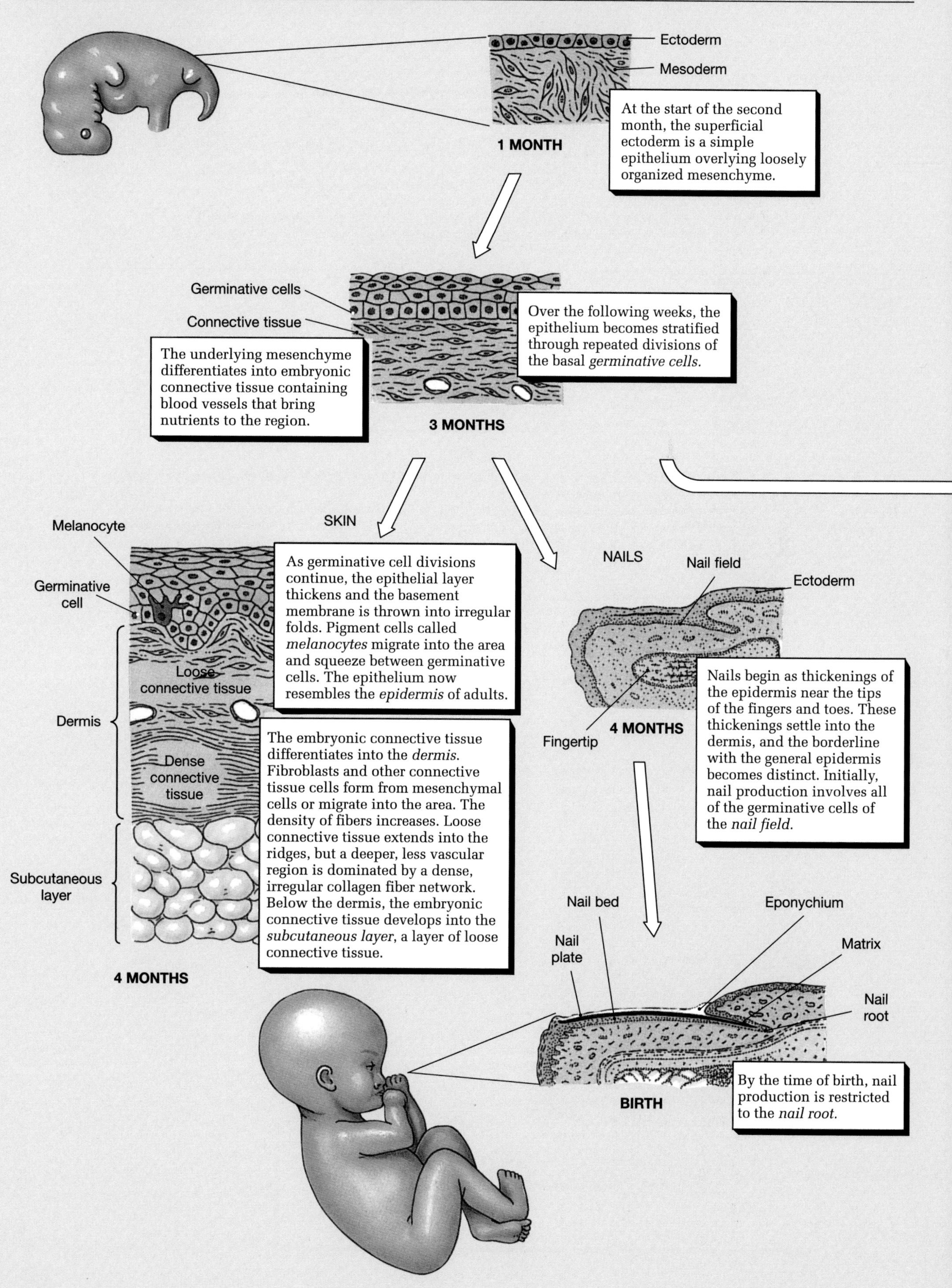
Ectoderm
Mesoderm
At the start of the second month, the superficial ectoderm is a simple epithelium overlying loosely organized mesenchyme.
1 MONTH
Germinative cells
Connective tissue
Over the following weeks, the epithelium becomes stratified through repeated divisions of the basal *germinative cells.*
The underlying mesenchyme differentiates into embryonic connective tissue containing blood vessels that bring nutrients to the region.
3 MONTHS
SKIN
Melanocyte
Germinative cell
As germinative cell divisions continue, the epithelial layer thickens and the basement membrane is thrown into irregular folds. Pigment cells called *melanocytes* migrate into the area and squeeze between germinative cells. The epithelium now resembles the *epidermis* of adults.
Loose connective tissue
Dermis
Dense connective tissue
The embryonic connective tissue differentiates into the *dermis.* Fibroblasts and other connective tissue cells form from mesenchymal cells or migrate into the area. The density of fibers increases. Loose connective tissue extends into the ridges, but a deeper, less vascular region is dominated by a dense, irregular collagen fiber network. Below the dermis, the embryonic connective tissue develops into the *subcutaneous layer*, a layer of loose connective tissue.
Subcutaneous layer
4 MONTHS
NAILS
Nail field
Ectoderm
Nails begin as thickenings of the epidermis near the tips of the fingers and toes. These thickenings settle into the dermis, and the borderline with the general epidermis becomes distinct. Initially, nail production involves all of the germinative cells of the *nail field.*
Fingertip
4 MONTHS
Nail bed
Eponychium
Nail plate
Matrix
Nail root
By the time of birth, nail production is restricted to the *nail root.*
BIRTH

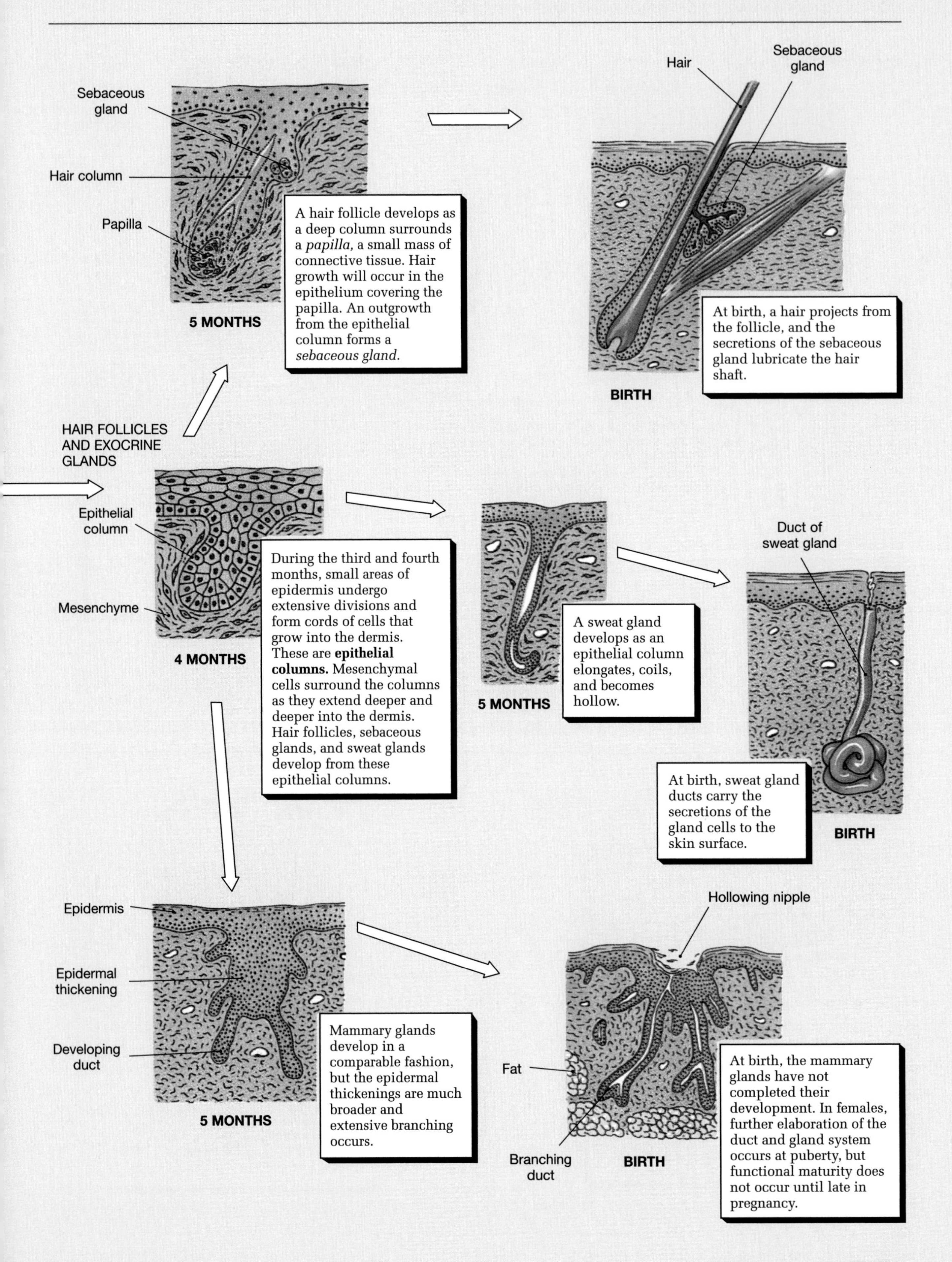
Sebaceous gland
Hair column
Papilla
5 MONTHS
A hair follicle develops as a deep column surrounds a *papilla,* a small mass of connective tissue. Hair growth will occur in the epithelium covering the papilla. An outgrowth from the epithelial column forms a *sebaceous gland.*
Hair
Sebaceous gland
At birth, a hair projects from the follicle, and the secretions of the sebaceous gland lubricate the hair shaft.
BIRTH
HAIR FOLLICLES AND EXOCRINE GLANDS
Epithelial column
Mesenchyme
4 MONTHS
During the third and fourth months, small areas of epidermis undergo extensive divisions and form cords of cells that grow into the dermis. These are **epithelial columns.** Mesenchymal cells surround the columns as they extend deeper and deeper into the dermis. Hair follicles, sebaceous glands, and sweat glands develop from these epithelial columns.
5 MONTHS
A sweat gland develops as an epithelial column elongates, coils, and becomes hollow.
Duct of sweat gland
At birth, sweat gland ducts carry the secretions of the gland cells to the skin surface.
BIRTH
Epidermis
Epidermal thickening
Developing duct
5 MONTHS
Mammary glands develop in a comparable fashion, but the epidermal thickenings are much broader and extensive branching occurs.
Hollowing nipple
Fat
Branching duct
BIRTH
At birth, the mammary glands have not completed their development. In females, further elaboration of the duct and gland system occurs at puberty, but functional maturity does not occur until late in pregnancy.

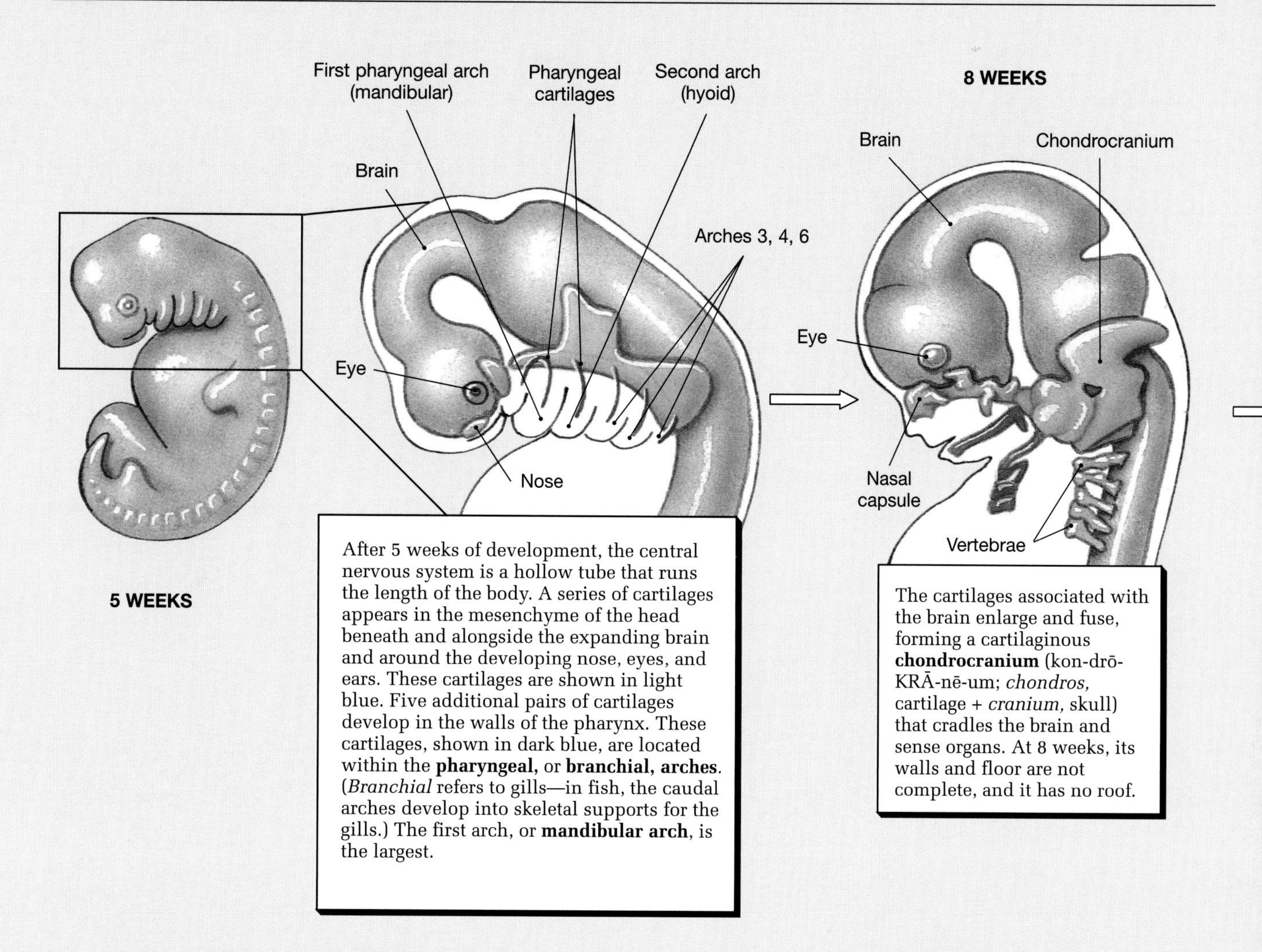

After 5 weeks of development, the central nervous system is a hollow tube that runs the length of the body. A series of cartilages appears in the mesenchyme of the head beneath and alongside the expanding brain and around the developing nose, eyes, and ears. These cartilages are shown in light blue. Five additional pairs of cartilages develop in the walls of the pharynx. These cartilages, shown in dark blue, are located within the **pharyngeal,** or **branchial, arches**. (*Branchial* refers to gills—in fish, the caudal arches develop into skeletal supports for the gills.) The first arch, or **mandibular arch**, is the largest.

The cartilages associated with the brain enlarge and fuse, forming a cartilaginous **chondrocranium** (kon-drō-KRĀ-nē-um; *chondros,* cartilage + *cranium,* skull) that cradles the brain and sense organs. At 8 weeks, its walls and floor are not complete, and it has no roof.

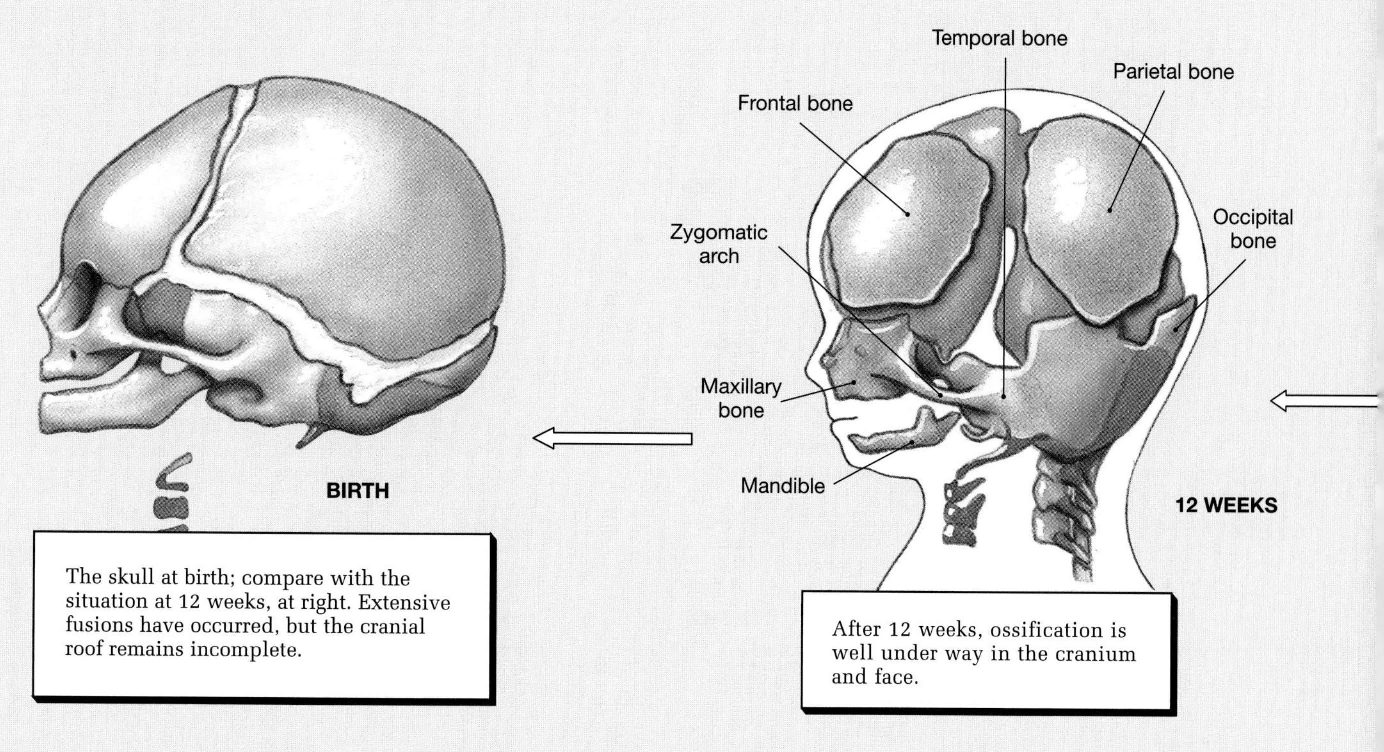

The skull at birth; compare with the situation at 12 weeks, at right. Extensive fusions have occurred, but the cranial roof remains incomplete.

After 12 weeks, ossification is well under way in the cranium and face.

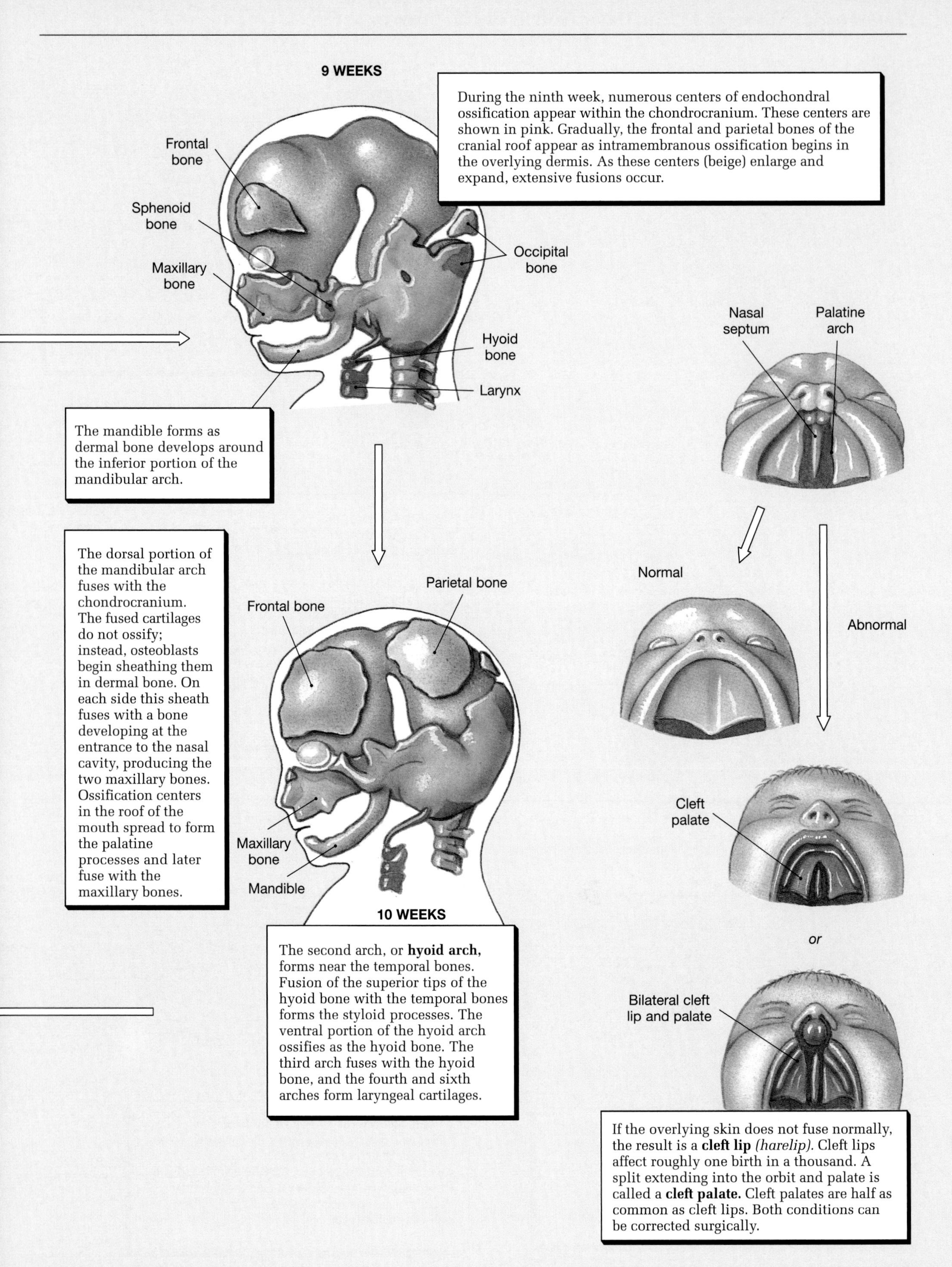

During the ninth week, numerous centers of endochondral ossification appear within the chondrocranium. These centers are shown in pink. Gradually, the frontal and parietal bones of the cranial roof appear as intramembranous ossification begins in the overlying dermis. As these centers (beige) enlarge and expand, extensive fusions occur.

The mandible forms as dermal bone develops around the inferior portion of the mandibular arch.

The dorsal portion of the mandibular arch fuses with the chondrocranium. The fused cartilages do not ossify; instead, osteoblasts begin sheathing them in dermal bone. On each side this sheath fuses with a bone developing at the entrance to the nasal cavity, producing the two maxillary bones. Ossification centers in the roof of the mouth spread to form the palatine processes and later fuse with the maxillary bones.

The second arch, or **hyoid arch,** forms near the temporal bones. Fusion of the superior tips of the hyoid bone with the temporal bones forms the styloid processes. The ventral portion of the hyoid arch ossifies as the hyoid bone. The third arch fuses with the hyoid bone, and the fourth and sixth arches form laryngeal cartilages.

If the overlying skin does not fuse normally, the result is a **cleft lip** *(harelip).* Cleft lips affect roughly one birth in a thousand. A split extending into the orbit and palate is called a **cleft palate.** Cleft palates are half as common as cleft lips. Both conditions can be corrected surgically.

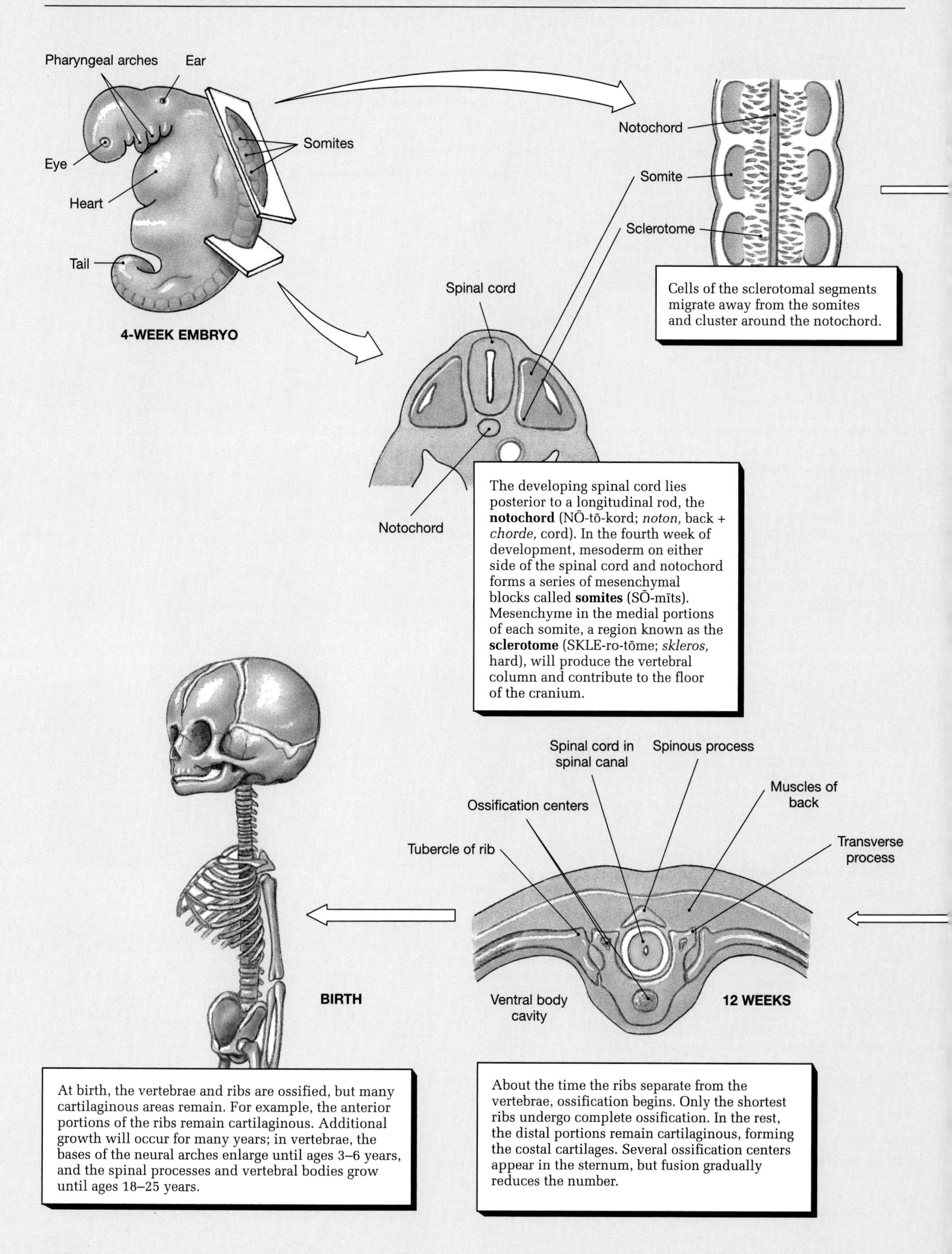

Cells of the sclerotomal segments migrate away from the somites and cluster around the notochord.

The developing spinal cord lies posterior to a longitudinal rod, the **notochord** (NŌ-tō-kord; *noton,* back + *chorde,* cord). In the fourth week of development, mesoderm on either side of the spinal cord and notochord forms a series of mesenchymal blocks called **somites** (SŌ-mīts). Mesenchyme in the medial portions of each somite, a region known as the **sclerotome** (SKLE-ro-tōme; *skleros,* hard), will produce the vertebral column and contribute to the floor of the cranium.

About the time the ribs separate from the vertebrae, ossification begins. Only the shortest ribs undergo complete ossification. In the rest, the distal portions remain cartilaginous, forming the costal cartilages. Several ossification centers appear in the sternum, but fusion gradually reduces the number.

At birth, the vertebrae and ribs are ossified, but many cartilaginous areas remain. For example, the anterior portions of the ribs remain cartilaginous. Additional growth will occur for many years; in vertebrae, the bases of the neural arches enlarge until ages 3–6 years, and the spinal processes and vertebral bodies grow until ages 18–25 years.

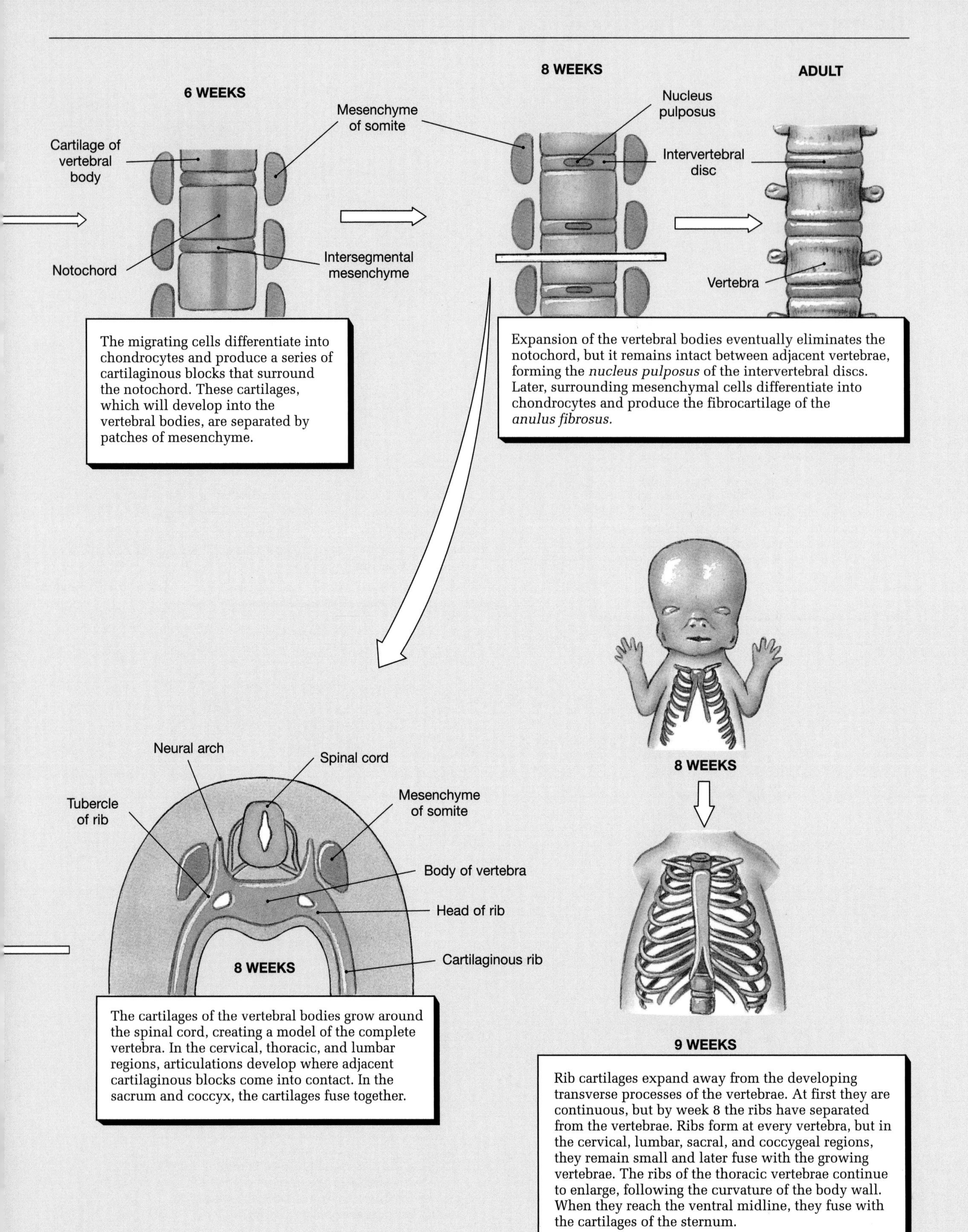
6 WEEKS
Cartilage of vertebral body
Mesenchyme of somite
Notochord
Intersegmental mesenchyme
The migrating cells differentiate into chondrocytes and produce a series of cartilaginous blocks that surround the notochord. These cartilages, which will develop into the vertebral bodies, are separated by patches of mesenchyme.
8 WEEKS
Nucleus pulposus
Intervertebral disc
ADULT
Vertebra
Expansion of the vertebral bodies eventually eliminates the notochord, but it remains intact between adjacent vertebrae, forming the *nucleus pulposus* of the intervertebral discs. Later, surrounding mesenchymal cells differentiate into chondrocytes and produce the fibrocartilage of the *anulus fibrosus.*
Neural arch
Spinal cord
Tubercle of rib
Mesenchyme of somite
Body of vertebra
Head of rib
Cartilaginous rib
8 WEEKS
The cartilages of the vertebral bodies grow around the spinal cord, creating a model of the complete vertebra. In the cervical, thoracic, and lumbar regions, articulations develop where adjacent cartilaginous blocks come into contact. In the sacrum and coccyx, the cartilages fuse together.
8 WEEKS
9 WEEKS
Rib cartilages expand away from the developing transverse processes of the vertebrae. At first they are continuous, but by week 8 the ribs have separated from the vertebrae. Ribs form at every vertebra, but in the cervical, lumbar, sacral, and coccygeal regions, they remain small and later fuse with the growing vertebrae. The ribs of the thoracic vertebrae continue to enlarge, following the curvature of the body wall. When they reach the ventral midline, they fuse with the cartilages of the sternum.

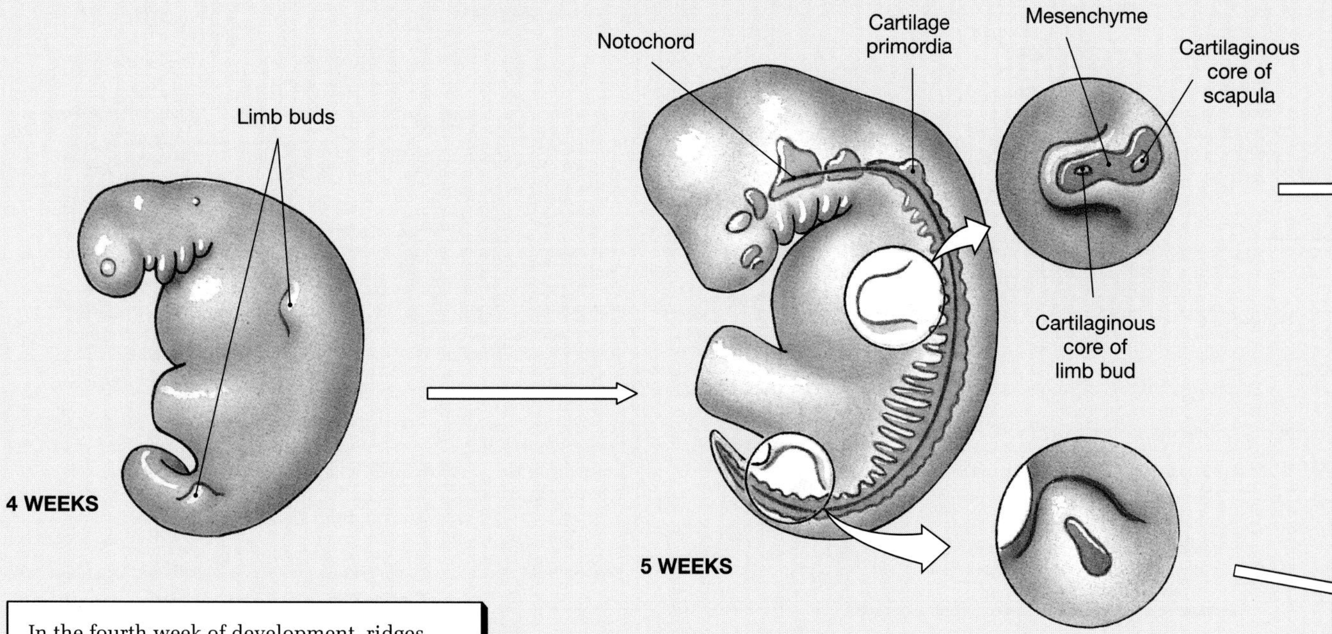

In the fourth week of development, ridges appear along the flanks of the embryo, extending from the caudal limits of the pharynx to the region of the anus. These ridges form as mesodermal cells congregate beneath the ectoderm of the flank. Mesoderm gradually accumulates at the end of each ridge, forming two pairs of limb buds.

After 5 weeks of development, the pectoral limb buds have a cartilaginous core and scapular cartilages are developing in the mesenchyme of the trunk.

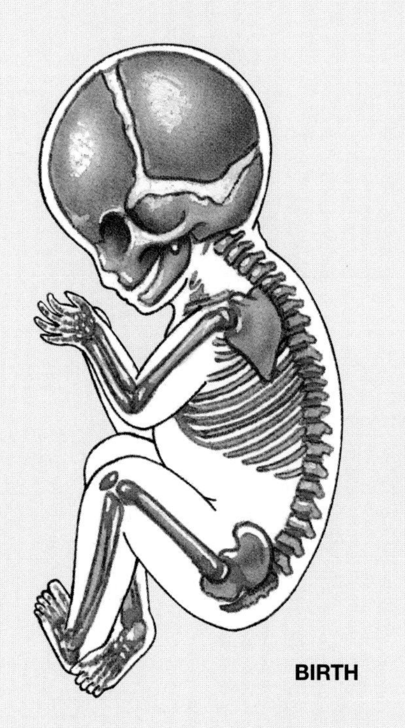

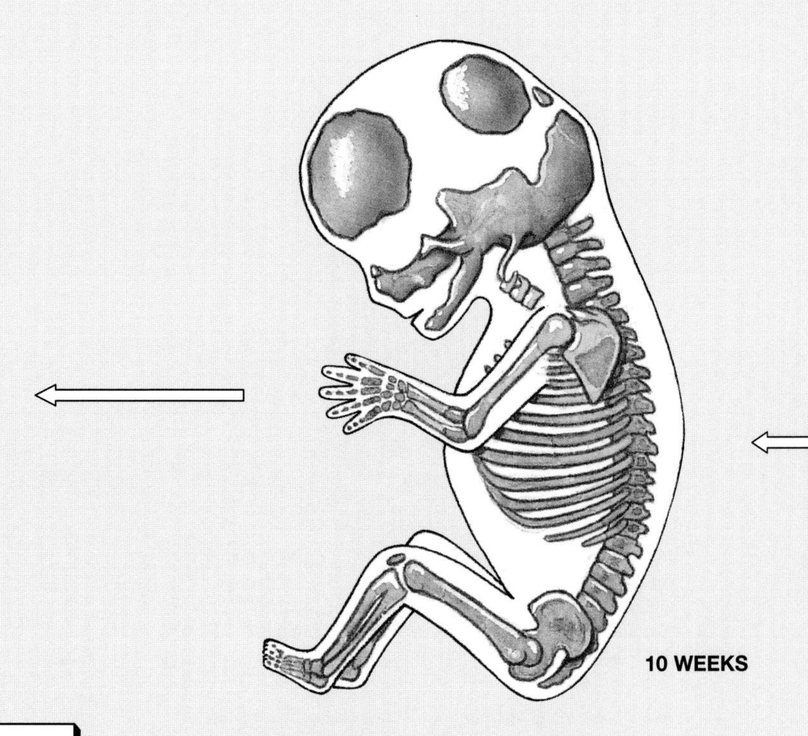

At birth, there are extensive areas of cartilage (blue) in the humeral head, in the wrist, between the bones of the palm and fingers, and in the ossa coxae. Notice the appearance of the axial skeleton, with reference to Embryology Summary 7.

After approximately 10 weeks of development, the shafts of the limb bones are undergoing rapid ossification, but the distal bones of the carpus and tarsus remain cartilaginous.

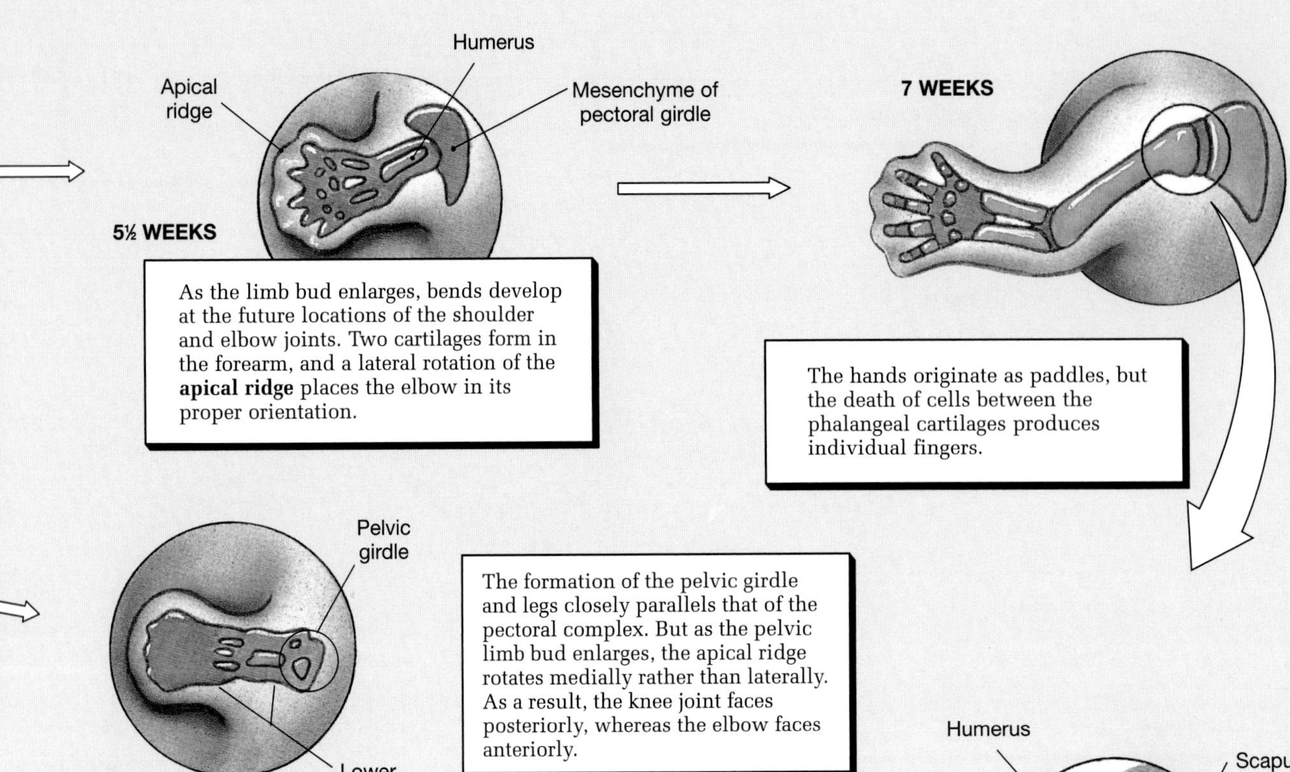

As the limb bud enlarges, bends develop at the future locations of the shoulder and elbow joints. Two cartilages form in the forearm, and a lateral rotation of the **apical ridge** places the elbow in its proper orientation.

The hands originate as paddles, but the death of cells between the phalangeal cartilages produces individual fingers.

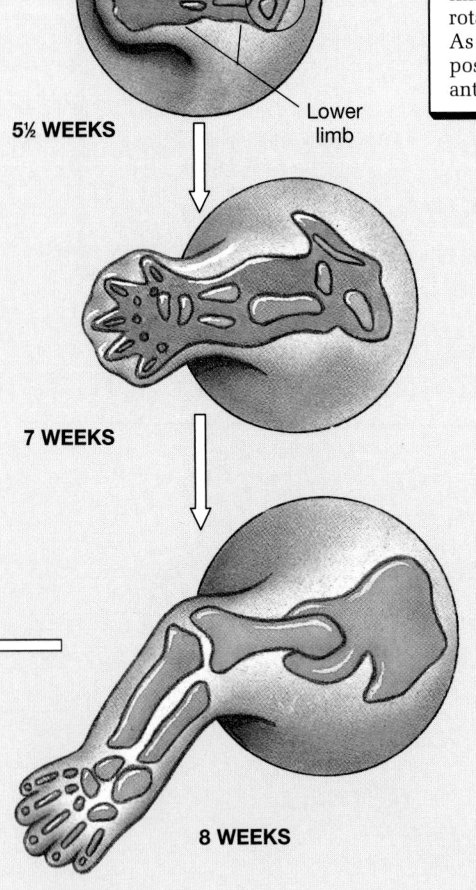

The formation of the pelvic girdle and legs closely parallels that of the pectoral complex. But as the pelvic limb bud enlarges, the apical ridge rotates medially rather than laterally. As a result, the knee joint faces posteriorly, whereas the elbow faces anteriorly.

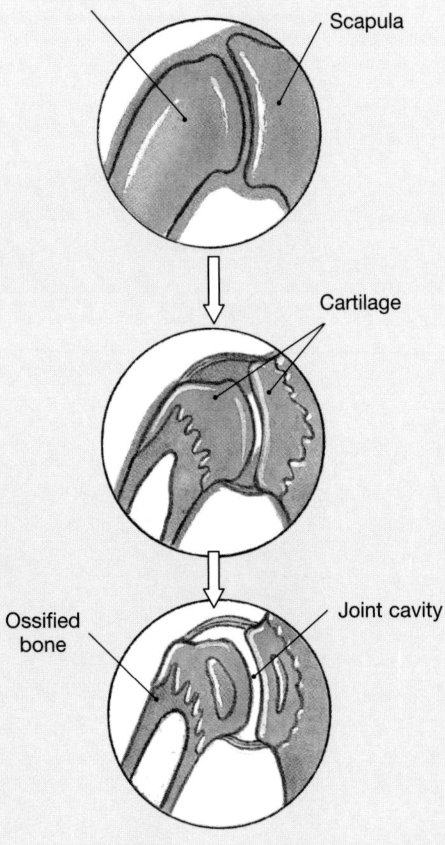

By week 8, cartilaginous models of all of the major skeletal components are well formed, and endochondral ossification begins in the future limb bones. Ossification of the coxal bones begins at three separate centers that gradually enlarge.

Joints form where two cartilages are in contact. The surfaces within the joint cavity remain cartilaginous; the rest of the bones undergo ossification.

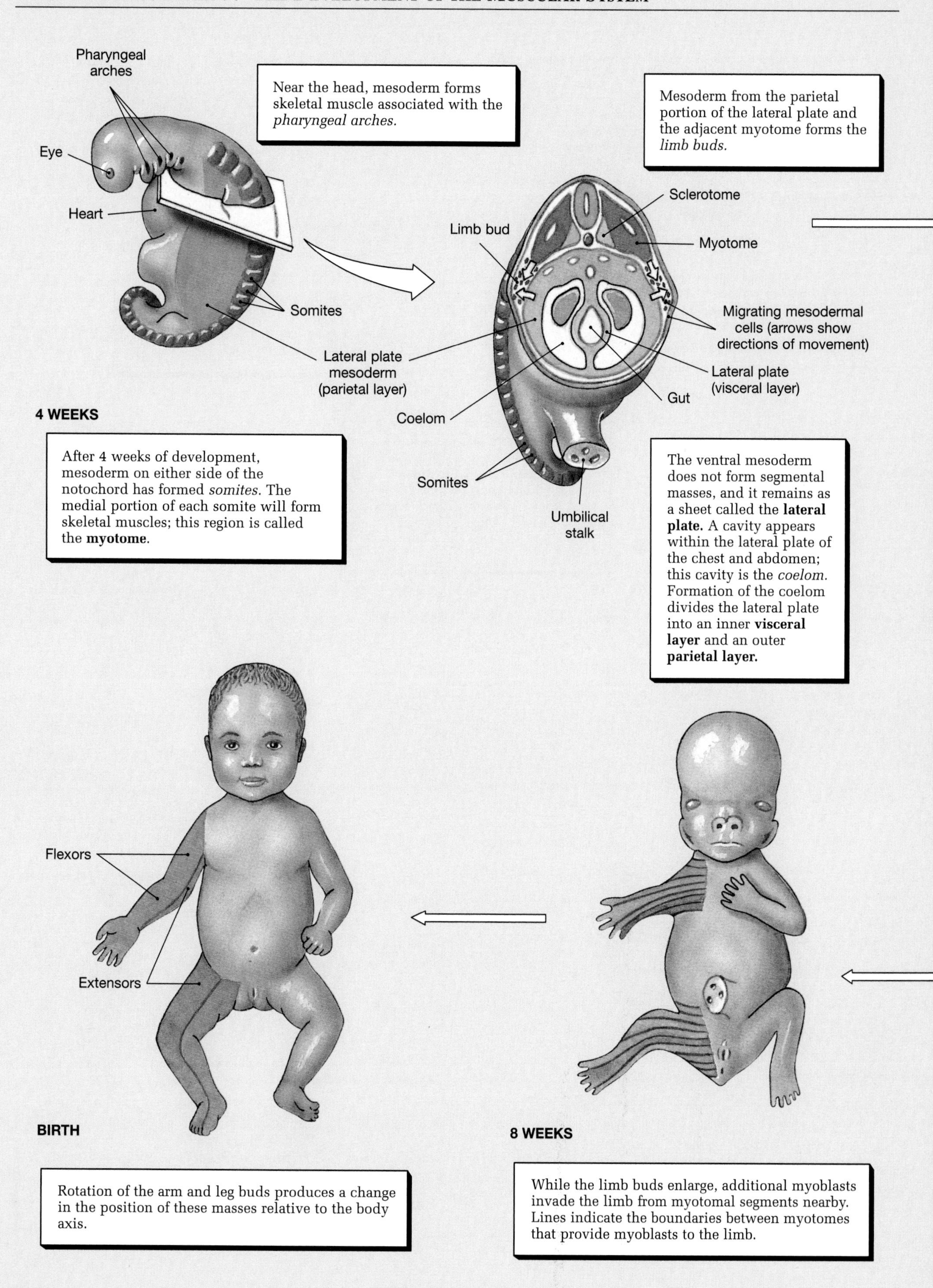

Pharyngeal arches
Eye
Heart
Somites
Lateral plate mesoderm (parietal layer)
4 WEEKS
Near the head, mesoderm forms skeletal muscle associated with the *pharyngeal arches.*
Mesoderm from the parietal portion of the lateral plate and the adjacent myotome forms the *limb buds.*
Sclerotome
Limb bud
Myotome
Migrating mesodermal cells (arrows show directions of movement)
Lateral plate (visceral layer)
Gut
Coelom
Somites
Umbilical stalk
After 4 weeks of development, mesoderm on either side of the notochord has formed *somites.* The medial portion of each somite will form skeletal muscles; this region is called the **myotome**.
The ventral mesoderm does not form segmental masses, and it remains as a sheet called the **lateral plate.** A cavity appears within the lateral plate of the chest and abdomen; this cavity is the *coelom.* Formation of the coelom divides the lateral plate into an inner **visceral layer** and an outer **parietal layer.**
Flexors
Extensors
BIRTH
8 WEEKS
Rotation of the arm and leg buds produces a change in the position of these masses relative to the body axis.
While the limb buds enlarge, additional myoblasts invade the limb from myotomal segments nearby. Lines indicate the boundaries between myotomes that provide myoblasts to the limb.

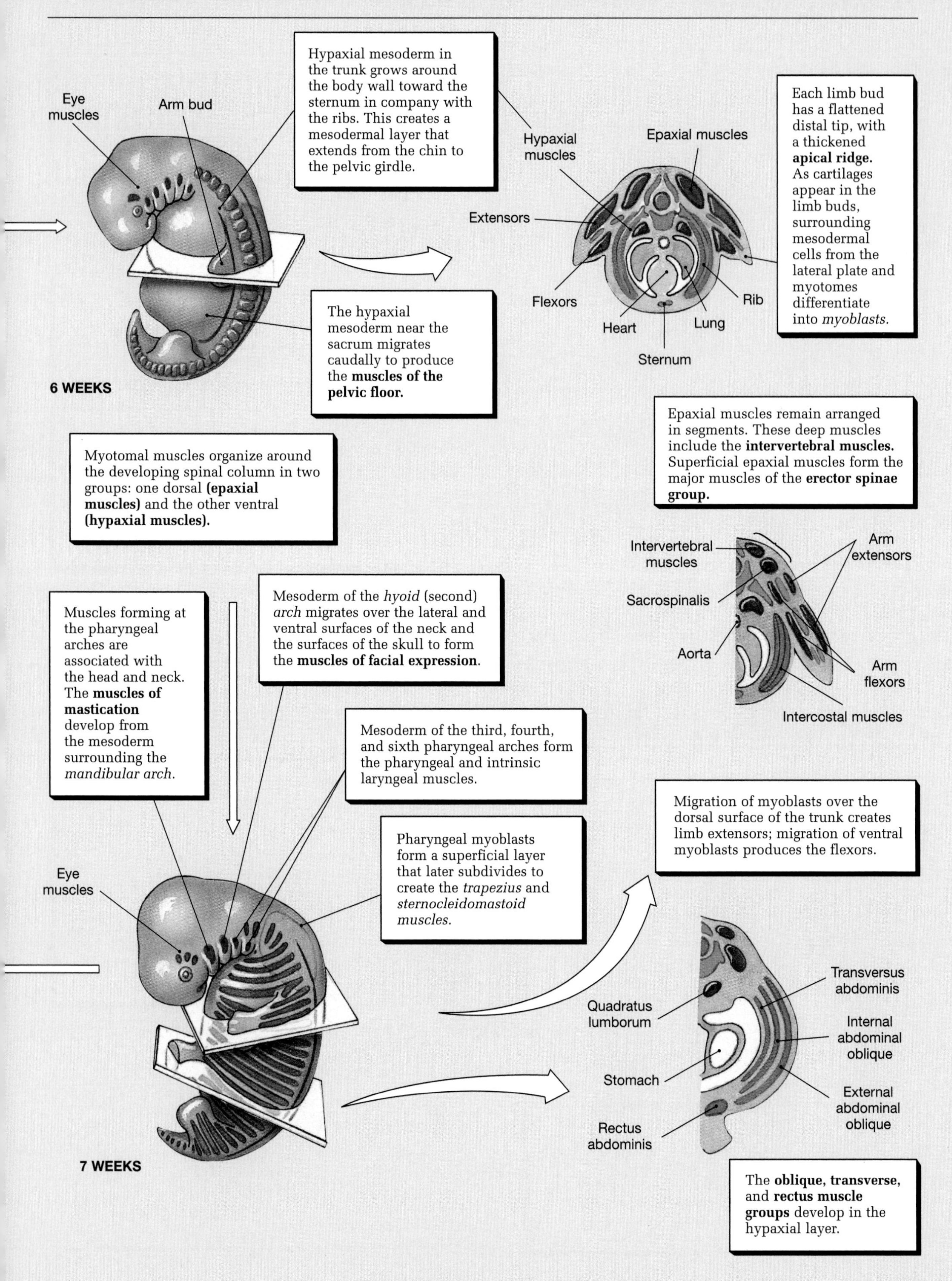
Eye muscles
Arm bud
Hypaxial mesoderm in the trunk grows around the body wall toward the sternum in company with the ribs. This creates a mesodermal layer that extends from the chin to the pelvic girdle.
Hypaxial muscles
Epaxial muscles
Each limb bud has a flattened distal tip, with a thickened **apical ridge.** As cartilages appear in the limb buds, surrounding mesodermal cells from the lateral plate and myotomes differentiate into *myoblasts.*
Extensors
Flexors
Heart
Lung
Rib
Sternum
The hypaxial mesoderm near the sacrum migrates caudally to produce the **muscles of the pelvic floor.**
6 WEEKS
Epaxial muscles remain arranged in segments. These deep muscles include the **intervertebral muscles.** Superficial epaxial muscles form the major muscles of the **erector spinae group.**
Myotomal muscles organize around the developing spinal column in two groups: one dorsal **(epaxial muscles)** and the other ventral **(hypaxial muscles).**
Intervertebral muscles
Arm extensors
Sacrospinalis
Aorta
Arm flexors
Intercostal muscles
Muscles forming at the pharyngeal arches are associated with the head and neck. The **muscles of mastication** develop from the mesoderm surrounding the *mandibular arch.*
Mesoderm of the *hyoid* (second) *arch* migrates over the lateral and ventral surfaces of the neck and the surfaces of the skull to form the **muscles of facial expression**.
Mesoderm of the third, fourth, and sixth pharyngeal arches form the pharyngeal and intrinsic laryngeal muscles.
Pharyngeal myoblasts form a superficial layer that later subdivides to create the *trapezius* and *sternocleidomastoid muscles.*
Migration of myoblasts over the dorsal surface of the trunk creates limb extensors; migration of ventral myoblasts produces the flexors.
Eye muscles
Transversus abdominis
Quadratus lumborum
Internal abdominal oblique
Stomach
External abdominal oblique
Rectus abdominis
7 WEEKS
The **oblique, transverse,** and **rectus muscle groups** develop in the hypaxial layer.

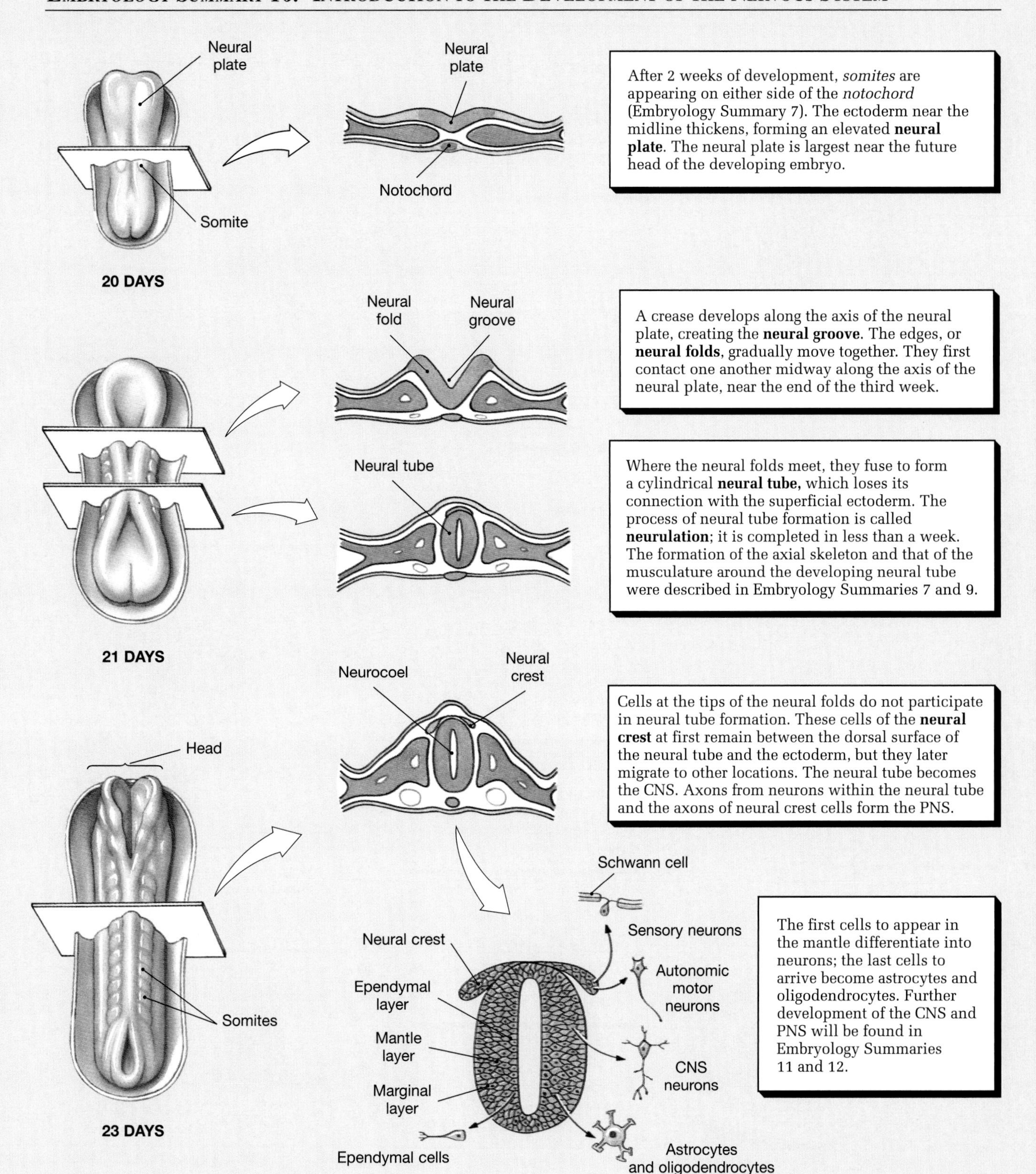

After 2 weeks of development, *somites* are appearing on either side of the *notochord* (Embryology Summary 7). The ectoderm near the midline thickens, forming an elevated **neural plate**. The neural plate is largest near the future head of the developing embryo.

A crease develops along the axis of the neural plate, creating the **neural groove**. The edges, or **neural folds**, gradually move together. They first contact one another midway along the axis of the neural plate, near the end of the third week.

Where the neural folds meet, they fuse to form a cylindrical **neural tube,** which loses its connection with the superficial ectoderm. The process of neural tube formation is called **neurulation**; it is completed in less than a week. The formation of the axial skeleton and that of the musculature around the developing neural tube were described in Embryology Summaries 7 and 9.

Cells at the tips of the neural folds do not participate in neural tube formation. These cells of the **neural crest** at first remain between the dorsal surface of the neural tube and the ectoderm, but they later migrate to other locations. The neural tube becomes the CNS. Axons from neurons within the neural tube and the axons of neural crest cells form the PNS.

The first cells to appear in the mantle differentiate into neurons; the last cells to arrive become astrocytes and oligodendrocytes. Further development of the CNS and PNS will be found in Embryology Summaries 11 and 12.

The neural tube increases in thickness as its epithelial lining undergoes repeated mitoses. By the middle of the fifth developmental week, there are three distinct layers. The **ependymal layer** lines the enclosed cavity, or **neurocoel**. The ependymal cells continue their mitotic activities, and daughter cells create the surrounding **mantle layer**. Axons from developing neurons form a superficial **marginal layer**.

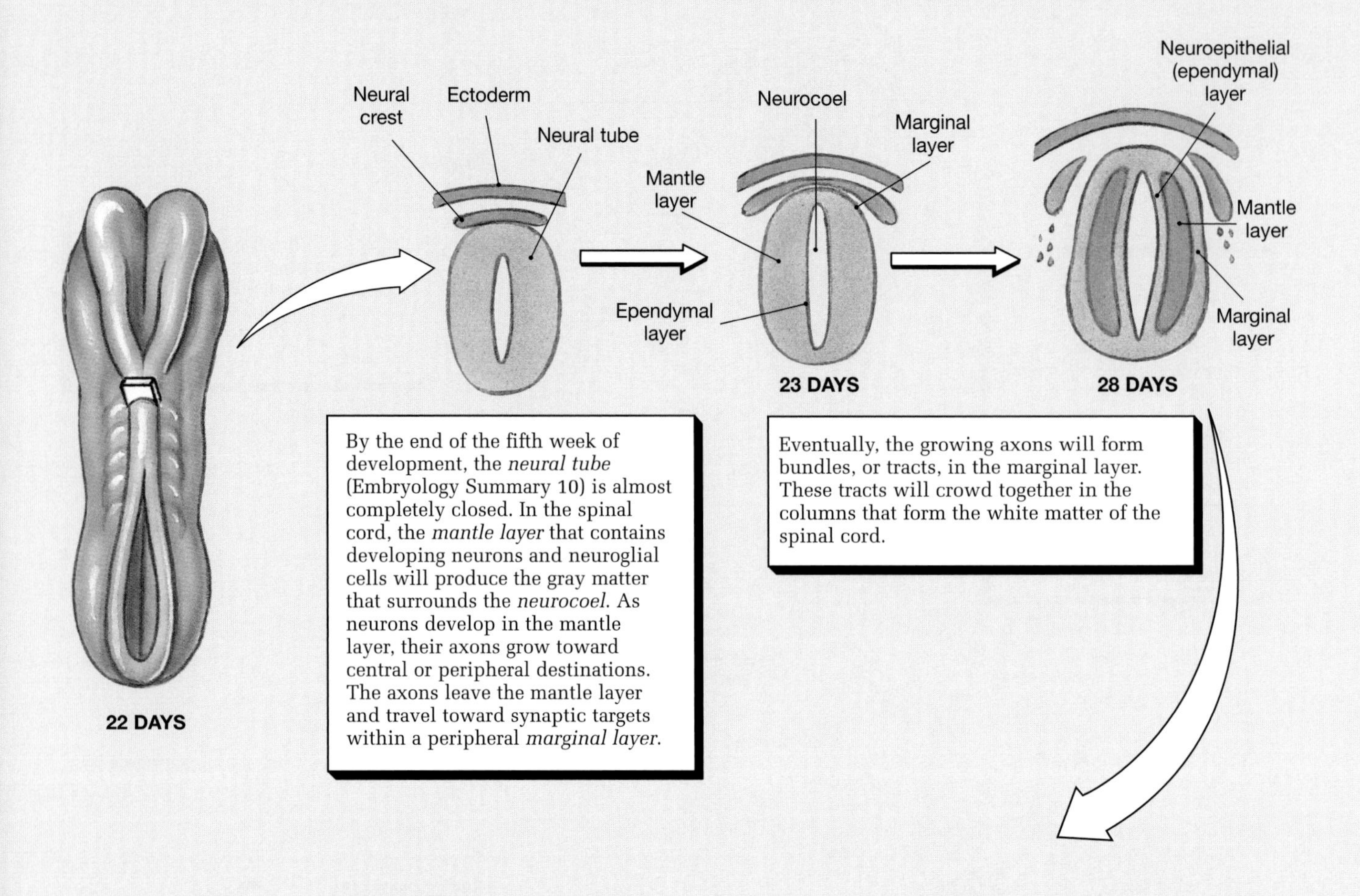

By the end of the fifth week of development, the *neural tube* (Embryology Summary 10) is almost completely closed. In the spinal cord, the *mantle layer* that contains developing neurons and neuroglial cells will produce the gray matter that surrounds the *neurocoel*. As neurons develop in the mantle layer, their axons grow toward central or peripheral destinations. The axons leave the mantle layer and travel toward synaptic targets within a peripheral *marginal layer*.

Eventually, the growing axons will form bundles, or tracts, in the marginal layer. These tracts will crowd together in the columns that form the white matter of the spinal cord.

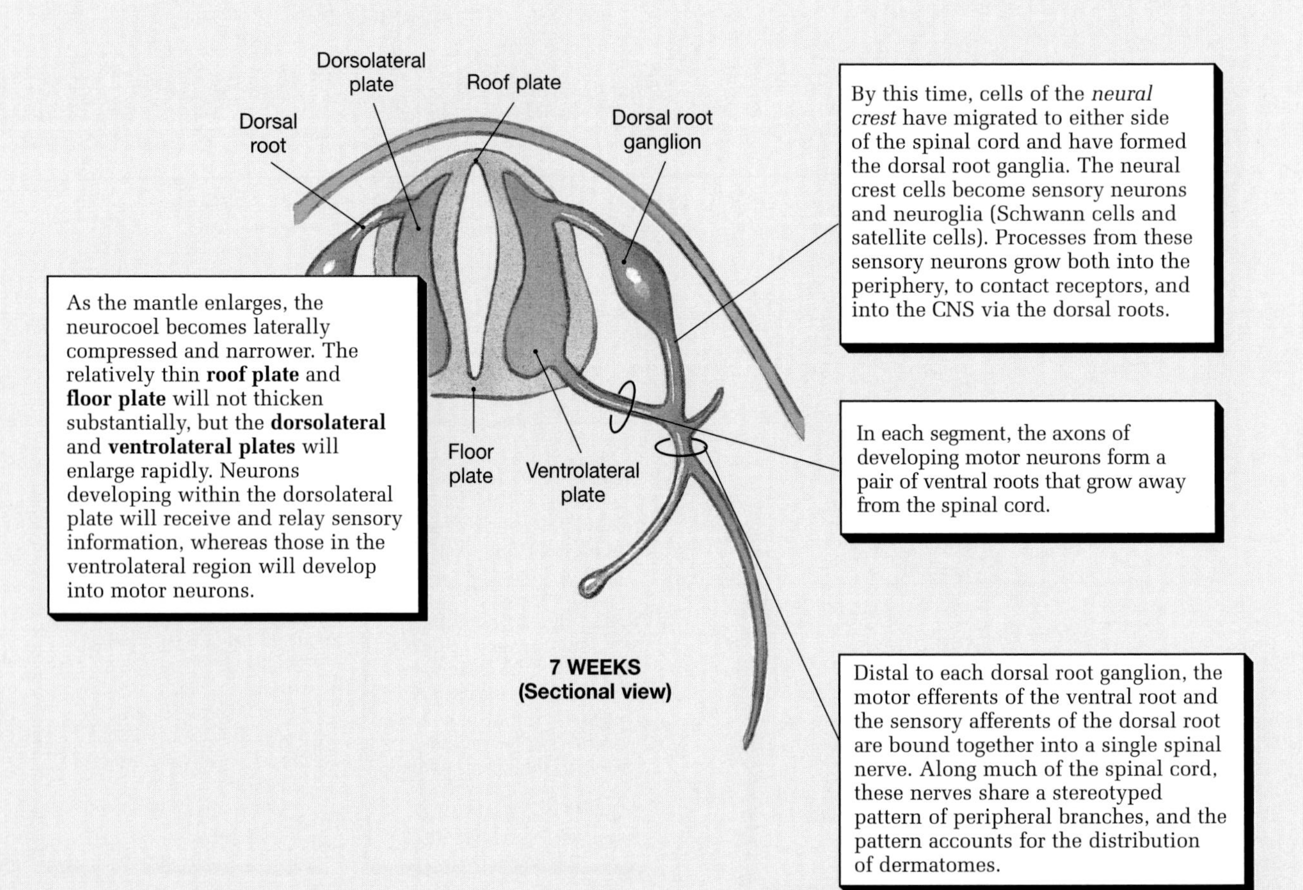

As the mantle enlarges, the neurocoel becomes laterally compressed and narrower. The relatively thin **roof plate** and **floor plate** will not thicken substantially, but the **dorsolateral** and **ventrolateral plates** will enlarge rapidly. Neurons developing within the dorsolateral plate will receive and relay sensory information, whereas those in the ventrolateral region will develop into motor neurons.

By this time, cells of the *neural crest* have migrated to either side of the spinal cord and have formed the dorsal root ganglia. The neural crest cells become sensory neurons and neuroglia (Schwann cells and satellite cells). Processes from these sensory neurons grow both into the periphery, to contact receptors, and into the CNS via the dorsal roots.

In each segment, the axons of developing motor neurons form a pair of ventral roots that grow away from the spinal cord.

Distal to each dorsal root ganglion, the motor efferents of the ventral root and the sensory afferents of the dorsal root are bound together into a single spinal nerve. Along much of the spinal cord, these nerves share a stereotyped pattern of peripheral branches, and the pattern accounts for the distribution of dermatomes.

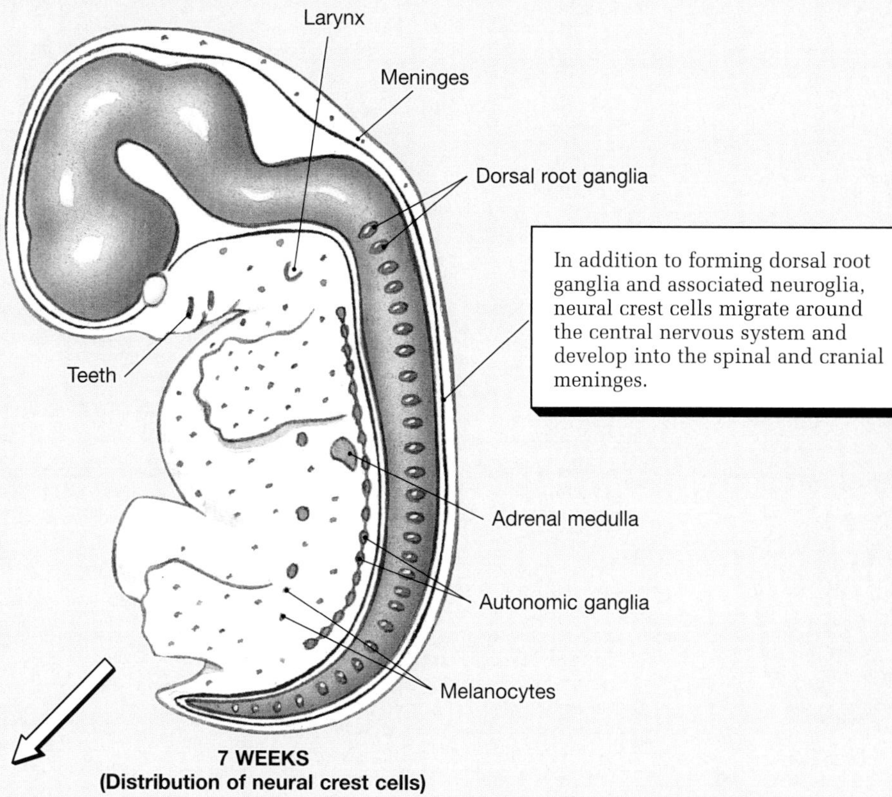

7 WEEKS
(Distribution of neural crest cells)

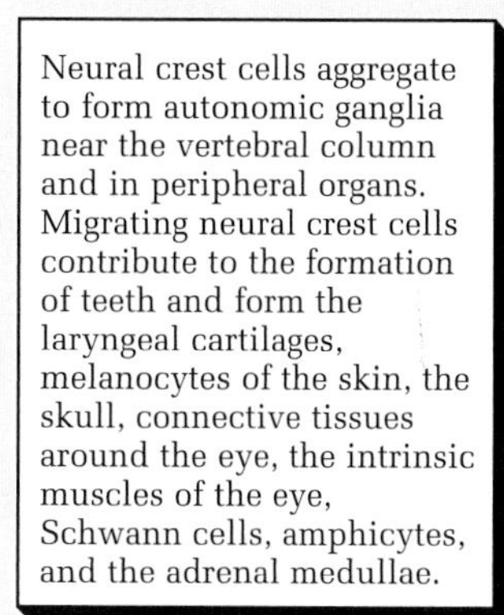
Neural crest cells aggregate to form autonomic ganglia near the vertebral column and in peripheral organs. Migrating neural crest cells contribute to the formation of teeth and form the laryngeal cartilages, melanocytes of the skin, the skull, connective tissues around the eye, the intrinsic muscles of the eye, Schwann cells, amphicytes, and the adrenal medullae.

In addition to forming dorsal root ganglia and associated neuroglia, neural crest cells migrate around the central nervous system and develop into the spinal and cranial meninges.

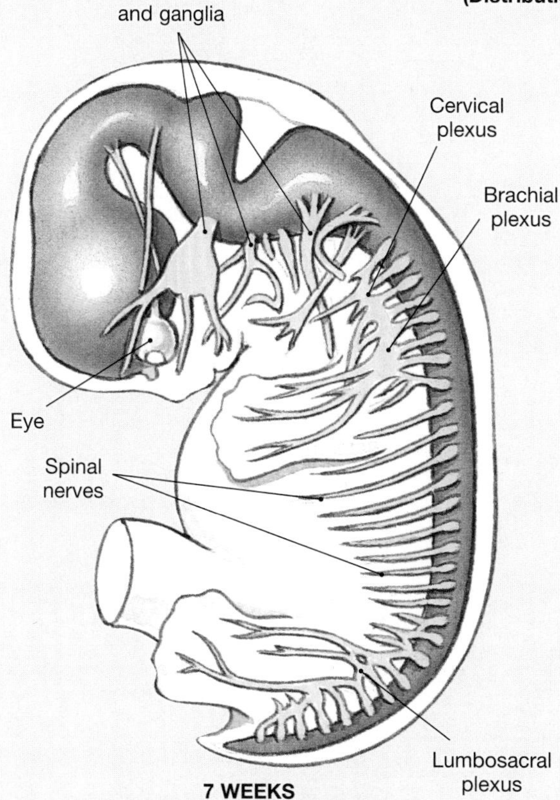

7 WEEKS
(Peripheral nerve distribution)

Several spinal nerves innervate each developing limb. When embryonic muscle cells migrate away from the myotome, the nerves grow along with them. If a large muscle in the adult is derived from several myotomal blocks, connective tissue partitions typically mark the original boundaries, and the innervation always involves more than one spinal nerve.

DEVELOPMENTAL ABNORMALITIES

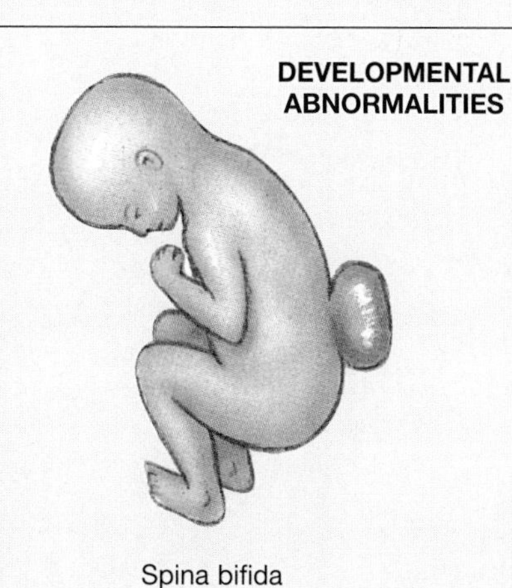

Spina bifida

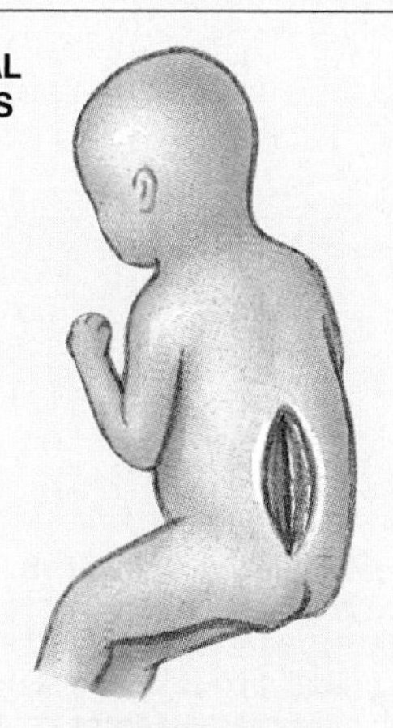

Neural tube defect

Spina bifida (BI-fi-da) results when the developing vertebral laminae fail to unite. The neural arch is incomplete, and the meninges bulge outward beneath the skin of the back. The extent of the abnormality determines the severity of the defects. In mild cases, the condition may pass unnoticed; extreme cases involve much of the length of the spinal column and are usually associated with abnormal nerve function.

A **neural tube defect (NTD)** is a condition that is secondary to a developmental error in the formation of the spinal cord. Instead of forming a hollow tube, a portion of the spinal cord develops as a broad plate. This is often associated with spina bifida. Neural tube defects affect roughly one individual in 1000; prenatal testing can detect the existence of these defects with an 80–85% success rate.

Before proceeding, briefly review Embryology Summaries 6, on skull formation, and 11, on spinal cord development.

The initial expansion occurs as the neurocoel enlarges, forming three distinct *brain vesicles:* (1) the *prosencephalon,* or "forebrain," (2) the *mesencephalon,* or "midbrain," and (3) the *rhombencephalon,* or "hindbrain." The prosencephalon and rhombencephalon will be subdivided further as development proceeds.

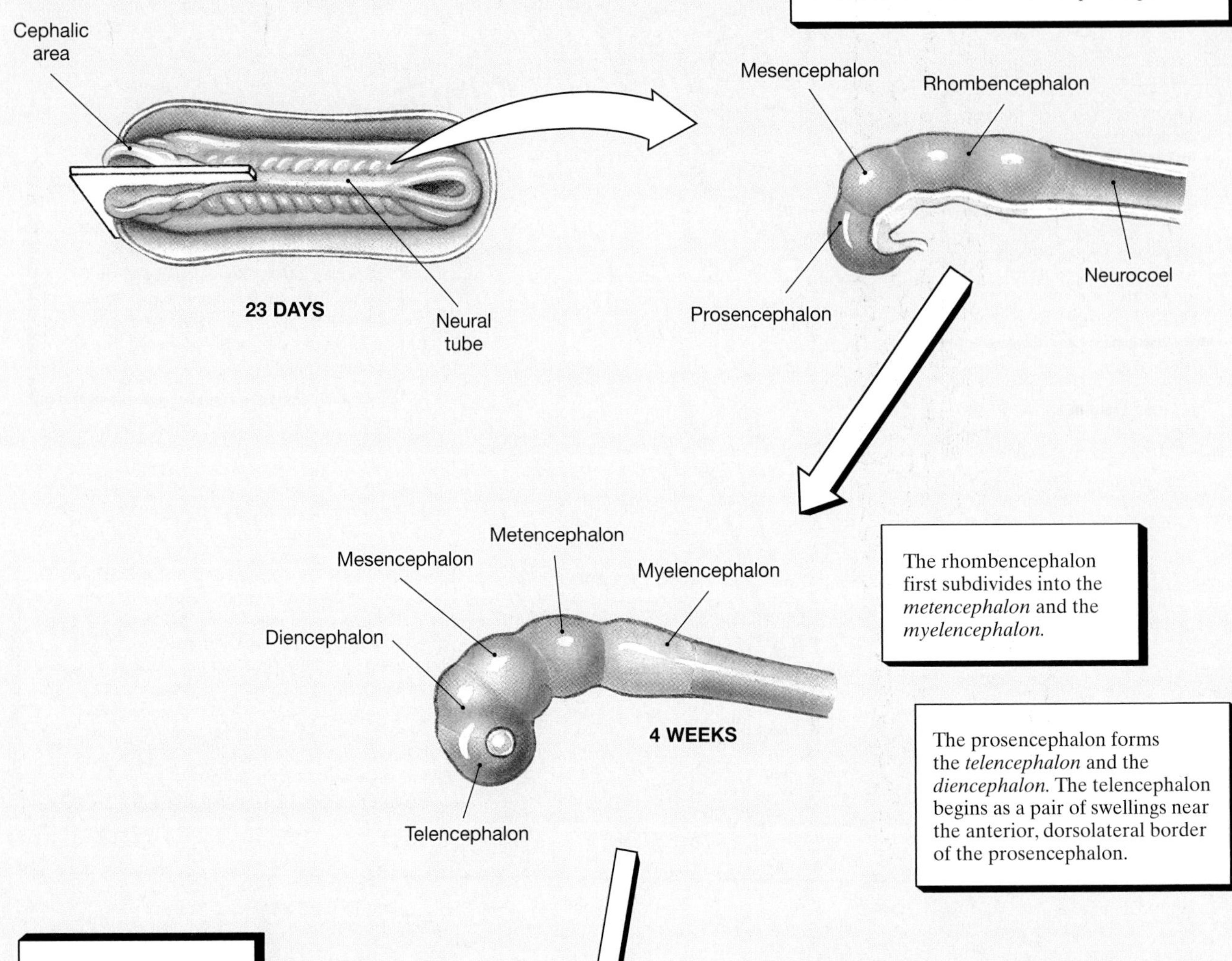

The rhombencephalon first subdivides into the *metencephalon* and the *myelencephalon.*

The prosencephalon forms the *telencephalon* and the *diencephalon.* The telencephalon begins as a pair of swellings near the anterior, dorsolateral border of the prosencephalon.

Development of the mesencephalon produces a small mass of neural tissue with a constricted neurocoel, the mesencephalic (cerebral) aqueduct.

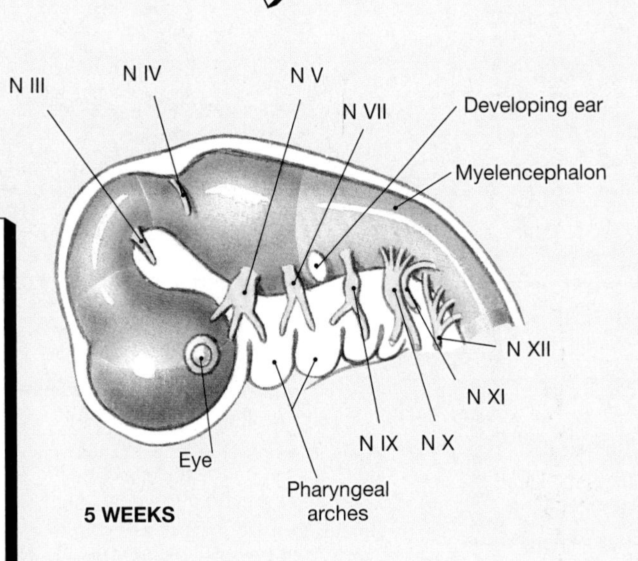

Cranial nerves develop as sensory ganglia link peripheral receptors with the brain and motor fibers grow out of developing cranial nuclei. Special sensory neurons of cranial nerves I, II, and VIII develop in association with the developing receptors. The somatic motor nerves III, IV, and VI grow to the eye muscles; the mixed nerves (IV, V, VII, IX, and X) innervate the *pharyngeal arches* (Embryology Summary 6).

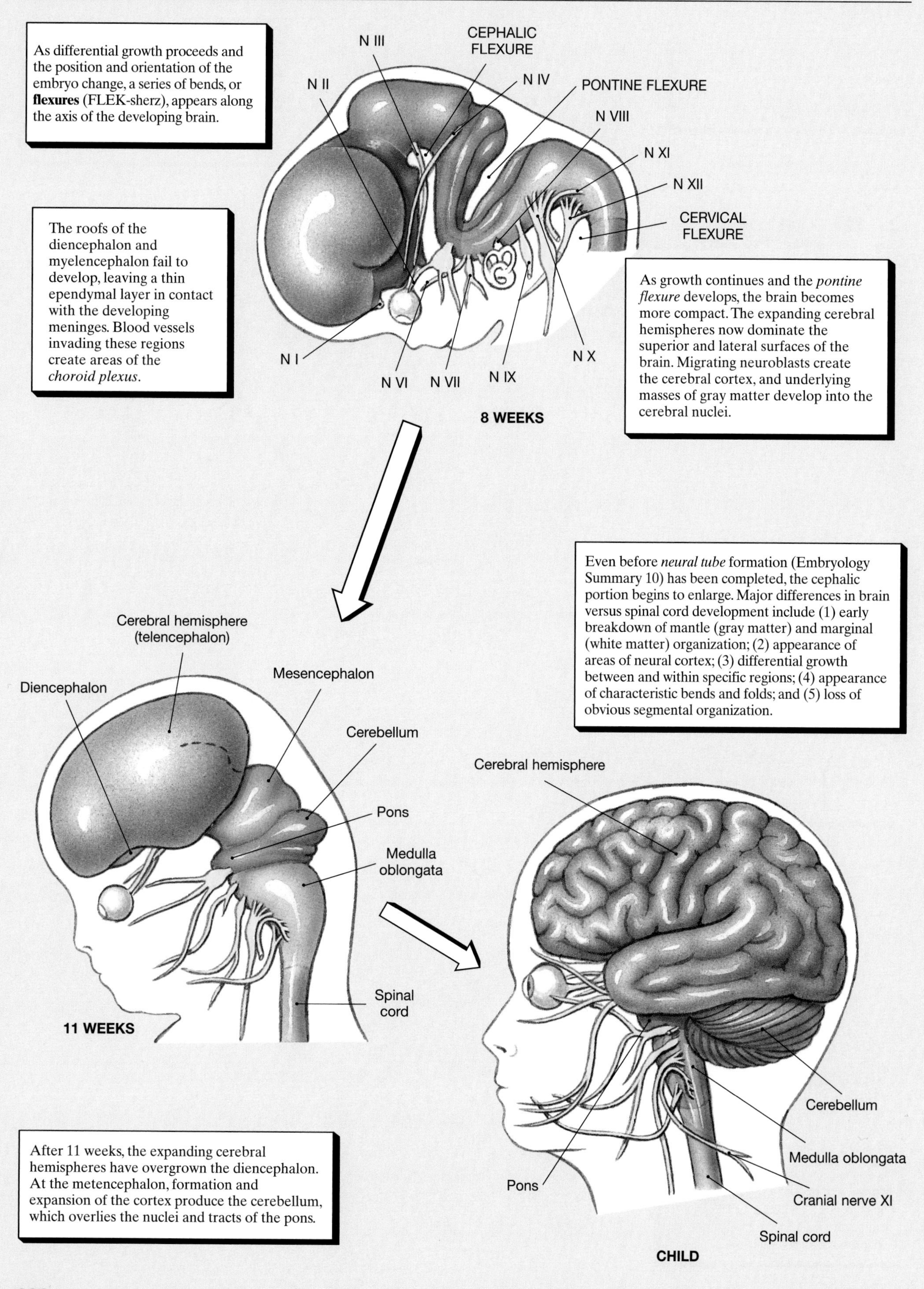

As differential growth proceeds and the position and orientation of the embryo change, a series of bends, or **flexures** (FLEK-sherz), appears along the axis of the developing brain.
N III
CEPHALIC FLEXURE
N II
N IV
PONTINE FLEXURE
N VIII
N XI
N XII
CERVICAL FLEXURE
The roofs of the diencephalon and myelencephalon fail to develop, leaving a thin ependymal layer in contact with the developing meninges. Blood vessels invading these regions create areas of the *choroid plexus*.
N I
N VI
N VII
N IX
N X
8 WEEKS
As growth continues and the *pontine flexure* develops, the brain becomes more compact. The expanding cerebral hemispheres now dominate the superior and lateral surfaces of the brain. Migrating neuroblasts create the cerebral cortex, and underlying masses of gray matter develop into the cerebral nuclei.
Even before *neural tube* formation (Embryology Summary 10) has been completed, the cephalic portion begins to enlarge. Major differences in brain versus spinal cord development include (1) early breakdown of mantle (gray matter) and marginal (white matter) organization; (2) appearance of areas of neural cortex; (3) differential growth between and within specific regions; (4) appearance of characteristic bends and folds; and (5) loss of obvious segmental organization.
Cerebral hemisphere (telencephalon)
Diencephalon
Mesencephalon
Cerebellum
Pons
Medulla oblongata
Spinal cord
11 WEEKS
Cerebral hemisphere
Cerebellum
Medulla oblongata
Pons
Cranial nerve XI
Spinal cord
CHILD
After 11 weeks, the expanding cerebral hemispheres have overgrown the diencephalon. At the metencephalon, formation and expansion of the cortex produce the cerebellum, which overlies the nuclei and tracts of the pons.

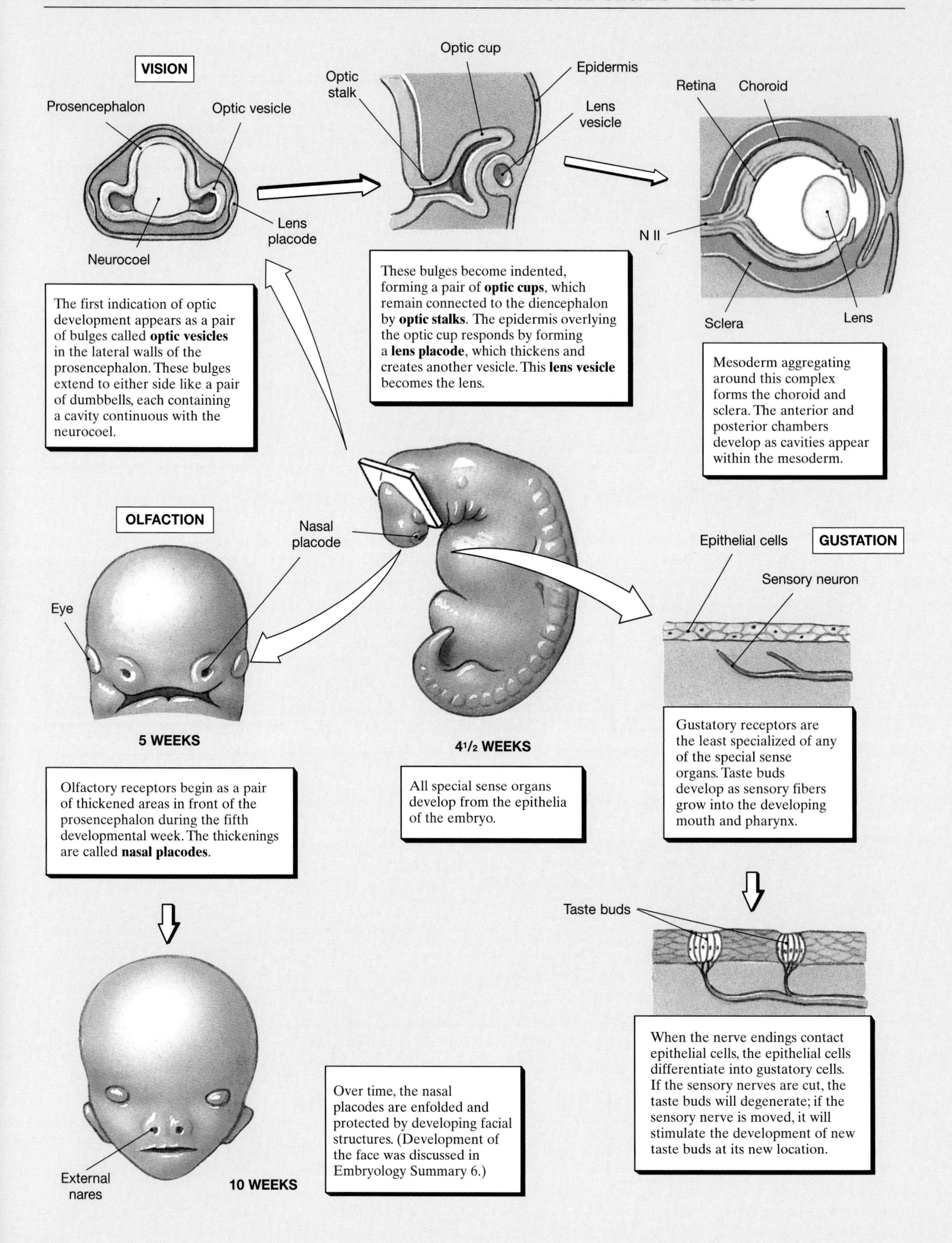

The first indication of optic development appears as a pair of bulges called **optic vesicles** in the lateral walls of the prosencephalon. These bulges extend to either side like a pair of dumbbells, each containing a cavity continuous with the neurocoel.

These bulges become indented, forming a pair of **optic cups**, which remain connected to the diencephalon by **optic stalks**. The epidermis overlying the optic cup responds by forming a **lens placode**, which thickens and creates another vesicle. This **lens vesicle** becomes the lens.

Mesoderm aggregating around this complex forms the choroid and sclera. The anterior and posterior chambers develop as cavities appear within the mesoderm.

Olfactory receptors begin as a pair of thickened areas in front of the prosencephalon during the fifth developmental week. The thickenings are called **nasal placodes**.

All special sense organs develop from the epithelia of the embryo.

Gustatory receptors are the least specialized of any of the special sense organs. Taste buds develop as sensory fibers grow into the developing mouth and pharynx.

Over time, the nasal placodes are enfolded and protected by developing facial structures. (Development of the face was discussed in Embryology Summary 6.)

When the nerve endings contact epithelial cells, the epithelial cells differentiate into gustatory cells. If the sensory nerves are cut, the taste buds will degenerate; if the sensory nerve is moved, it will stimulate the development of new taste buds at its new location.

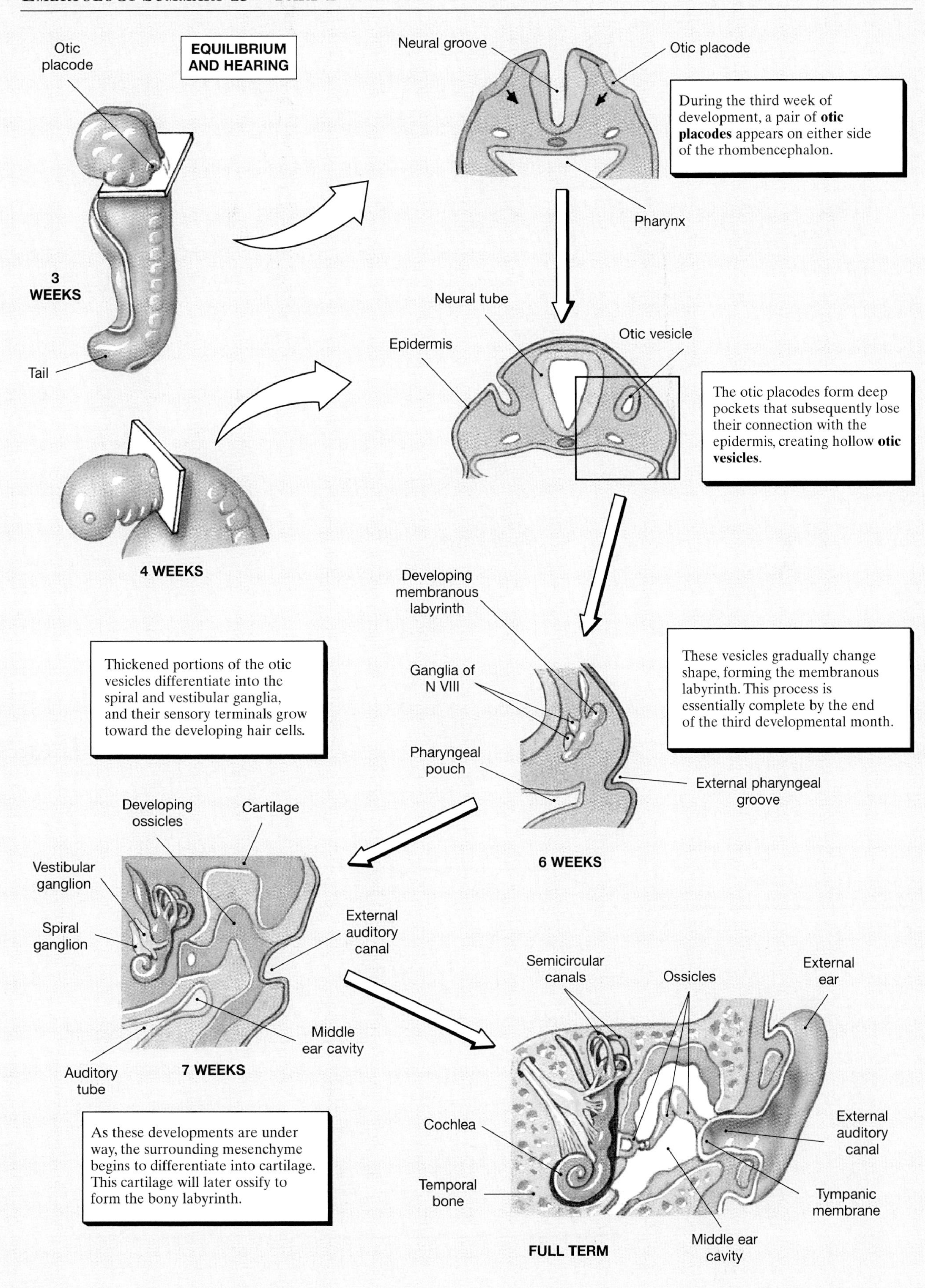
EQUILIBRIUM AND HEARING
Otic placode
3 WEEKS
Tail
4 WEEKS
Neural groove
Otic placode
Pharynx
During the third week of development, a pair of **otic placodes** appears on either side of the rhombencephalon.
Neural tube
Epidermis
Otic vesicle
The otic placodes form deep pockets that subsequently lose their connection with the epidermis, creating hollow **otic vesicles**.
Developing membranous labyrinth
Ganglia of N VIII
Pharyngeal pouch
External pharyngeal groove
6 WEEKS
These vesicles gradually change shape, forming the membranous labyrinth. This process is essentially complete by the end of the third developmental month.
Thickened portions of the otic vesicles differentiate into the spiral and vestibular ganglia, and their sensory terminals grow toward the developing hair cells.
Developing ossicles
Cartilage
Vestibular ganglion
Spiral ganglion
External auditory canal
Middle ear cavity
Auditory tube
7 WEEKS
As these developments are under way, the surrounding mesenchyme begins to differentiate into cartilage. This cartilage will later ossify to form the bony labyrinth.
Semicircular canals
Ossicles
External ear
Cochlea
Temporal bone
External auditory canal
Tympanic membrane
Middle ear cavity
FULL TERM

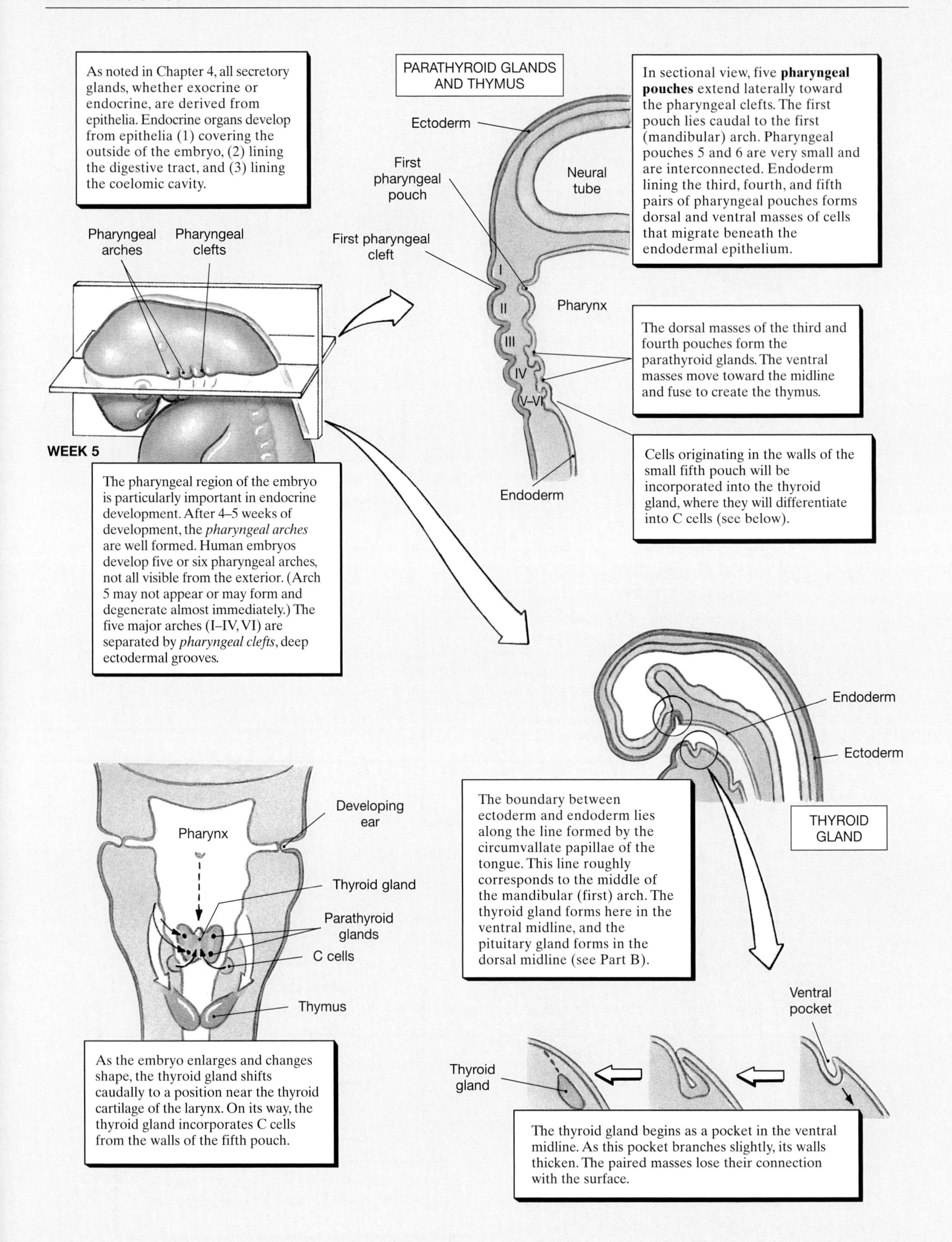

As noted in Chapter 4, all secretory glands, whether exocrine or endocrine, are derived from epithelia. Endocrine organs develop from epithelia (1) covering the outside of the embryo, (2) lining the digestive tract, and (3) lining the coelomic cavity.
PARATHYROID GLANDS AND THYMUS
In sectional view, five **pharyngeal pouches** extend laterally toward the pharyngeal clefts. The first pouch lies caudal to the first (mandibular) arch. Pharyngeal pouches 5 and 6 are very small and are interconnected. Endoderm lining the third, fourth, and fifth pairs of pharyngeal pouches forms dorsal and ventral masses of cells that migrate beneath the endodermal epithelium.
Ectoderm
First pharyngeal pouch
Neural tube
Pharyngeal arches
Pharyngeal clefts
First pharyngeal cleft
I
II
III
IV
V–VI
Pharynx
The dorsal masses of the third and fourth pouches form the parathyroid glands. The ventral masses move toward the midline and fuse to create the thymus.
WEEK 5
Cells originating in the walls of the small fifth pouch will be incorporated into the thyroid gland, where they will differentiate into C cells (see below).
Endoderm
The pharyngeal region of the embryo is particularly important in endocrine development. After 4–5 weeks of development, the *pharyngeal arches* are well formed. Human embryos develop five or six pharyngeal arches, not all visible from the exterior. (Arch 5 may not appear or may form and degenerate almost immediately.) The five major arches (I–IV, VI) are separated by *pharyngeal clefts*, deep ectodermal grooves.
Endoderm
Ectoderm
Developing ear
Pharynx
Thyroid gland
Parathyroid glands
C cells
Thymus
The boundary between ectoderm and endoderm lies along the line formed by the circumvallate papillae of the tongue. This line roughly corresponds to the middle of the mandibular (first) arch. The thyroid gland forms here in the ventral midline, and the pituitary gland forms in the dorsal midline (see Part B).
THYROID GLAND
Ventral pocket
Thyroid gland
As the embryo enlarges and changes shape, the thyroid gland shifts caudally to a position near the thyroid cartilage of the larynx. On its way, the thyroid gland incorporates C cells from the walls of the fifth pouch.
The thyroid gland begins as a pocket in the ventral midline. As this pocket branches slightly, its walls thicken. The paired masses lose their connection with the surface.

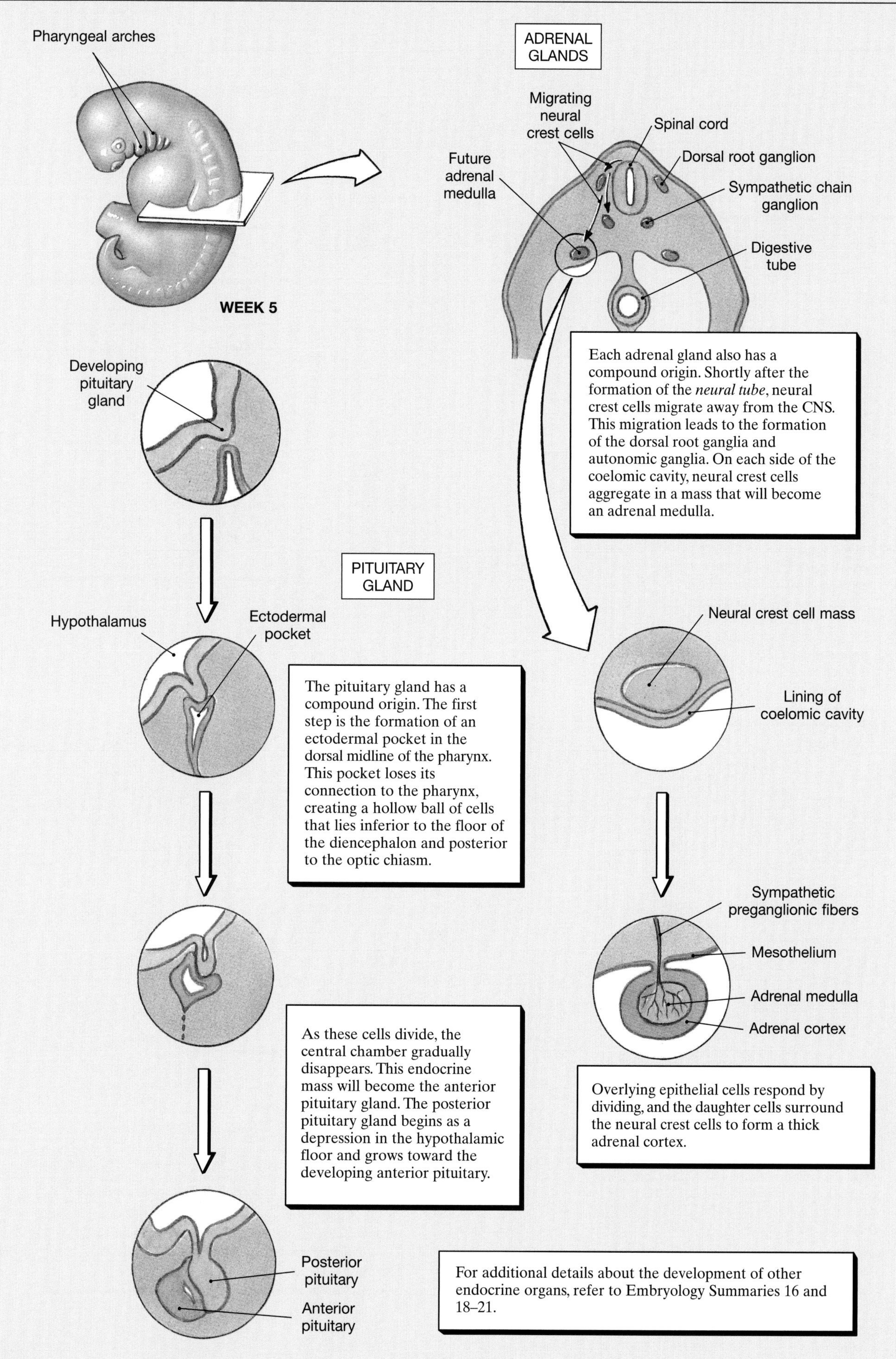
Pharyngeal arches
WEEK 5
ADRENAL GLANDS
Migrating neural crest cells
Spinal cord
Future adrenal medulla
Dorsal root ganglion
Sympathetic chain ganglion
Digestive tube
Each adrenal gland also has a compound origin. Shortly after the formation of the *neural tube*, neural crest cells migrate away from the CNS. This migration leads to the formation of the dorsal root ganglia and autonomic ganglia. On each side of the coelomic cavity, neural crest cells aggregate in a mass that will become an adrenal medulla.
Developing pituitary gland
PITUITARY GLAND
Hypothalamus
Ectodermal pocket
The pituitary gland has a compound origin. The first step is the formation of an ectodermal pocket in the dorsal midline of the pharynx. This pocket loses its connection to the pharynx, creating a hollow ball of cells that lies inferior to the floor of the diencephalon and posterior to the optic chiasm.
Neural crest cell mass
Lining of coelomic cavity
Sympathetic preganglionic fibers
Mesothelium
Adrenal medulla
Adrenal cortex
As these cells divide, the central chamber gradually disappears. This endocrine mass will become the anterior pituitary gland. The posterior pituitary gland begins as a depression in the hypothalamic floor and grows toward the developing anterior pituitary.
Overlying epithelial cells respond by dividing, and the daughter cells surround the neural crest cells to form a thick adrenal cortex.
Posterior pituitary
Anterior pituitary
For additional details about the development of other endocrine organs, refer to Embryology Summaries 16 and 18–21.

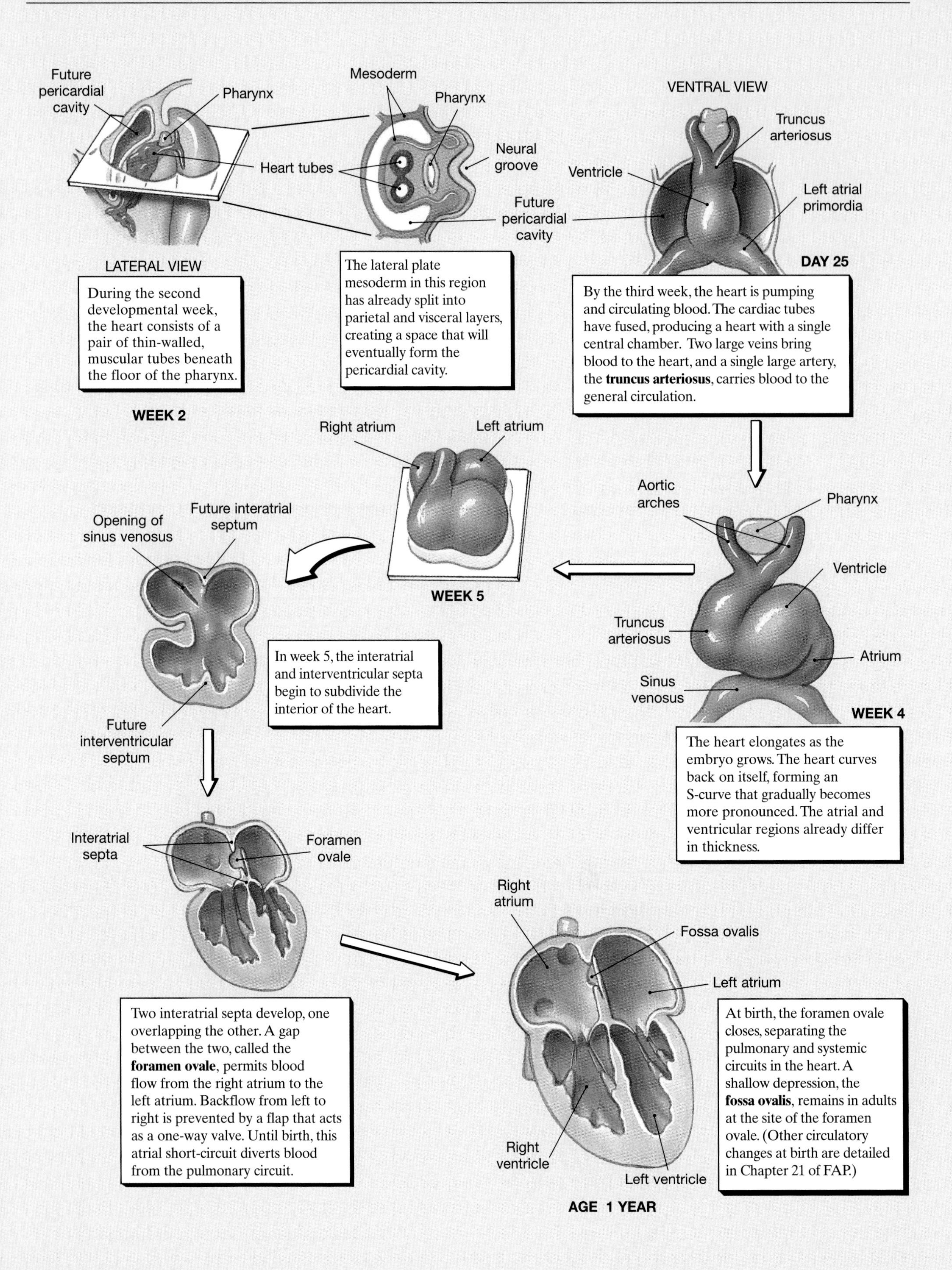
Future pericardial cavity
Pharynx
Heart tubes
LATERAL VIEW
During the second developmental week, the heart consists of a pair of thin-walled, muscular tubes beneath the floor of the pharynx.
WEEK 2
Mesoderm
Pharynx
Neural groove
Future pericardial cavity
The lateral plate mesoderm in this region has already split into parietal and visceral layers, creating a space that will eventually form the pericardial cavity.
VENTRAL VIEW
Truncus arteriosus
Ventricle
Left atrial primordia
DAY 25
By the third week, the heart is pumping and circulating blood. The cardiac tubes have fused, producing a heart with a single central chamber. Two large veins bring blood to the heart, and a single large artery, the **truncus arteriosus**, carries blood to the general circulation.
Aortic arches
Pharynx
Ventricle
Truncus arteriosus
Atrium
Sinus venosus
WEEK 4
The heart elongates as the embryo grows. The heart curves back on itself, forming an S-curve that gradually becomes more pronounced. The atrial and ventricular regions already differ in thickness.
Right atrium
Left atrium
WEEK 5
Opening of sinus venosus
Future interatrial septum
Future interventricular septum
In week 5, the interatrial and interventricular septa begin to subdivide the interior of the heart.
Interatrial septa
Foramen ovale
Two interatrial septa develop, one overlapping the other. A gap between the two, called the **foramen ovale**, permits blood flow from the right atrium to the left atrium. Backflow from left to right is prevented by a flap that acts as a one-way valve. Until birth, this atrial short-circuit diverts blood from the pulmonary circuit.
Right atrium
Fossa ovalis
Left atrium
Right ventricle
Left ventricle
AGE 1 YEAR
At birth, the foramen ovale closes, separating the pulmonary and systemic circuits in the heart. A shallow depression, the **fossa ovalis**, remains in adults at the site of the foramen ovale. (Other circulatory changes at birth are detailed in Chapter 21 of FAP.)

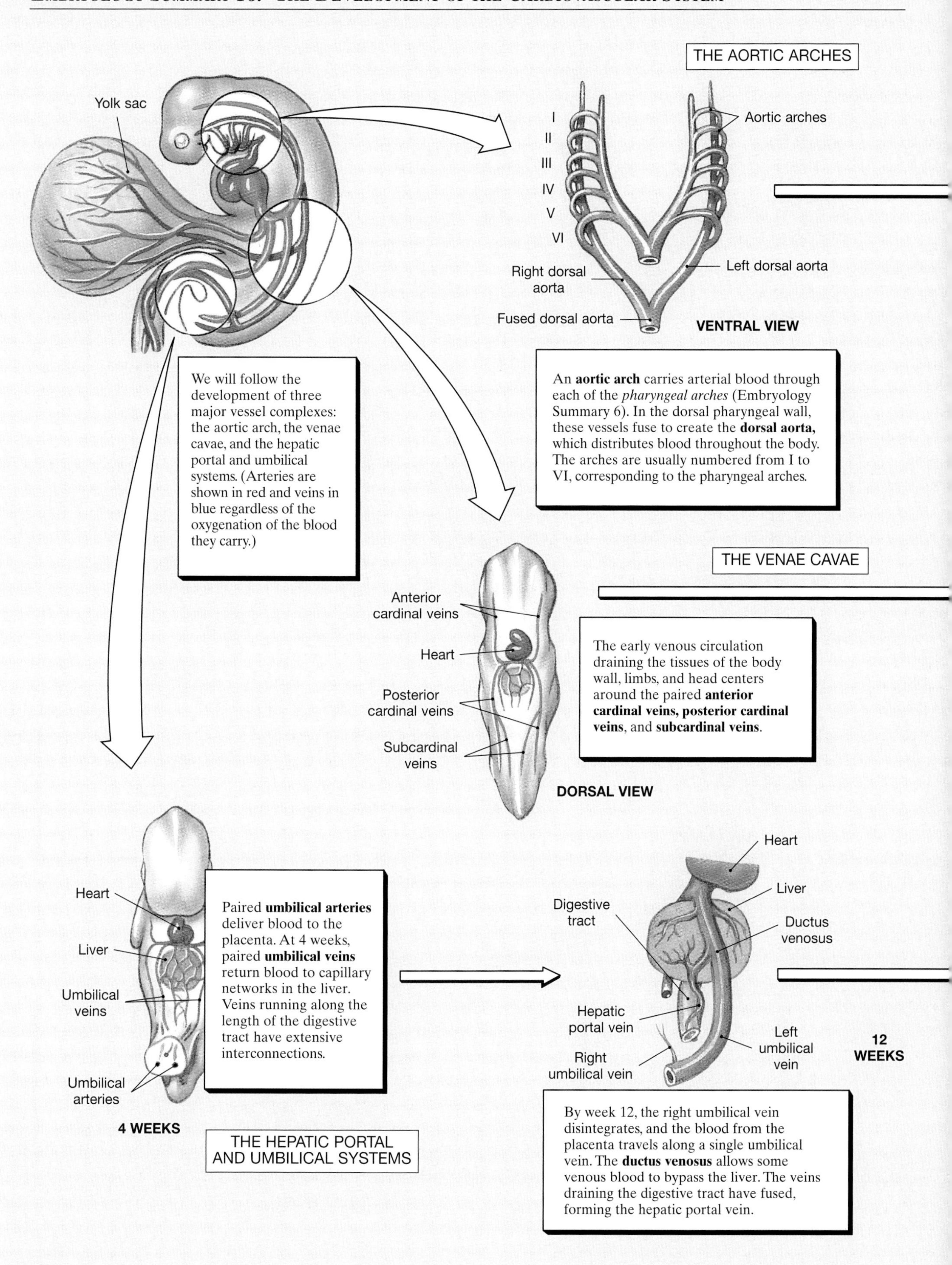
THE AORTIC ARCHES
Yolk sac
I
II
III
IV
V
VI
Aortic arches
Right dorsal aorta
Left dorsal aorta
Fused dorsal aorta
VENTRAL VIEW
We will follow the development of three major vessel complexes: the aortic arch, the venae cavae, and the hepatic portal and umbilical systems. (Arteries are shown in red and veins in blue regardless of the oxygenation of the blood they carry.)
An aortic arch carries arterial blood through each of the pharyngeal arches (Embryology Summary 6). In the dorsal pharyngeal wall, these vessels fuse to create the dorsal aorta, which distributes blood throughout the body. The arches are usually numbered from I to VI, corresponding to the pharyngeal arches.
THE VENAE CAVAE
Anterior cardinal veins
Heart
Posterior cardinal veins
Subcardinal veins
DORSAL VIEW
The early venous circulation draining the tissues of the body wall, limbs, and head centers around the paired anterior cardinal veins, posterior cardinal veins, and subcardinal veins.
Heart
Liver
Umbilical veins
Umbilical arteries
4 WEEKS
Paired umbilical arteries deliver blood to the placenta. At 4 weeks, paired umbilical veins return blood to capillary networks in the liver. Veins running along the length of the digestive tract have extensive interconnections.
THE HEPATIC PORTAL AND UMBILICAL SYSTEMS
Heart
Liver
Digestive tract
Ductus venosus
Hepatic portal vein
Right umbilical vein
Left umbilical vein
12 WEEKS
By week 12, the right umbilical vein disintegrates, and the blood from the placenta travels along a single umbilical vein. The ductus venosus allows some venous blood to bypass the liver. The veins draining the digestive tract have fused, forming the hepatic portal vein.

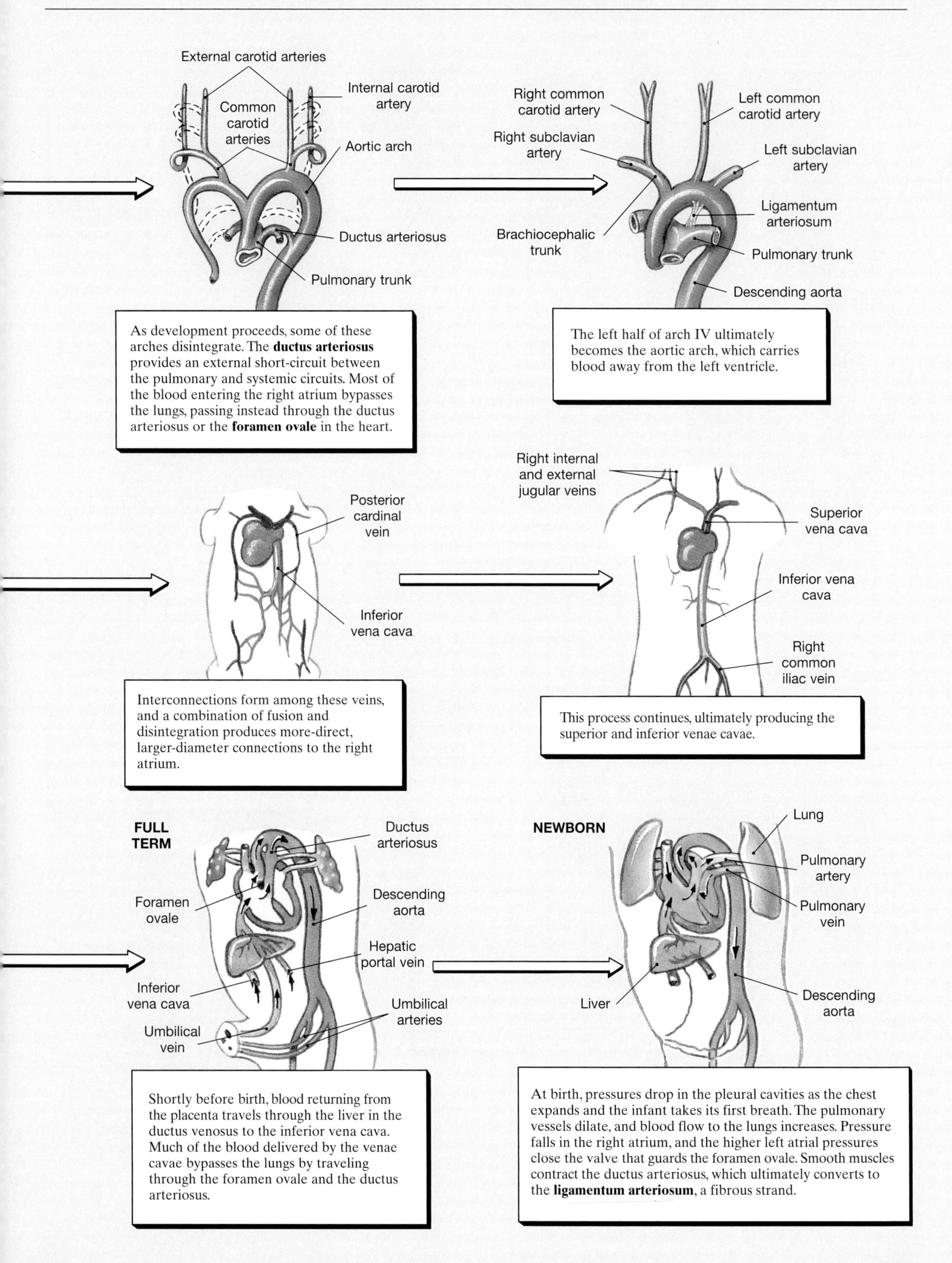

As development proceeds, some of these arches disintegrate. The **ductus arteriosus** provides an external short-circuit between the pulmonary and systemic circuits. Most of the blood entering the right atrium bypasses the lungs, passing instead through the ductus arteriosus or the **foramen ovale** in the heart.

The left half of arch IV ultimately becomes the aortic arch, which carries blood away from the left ventricle.

Interconnections form among these veins, and a combination of fusion and disintegration produces more-direct, larger-diameter connections to the right atrium.

This process continues, ultimately producing the superior and inferior venae cavae.

Shortly before birth, blood returning from the placenta travels through the liver in the ductus venosus to the inferior vena cava. Much of the blood delivered by the venae cavae bypasses the lungs by traveling through the foramen ovale and the ductus arteriosus.

At birth, pressures drop in the pleural cavities as the chest expands and the infant takes its first breath. The pulmonary vessels dilate, and blood flow to the lungs increases. Pressure falls in the right atrium, and the higher left atrial pressures close the valve that guards the foramen ovale. Smooth muscles contract the ductus arteriosus, which ultimately converts to the **ligamentum arteriosum**, a fibrous strand.

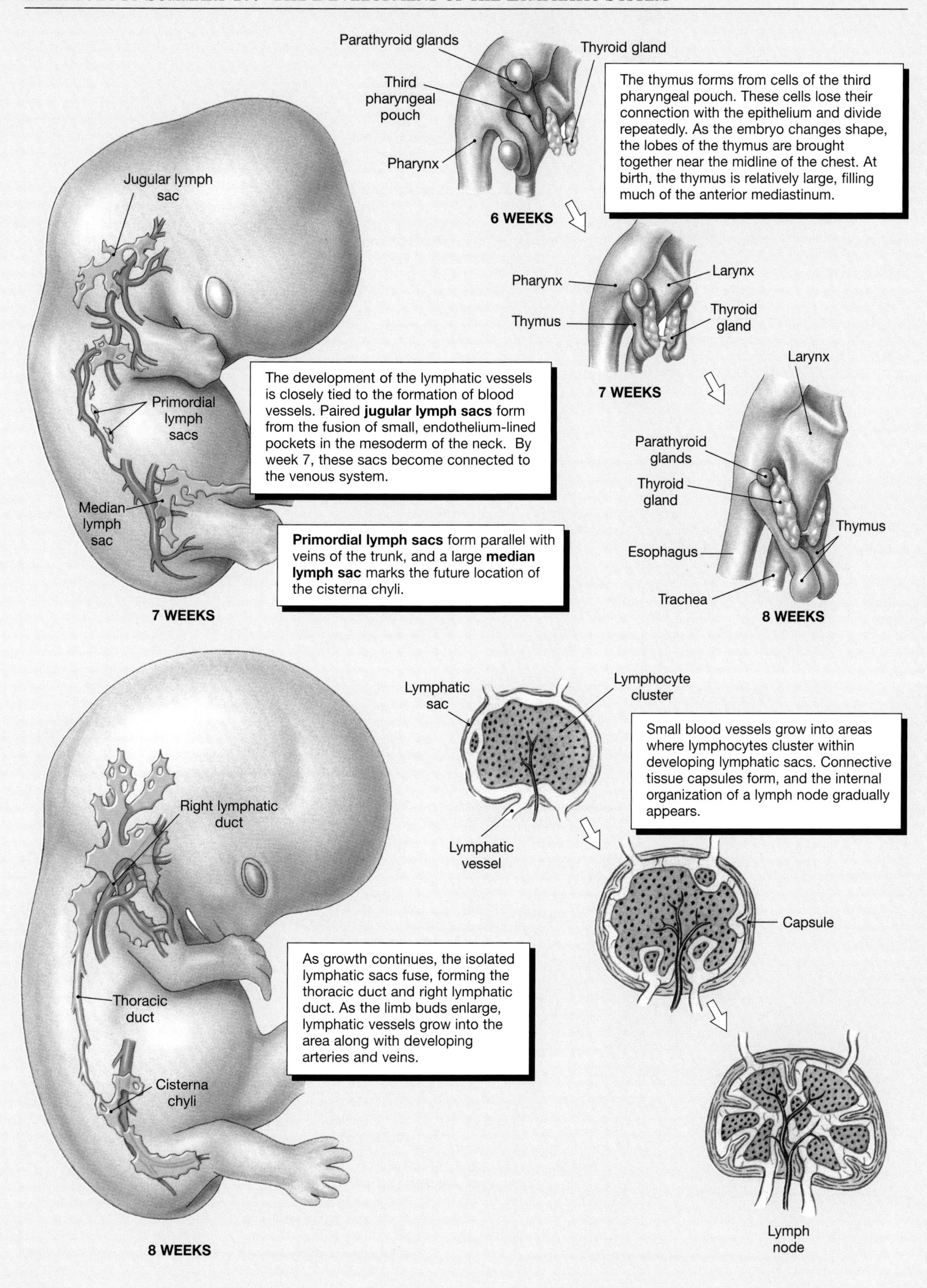

Parathyroid glands
Thyroid gland
Third pharyngeal pouch
Pharynx
6 WEEKS
The thymus forms from cells of the third pharyngeal pouch. These cells lose their connection with the epithelium and divide repeatedly. As the embryo changes shape, the lobes of the thymus are brought together near the midline of the chest. At birth, the thymus is relatively large, filling much of the anterior mediastinum.
Pharynx
Larynx
Thymus
Thyroid gland
7 WEEKS
Larynx
Parathyroid glands
Thyroid gland
Thymus
Esophagus
Trachea
8 WEEKS
Jugular lymph sac
Primordial lymph sacs
Median lymph sac
7 WEEKS
The development of the lymphatic vessels is closely tied to the formation of blood vessels. Paired **jugular lymph sacs** form from the fusion of small, endothelium-lined pockets in the mesoderm of the neck. By week 7, these sacs become connected to the venous system.
Primordial lymph sacs form parallel with veins of the trunk, and a large **median lymph sac** marks the future location of the cisterna chyli.
Lymphatic sac
Lymphocyte cluster
Lymphatic vessel
Small blood vessels grow into areas where lymphocytes cluster within developing lymphatic sacs. Connective tissue capsules form, and the internal organization of a lymph node gradually appears.
Capsule
Lymph node
Right lymphatic duct
Thoracic duct
Cisterna chyli
As growth continues, the isolated lymphatic sacs fuse, forming the thoracic duct and right lymphatic duct. As the limb buds enlarge, lymphatic vessels grow into the area along with developing arteries and veins.
8 WEEKS

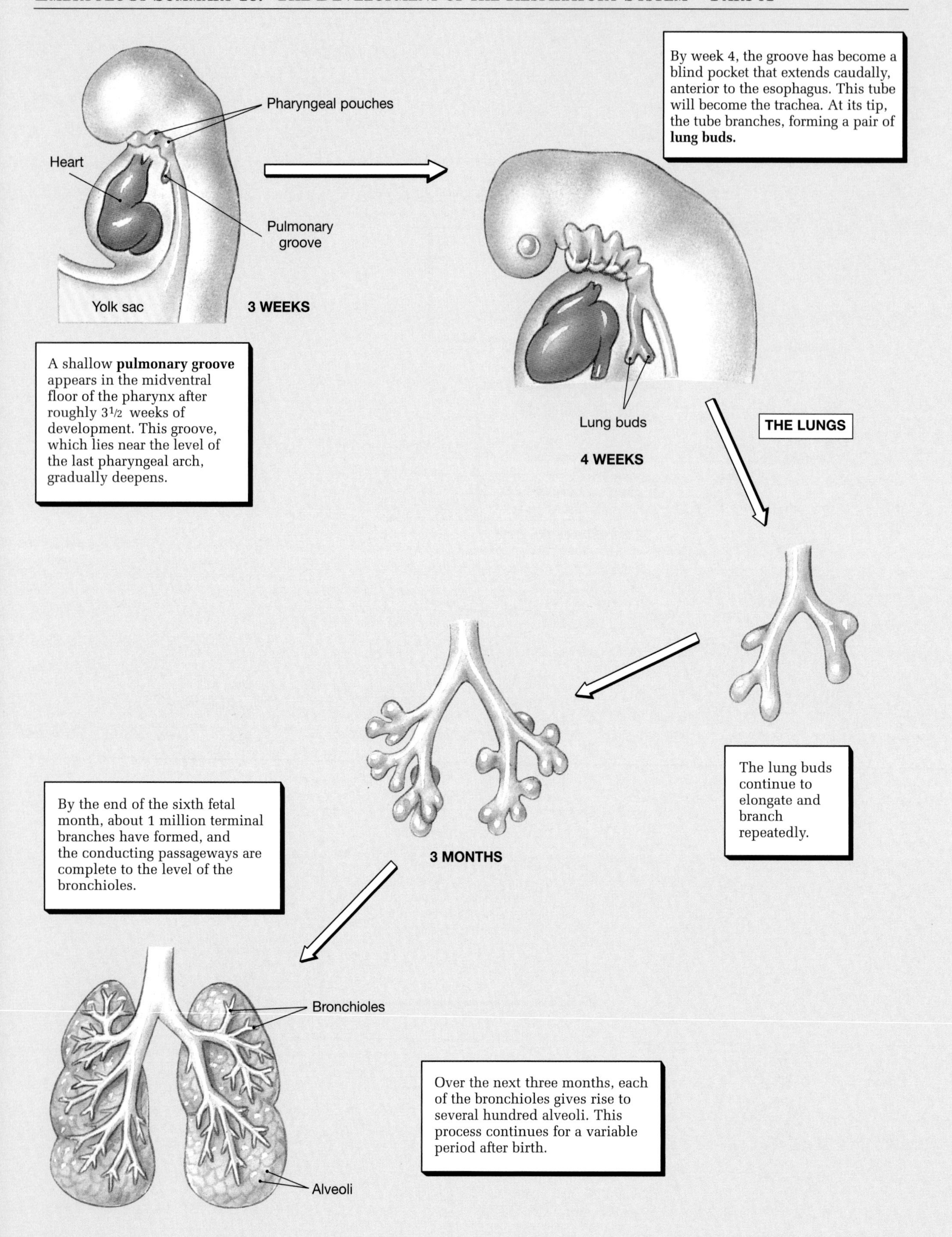

By week 4, the groove has become a blind pocket that extends caudally, anterior to the esophagus. This tube will become the trachea. At its tip, the tube branches, forming a pair of **lung buds.**

A shallow **pulmonary groove** appears in the midventral floor of the pharynx after roughly 3½ weeks of development. This groove, which lies near the level of the last pharyngeal arch, gradually deepens.

The lung buds continue to elongate and branch repeatedly.

By the end of the sixth fetal month, about 1 million terminal branches have formed, and the conducting passageways are complete to the level of the bronchioles.

Over the next three months, each of the bronchioles gives rise to several hundred alveoli. This process continues for a variable period after birth.

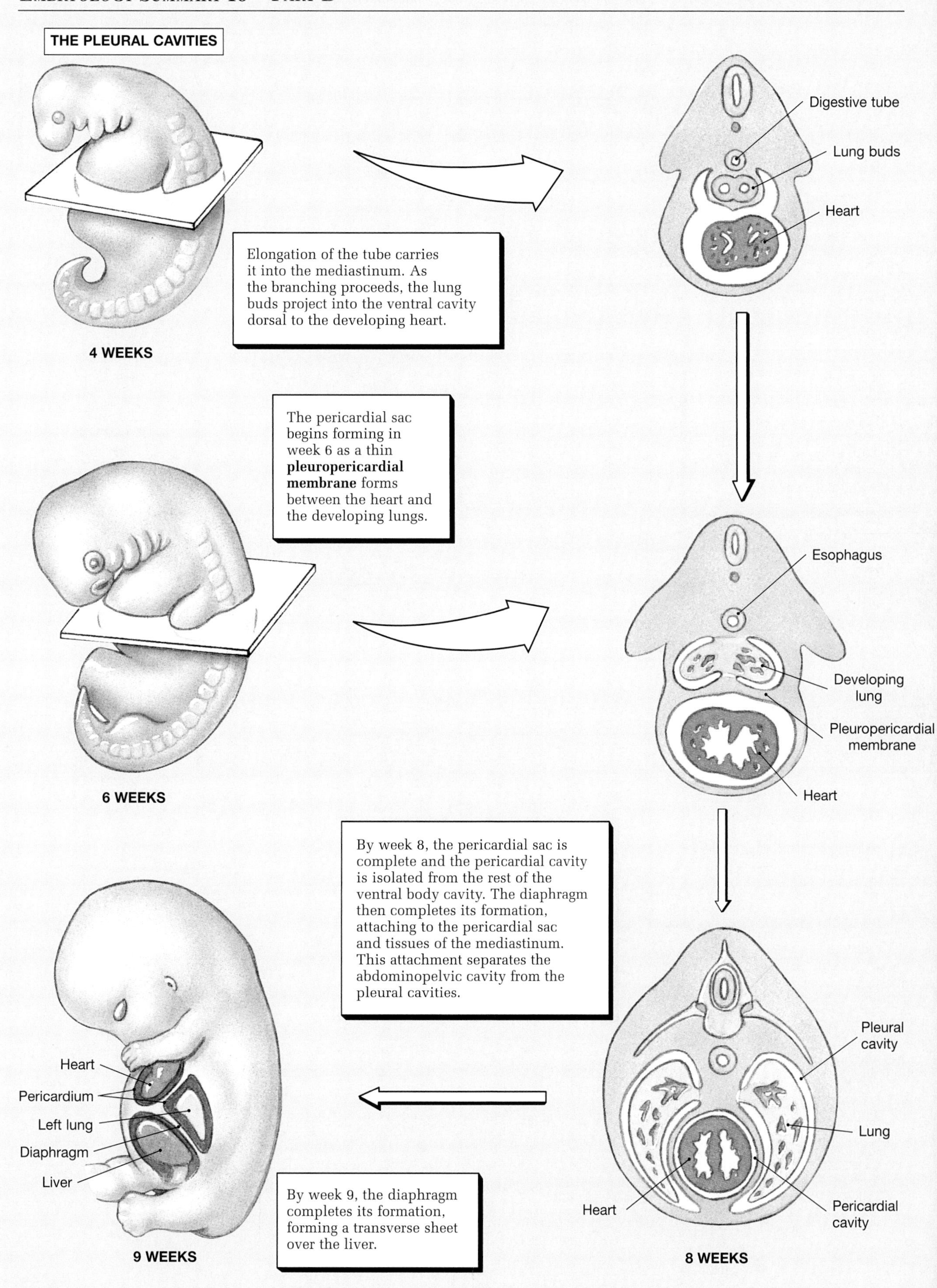
THE PLEURAL CAVITIES
4 WEEKS
Elongation of the tube carries it into the mediastinum. As the branching proceeds, the lung buds project into the ventral cavity dorsal to the developing heart.
Digestive tube
Lung buds
Heart
The pericardial sac begins forming in week 6 as a thin **pleuropericardial membrane** forms between the heart and the developing lungs.
6 WEEKS
Esophagus
Developing lung
Pleuropericardial membrane
Heart
By week 8, the pericardial sac is complete and the pericardial cavity is isolated from the rest of the ventral body cavity. The diaphragm then completes its formation, attaching to the pericardial sac and tissues of the mediastinum. This attachment separates the abdominopelvic cavity from the pleural cavities.
Pleural cavity
Lung
Heart
Pericardial cavity
8 WEEKS
Heart
Pericardium
Left lung
Diaphragm
Liver
9 WEEKS
By week 9, the diaphragm completes its formation, forming a transverse sheet over the liver.

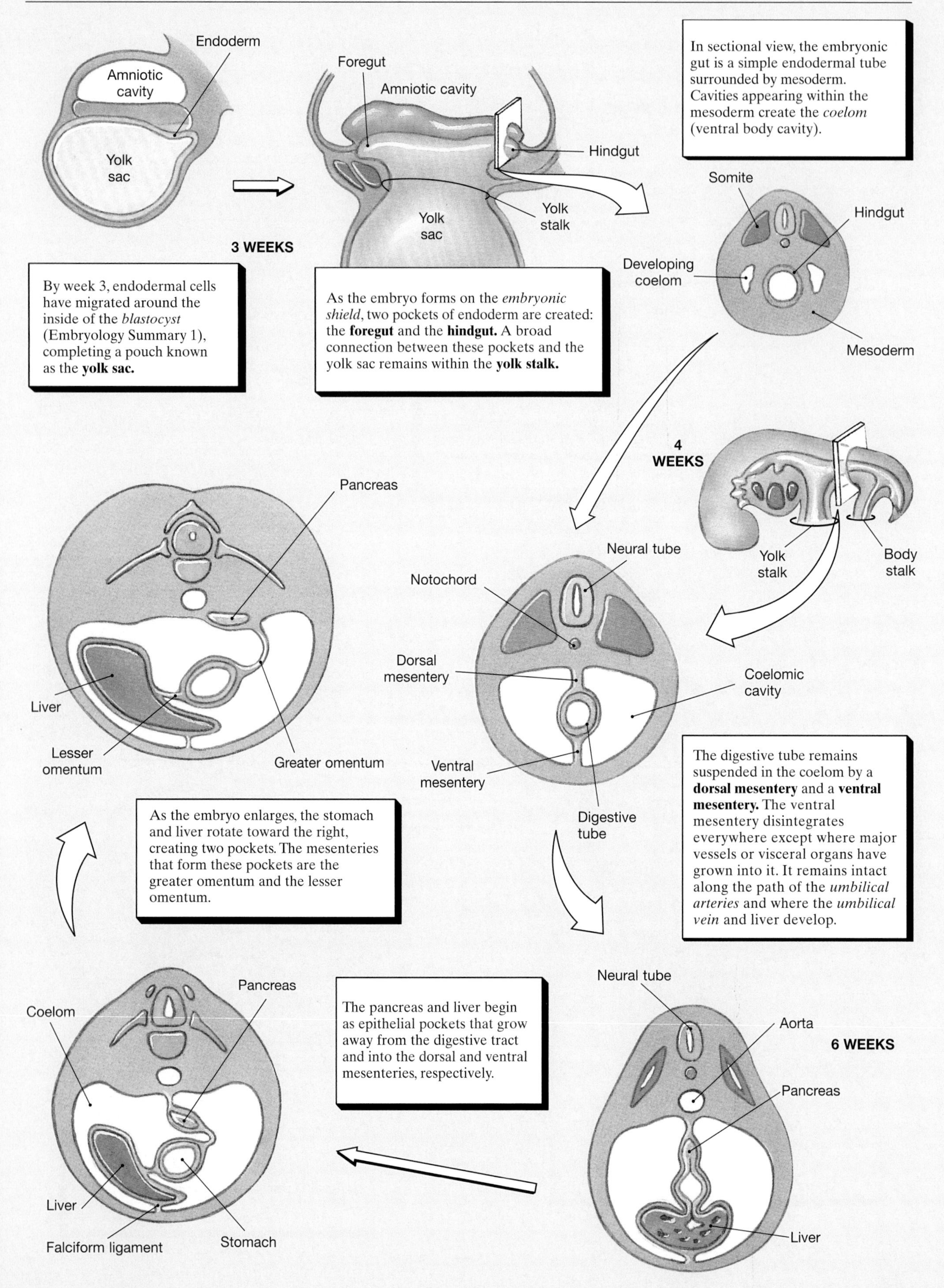
Endoderm
Amniotic cavity
Yolk sac
Foregut
Amniotic cavity
Hindgut
Yolk stalk
Yolk sac
3 WEEKS
In sectional view, the embryonic gut is a simple endodermal tube surrounded by mesoderm. Cavities appearing within the mesoderm create the coelom (ventral body cavity).
Somite
Hindgut
Developing coelom
Mesoderm
By week 3, endodermal cells have migrated around the inside of the blastocyst (Embryology Summary 1), completing a pouch known as the yolk sac.
As the embryo forms on the embryonic shield, two pockets of endoderm are created: the foregut and the hindgut. A broad connection between these pockets and the yolk sac remains within the yolk stalk.
4 WEEKS
Yolk stalk
Body stalk
Pancreas
Liver
Lesser omentum
Greater omentum
Neural tube
Notochord
Dorsal mesentery
Coelomic cavity
Ventral mesentery
Digestive tube
The digestive tube remains suspended in the coelom by a dorsal mesentery and a ventral mesentery. The ventral mesentery disintegrates everywhere except where major vessels or visceral organs have grown into it. It remains intact along the path of the umbilical arteries and where the umbilical vein and liver develop.
As the embryo enlarges, the stomach and liver rotate toward the right, creating two pockets. The mesenteries that form these pockets are the greater omentum and the lesser omentum.
Coelom
Pancreas
The pancreas and liver begin as epithelial pockets that grow away from the digestive tract and into the dorsal and ventral mesenteries, respectively.
Liver
Falciform ligament
Stomach
Neural tube
Aorta
6 WEEKS
Pancreas
Liver

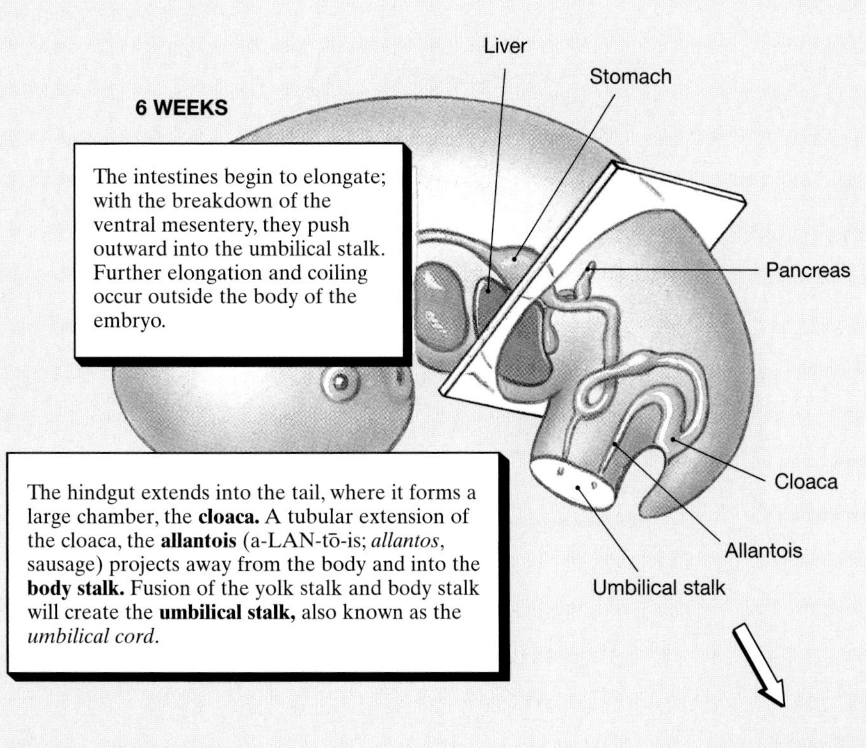

The intestines begin to elongate; with the breakdown of the ventral mesentery, they push outward into the umbilical stalk. Further elongation and coiling occur outside the body of the embryo.

The hindgut extends into the tail, where it forms a large chamber, the **cloaca.** A tubular extension of the cloaca, the **allantois** (a-LAN-tō-is; *allantos*, sausage) projects away from the body and into the **body stalk.** Fusion of the yolk stalk and body stalk will create the **umbilical stalk,** also known as the *umbilical cord*.

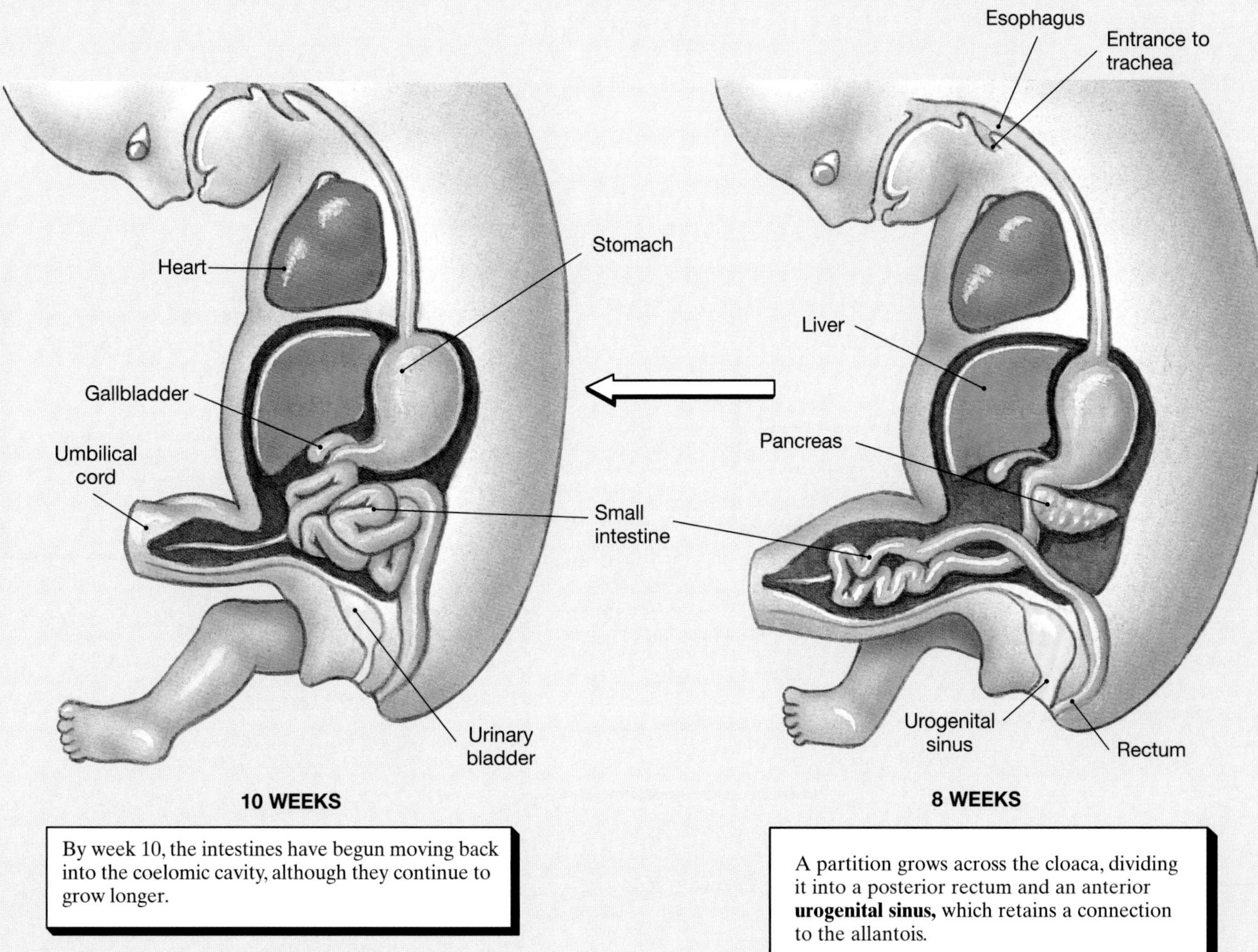

A partition grows across the cloaca, dividing it into a posterior rectum and an anterior **urogenital sinus,** which retains a connection to the allantois.

By week 10, the intestines have begun moving back into the coelomic cavity, although they continue to grow longer.

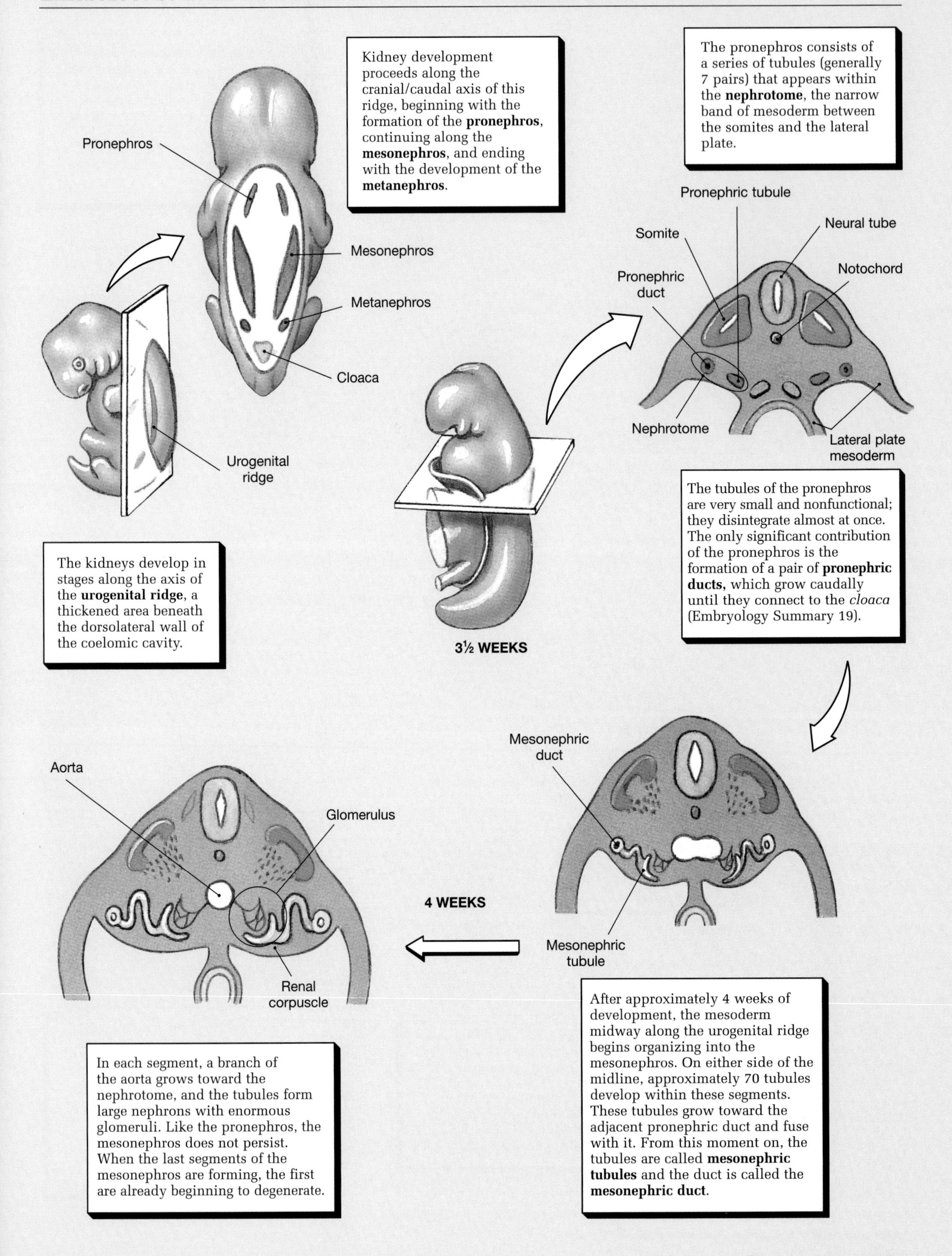

Kidney development proceeds along the cranial/caudal axis of this ridge, beginning with the formation of the **pronephros**, continuing along the **mesonephros**, and ending with the development of the **metanephros**.

The pronephros consists of a series of tubules (generally 7 pairs) that appears within the **nephrotome**, the narrow band of mesoderm between the somites and the lateral plate.

The kidneys develop in stages along the axis of the **urogenital ridge**, a thickened area beneath the dorsolateral wall of the coelomic cavity.

The tubules of the pronephros are very small and nonfunctional; they disintegrate almost at once. The only significant contribution of the pronephros is the formation of a pair of **pronephric ducts,** which grow caudally until they connect to the *cloaca* (Embryology Summary 19).

After approximately 4 weeks of development, the mesoderm midway along the urogenital ridge begins organizing into the mesonephros. On either side of the midline, approximately 70 tubules develop within these segments. These tubules grow toward the adjacent pronephric duct and fuse with it. From this moment on, the tubules are called **mesonephric tubules** and the duct is called the **mesonephric duct**.

In each segment, a branch of the aorta grows toward the nephrotome, and the tubules form large nephrons with enormous glomeruli. Like the pronephros, the mesonephros does not persist. When the last segments of the mesonephros are forming, the first are already beginning to degenerate.

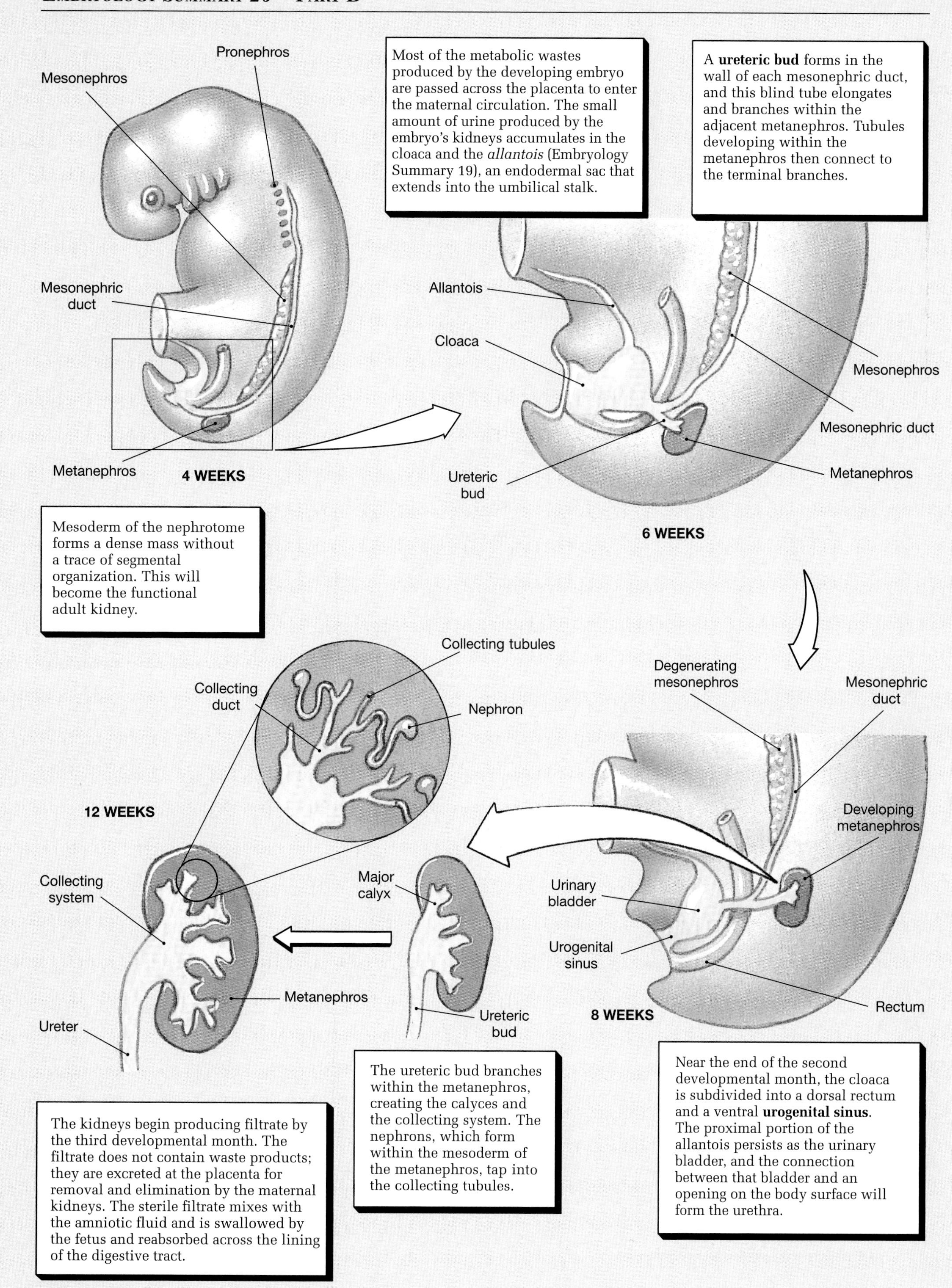
Pronephros
Mesonephros
Mesonephric duct
Metanephros
4 WEEKS
Most of the metabolic wastes produced by the developing embryo are passed across the placenta to enter the maternal circulation. The small amount of urine produced by the embryo's kidneys accumulates in the cloaca and the *allantois* (Embryology Summary 19), an endodermal sac that extends into the umbilical stalk.
A **ureteric bud** forms in the wall of each mesonephric duct, and this blind tube elongates and branches within the adjacent metanephros. Tubules developing within the metanephros then connect to the terminal branches.
Allantois
Cloaca
Mesonephros
Mesonephric duct
Metanephros
Ureteric bud
6 WEEKS
Mesoderm of the nephrotome forms a dense mass without a trace of segmental organization. This will become the functional adult kidney.
Collecting tubules
Collecting duct
Nephron
12 WEEKS
Collecting system
Metanephros
Ureter
Major calyx
Ureteric bud
Degenerating mesonephros
Mesonephric duct
Developing metanephros
Urinary bladder
Urogenital sinus
Rectum
8 WEEKS
The kidneys begin producing filtrate by the third developmental month. The filtrate does not contain waste products; they are excreted at the placenta for removal and elimination by the maternal kidneys. The sterile filtrate mixes with the amniotic fluid and is swallowed by the fetus and reabsorbed across the lining of the digestive tract.
The ureteric bud branches within the metanephros, creating the calyces and the collecting system. The nephrons, which form within the mesoderm of the metanephros, tap into the collecting tubules.
Near the end of the second developmental month, the cloaca is subdivided into a dorsal rectum and a ventral **urogenital sinus**. The proximal portion of the allantois persists as the urinary bladder, and the connection between that bladder and an opening on the body surface will form the urethra.

GENDER-INDIFFERENT STAGES

(WEEKS 3–6)

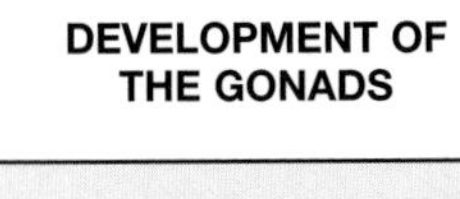

DEVELOPMENT OF THE GONADS

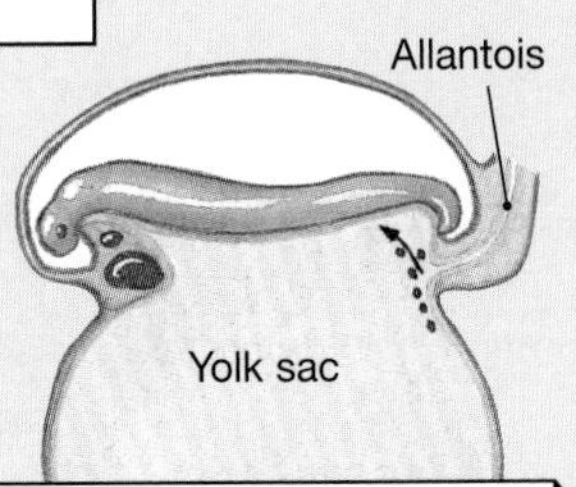

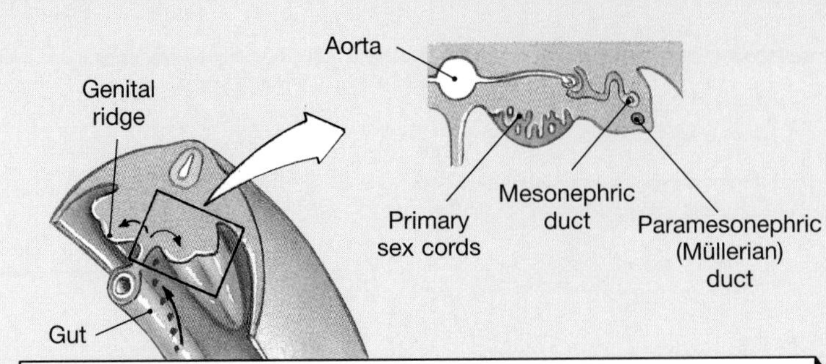

During the third week, endodermal cells migrate from the wall of the yolk sac near the allantois to the dorsal wall of the abdominal cavity. These primordial germ cells enter the **genital ridges,** which parallel the mesonephros.

3 WEEKS

Each ridge has a thick epithelium that is continuous with columns of cells, the **primary sex cords.** They extend into the center (medulla) of the ridge. Anterior to each mesonephric duct, a duct forms that has no connection to the kidneys. This is the **paramesonephric** (Müllerian) **duct;** it extends along the genital ridge and continues toward the cloaca. At this gender-indifferent stage, male embryos cannot be distinguished from female embryos.

DEVELOPMENT OF DUCTS AND ACCESSORY ORGANS

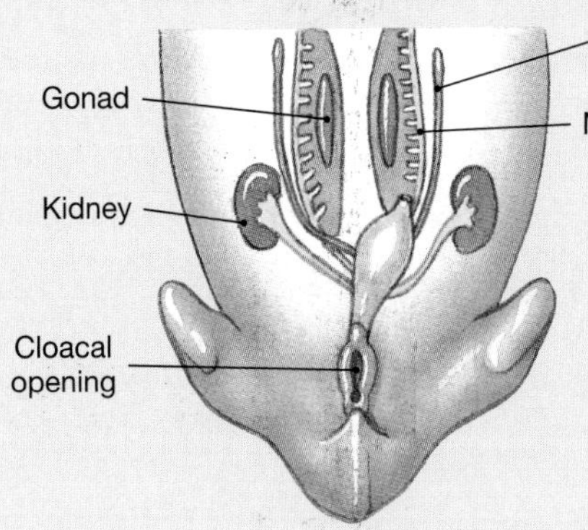

Both genders have mesonephric and paramesonephric ducts at this stage. Unless exposed to androgens, the embryo—regardless of its genetic gender—will develop into a female. In a normal male embyro, cells in the core (medulla) of the genital ridge begin producing testosterone sometime after week 6. Testosterone triggers the changes in the duct system and external genitalia that are detailed in Part B.

DEVELOPMENT OF EXTERNAL GENITALIA

4 WEEKS

Genital tubercle

Cloacal fold

Cloacal membrane

6 WEEKS

Urogenital membrane

Urethral fold

Genital swelling

Anal fold

After 4 weeks of development, mesenchymal swellings called **cloacal folds** are around the **cloacal membrane.** (The cloaca does not open to the exterior.) The **genital tubercle** forms the glans of the penis in males and the clitoris in females.

Two weeks later, the cloaca has subdivided, separating the cloacal membrane into a posterior *anal membrane*, bounded by the **anal folds,** and an anterior **urogenital membrane,** bounded by the **urethral folds.** A prominent **genital swelling** forms lateral to each urethral fold.

THE DEVELOPMENT OF THE MALE REPRODUCTIVE SYSTEM

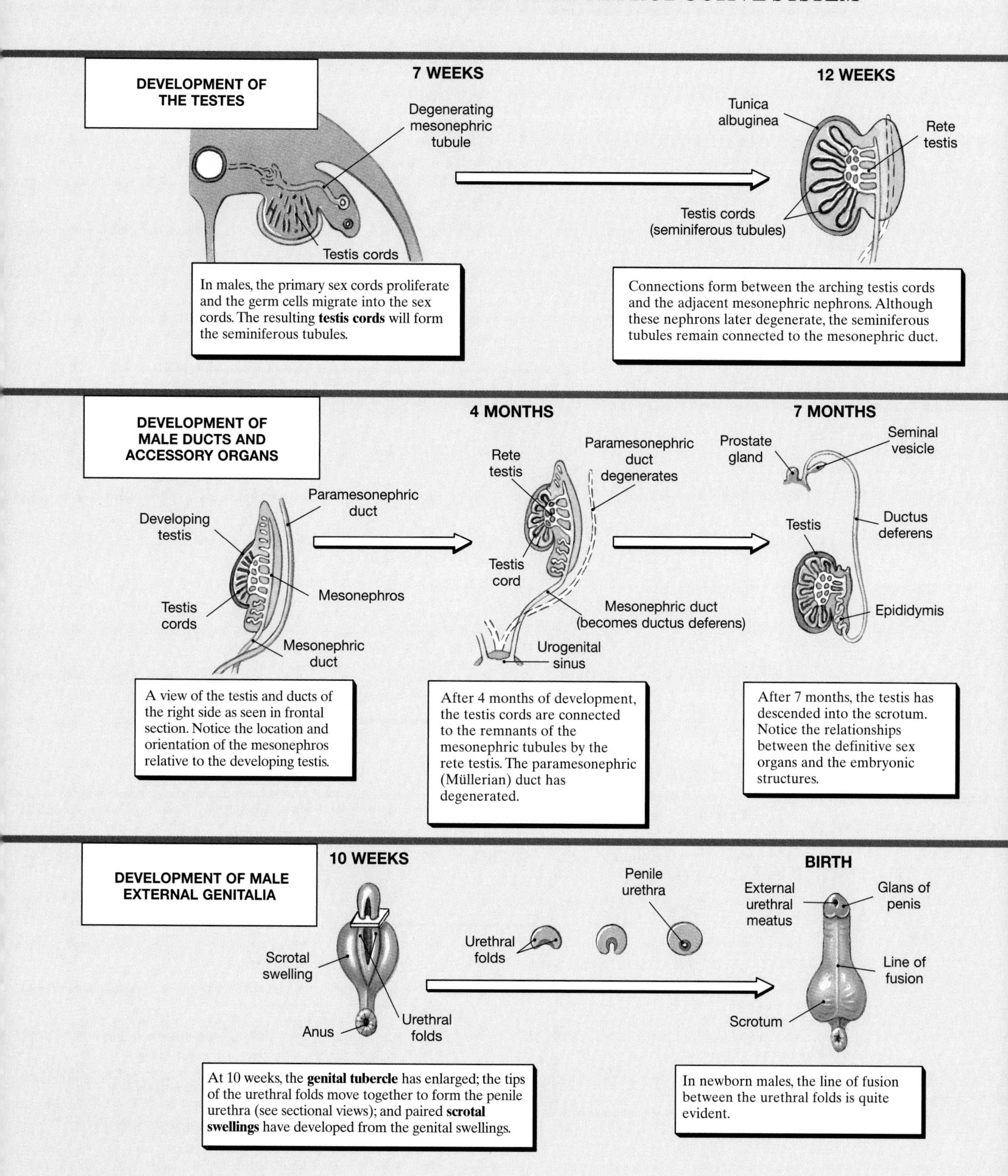

In males, the primary sex cords proliferate and the germ cells migrate into the sex cords. The resulting **testis cords** will form the seminiferous tubules.

Connections form between the arching testis cords and the adjacent mesonephric nephrons. Although these nephrons later degenerate, the seminiferous tubules remain connected to the mesonephric duct.

A view of the testis and ducts of the right side as seen in frontal section. Notice the location and orientation of the mesonephros relative to the developing testis.

After 4 months of development, the testis cords are connected to the remnants of the mesonephric tubules by the rete testis. The paramesonephric (Müllerian) duct has degenerated.

After 7 months, the testis has descended into the scrotum. Notice the relationships between the definitive sex organs and the embryonic structures.

At 10 weeks, the **genital tubercle** has enlarged; the tips of the urethral folds move together to form the penile urethra (see sectional views); and paired **scrotal swellings** have developed from the genital swellings.

In newborn males, the line of fusion between the urethral folds is quite evident.

THE DEVELOPMENT OF THE FEMALE REPRODUCTIVE SYSTEM

DEVELOPMENT OF THE OVARIES

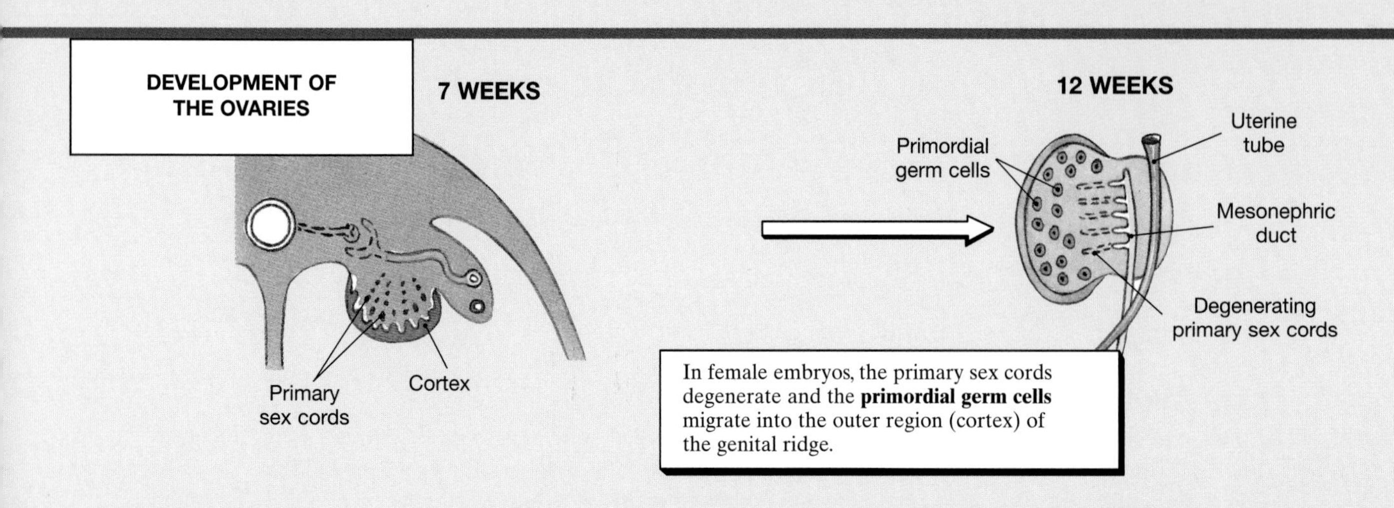

In female embryos, the primary sex cords degenerate and the **primordial germ cells** migrate into the outer region (cortex) of the genital ridge.

DEVELOPMENT OF FEMALE DUCTS AND ACCESSORY ORGANS

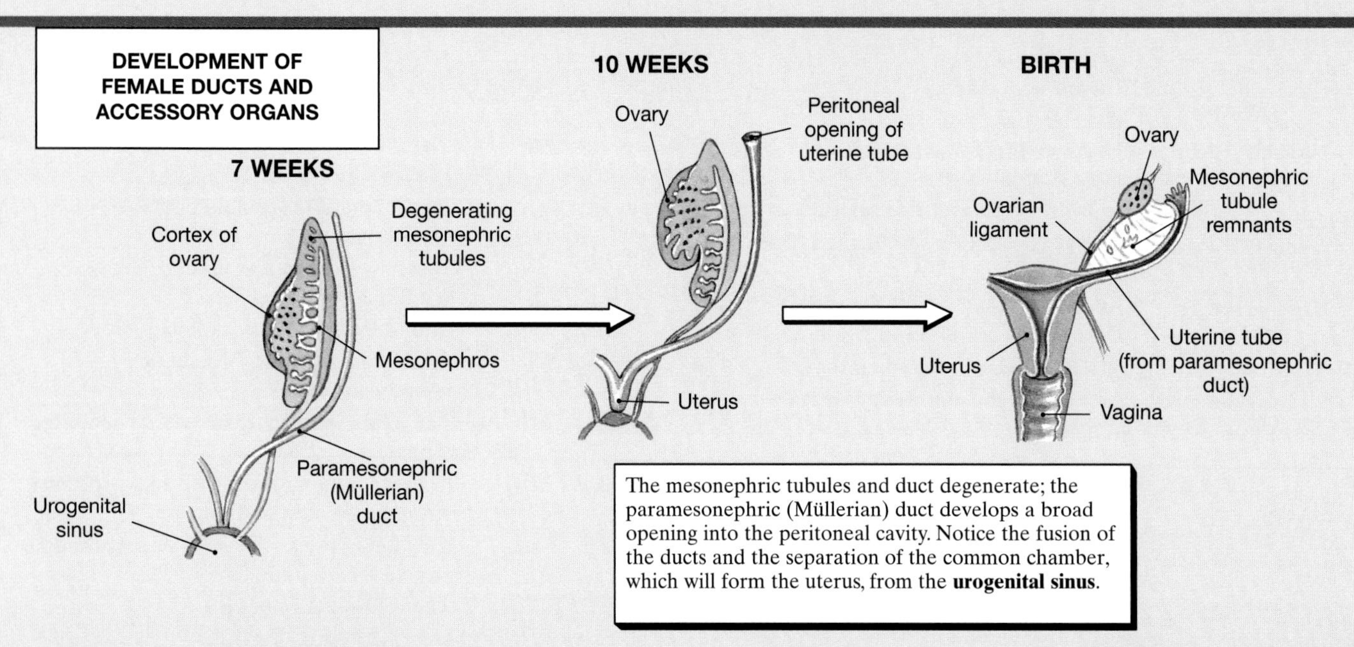

The mesonephric tubules and duct degenerate; the paramesonephric (Müllerian) duct develops a broad opening into the peritoneal cavity. Notice the fusion of the ducts and the separation of the common chamber, which will form the uterus, from the **urogenital sinus**.

DEVELOPMENT OF FEMALE EXTERNAL GENITALIA

Comparison of Male and Female External Genitalia

Males	Females
Penis	Clitoris
Corpora cavernosa	Erectile tissue
Corpus spongiosum	Vestibular bulbs
Proximal shaft of penis	Labia minora
Penile urethra	Vestibule
Bulbourethral glands	Greater vestibular glands
Scrotum	Labia majora

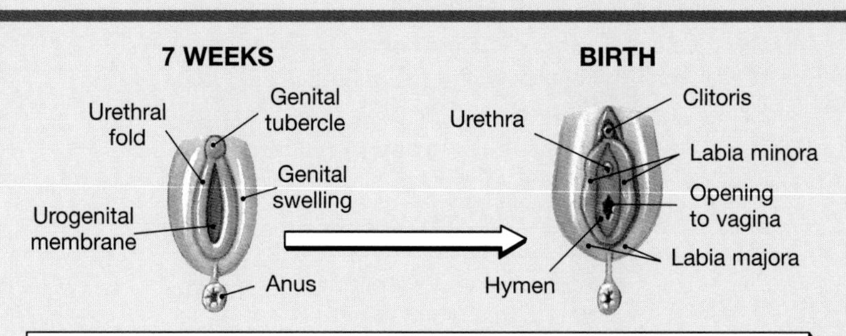

In females, the urethral folds do not fuse; they develop into the labia minora. The genital swellings form the labia majora. The genital tubercle develops into the clitoris. The urethra opens to the exterior immediately posterior to the clitoris. The hymen remains as an elaboration of the urogenital membrane.

Surface Anatomy and Cadaver Atlas

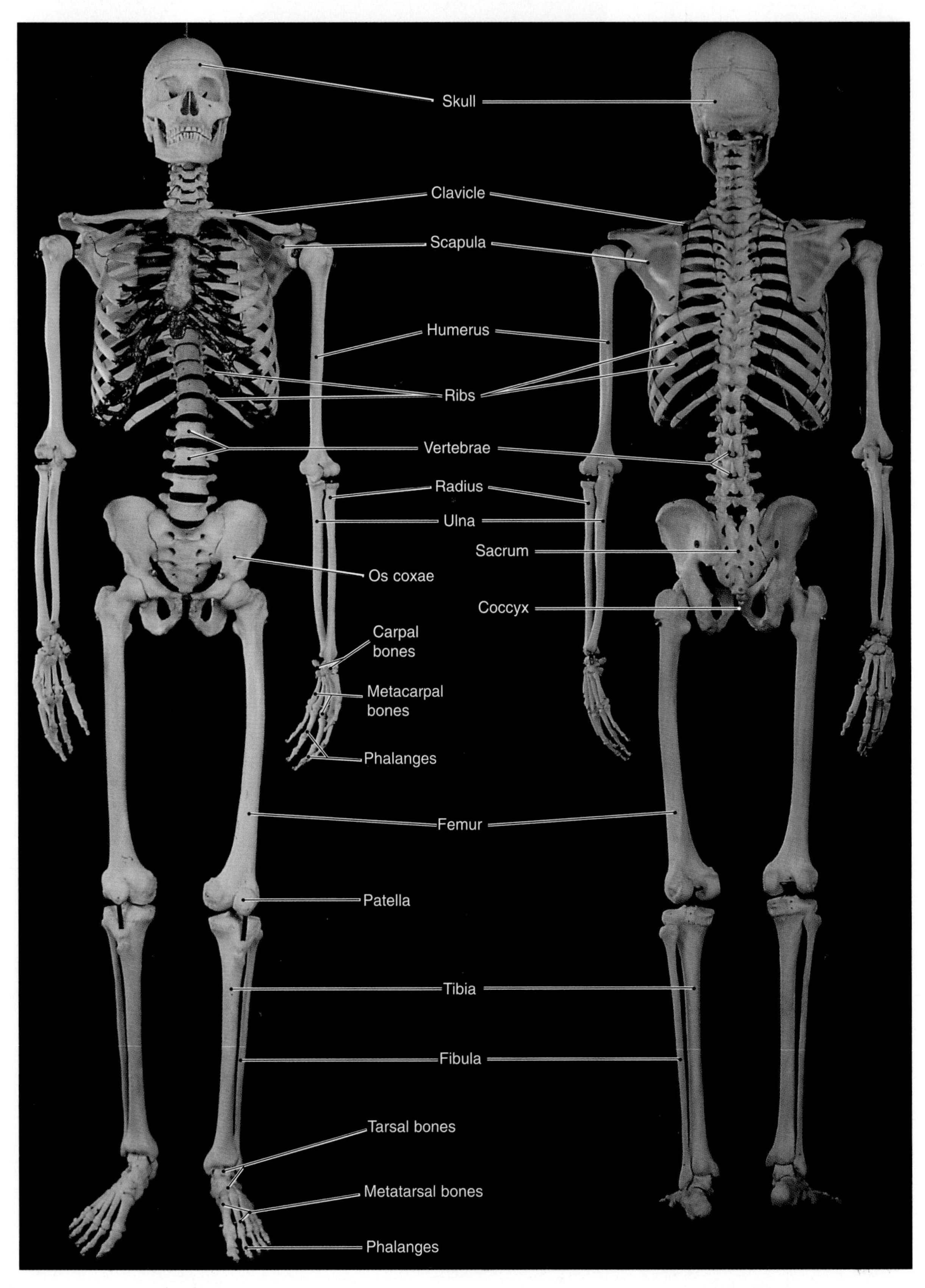

1.1 Anterior view **1.2 Posterior view**

PLATE 2 THE SKULL
2.1 Painted Skull Series

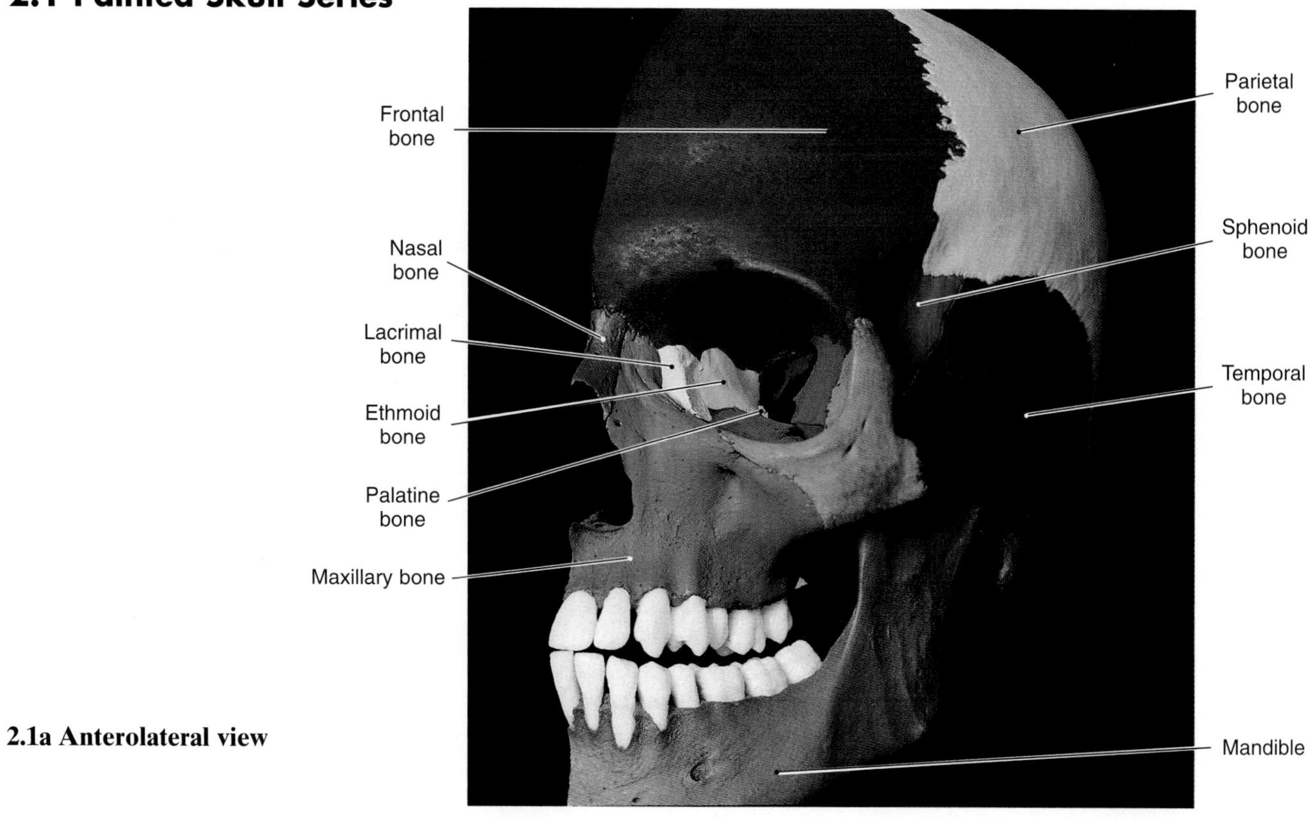

2.1a Anterolateral view

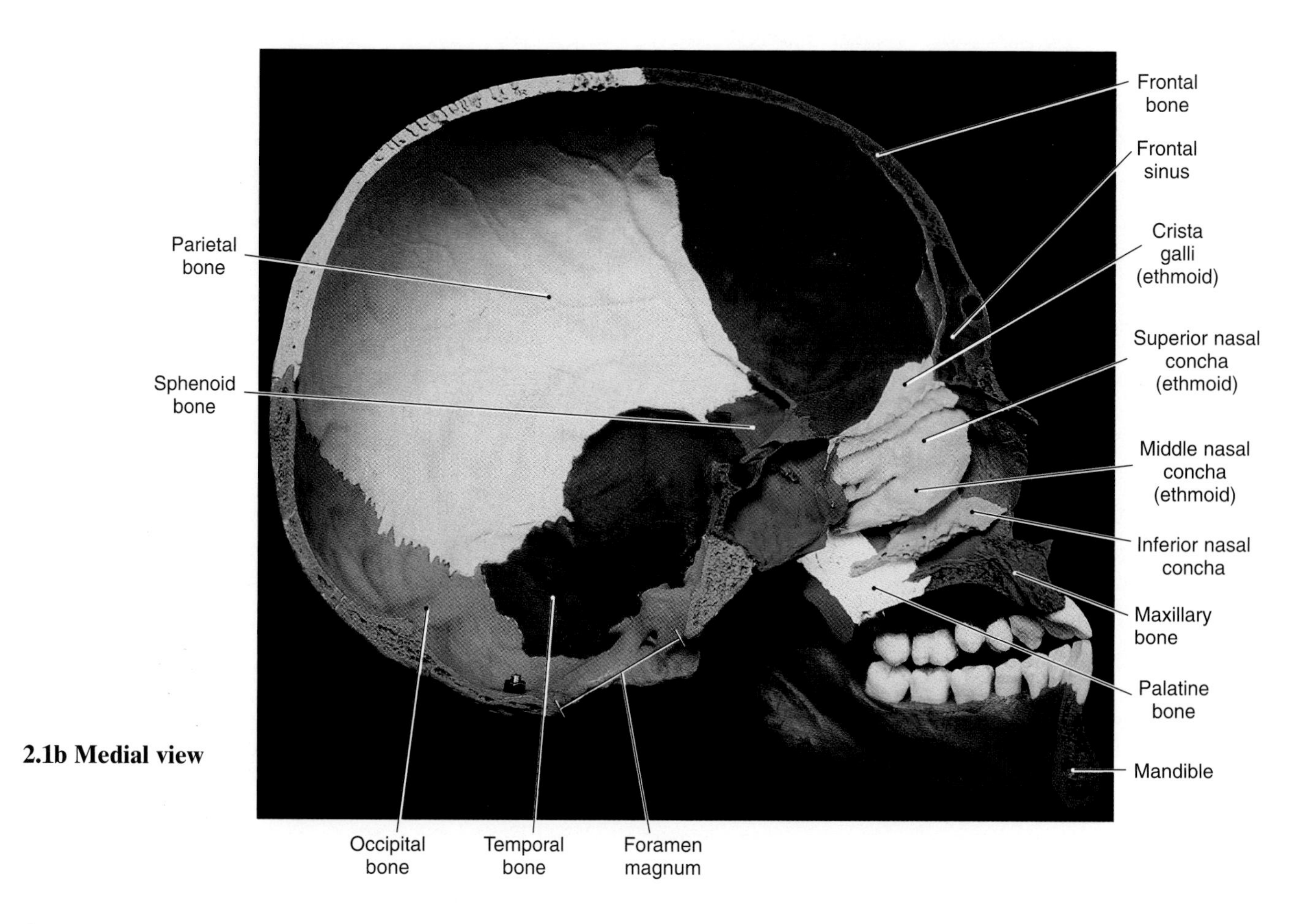

2.1b Medial view

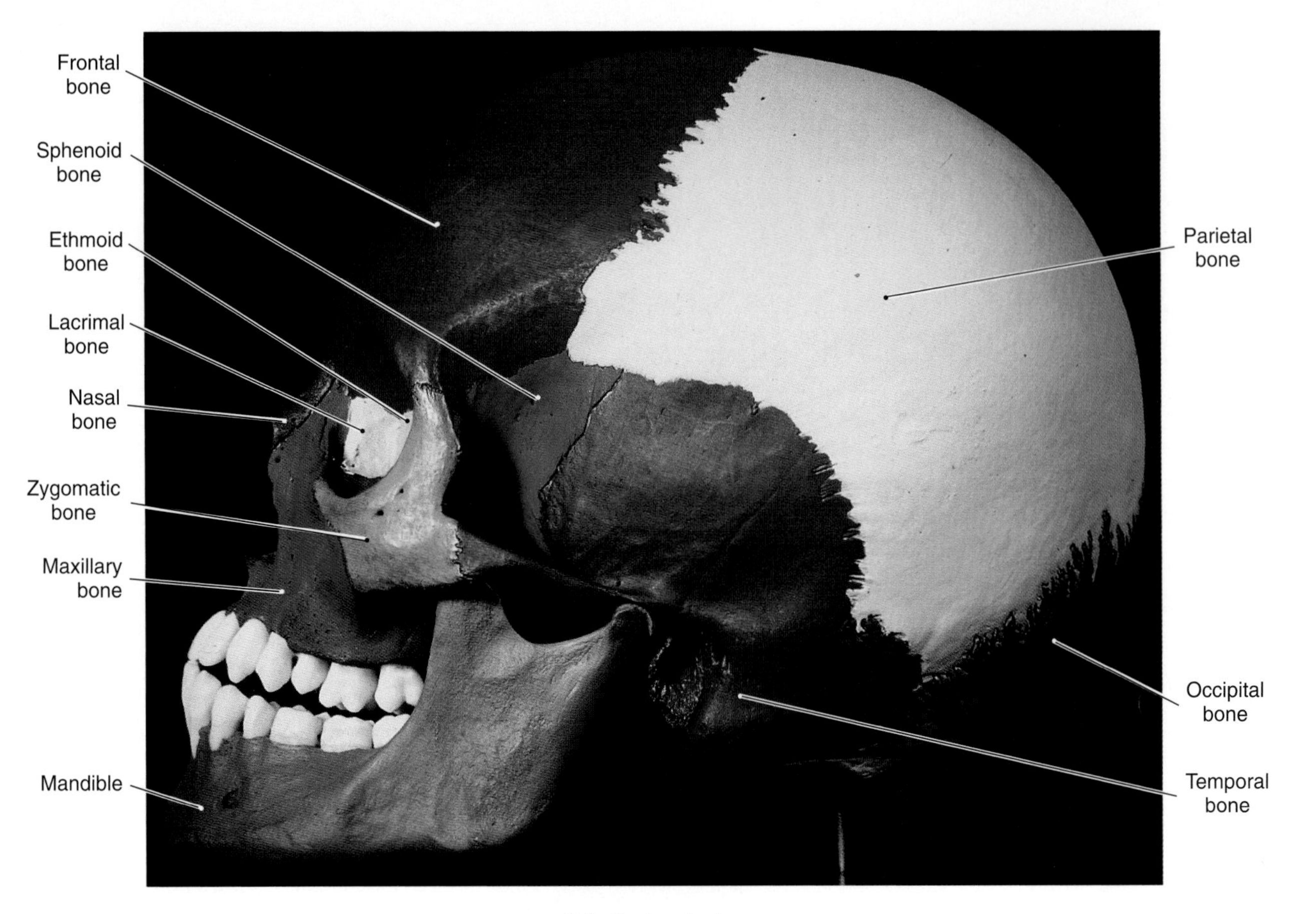

2.1c Lateral view

2.2, 2.3 Natural Skull Series

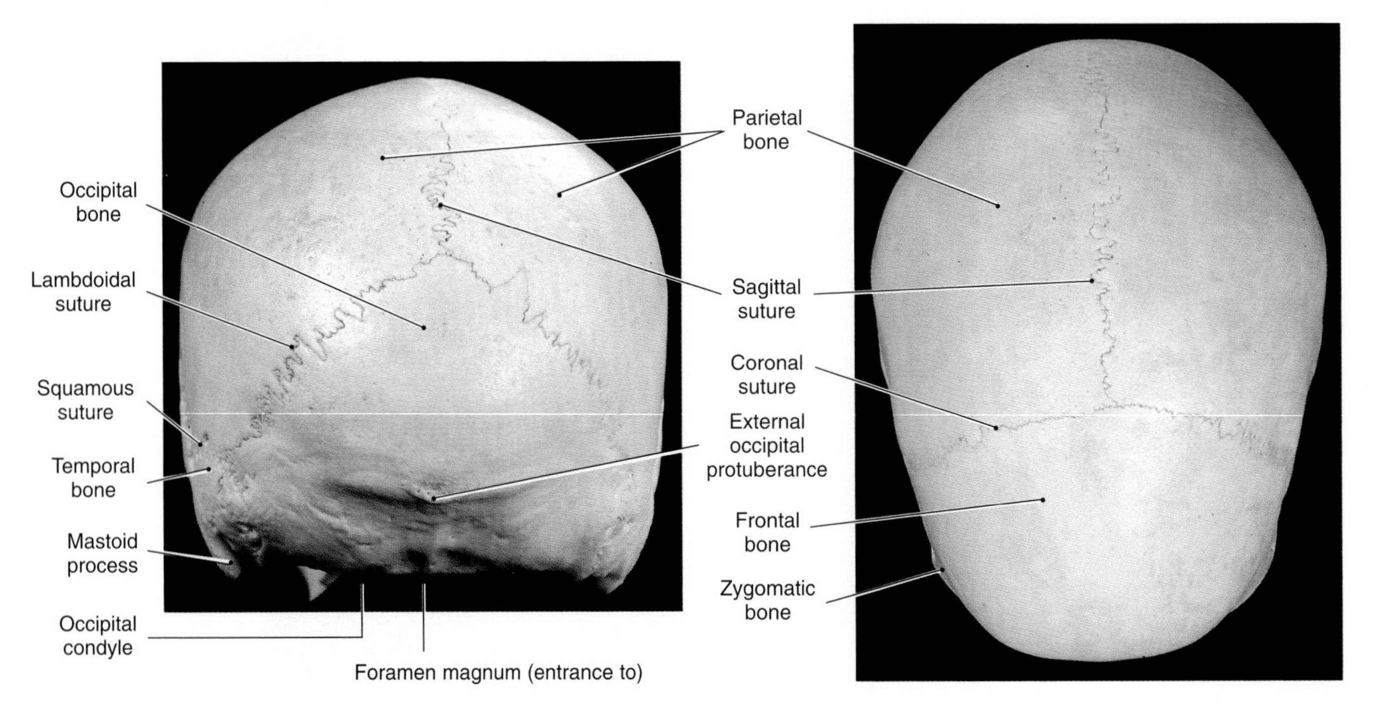

2.2a Posterior view

2.2b Superior view

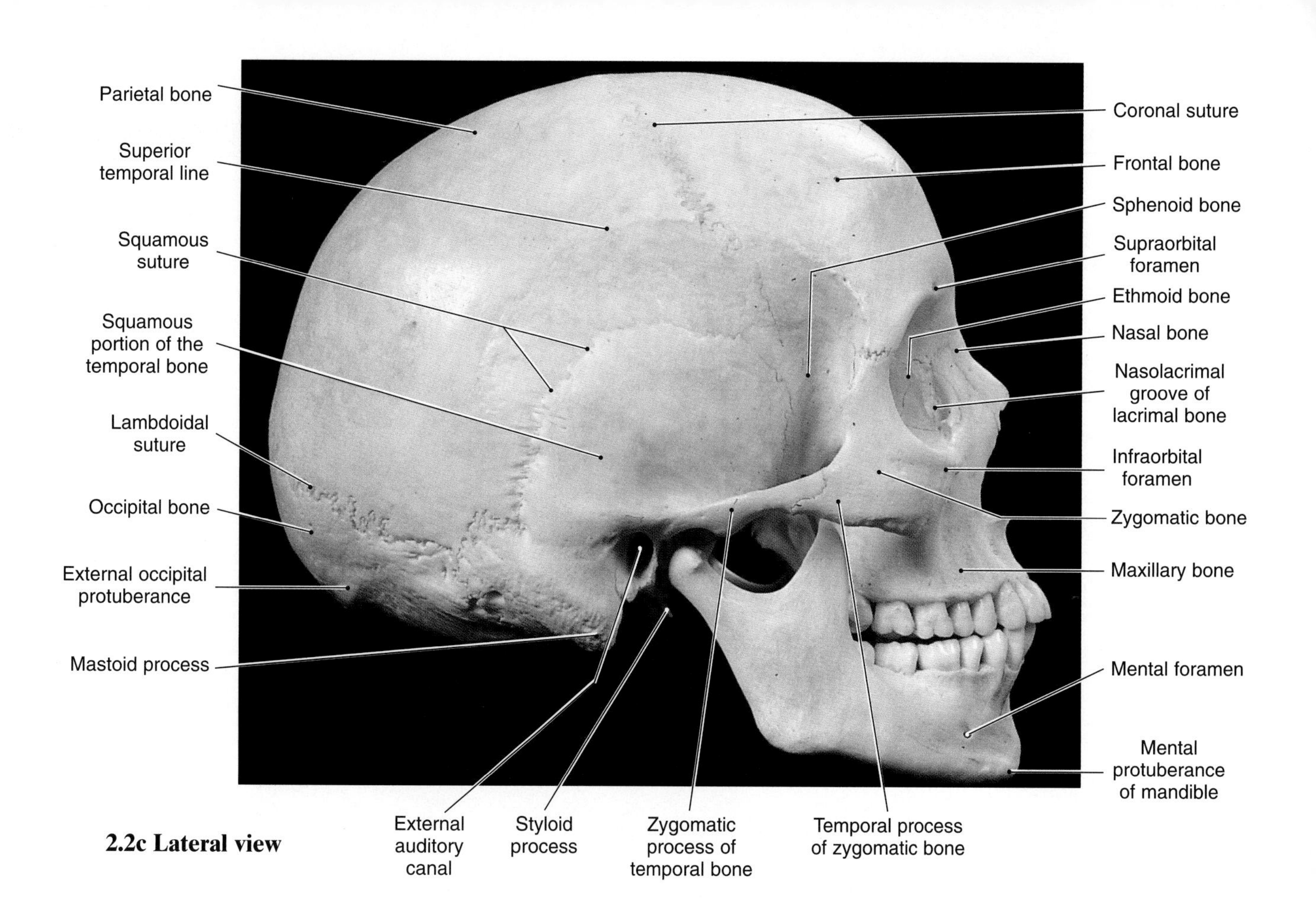
Parietal bone
Superior temporal line
Squamous suture
Squamous portion of the temporal bone
Lambdoidal suture
Occipital bone
External occipital protuberance
Mastoid process
Coronal suture
Frontal bone
Sphenoid bone
Supraorbital foramen
Ethmoid bone
Nasal bone
Nasolacrimal groove of lacrimal bone
Infraorbital foramen
Zygomatic bone
Maxillary bone
Mental foramen
Mental protuberance of mandible
External auditory canal
Styloid process
Zygomatic process of temporal bone
Temporal process of zygomatic bone

2.2c Lateral view

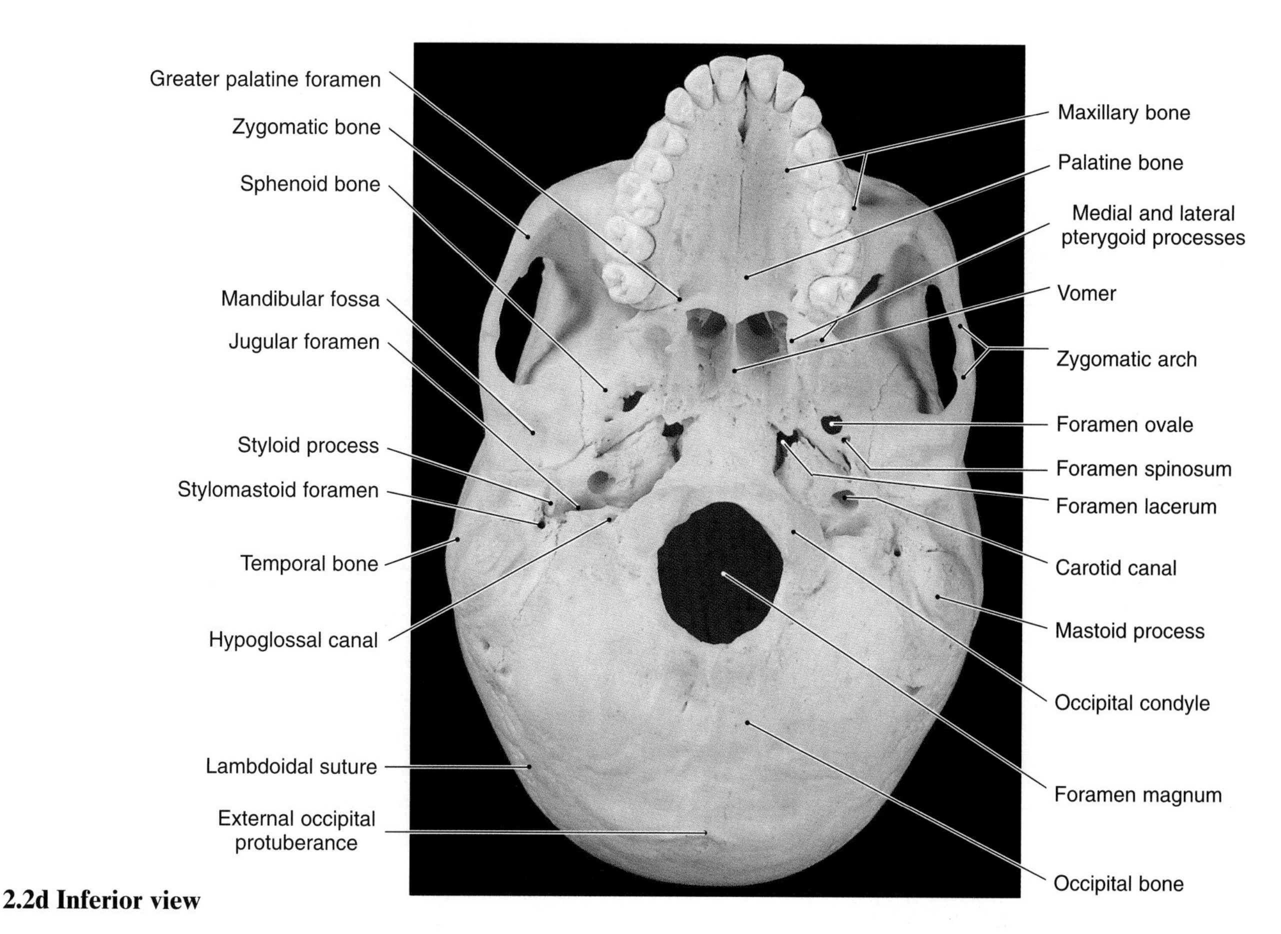
Greater palatine foramen
Zygomatic bone
Sphenoid bone
Mandibular fossa
Jugular foramen
Styloid process
Stylomastoid foramen
Temporal bone
Hypoglossal canal
Lambdoidal suture
External occipital protuberance
Maxillary bone
Palatine bone
Medial and lateral pterygoid processes
Vomer
Zygomatic arch
Foramen ovale
Foramen spinosum
Foramen lacerum
Carotid canal
Mastoid process
Occipital condyle
Foramen magnum
Occipital bone

2.2d Inferior view

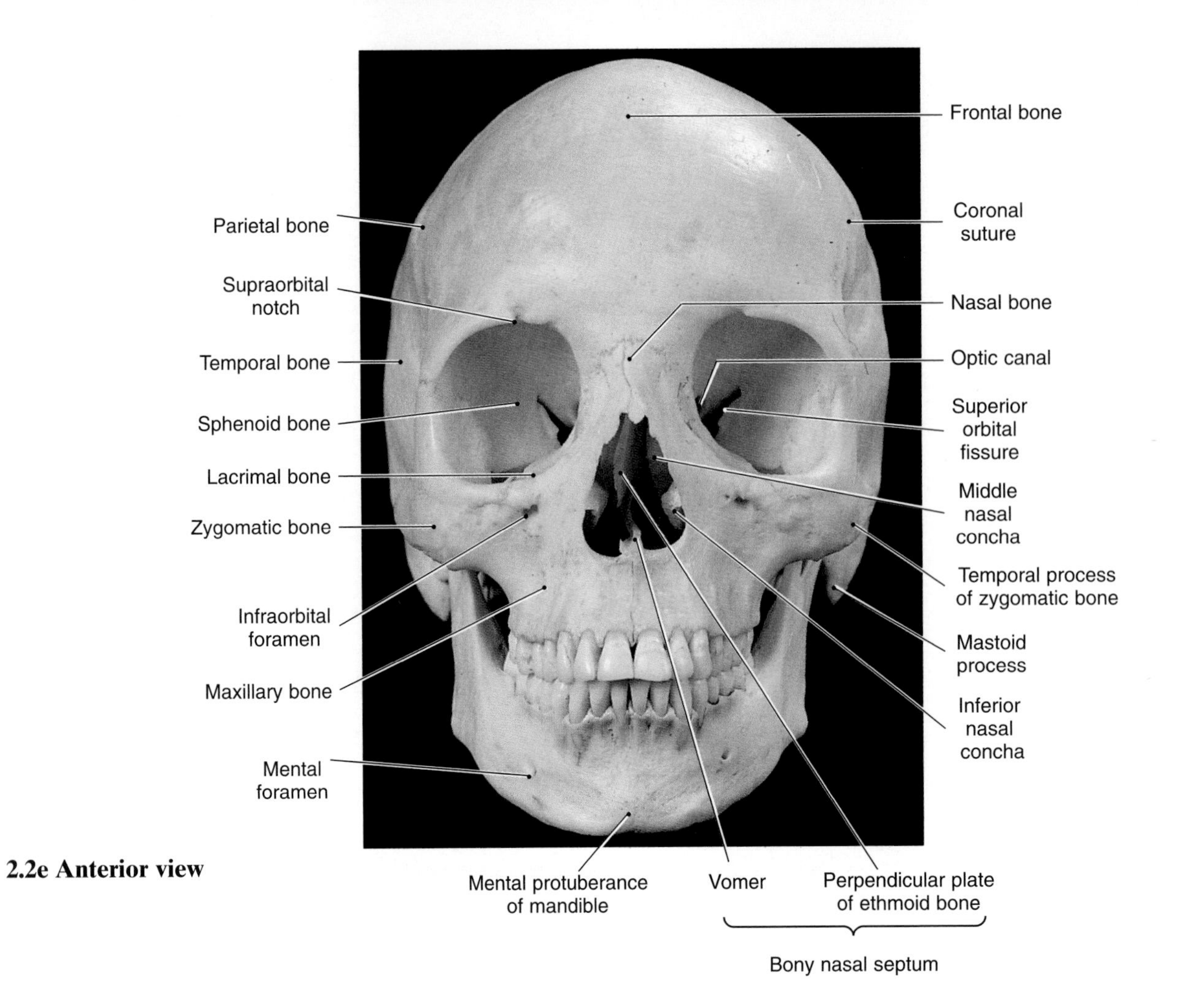
Frontal bone
Parietal bone
Coronal suture
Supraorbital notch
Nasal bone
Temporal bone
Optic canal
Sphenoid bone
Superior orbital fissure
Lacrimal bone
Middle nasal concha
Zygomatic bone
Temporal process of zygomatic bone
Infraorbital foramen
Mastoid process
Maxillary bone
Inferior nasal concha
Mental foramen
Mental protuberance of mandible
Vomer
Perpendicular plate of ethmoid bone
Bony nasal septum

2.2e Anterior view

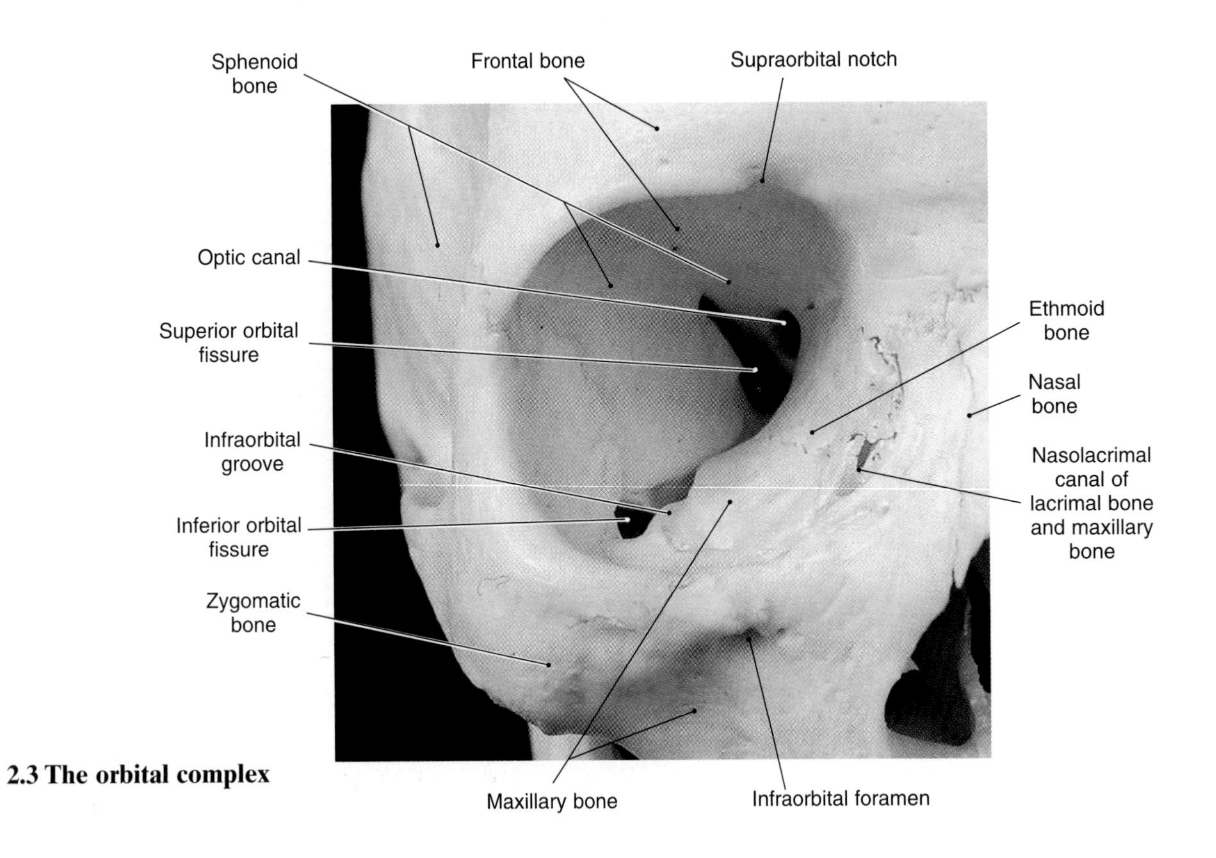
Sphenoid bone
Frontal bone
Supraorbital notch
Optic canal
Superior orbital fissure
Ethmoid bone
Nasal bone
Infraorbital groove
Nasolacrimal canal of lacrimal bone and maxillary bone
Inferior orbital fissure
Zygomatic bone
Maxillary bone
Infraorbital foramen

2.3 The orbital complex

2.4 The Cranium

Optic canal
Anterior clinoid process
Superior orbital fissure
Foramen rotundum
Sella turcica
Posterior clinoid process
Foramen ovale
Foramen spinosum
Foramen lacerum
Petrous portion of temporal bone
Internal auditory canal
Jugular foramen
Crista galli of ethmoid bone
Anterior cranial fossa
Middle cranial fossa
Hypoglossal canal
Foramen magnum
Posterior cranial fossa

2.4a The cranial floor, diagrammatic superior view

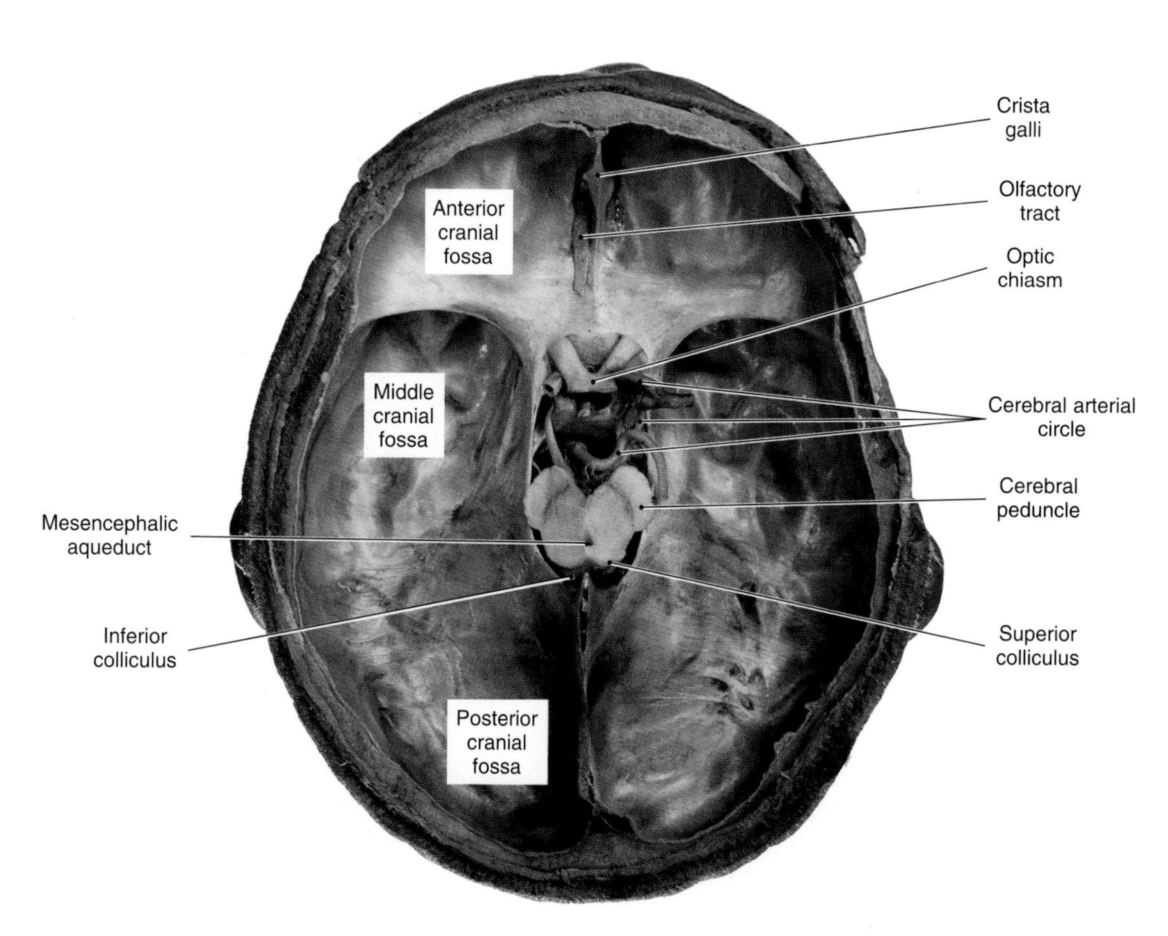

2.4b The cranial floor, superior view

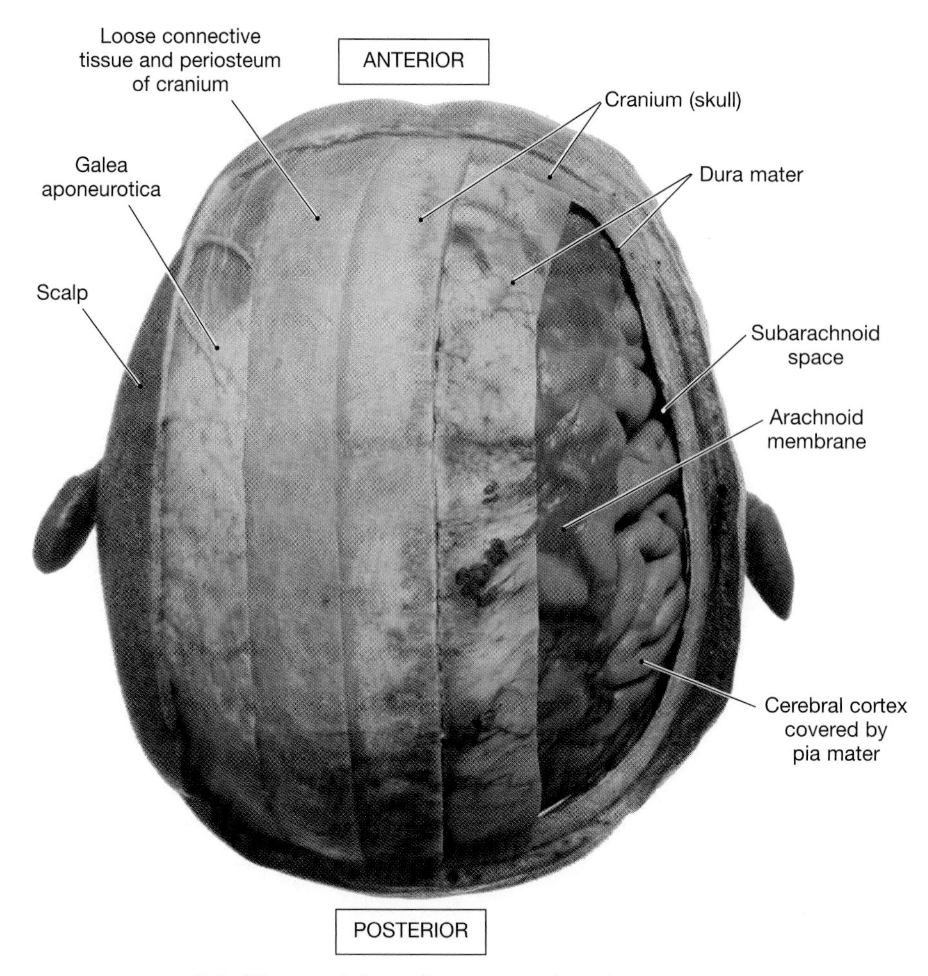

2.4c The cranial meninges, superior view

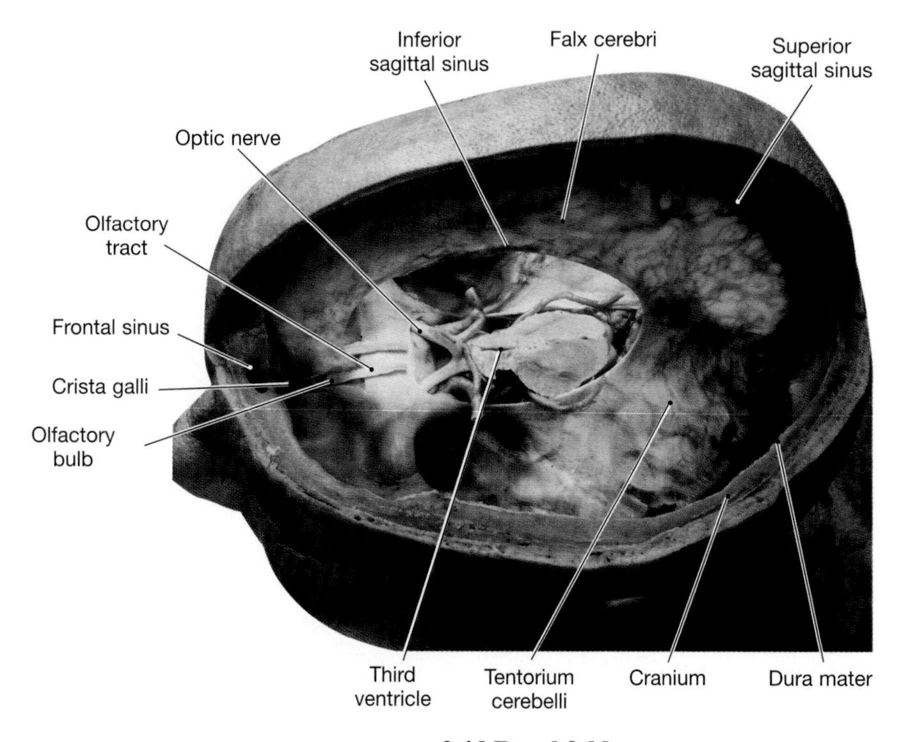

2.4d Dural folds

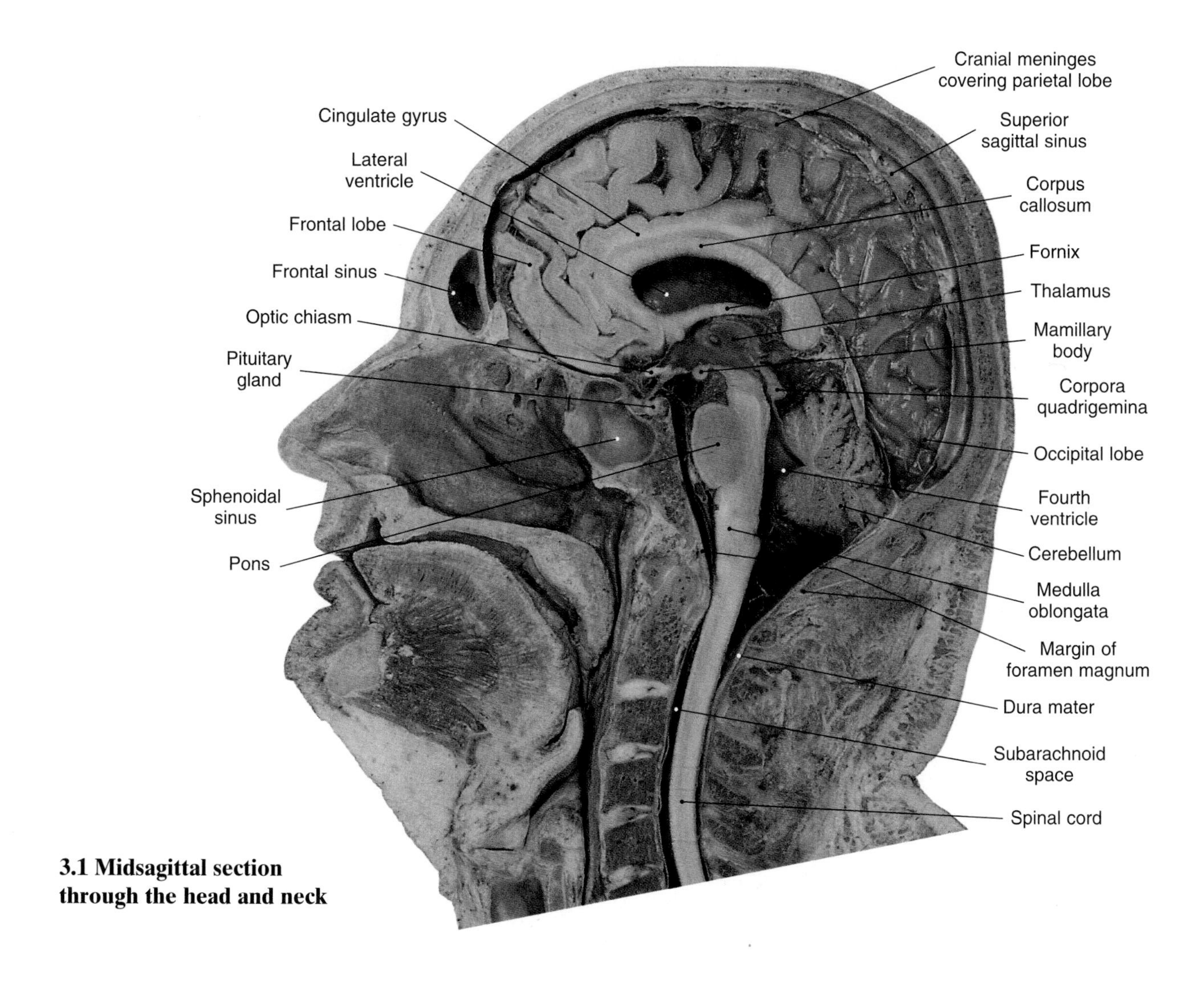

3.1 Midsagittal section through the head and neck

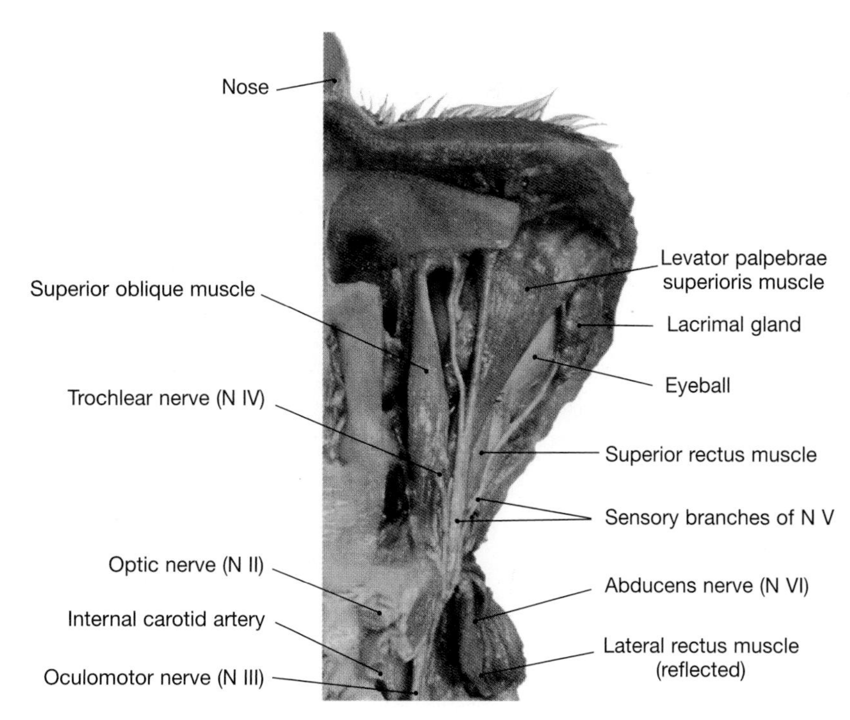

3.2 Accessory structures of the eye, superior view

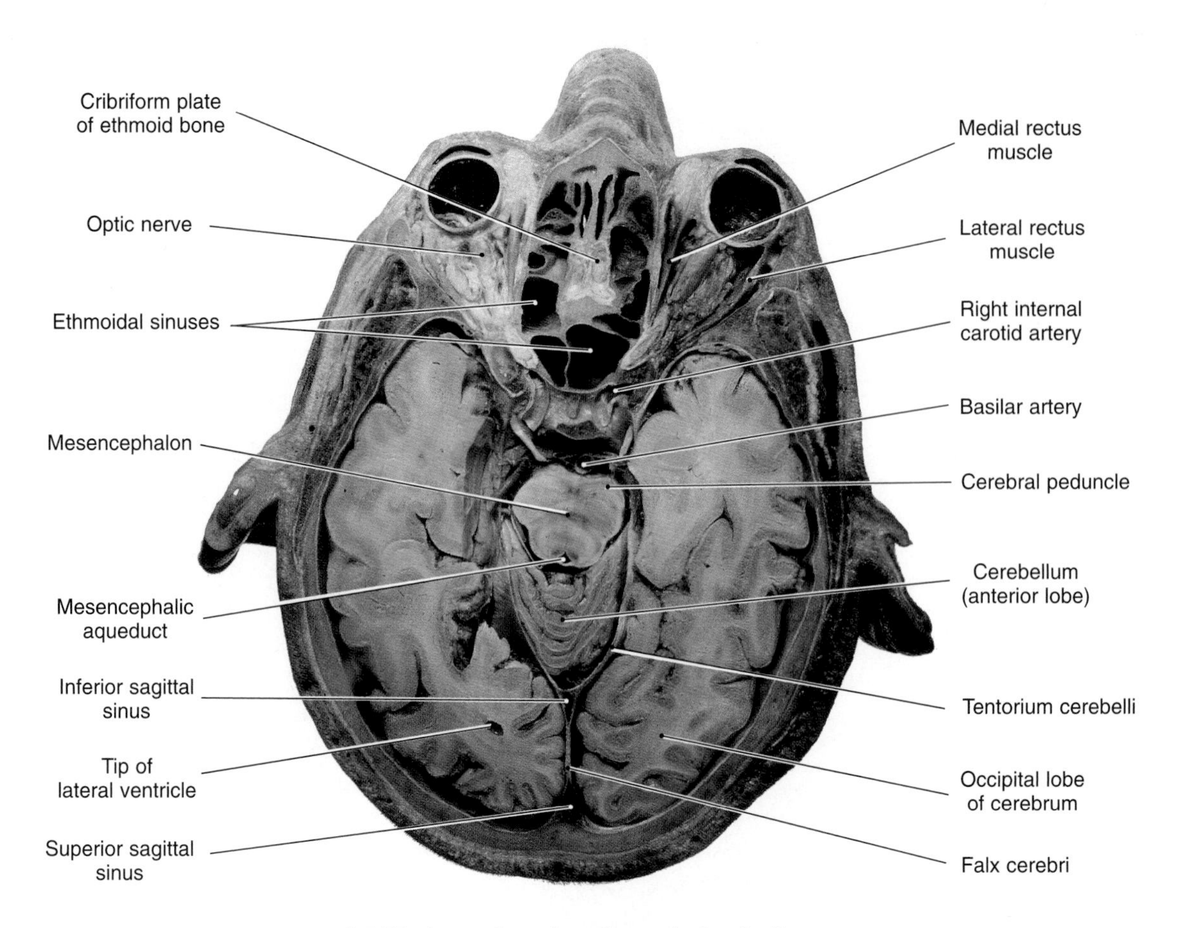

3.3 Horizontal section through the skull

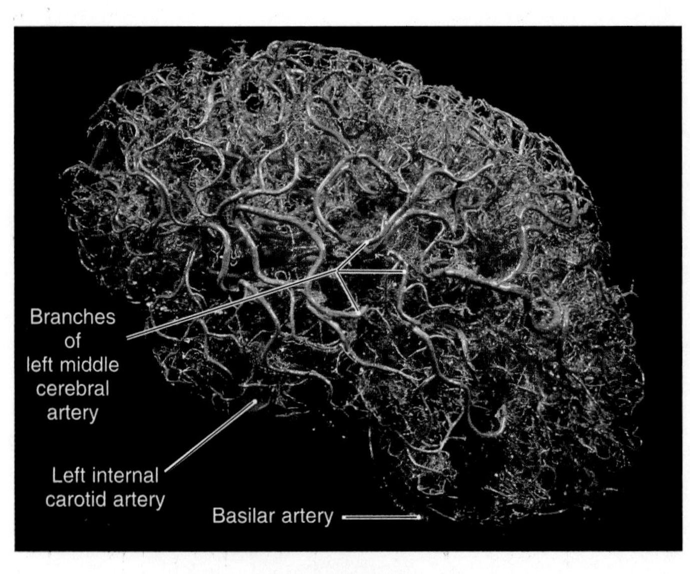

3.4 Arterial circulation to the brain, lateral view of corrosion cast

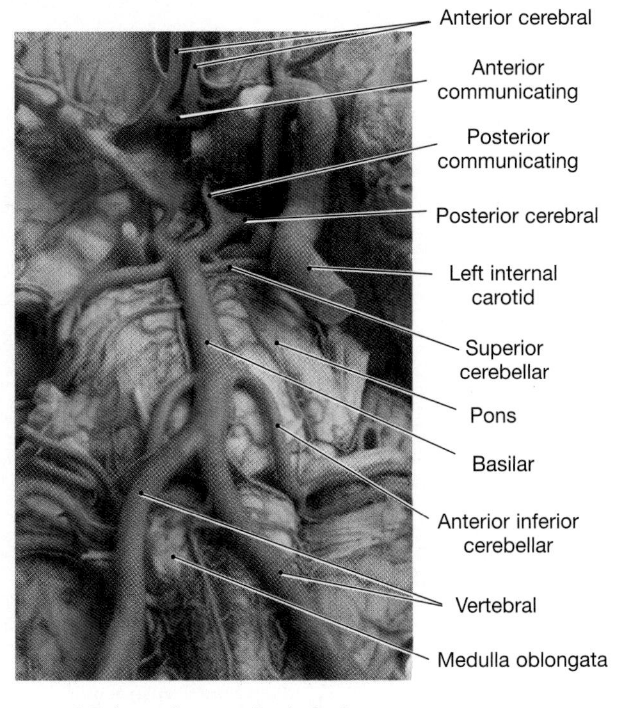

3.5 Arteries on the inferior surface of the brain

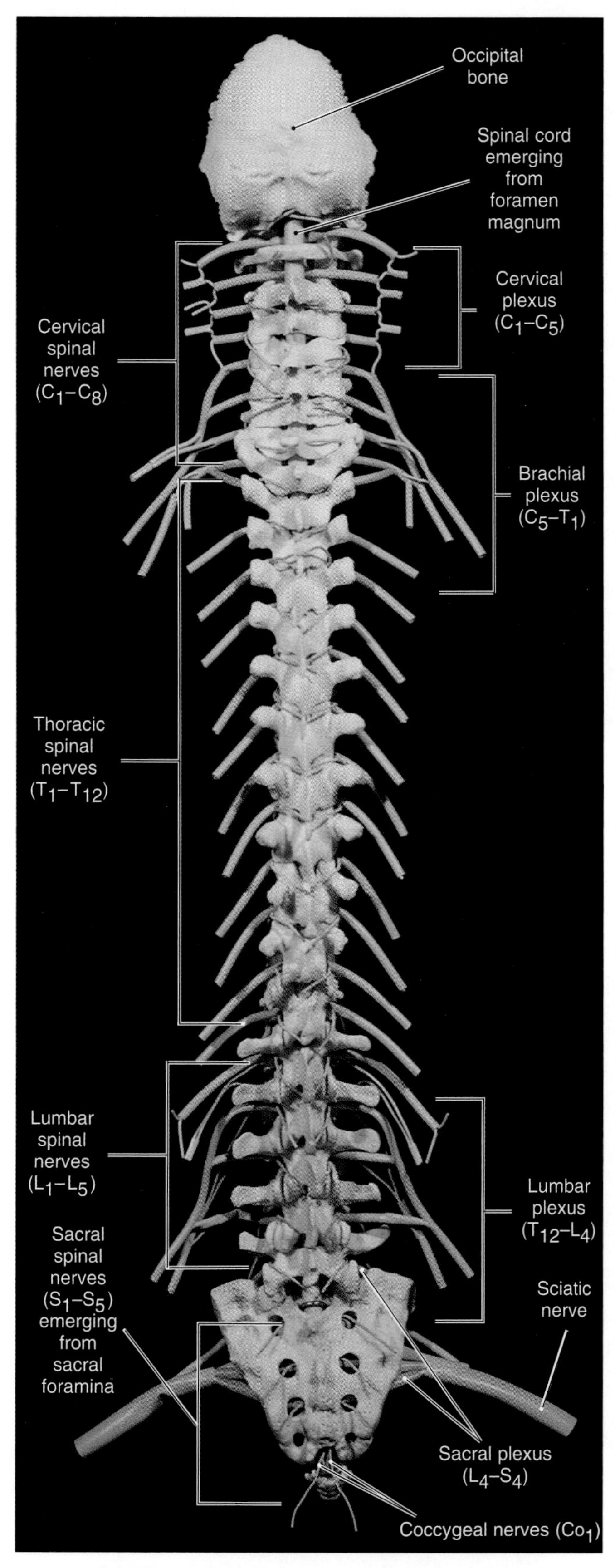

4.1 The vertebral column and spinal nerves

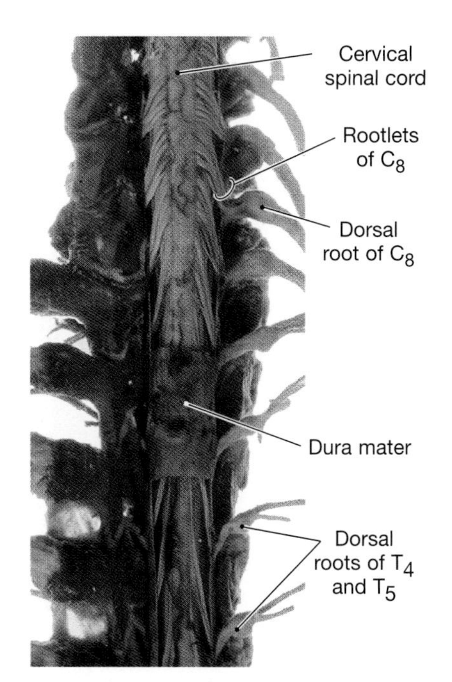

4.2 The cervical and thoracic regions of the spinal cord, posterior view

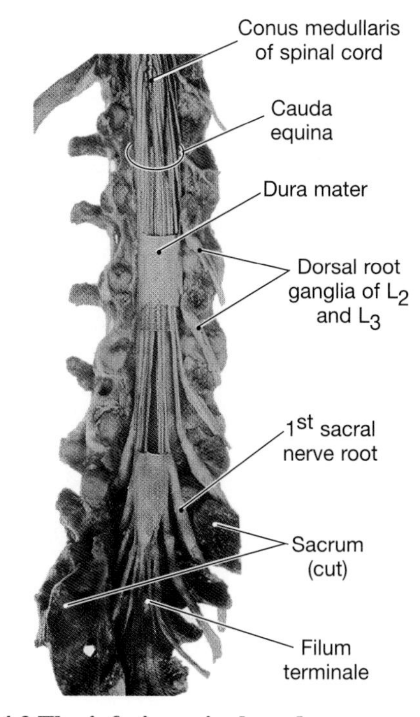

4.3 The inferior spinal cord and cauda equina, posterior view

PLATE 5 THE HEAD AND NECK
5.1 Surface Anatomy of the Head and Neck

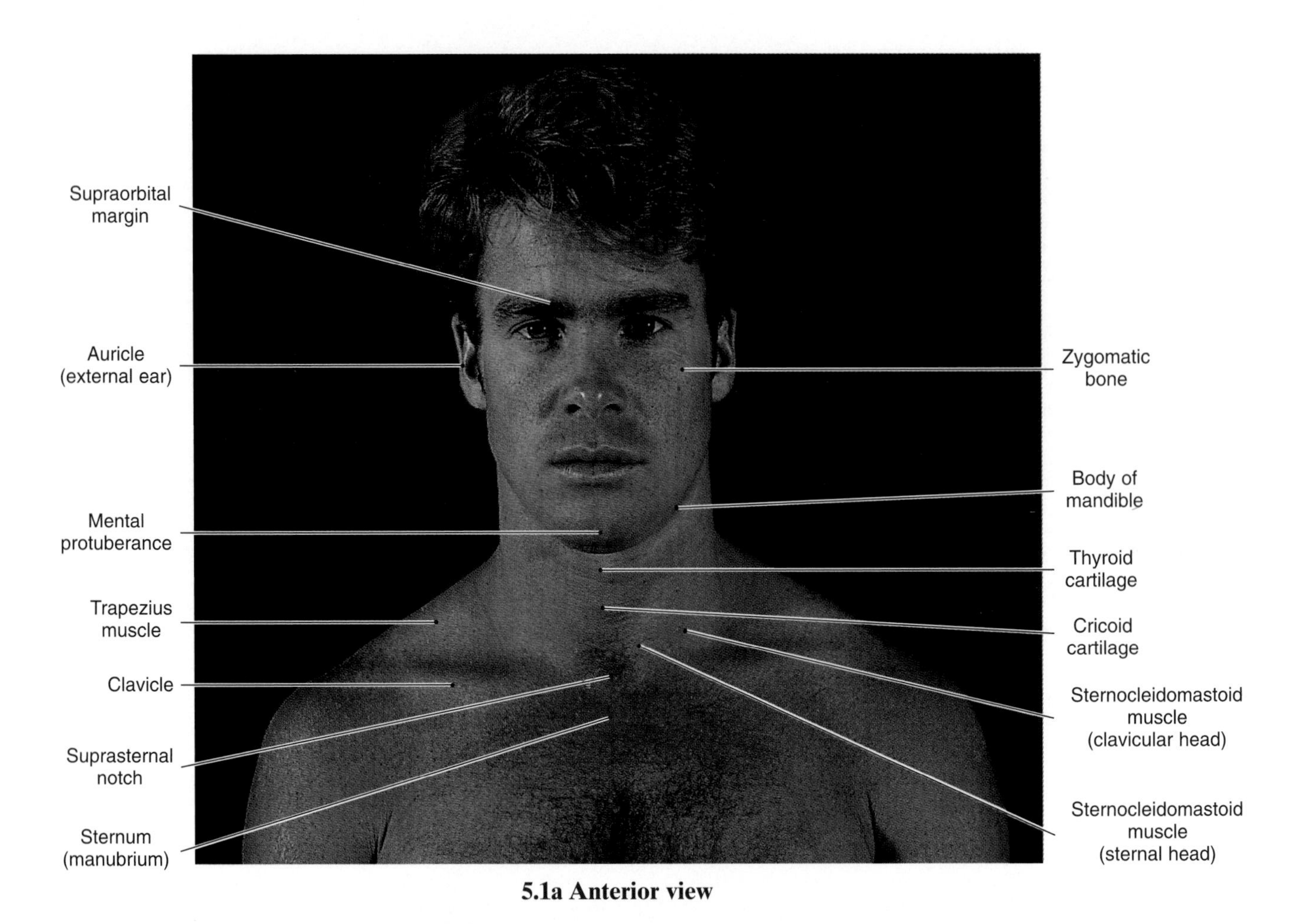

5.1a Anterior view

Angle of mandible

Site for palpation of submandibular gland and submandibular lymph nodes

Hyoid bone

Trapezius muscle

Thyroid cartilage

Supraclavicular fossa

Anterior cervical triangle

Suprasternal notch

Key to Divisions of the Anterior Cervical Triangle:

SHT= Suprahyoid triangle

SMT= Submandibular triangle

SCT= Superior carotid triangle

ICT= Inferior carotid triangle

SHT

SMT

SCT

ICT

Mastoid process

Occipital triangle

Site for palpation of carotid pulse

External jugular vein beneath platysma muscle

Posterior cervical triangle

Origin of brachial plexus

Acromion

Clavicle

Subclavian triangle

Sternocleidomastoid muscle (clavicular and sternal heads)

5.1b The cervical triangles

5.2 Superficial Dissection of the Face

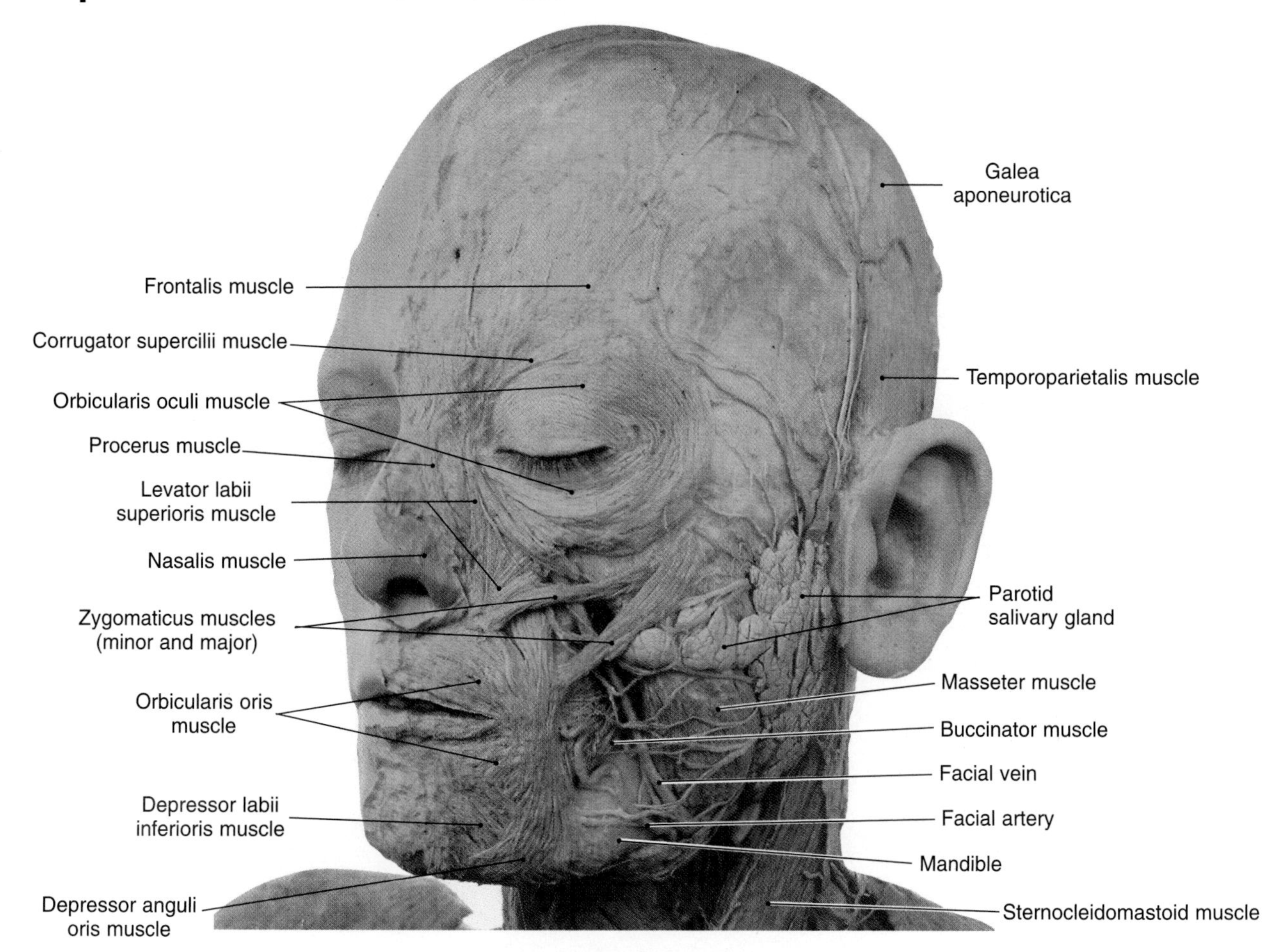
Galea aponeurotica
Frontalis muscle
Corrugator supercilii muscle
Orbicularis oculi muscle
Temporoparietalis muscle
Procerus muscle
Levator labii superioris muscle
Nasalis muscle
Zygomaticus muscles (minor and major)
Parotid salivary gland
Orbicularis oris muscle
Masseter muscle
Buccinator muscle
Facial vein
Depressor labii inferioris muscle
Facial artery
Mandible
Depressor anguli oris muscle
Sternocleidomastoid muscle

5.2a Anterolateral view

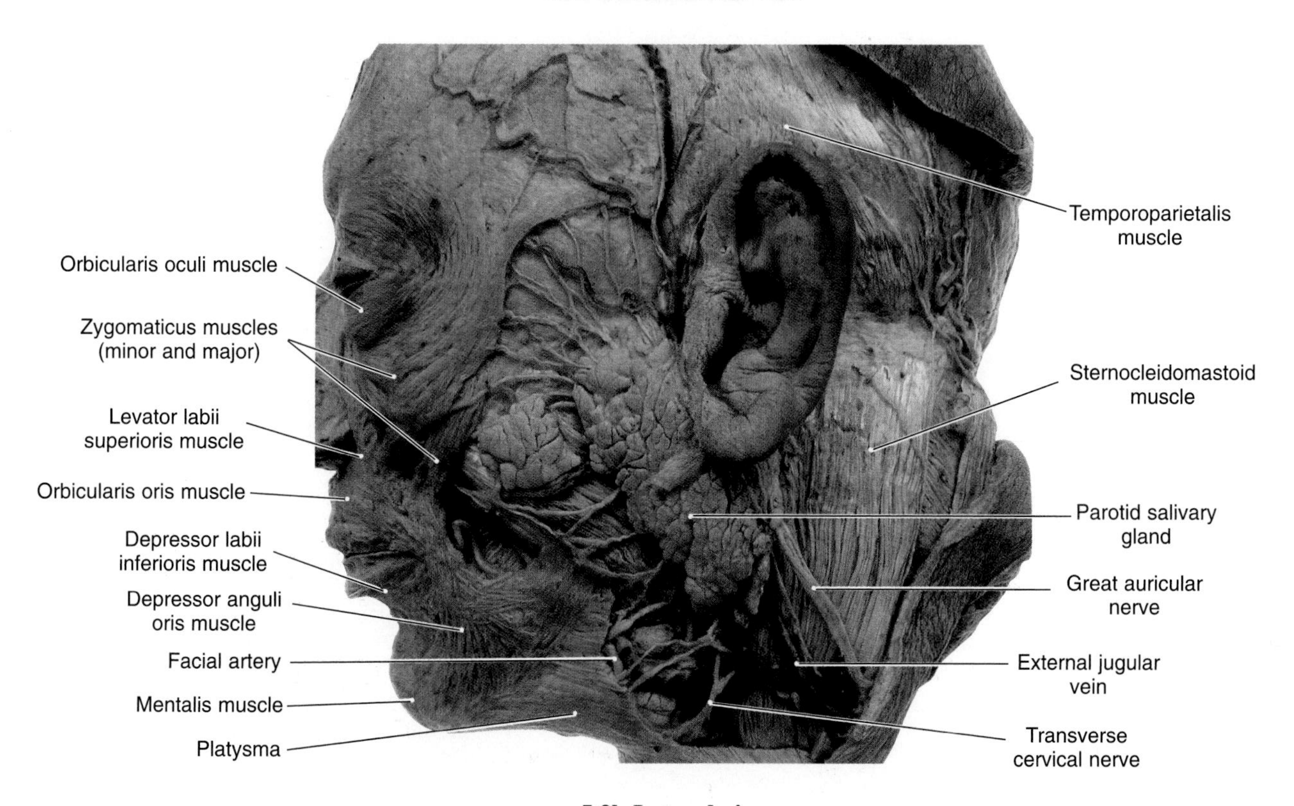
Temporoparietalis muscle
Orbicularis oculi muscle
Zygomaticus muscles (minor and major)
Sternocleidomastoid muscle
Levator labii superioris muscle
Orbicularis oris muscle
Parotid salivary gland
Depressor labii inferioris muscle
Great auricular nerve
Depressor anguli oris muscle
Facial artery
External jugular vein
Mentalis muscle
Platysma
Transverse cervical nerve

5.2b Lateral view

5.3 Dissection of the Neck

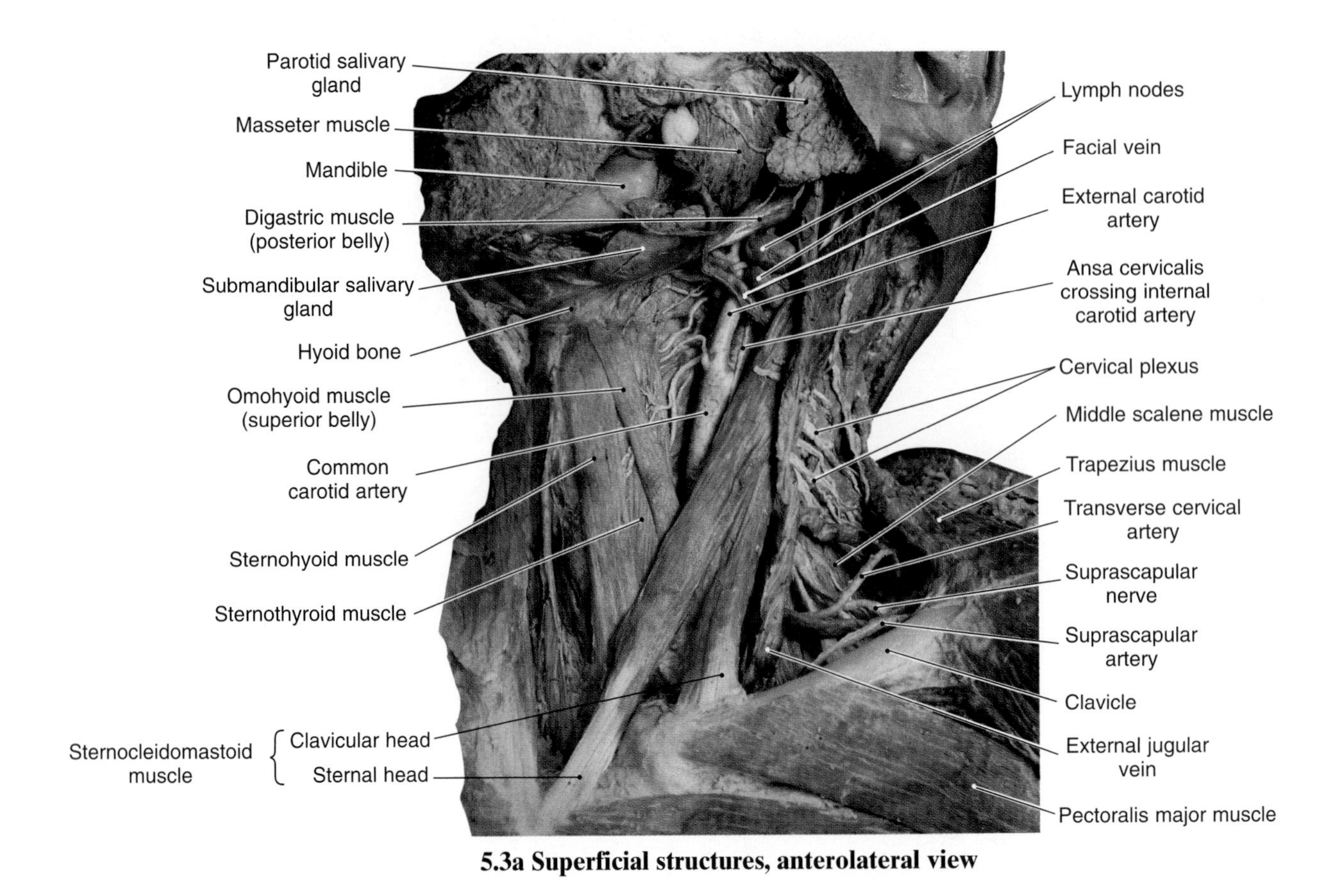

Parotid salivary gland
Masseter muscle
Mandible
Digastric muscle (posterior belly)
Submandibular salivary gland
Hyoid bone
Omohyoid muscle (superior belly)
Common carotid artery
Sternohyoid muscle
Sternothyroid muscle
Sternocleidomastoid muscle
Clavicular head
Sternal head
Lymph nodes
Facial vein
External carotid artery
Ansa cervicalis crossing internal carotid artery
Cervical plexus
Middle scalene muscle
Trapezius muscle
Transverse cervical artery
Suprascapular nerve
Suprascapular artery
Clavicle
External jugular vein
Pectoralis major muscle

5.3a Superficial structures, anterolateral view

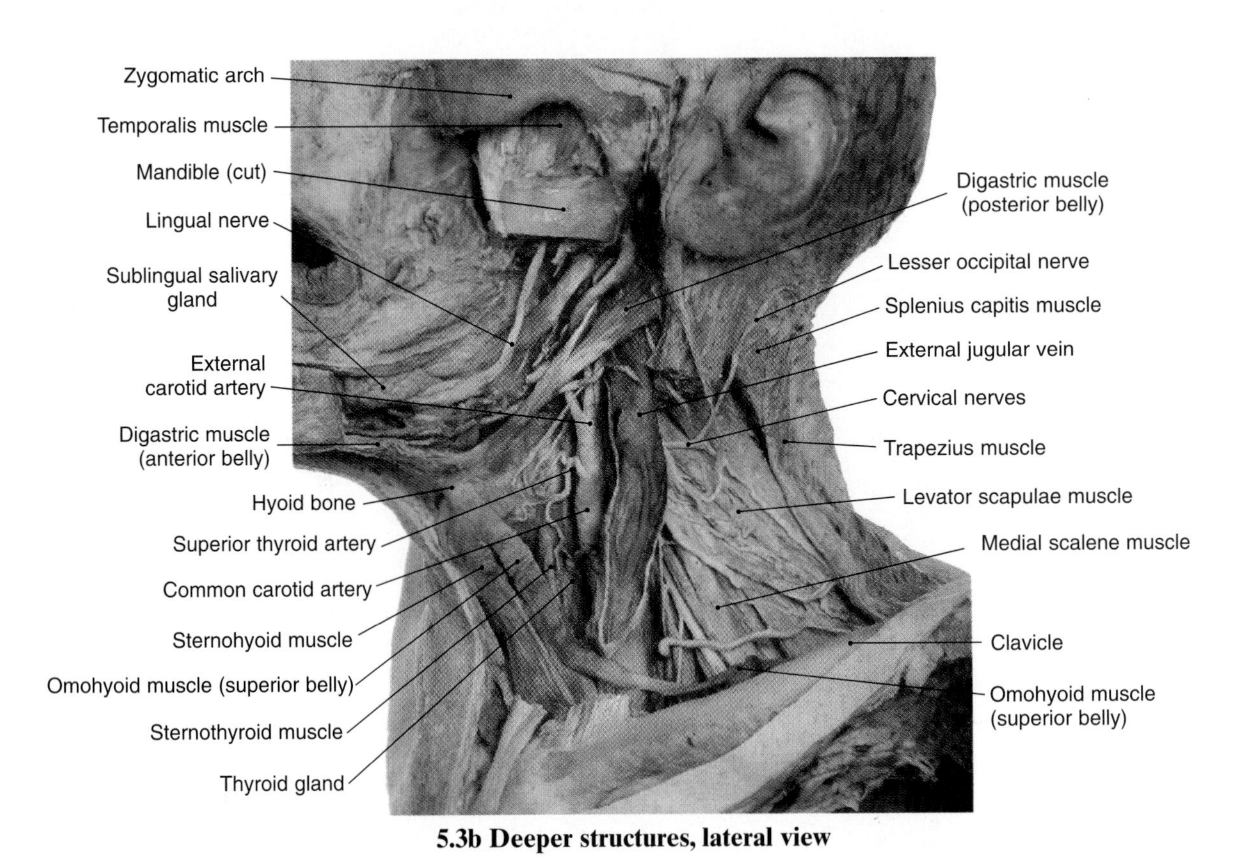

Zygomatic arch
Temporalis muscle
Mandible (cut)
Lingual nerve
Sublingual salivary gland
External carotid artery
Digastric muscle (anterior belly)
Hyoid bone
Superior thyroid artery
Common carotid artery
Sternohyoid muscle
Omohyoid muscle (superior belly)
Sternothyroid muscle
Thyroid gland
Digastric muscle (posterior belly)
Lesser occipital nerve
Splenius capitis muscle
External jugular vein
Cervical nerves
Trapezius muscle
Levator scapulae muscle
Medial scalene muscle
Clavicle
Omohyoid muscle (superior belly)

5.3b Deeper structures, lateral view

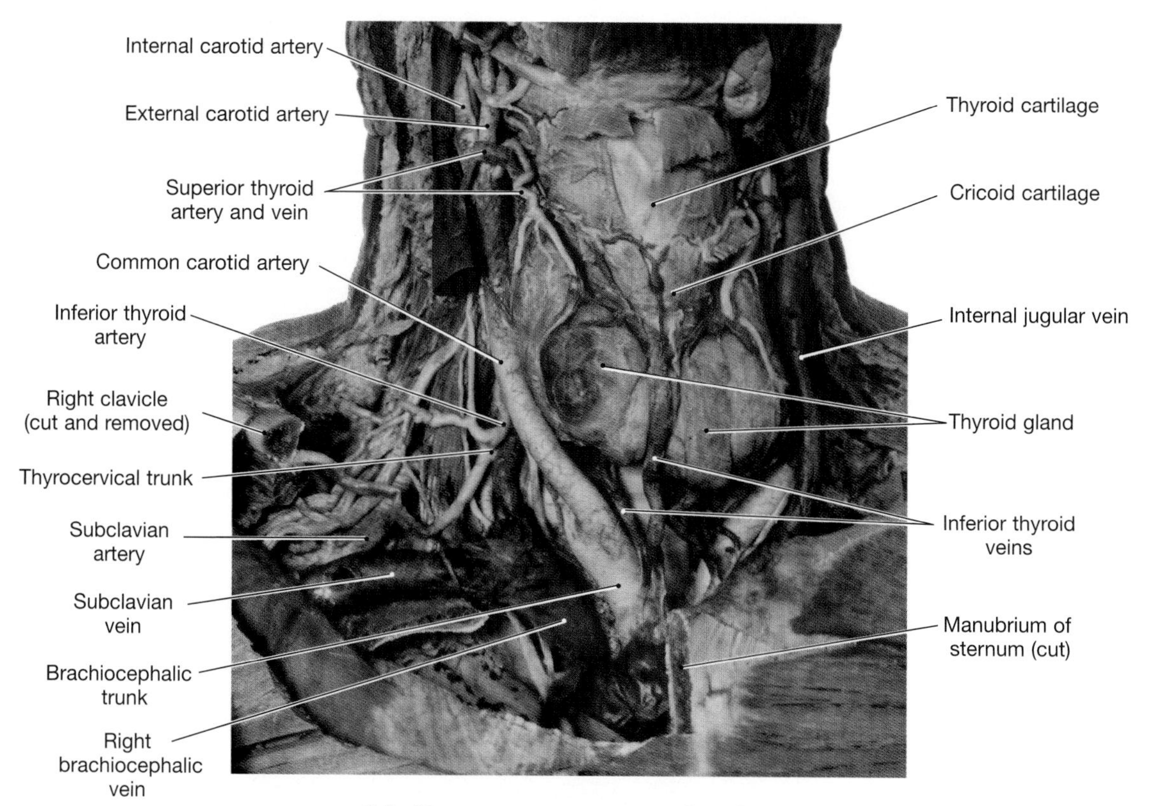

5.3c Deeper structures, anterior view

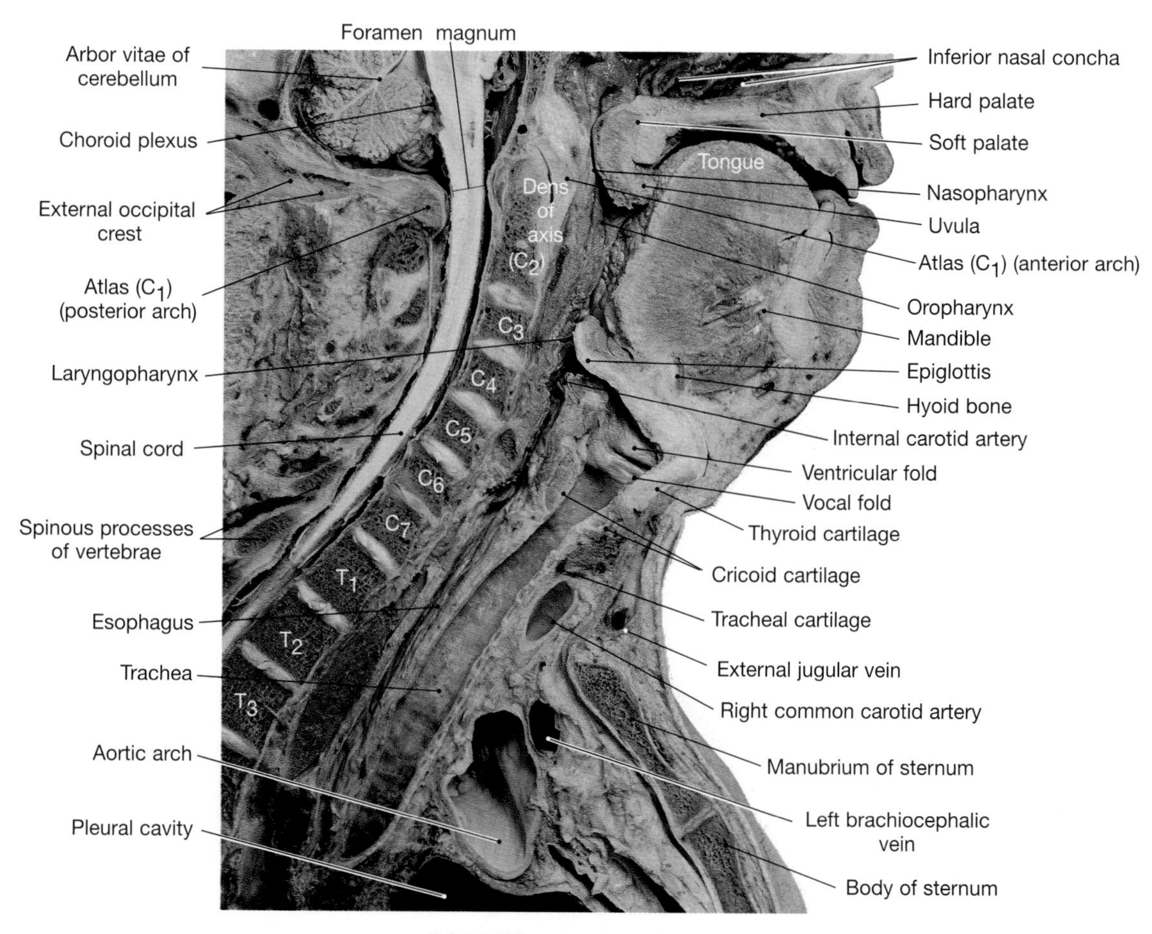

5.3d Midsagittal section

PLATE 6 THE UPPER LIMBS
6.1 Surface Anatomy

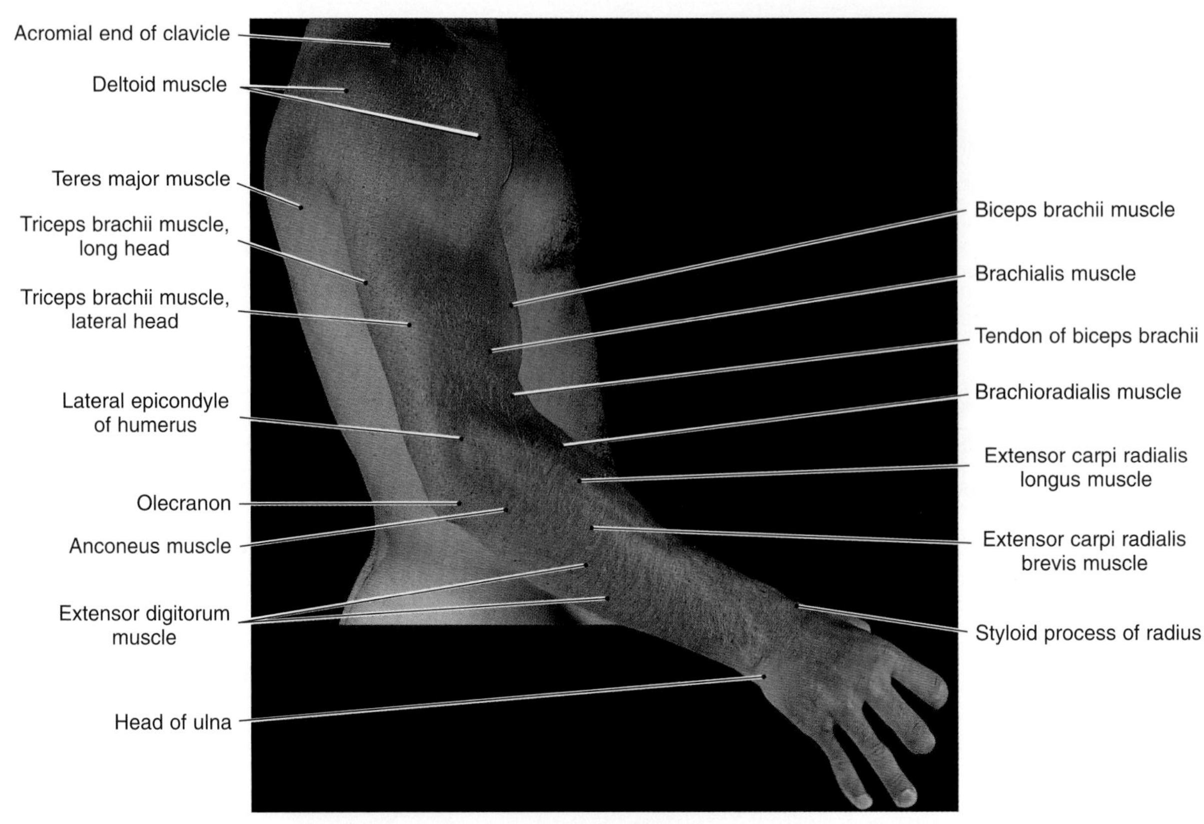

Acromial end of clavicle
Deltoid muscle
Teres major muscle
Triceps brachii muscle, long head
Triceps brachii muscle, lateral head
Lateral epicondyle of humerus
Olecranon
Anconeus muscle
Extensor digitorum muscle
Head of ulna
Biceps brachii muscle
Brachialis muscle
Tendon of biceps brachii
Brachioradialis muscle
Extensor carpi radialis longus muscle
Extensor carpi radialis brevis muscle
Styloid process of radius

6.1a Lateral view

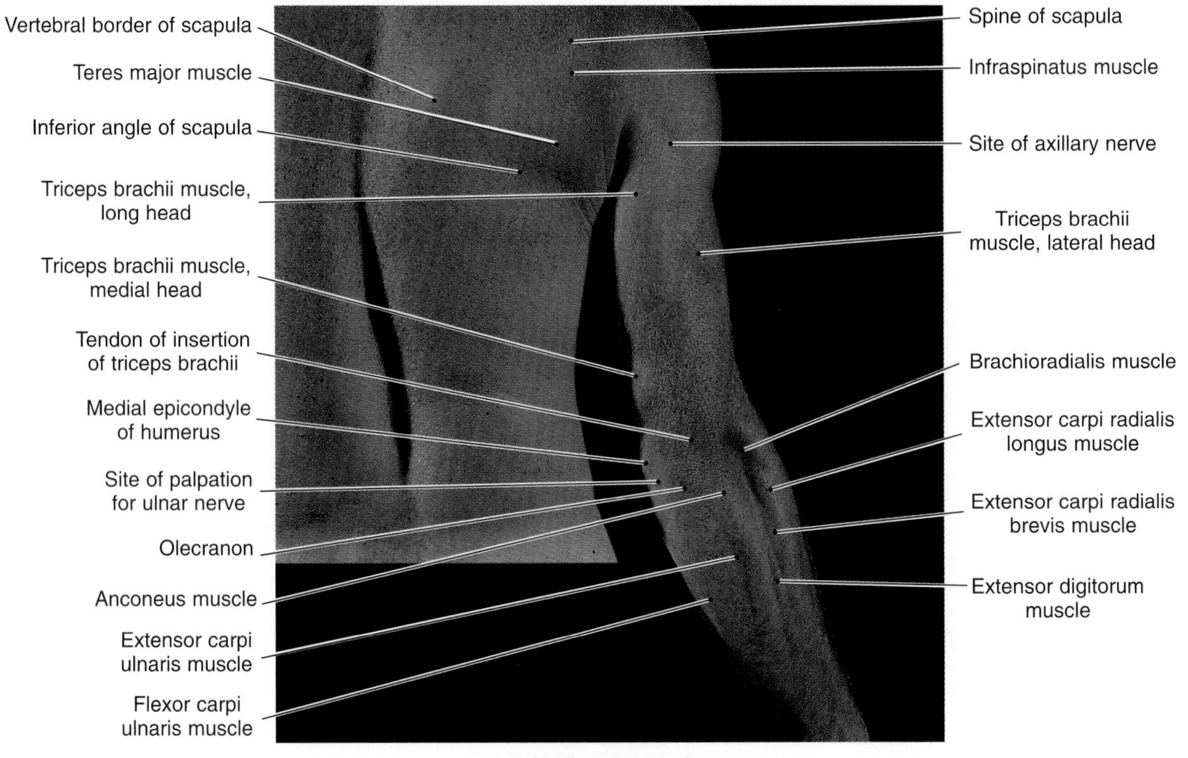

Vertebral border of scapula
Teres major muscle
Inferior angle of scapula
Triceps brachii muscle, long head
Triceps brachii muscle, medial head
Tendon of insertion of triceps brachii
Medial epicondyle of humerus
Site of palpation for ulnar nerve
Olecranon
Anconeus muscle
Extensor carpi ulnaris muscle
Flexor carpi ulnaris muscle
Spine of scapula
Infraspinatus muscle
Site of axillary nerve
Triceps brachii muscle, lateral head
Brachioradialis muscle
Extensor carpi radialis longus muscle
Extensor carpi radialis brevis muscle
Extensor digitorum muscle

6.1b Posterior view

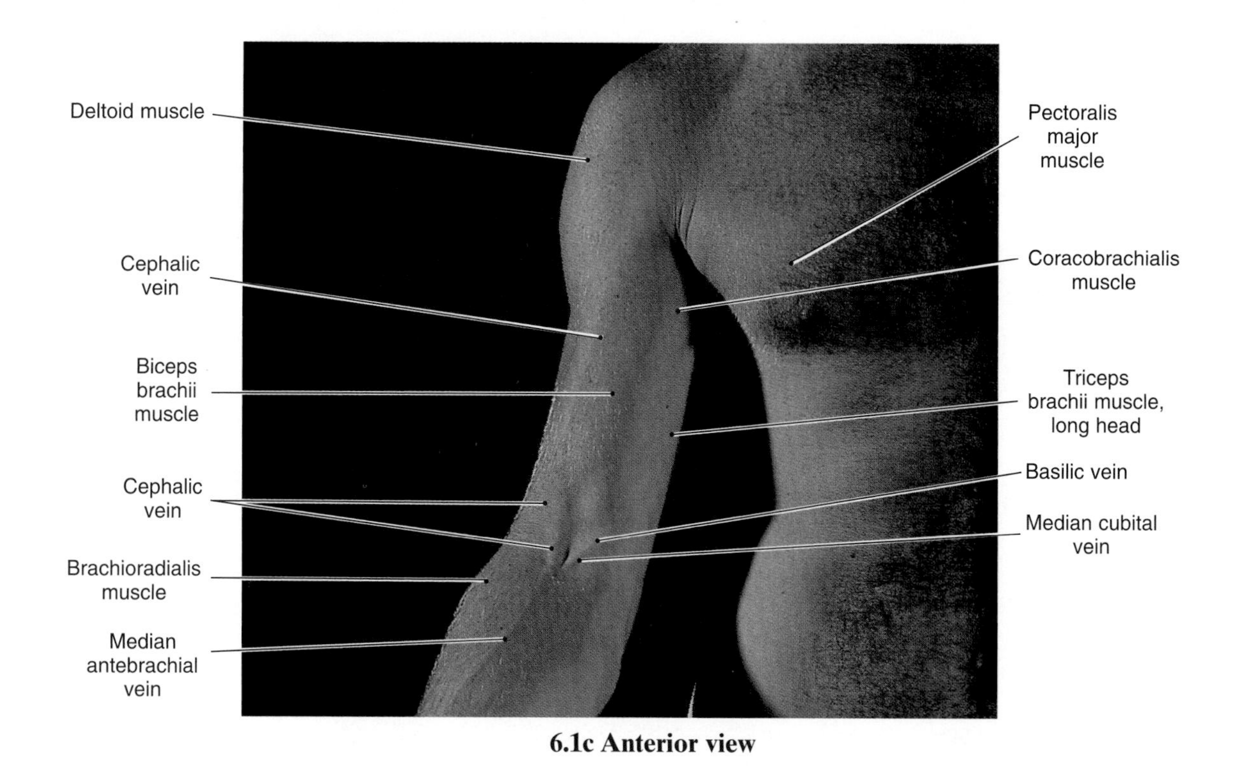
Deltoid muscle
Cephalic
vein
Biceps
brachii
muscle
Cephalic
vein
Brachioradialis
muscle
Median
antebrachial
vein
Pectoralis
major
muscle
Coracobrachialis
muscle
Triceps
brachii muscle,
long head
Basilic vein
Median cubital
vein

6.1c Anterior view

6.2 Nerves and Blood Vessels

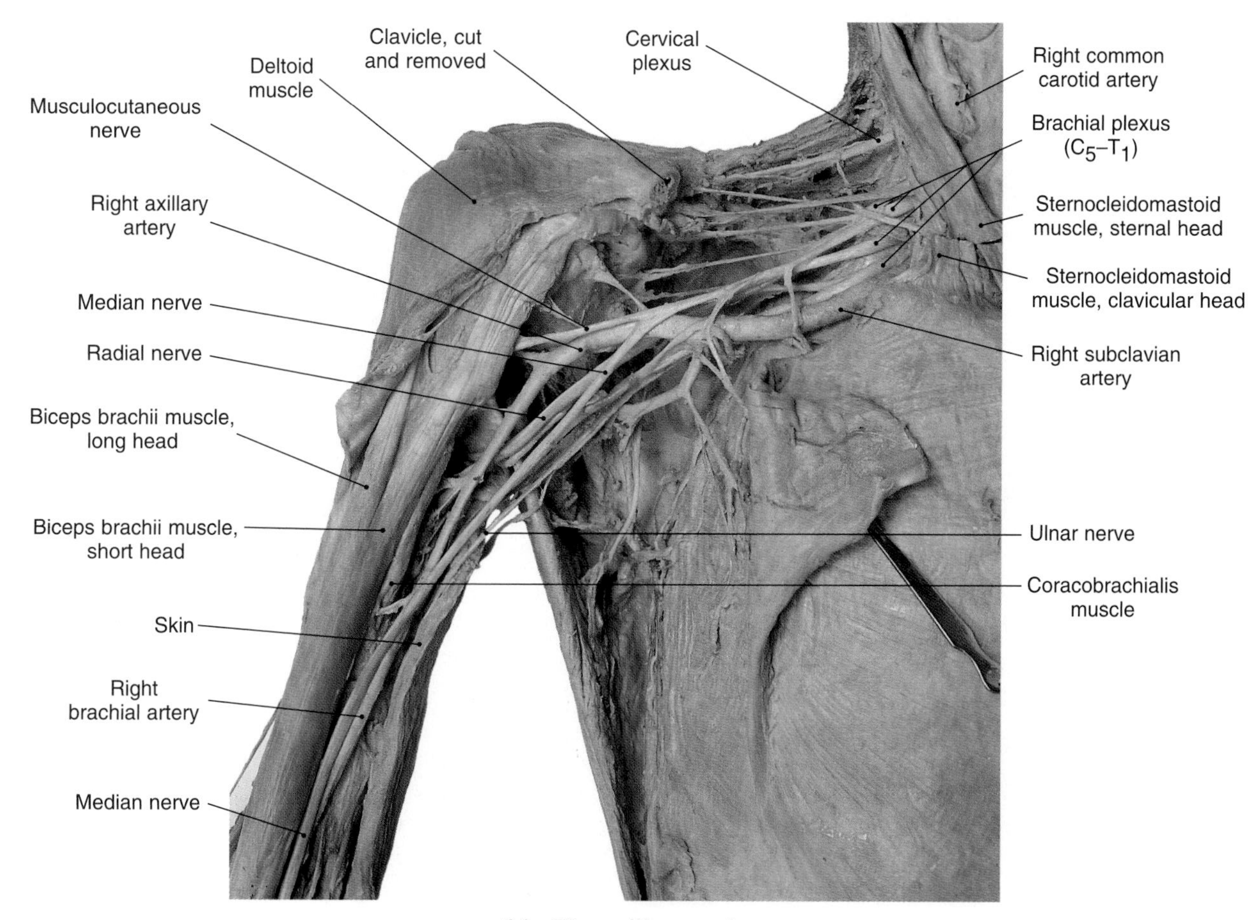
Deltoid
muscle
Clavicle, cut
and removed
Cervical
plexus
Musculocutaneous
nerve
Right axillary
artery
Median nerve
Radial nerve
Biceps brachii muscle,
long head
Biceps brachii muscle,
short head
Skin
Right
brachial artery
Median nerve
Right common
carotid artery
Brachial plexus
(C_5–T_1)
Sternocleidomastoid
muscle, sternal head
Sternocleidomastoid
muscle, clavicular head
Right subclavian
artery
Ulnar nerve
Coracobrachialis
muscle

6.2a The axillary region

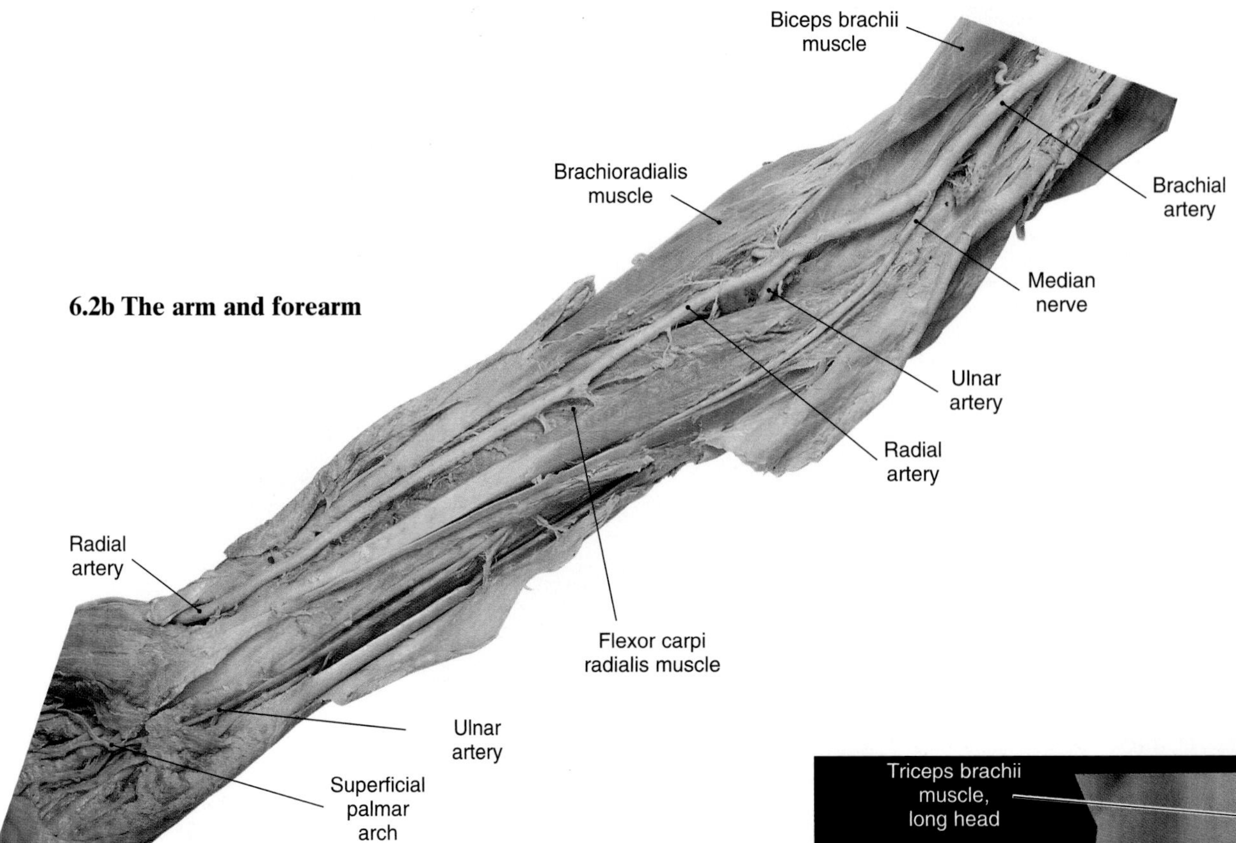

6.2b The arm and forearm

6.3 The Right Upper Limb, Anterior View

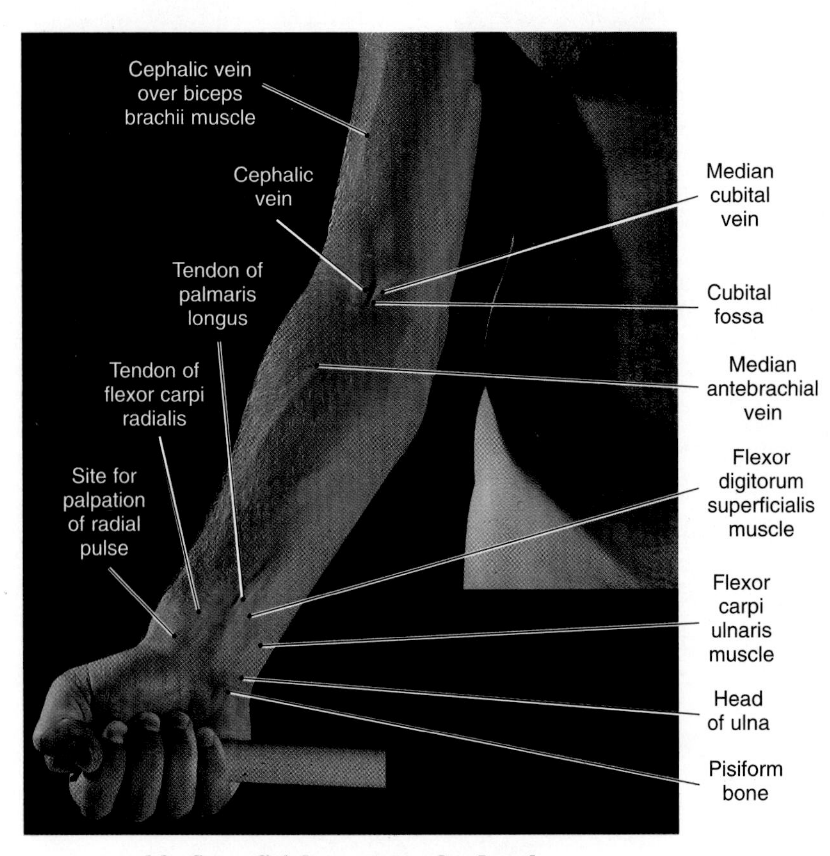

6.3a Superficial structures, landmarks

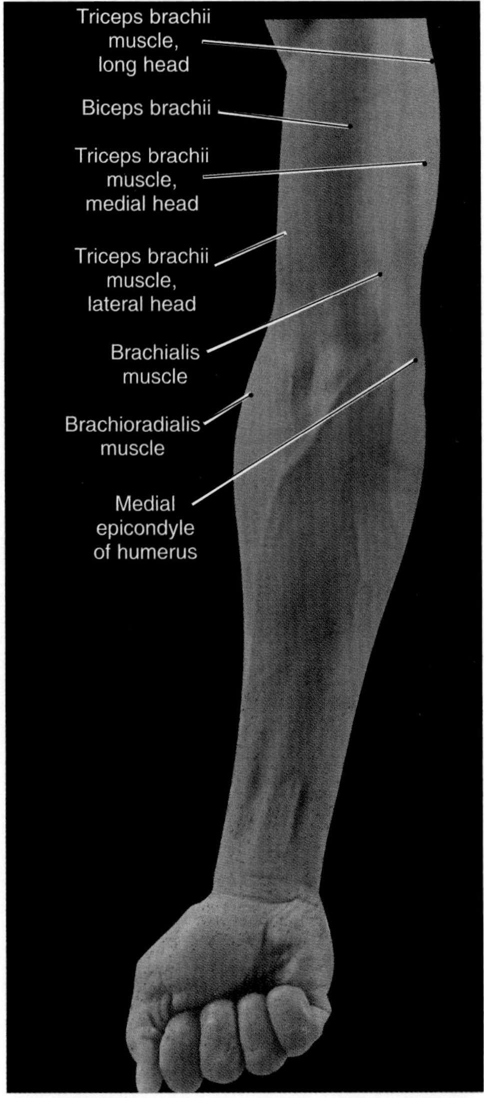

6.3b Superficial structures, muscles

6.4 The Right Upper Limb, Posterior View

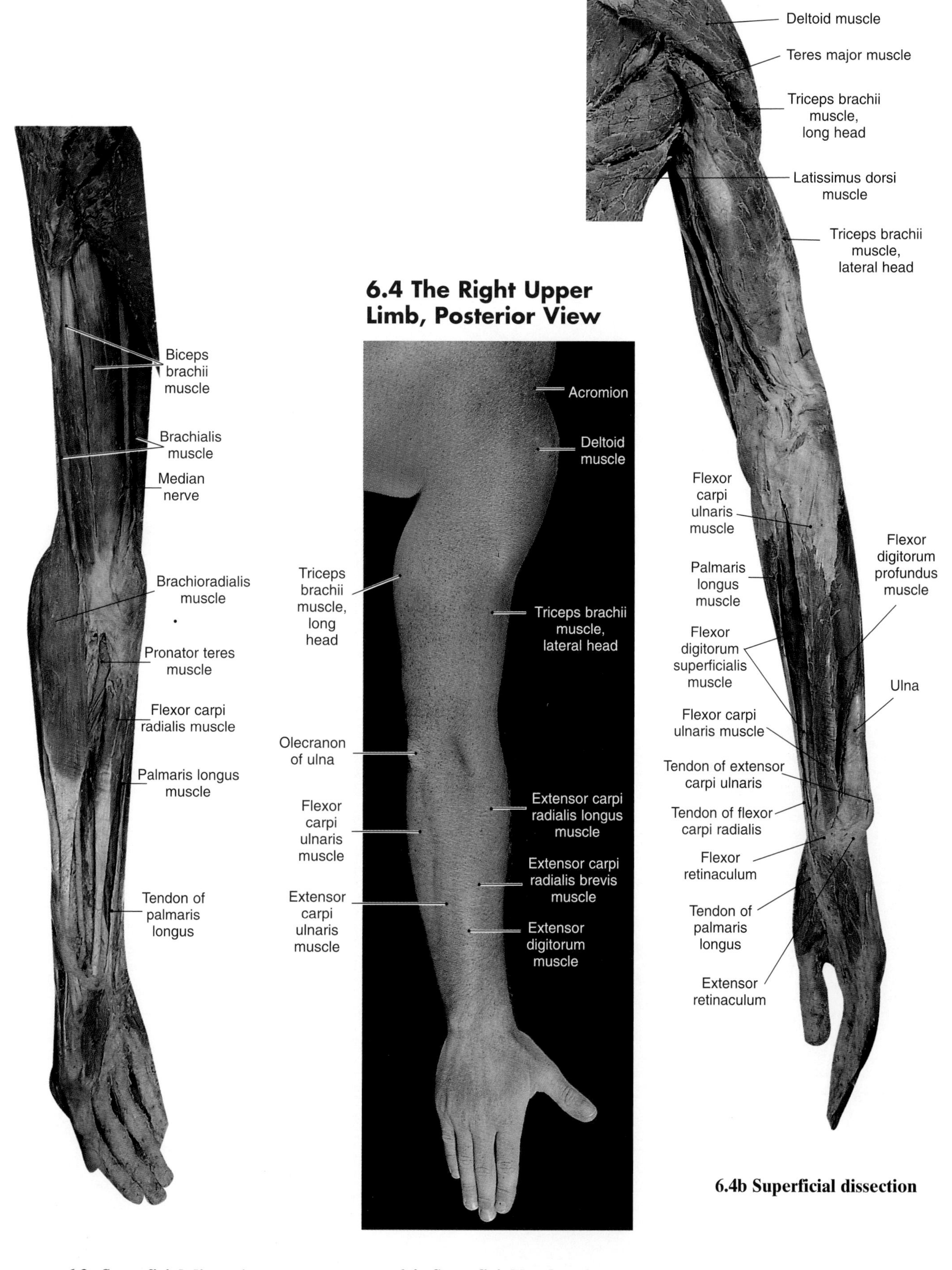

6.4b Superficial dissection

6.3c Superficial dissection

6.4a Superficial landmarks

6.6 The Right Hand

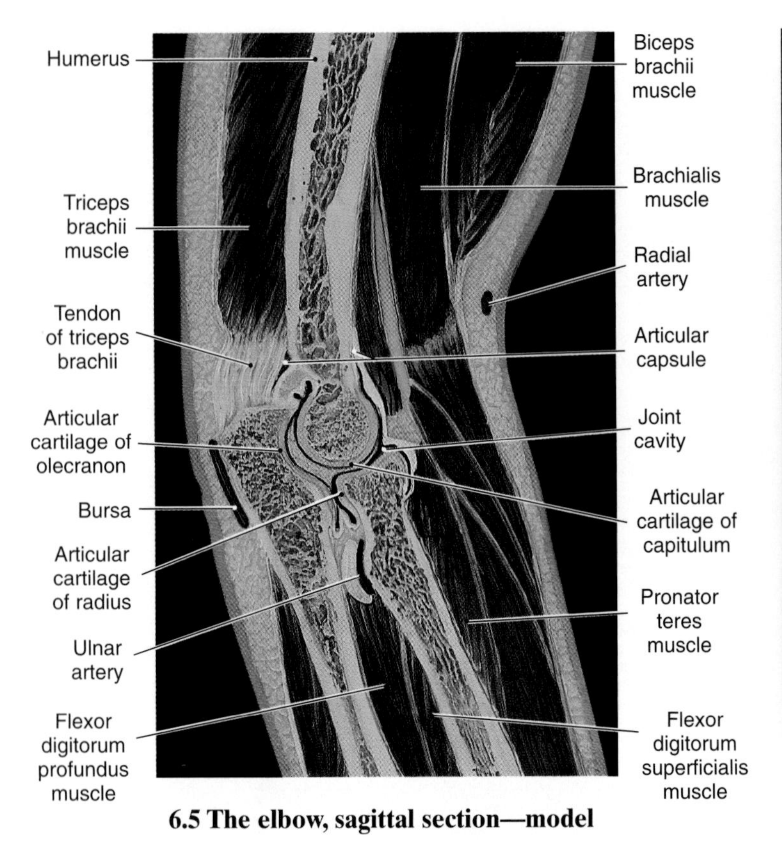
Humerus
Triceps brachii muscle
Tendon of triceps brachii
Articular cartilage of olecranon
Bursa
Articular cartilage of radius
Ulnar artery
Flexor digitorum profundus muscle
Biceps brachii muscle
Brachialis muscle
Radial artery
Articular capsule
Joint cavity
Articular cartilage of capitulum
Pronator teres muscle
Flexor digitorum superficialis muscle

6.5 The elbow, sagittal section—model

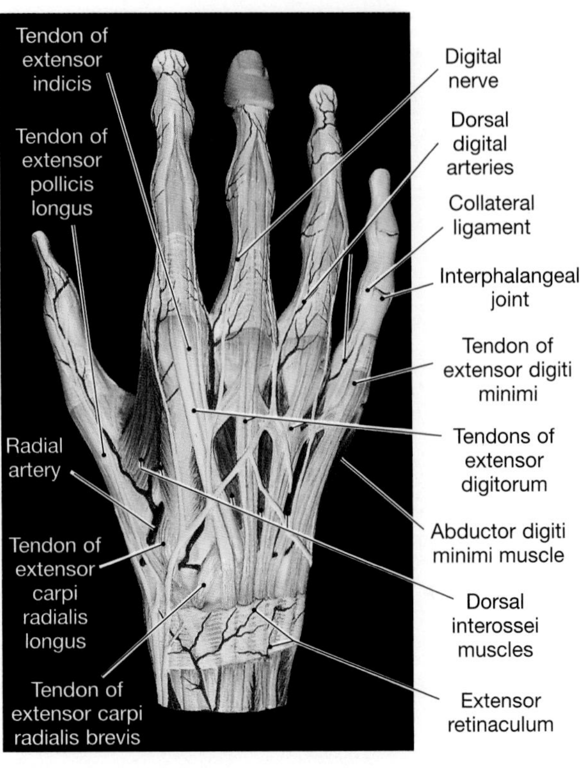
Tendon of extensor indicis
Tendon of extensor pollicis longus
Radial artery
Tendon of extensor carpi radialis longus
Tendon of extensor carpi radialis brevis
Digital nerve
Dorsal digital arteries
Collateral ligament
Interphalangeal joint
Tendon of extensor digiti minimi
Tendons of extensor digitorum
Abductor digiti minimi muscle
Dorsal interossei muscles
Extensor retinaculum

6.6a Posterior view—model

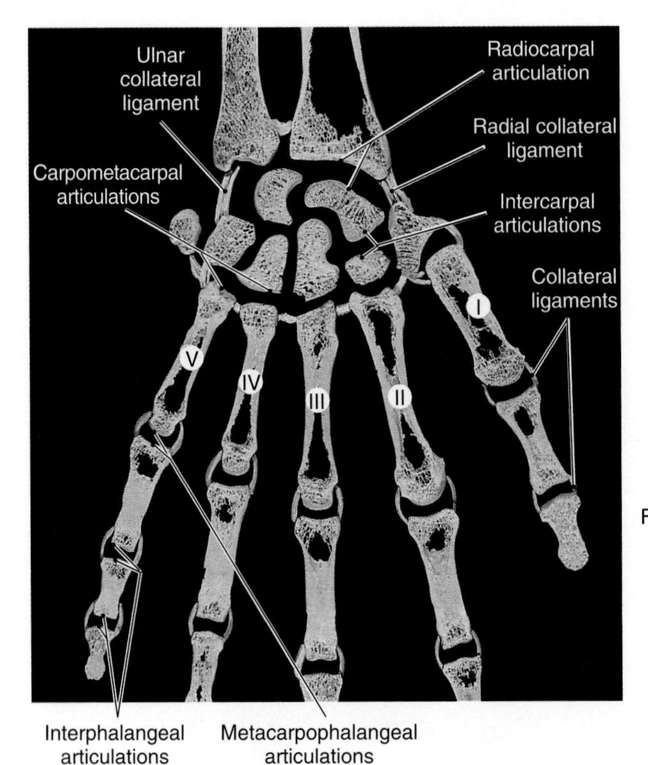
Ulnar collateral ligament
Carpometacarpal articulations
Radiocarpal articulation
Radial collateral ligament
Intercarpal articulations
Collateral ligaments
I
II
III
IV
V
Interphalangeal articulations
Metacarpophalangeal articulations

6.6b Joints of the wrist and hand, coronal section

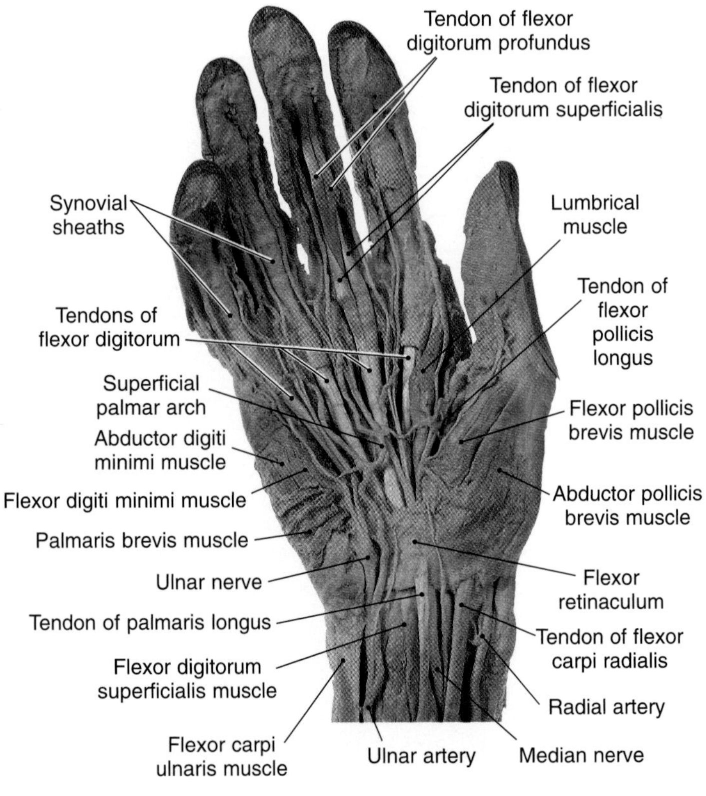
Tendon of flexor digitorum profundus
Tendon of flexor digitorum superficialis
Synovial sheaths
Lumbrical muscle
Tendon of flexor pollicis longus
Tendons of flexor digitorum
Superficial palmar arch
Abductor digiti minimi muscle
Flexor digiti minimi muscle
Palmaris brevis muscle
Ulnar nerve
Tendon of palmaris longus
Flexor digitorum superficialis muscle
Flexor carpi ulnaris muscle
Ulnar artery
Median nerve
Radial artery
Tendon of flexor carpi radialis
Flexor retinaculum
Abductor pollicis brevis muscle
Flexor pollicis brevis muscle

6.6c Superficial dissection, anterior view

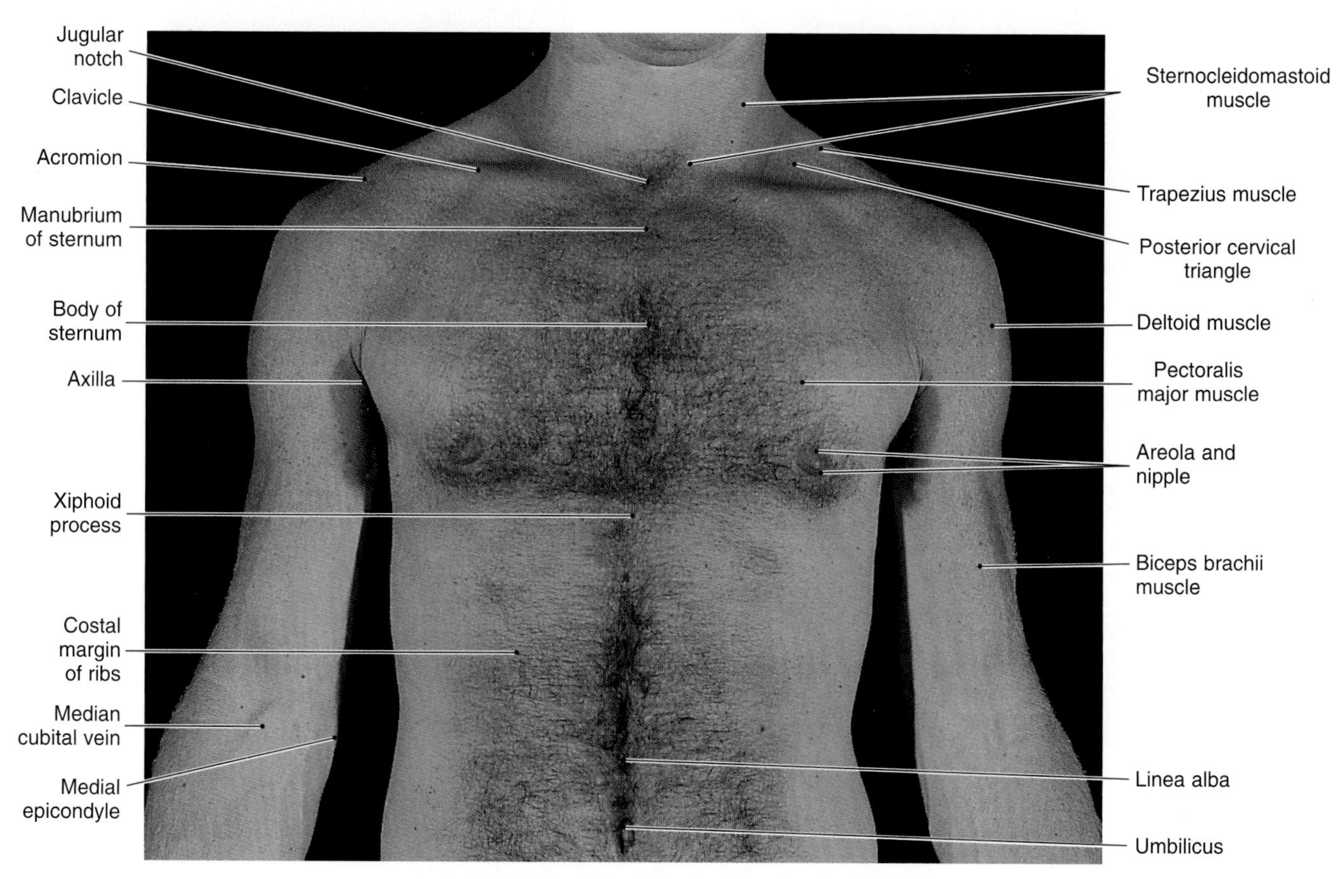

7.1 Surface anatomy of the thorax, anterior view

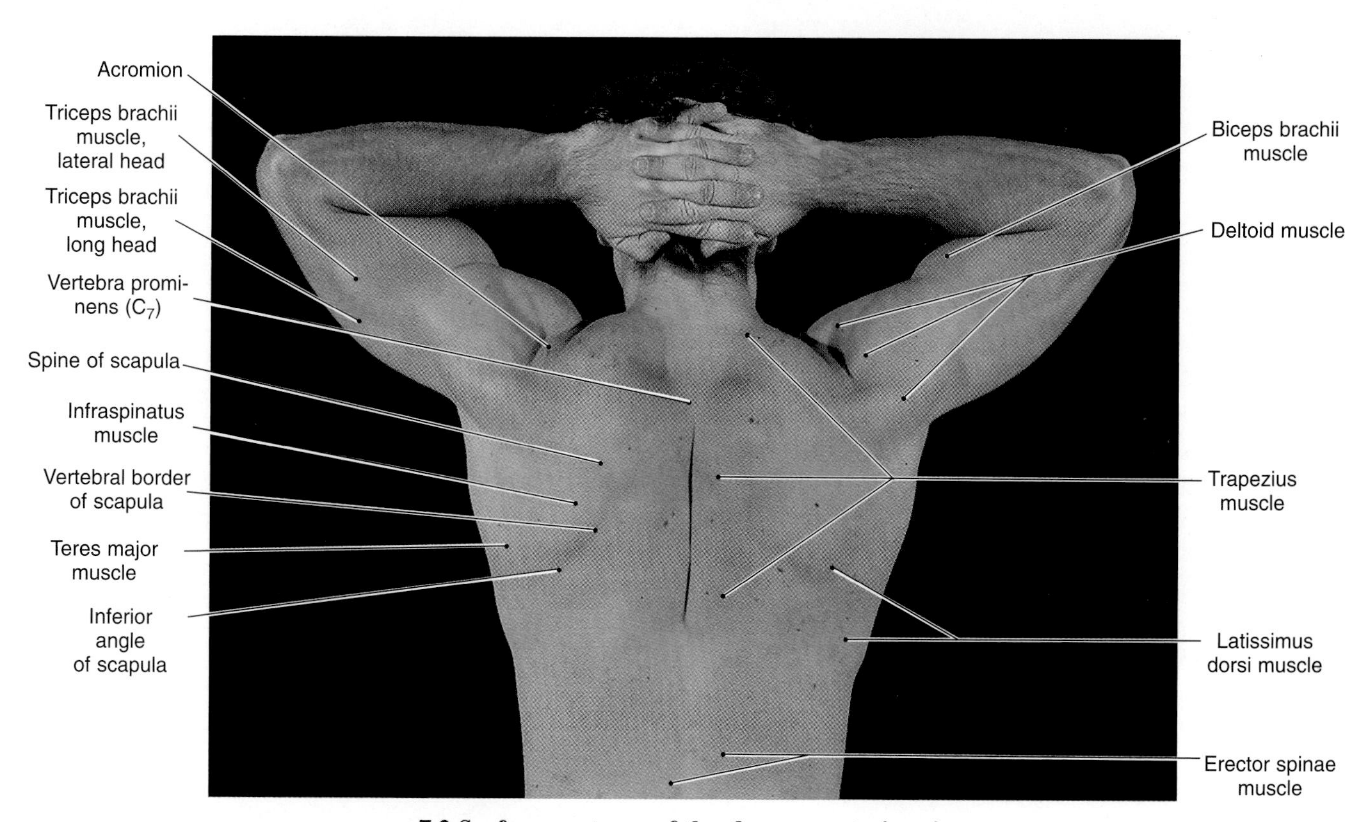

7.2 Surface anatomy of the thorax, posterior view

7.3 Surface Anatomy of the Abdomen

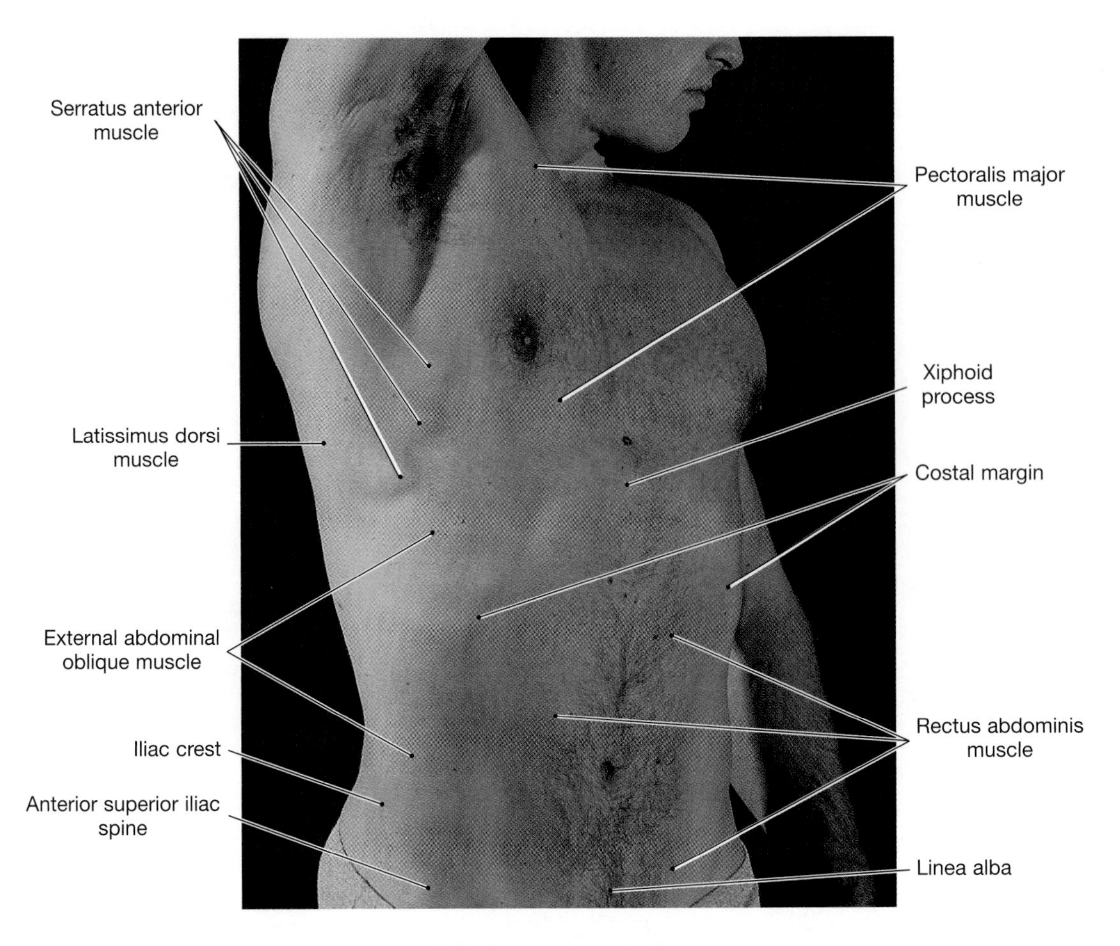

7.3a Anterolateral view

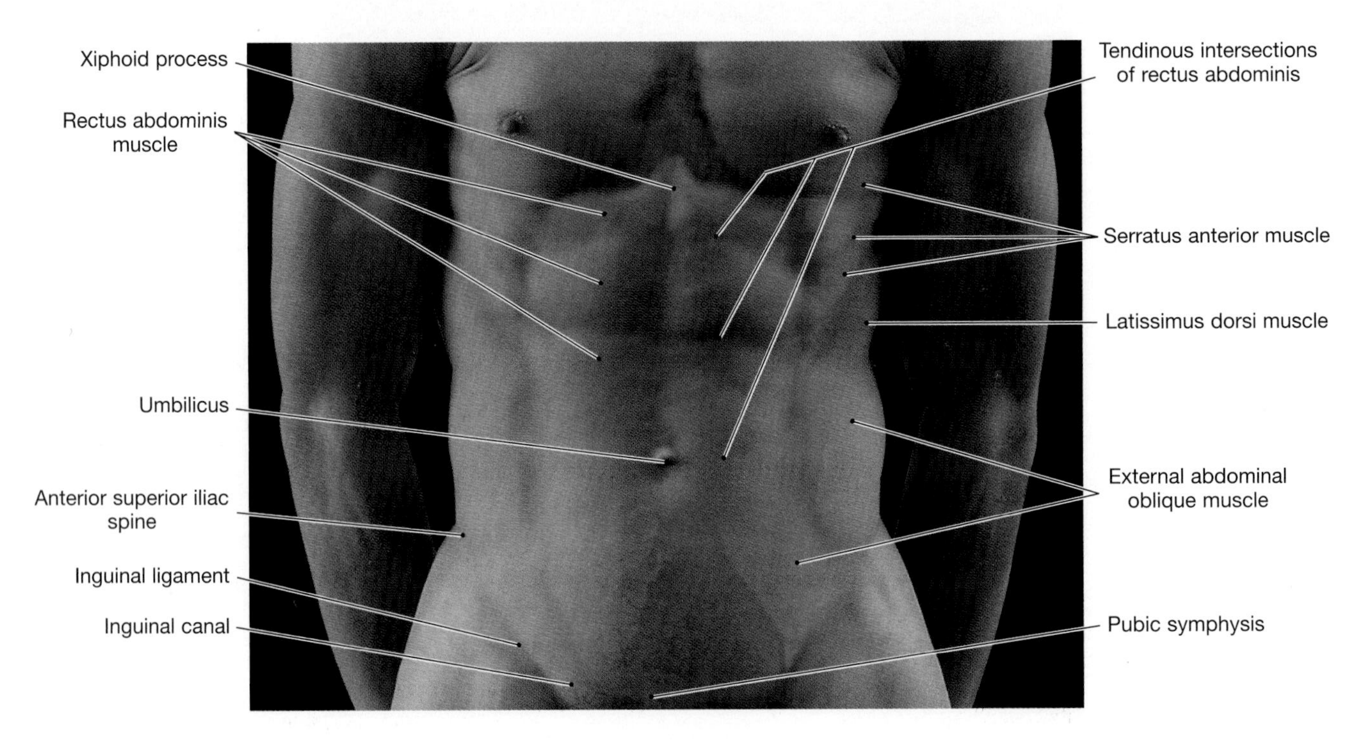

7.3b Anterior view

7.4 The Heart and Lungs

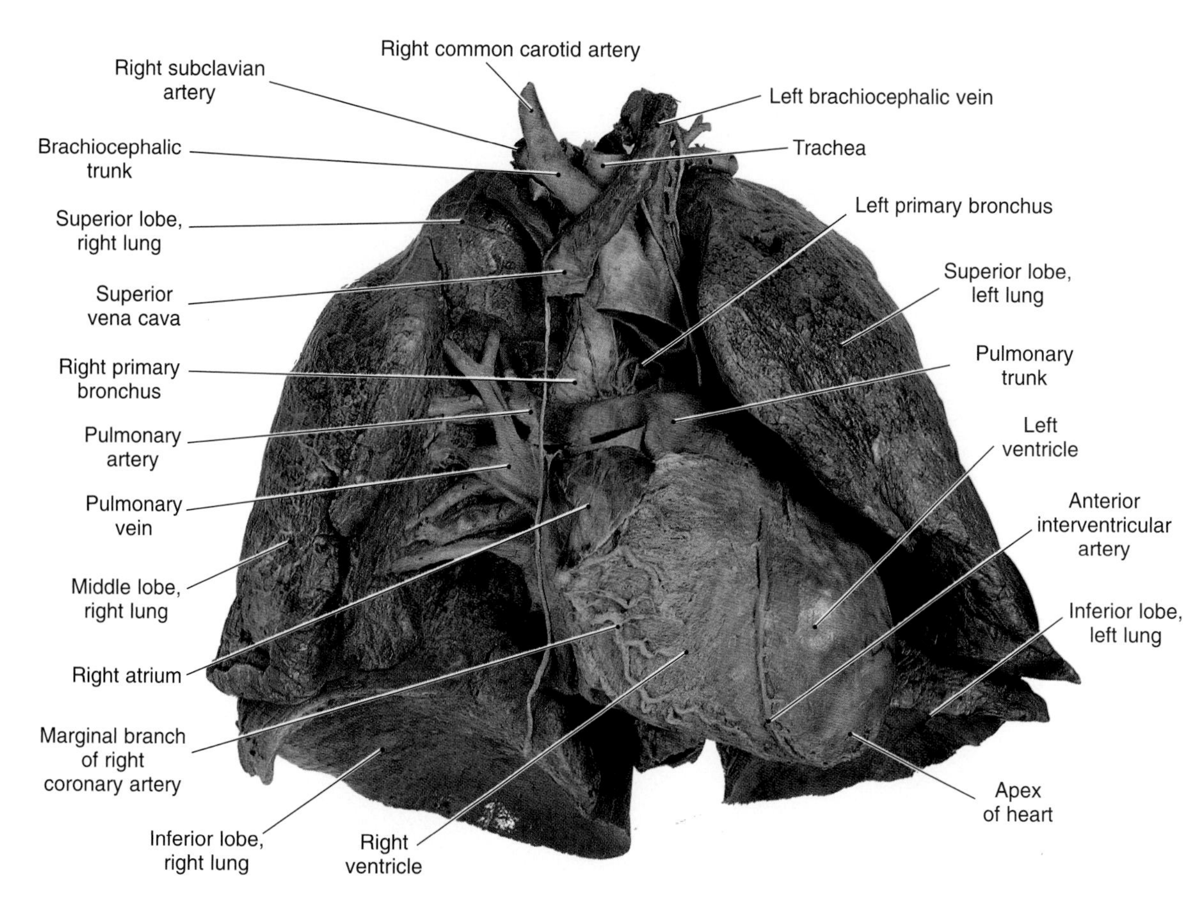

7.4a Anterior view

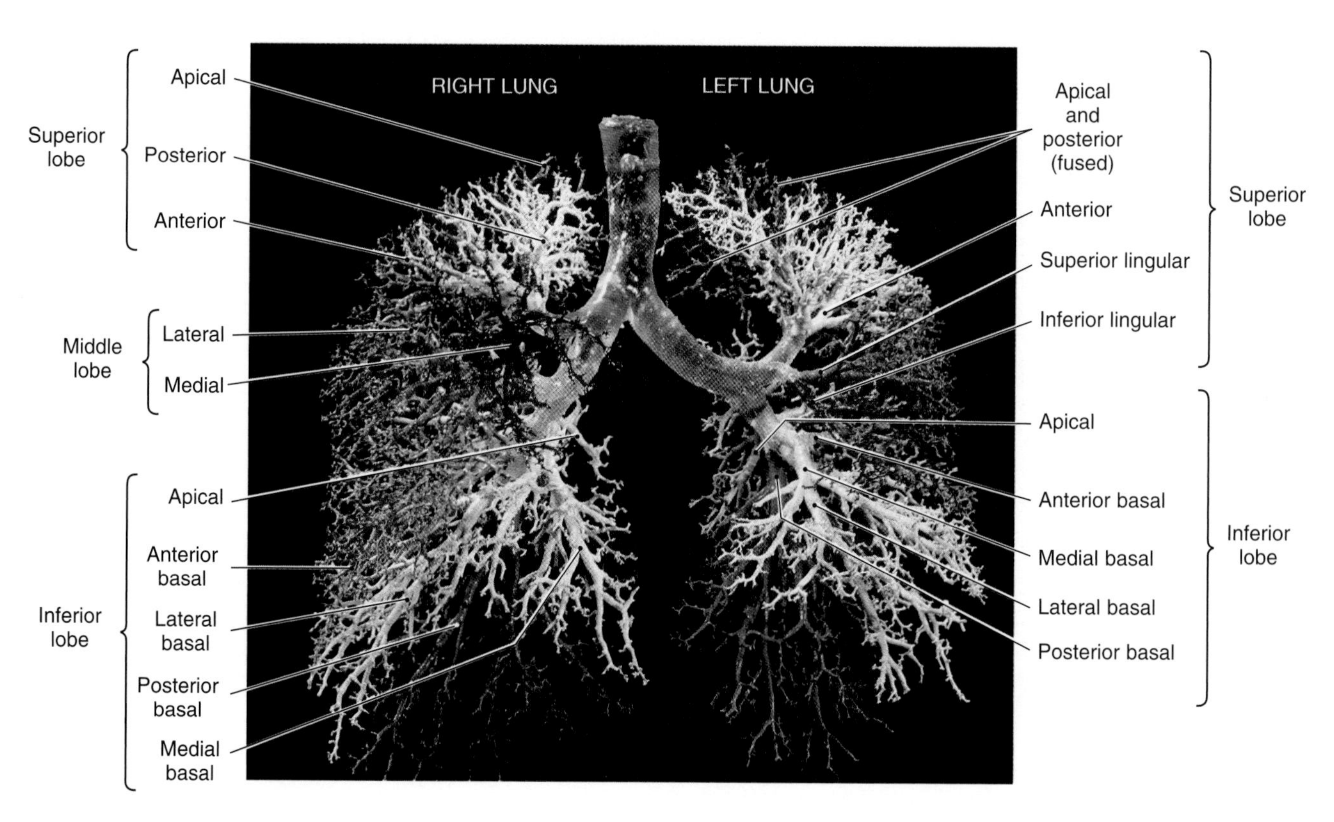

7.4b Color-coded corrosion cast of the bronchial tree, anterior view

7.5 The Lungs

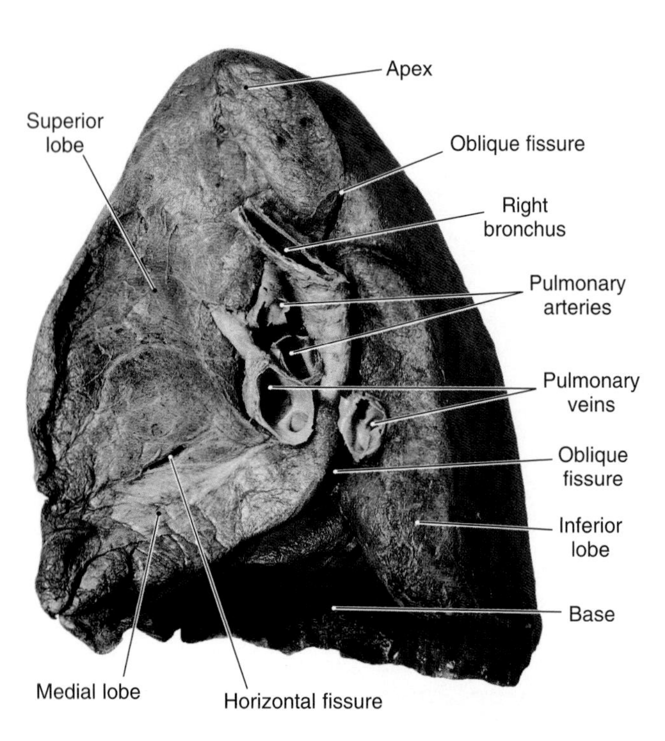

7.5a Medial surface of right lung

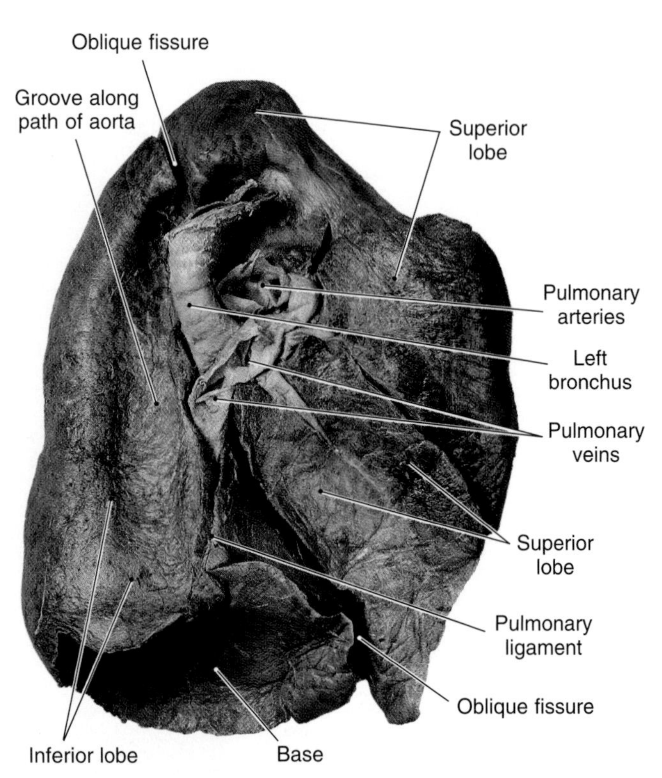

7.5b Medial surface of left lung

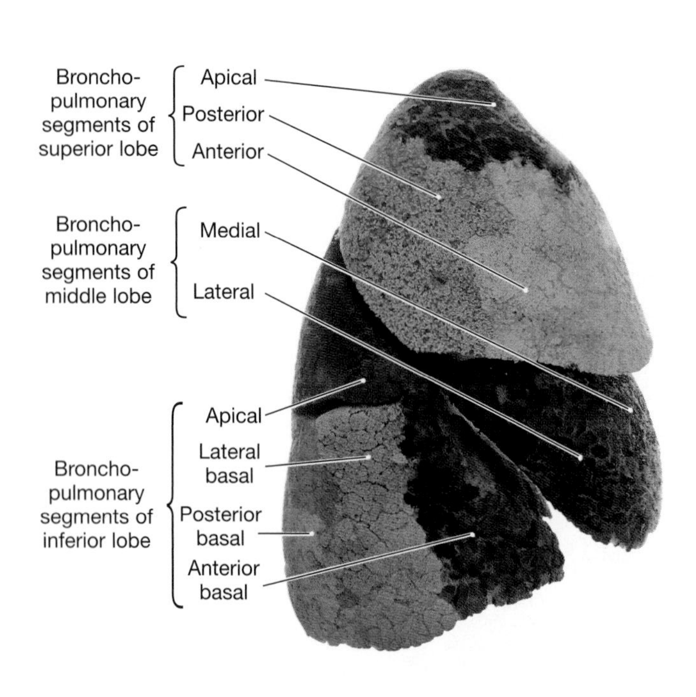

7.5c Lateral surface of right lung, color-coded for bronchopulmonary segments

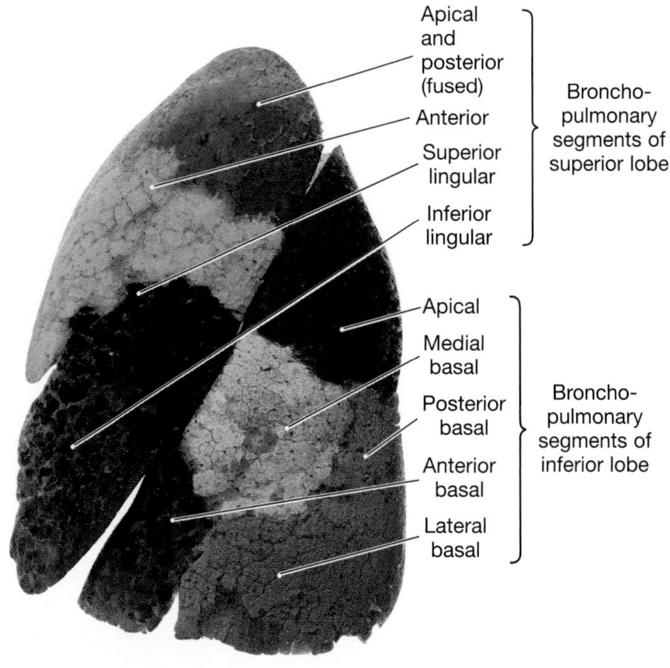

7.5d Lateral surface of left lung, color-coded for bronchopulmonary segments

7.6 The Thoracic and Abdominopelvic Viscera—Models

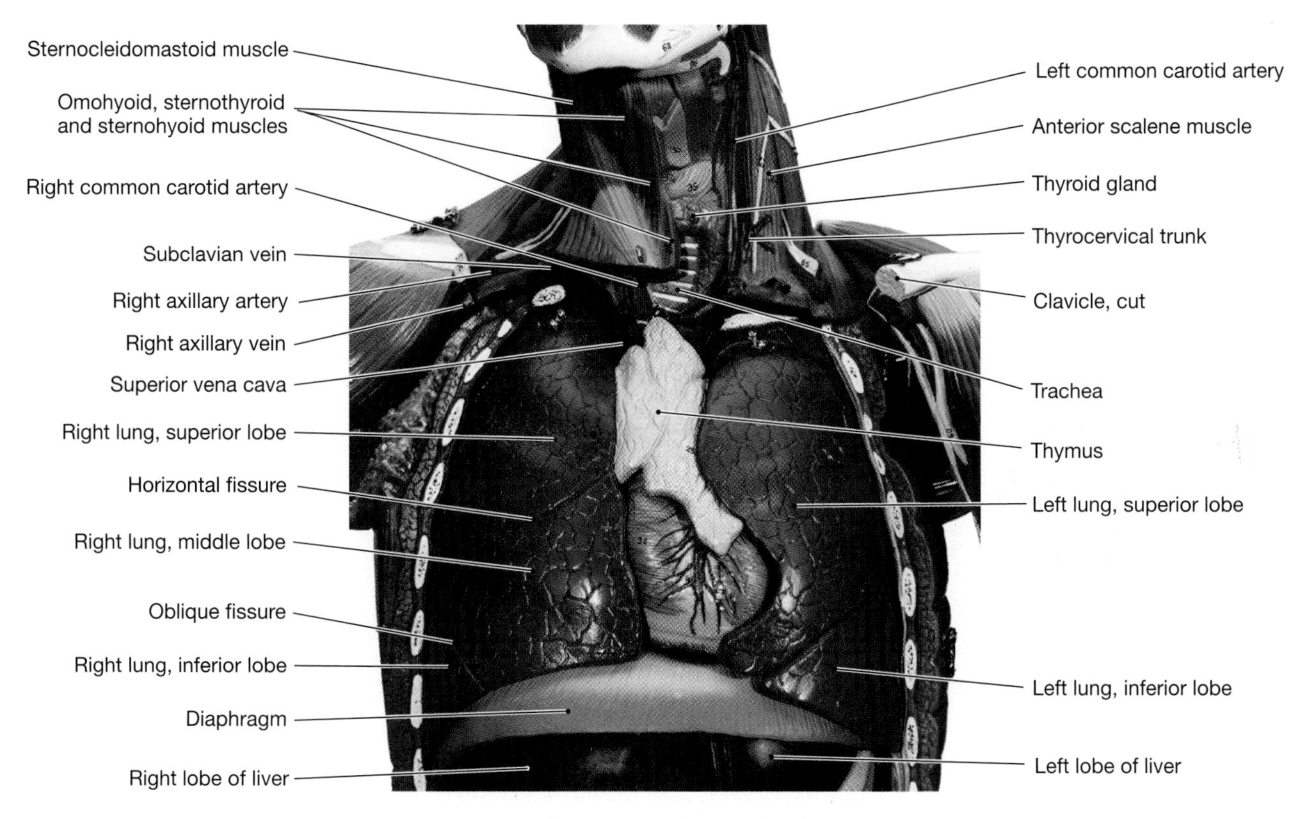

7.6a Thoracic organs, superficial view

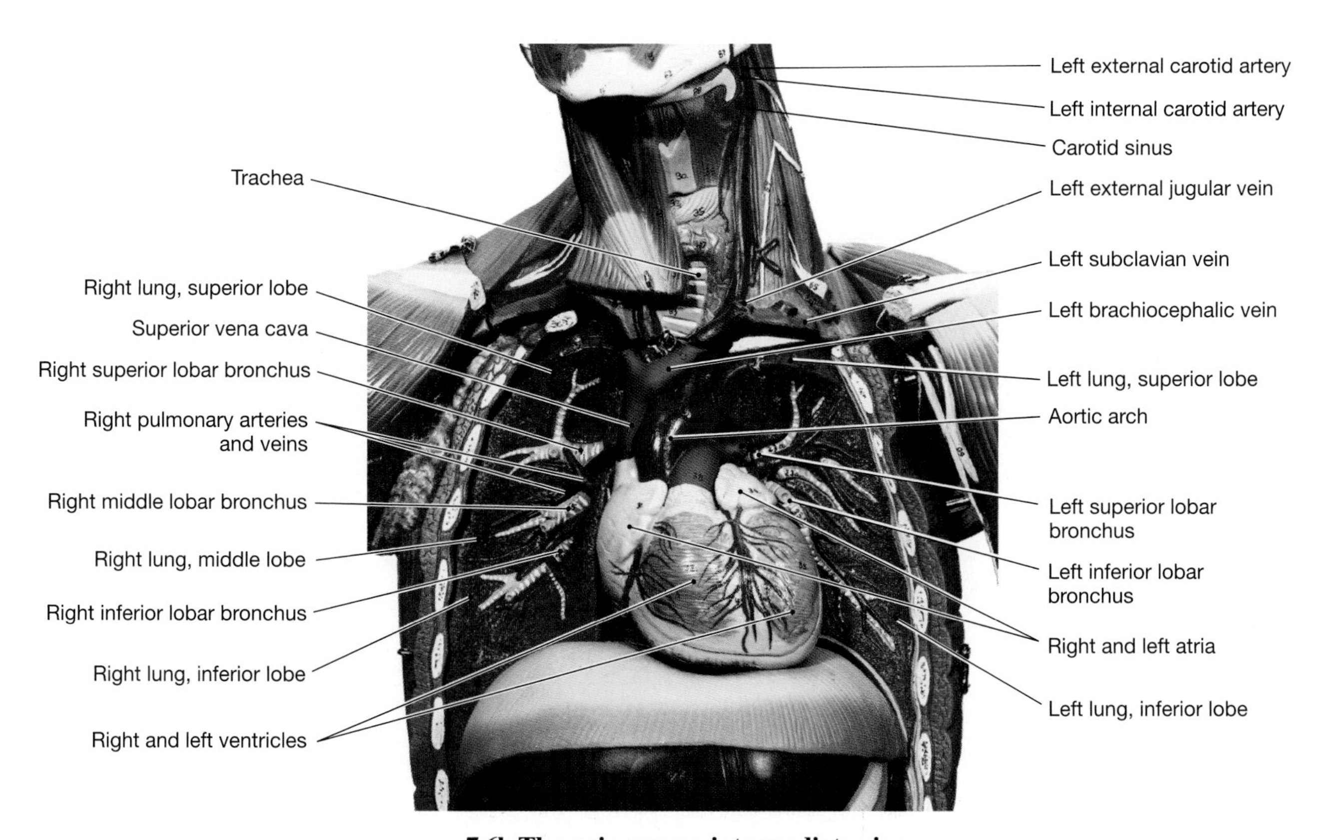

7.6b Thoracic organs, intermediate view

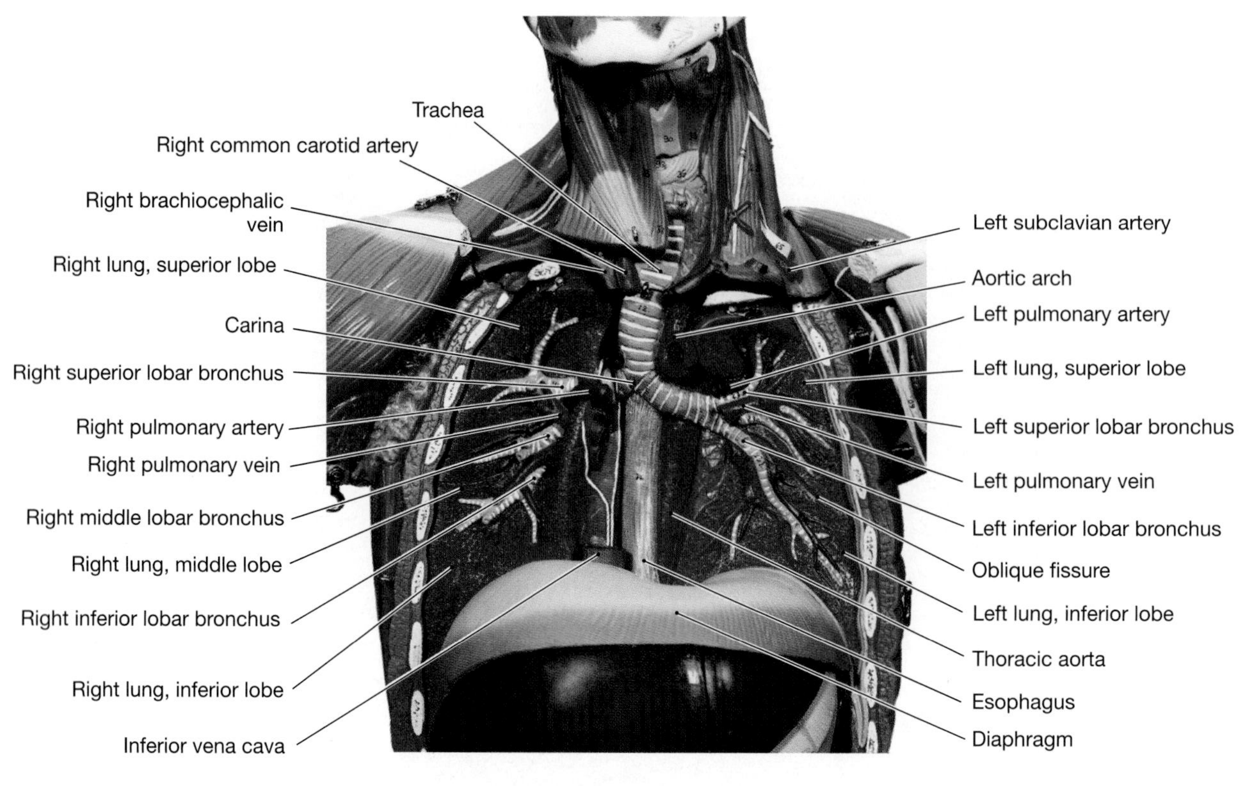

Trachea
Right common carotid artery
Right brachiocephalic vein
Right lung, superior lobe
Carina
Right superior lobar bronchus
Right pulmonary artery
Right pulmonary vein
Right middle lobar bronchus
Right lung, middle lobe
Right inferior lobar bronchus
Right lung, inferior lobe
Inferior vena cava
Left subclavian artery
Aortic arch
Left pulmonary artery
Left lung, superior lobe
Left superior lobar bronchus
Left pulmonary vein
Left inferior lobar bronchus
Oblique fissure
Left lung, inferior lobe
Thoracic aorta
Esophagus
Diaphragm

7.6c Thoracic organs, deeper view

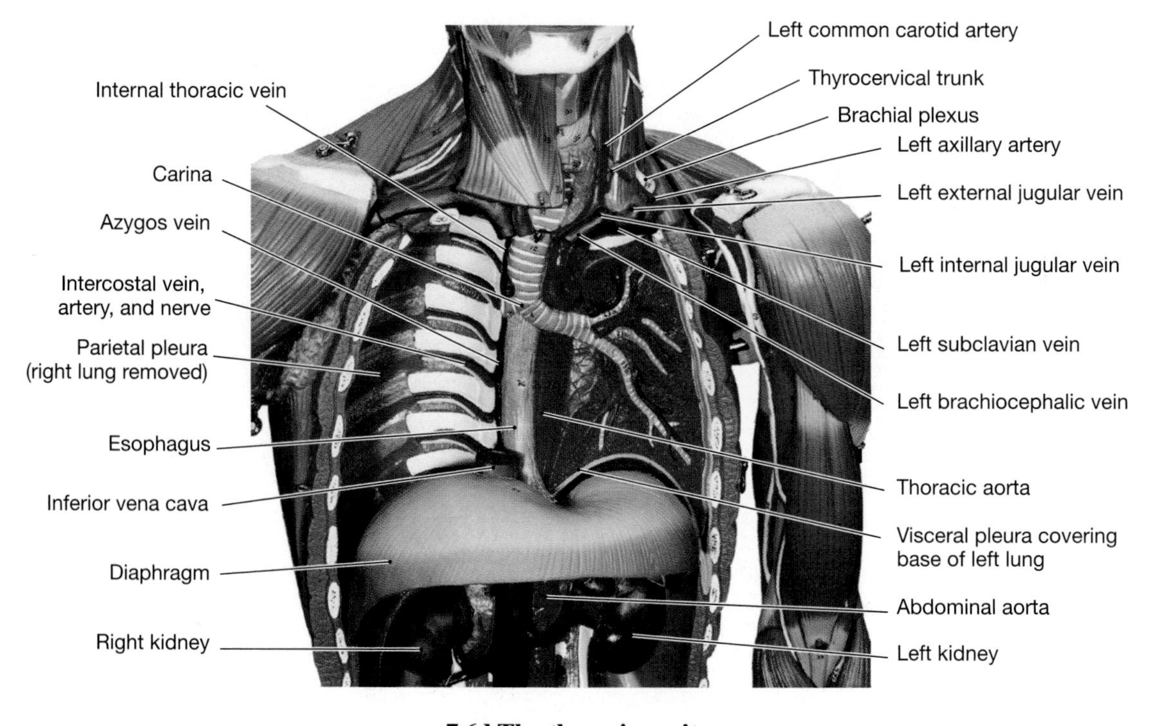

Internal thoracic vein
Carina
Azygos vein
Intercostal vein, artery, and nerve
Parietal pleura (right lung removed)
Esophagus
Inferior vena cava
Diaphragm
Right kidney
Left common carotid artery
Thyrocervical trunk
Brachial plexus
Left axillary artery
Left external jugular vein
Left internal jugular vein
Left subclavian vein
Left brachiocephalic vein
Thoracic aorta
Visceral pleura covering base of left lung
Abdominal aorta
Left kidney

7.6d The thoracic cavity

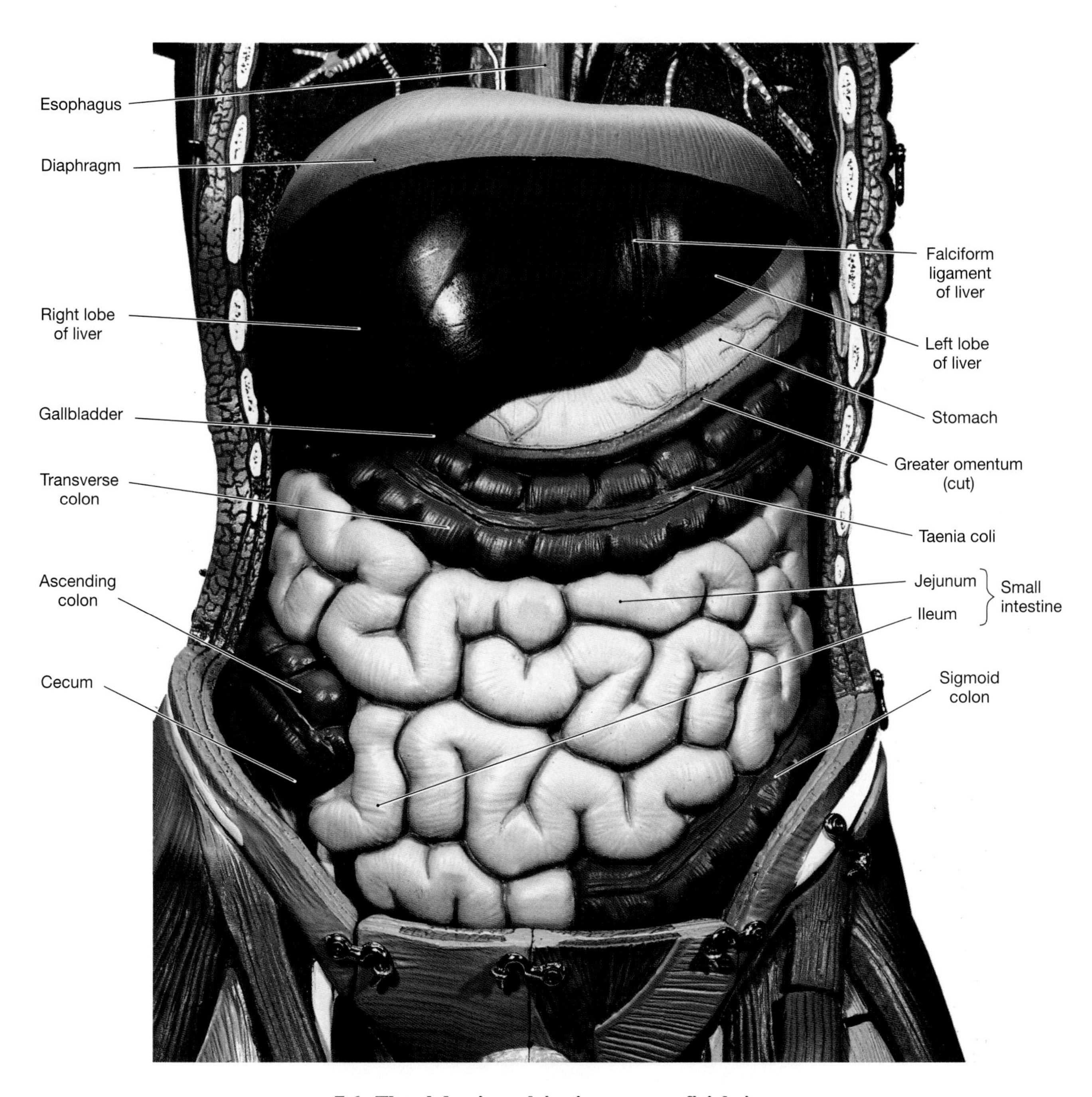

7.6e The abdominopelvic viscera, superficial view

7.7 Dissection of the Abdominopelvic Viscera

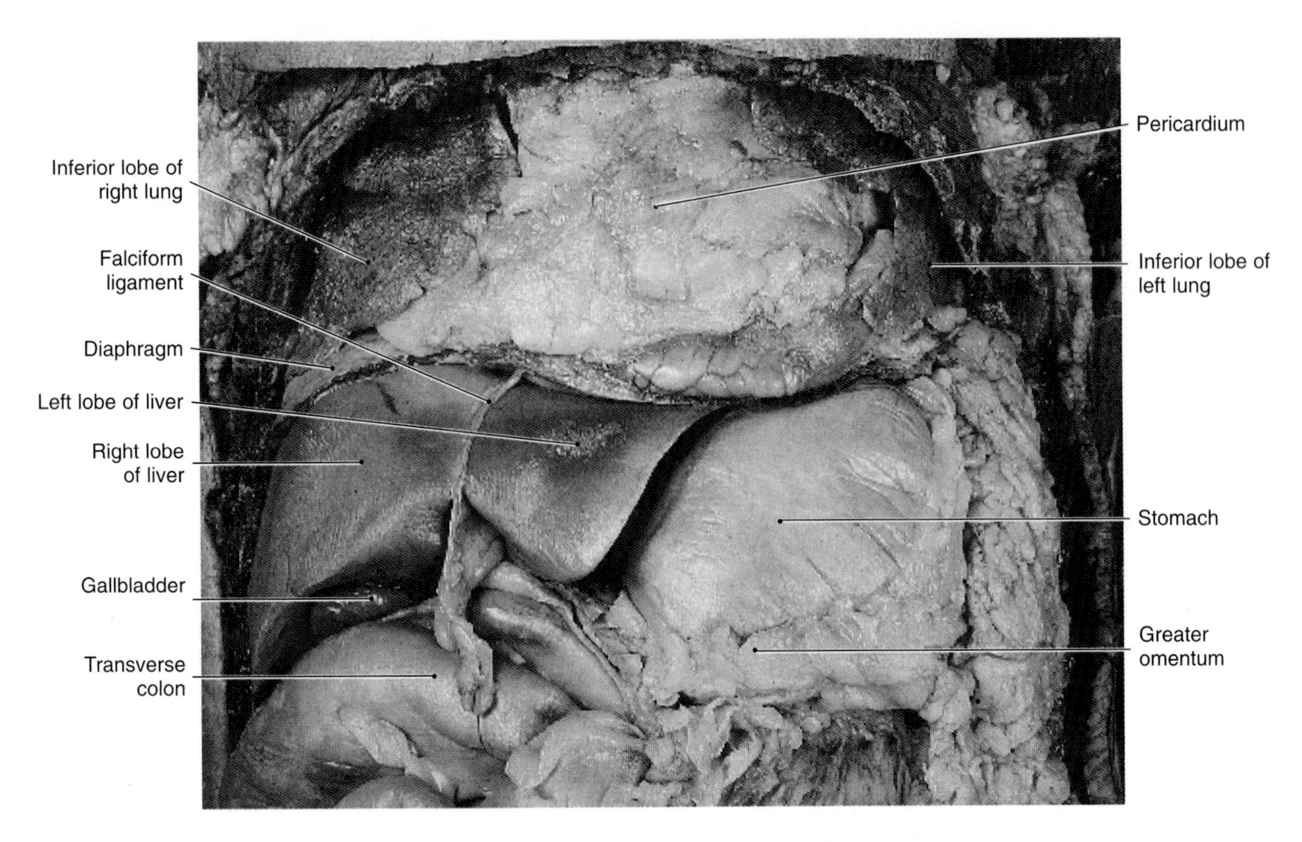

7.7a Superior portion, anterior view

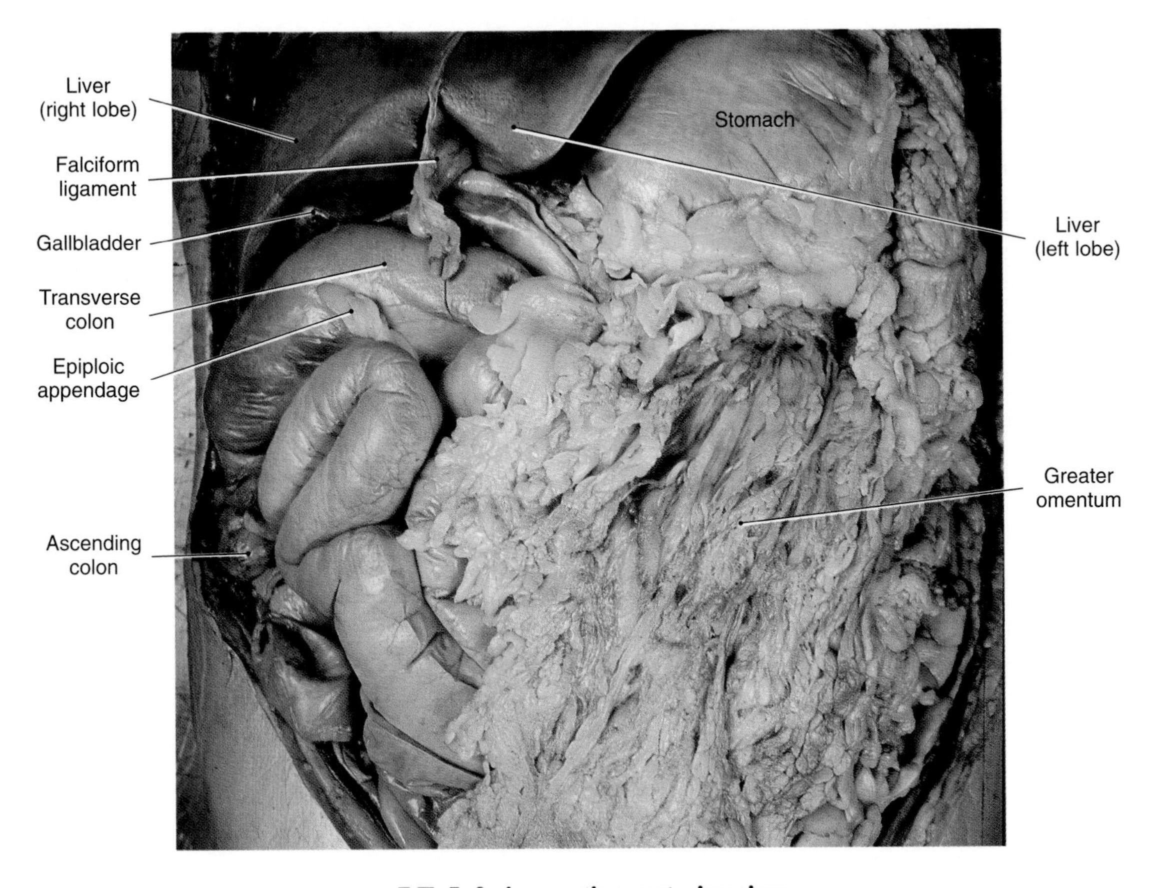

7.7b Inferior portion, anterior view

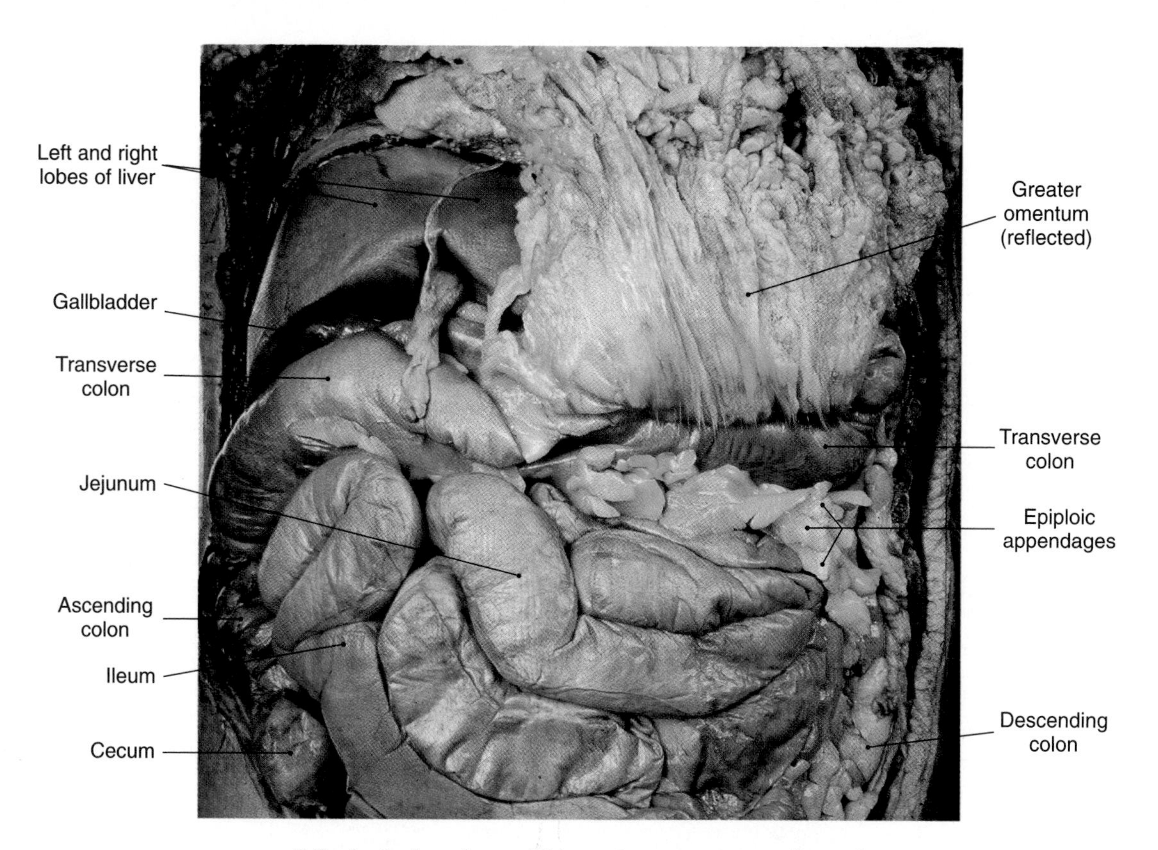

7.7c Anterior view, with greater omentum reflected

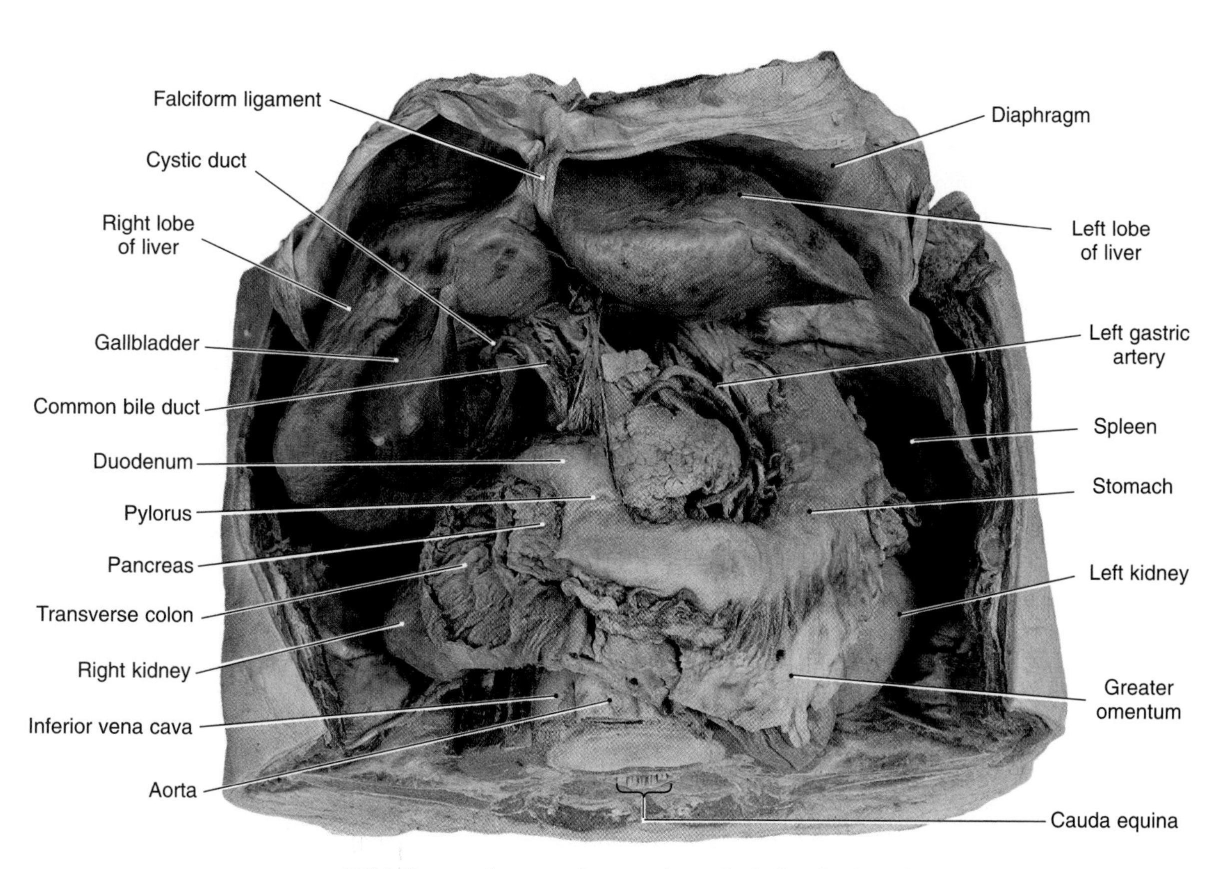

7.7d Organs in superior portion of abdominal cavity

7.8 The Isolated Liver and Gallbladder

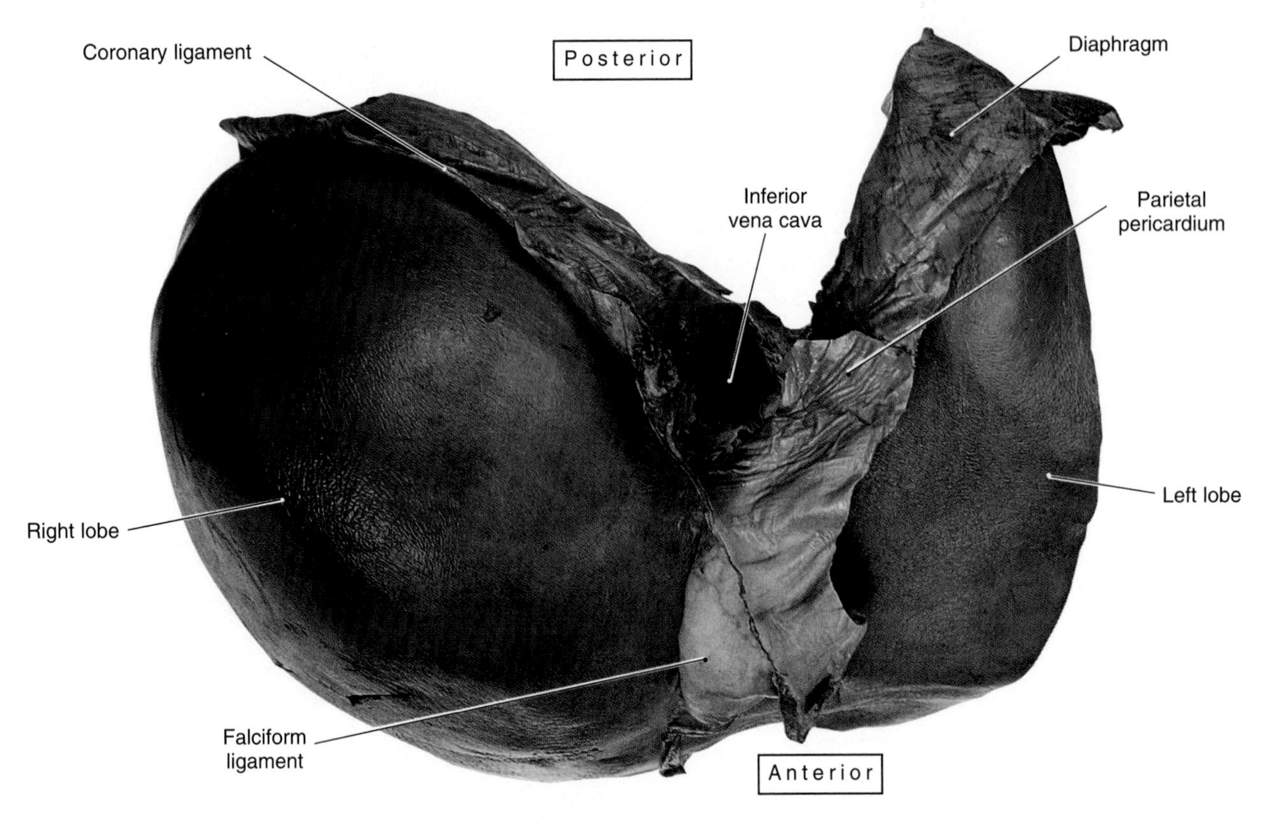

7.8a Superior view

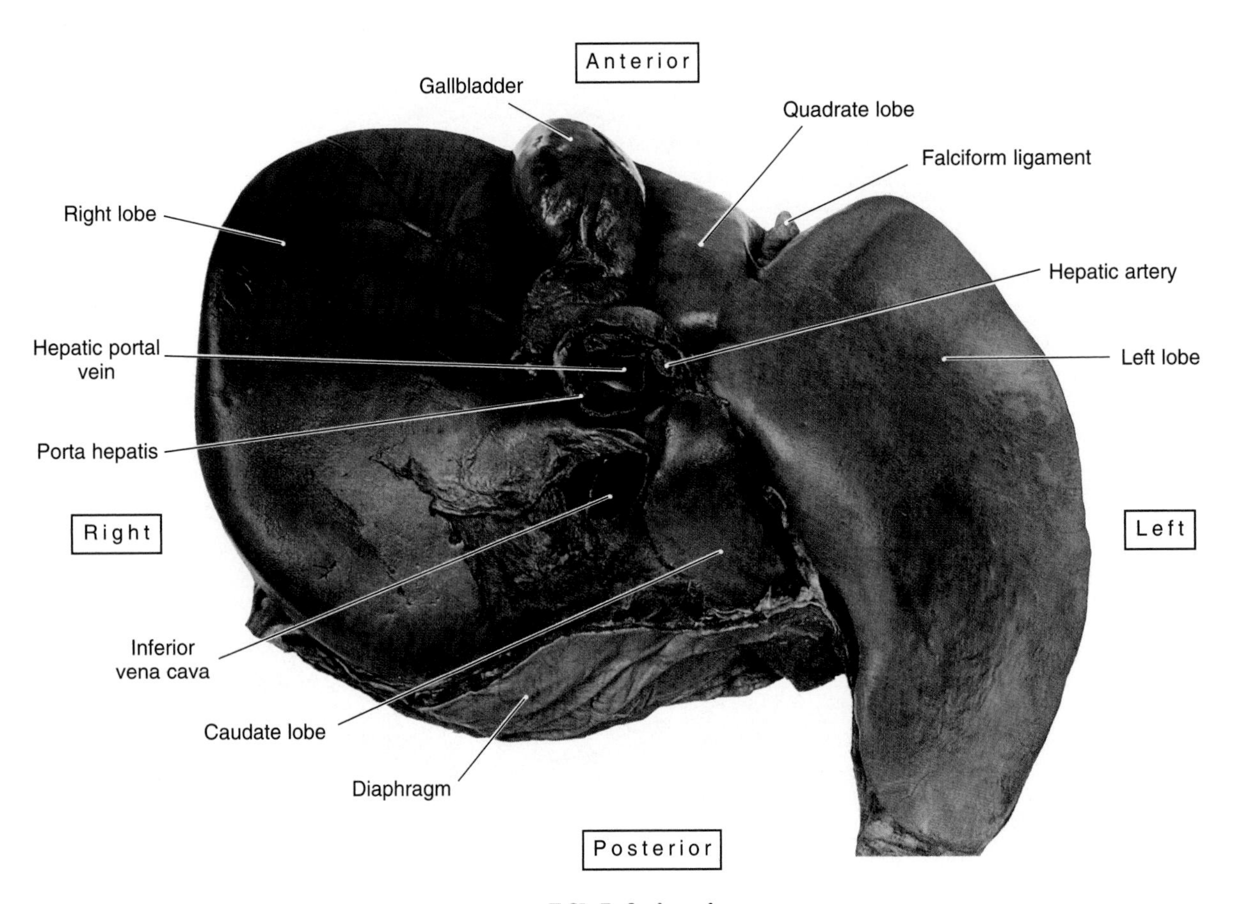

7.8b Inferior view

7.9 Horizontal Sections through the Abdominal Cavity

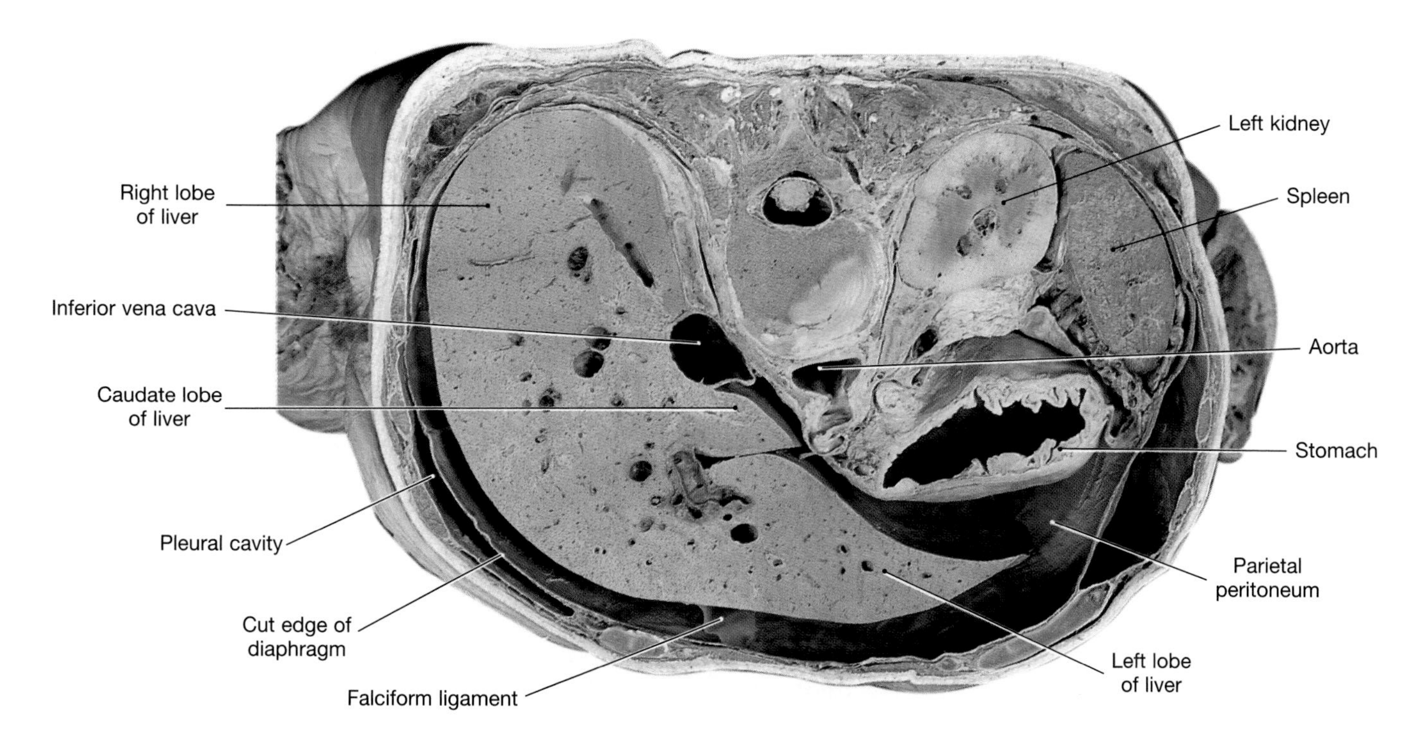

7.9a Horizontal section at T_{12}

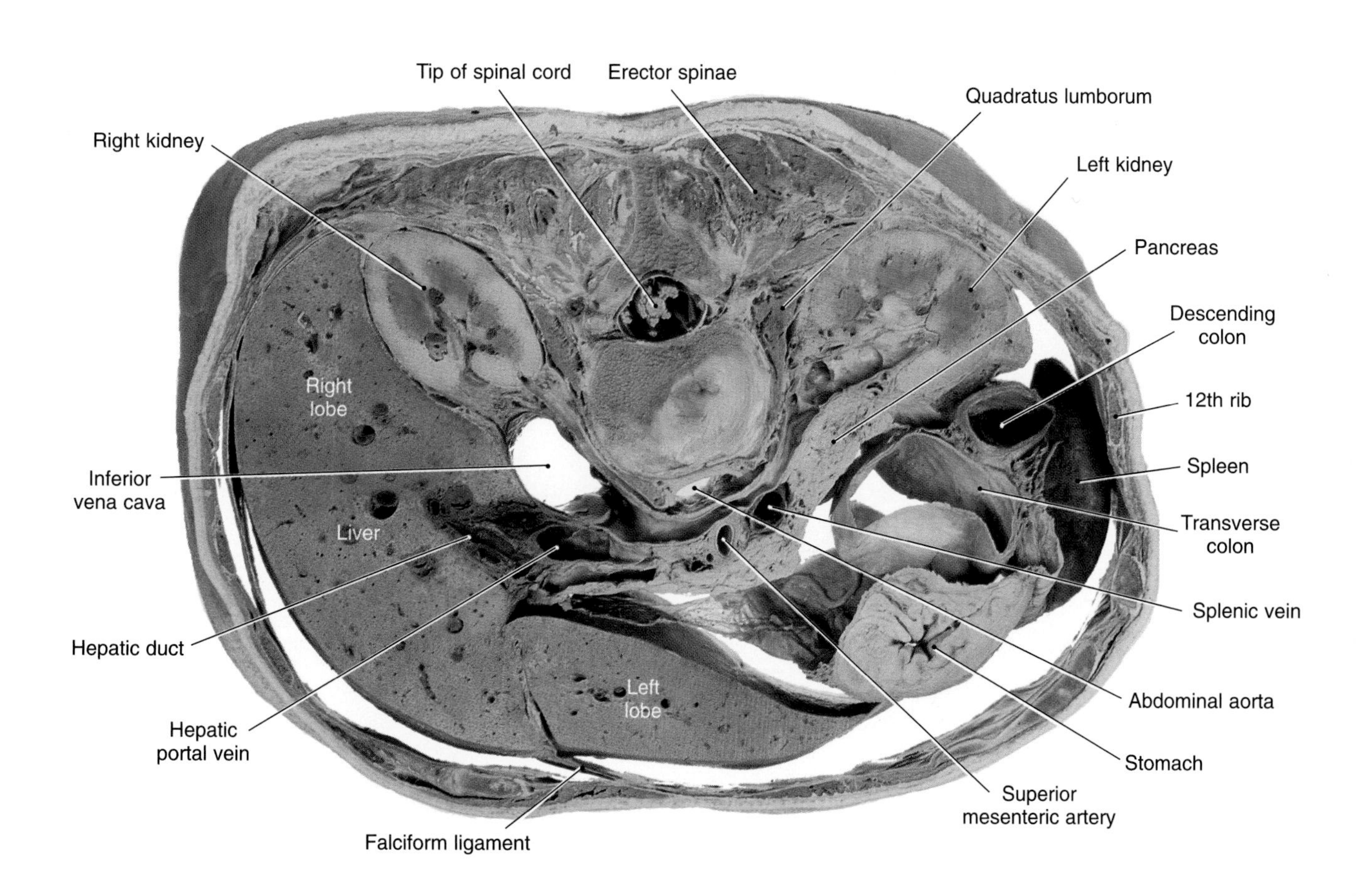

7.9b Horizontal section at L_1

7.10 The Abdominopelvic Cavity and Associated Structures

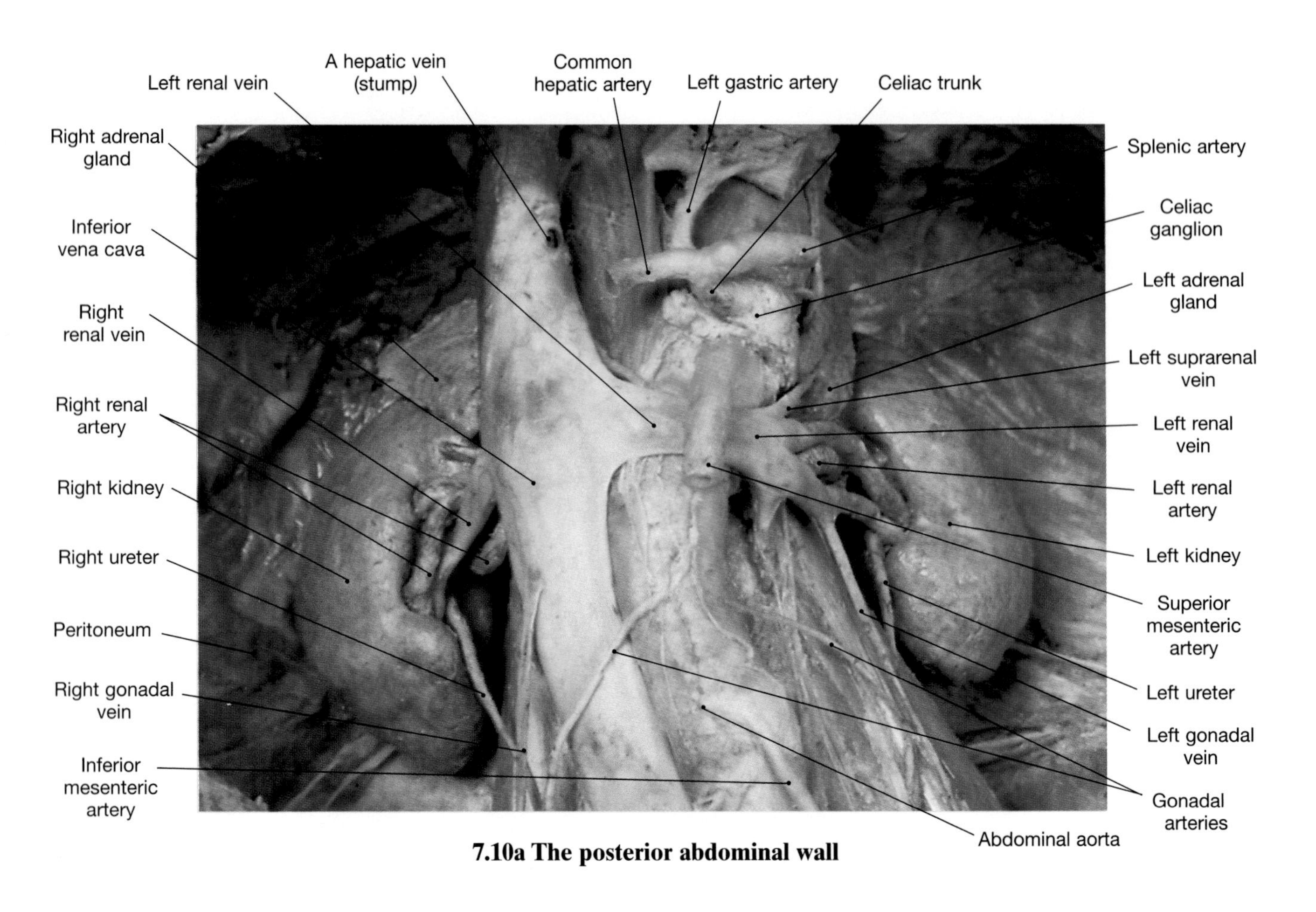

7.10a The posterior abdominal wall

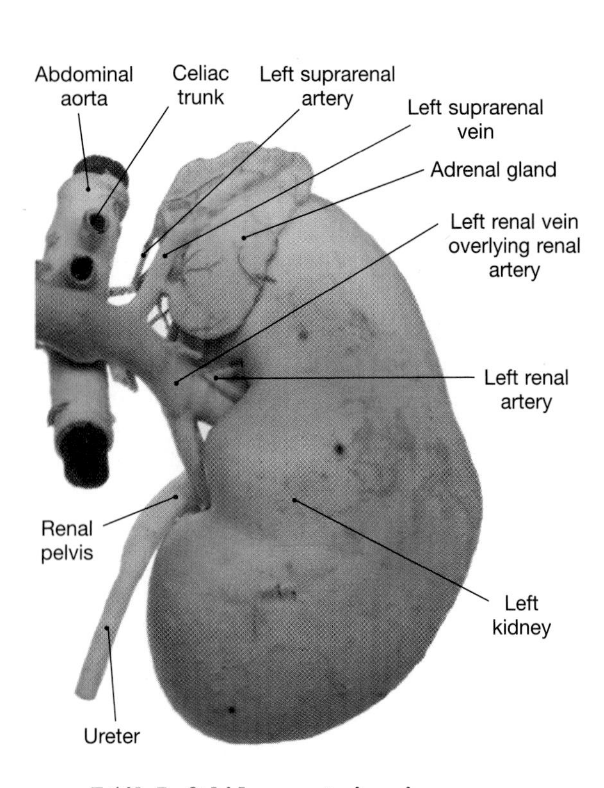

7.10b Left kidney, anterior view

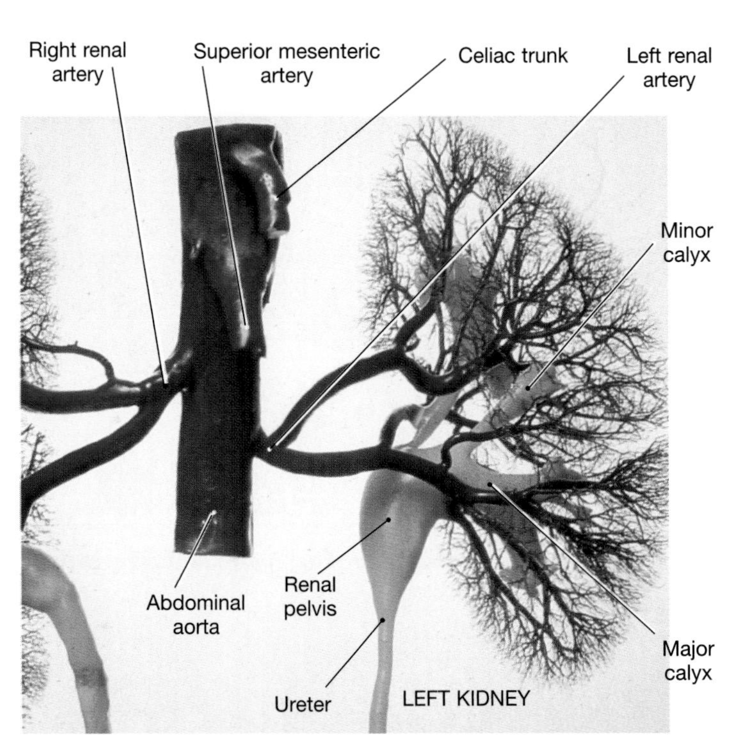

7.10c The renal arteries, ureters, and renal pelvis, corrosion cast

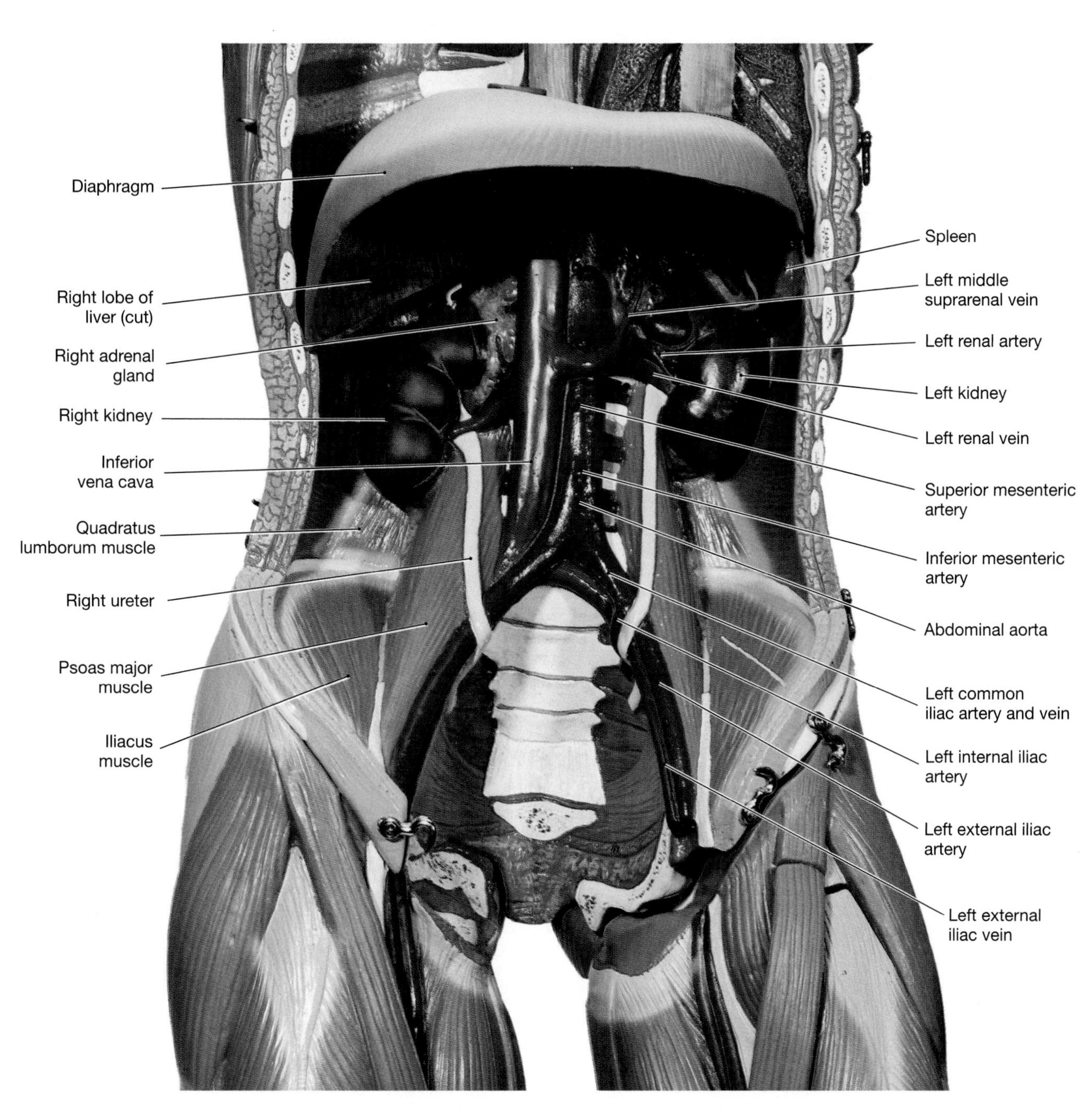

7.10d The abdominopelvic cavity, female, anterior view—model

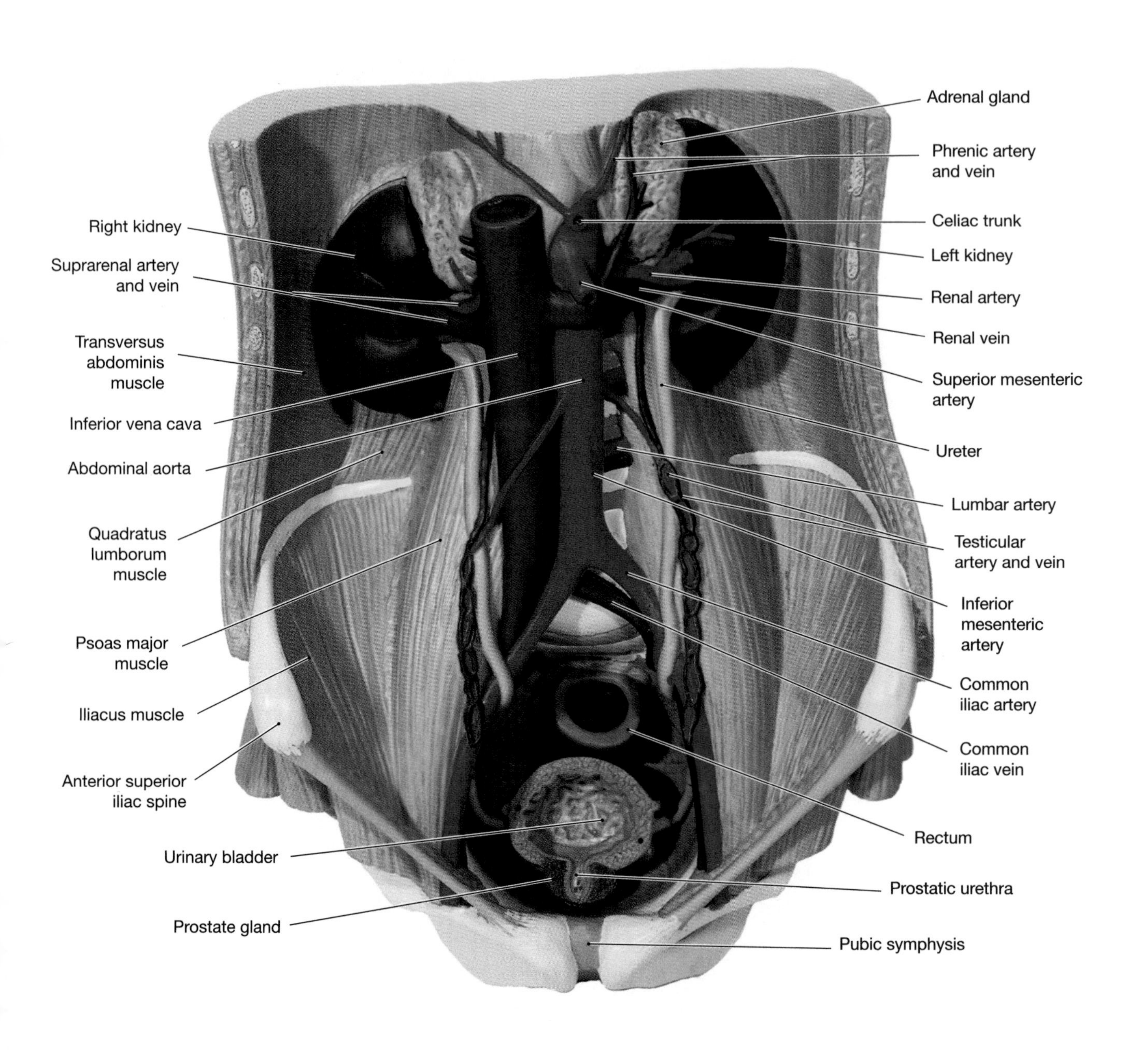

7.10e The abdominopelvic cavity, male, anterior view—model

7.11 The Reproductive Organs

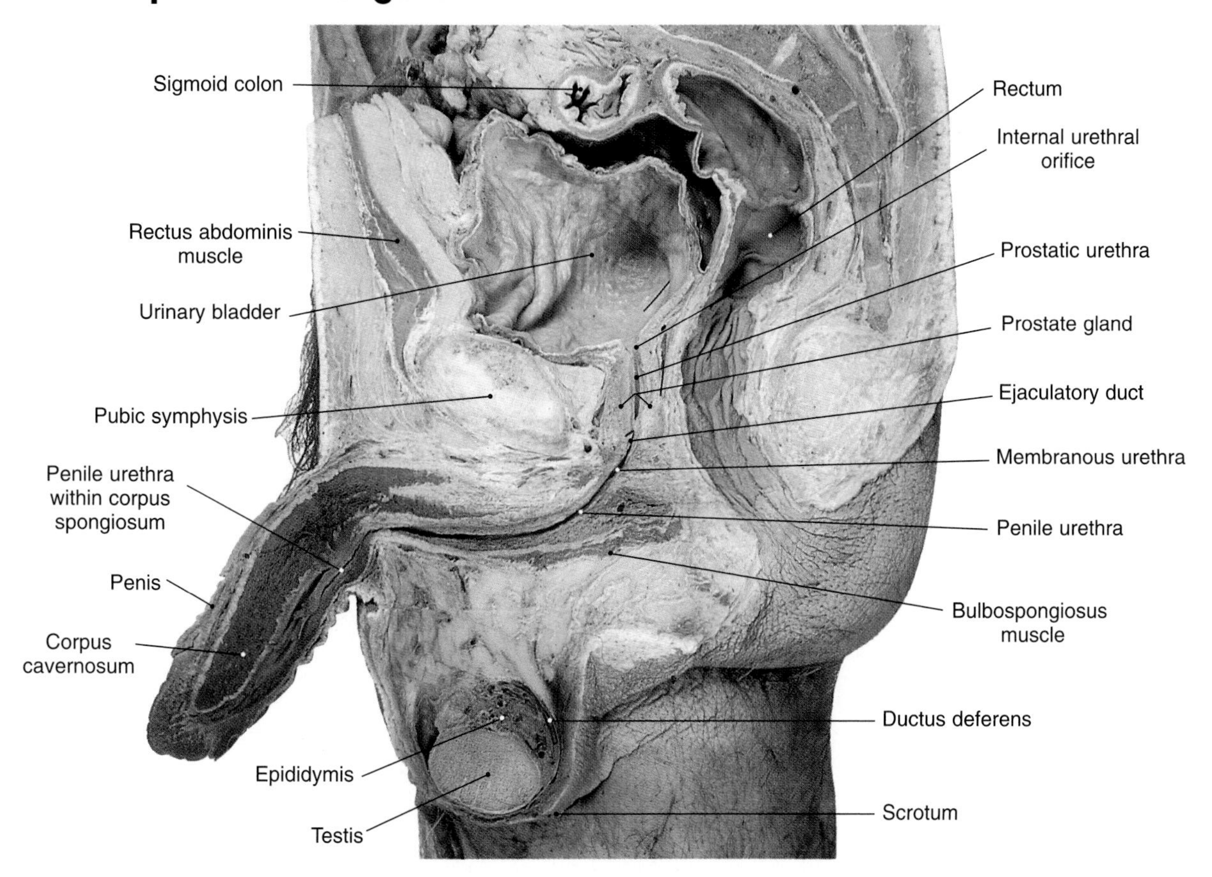

7.11a Pelvic region of a male, sagittal section

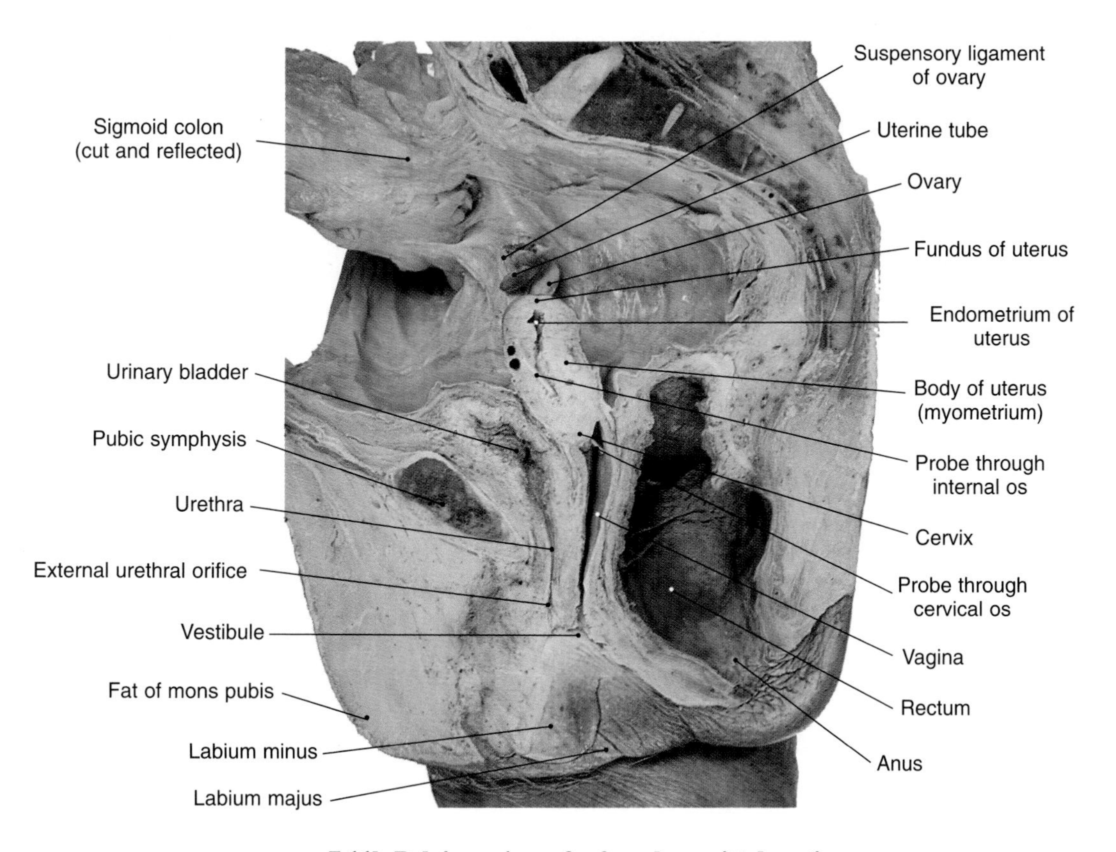

7.11b Pelvic region of a female, sagittal section

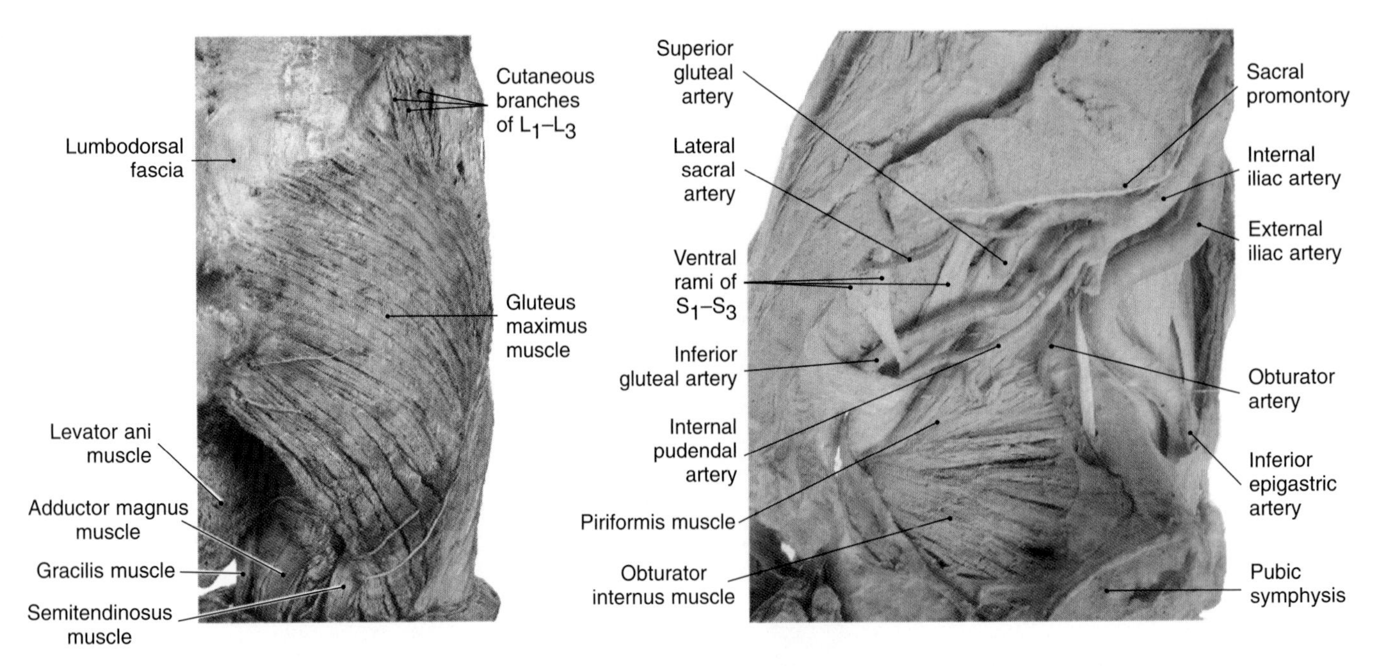

8.1 Superficial dissection of the right hip, posterior view

8.2 Dissection of blood vessels, nerves, and muscles in the left half of the pelvis

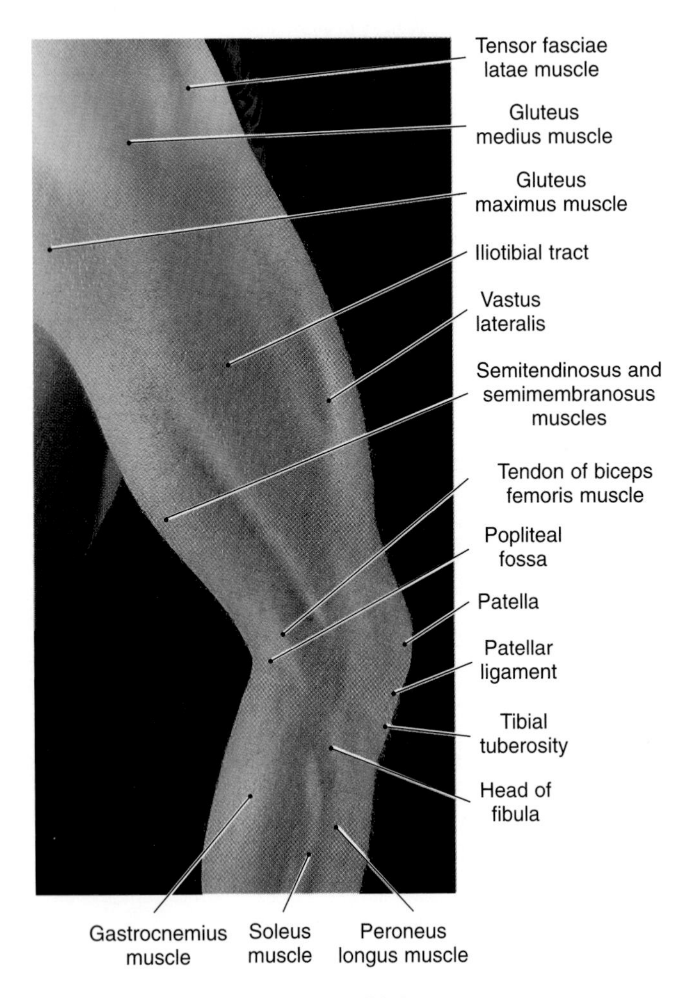

8.3 Surface anatomy of the thigh, lateral view

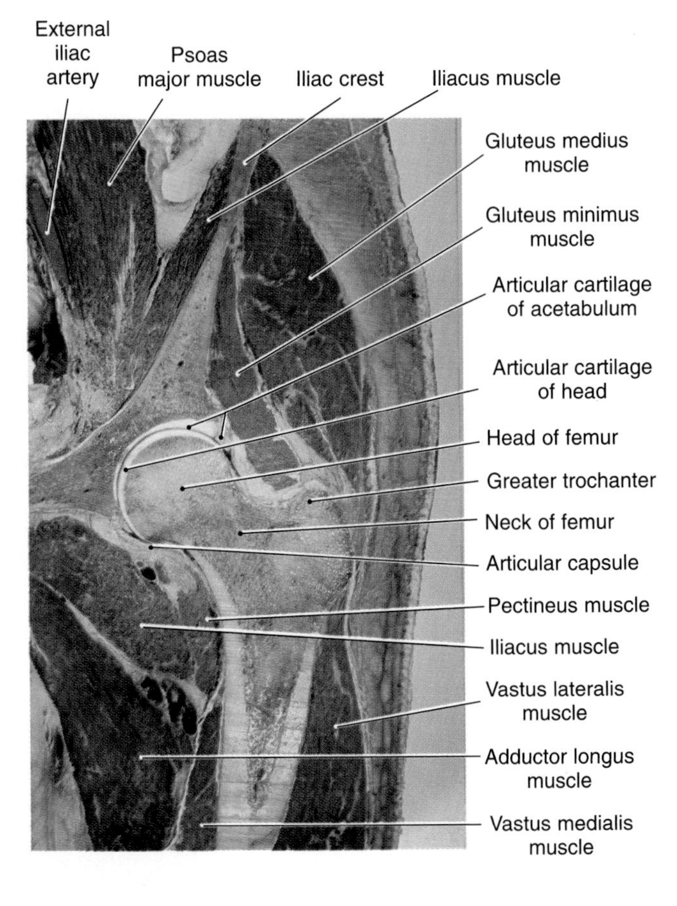

8.4 Coronal section through the left hip

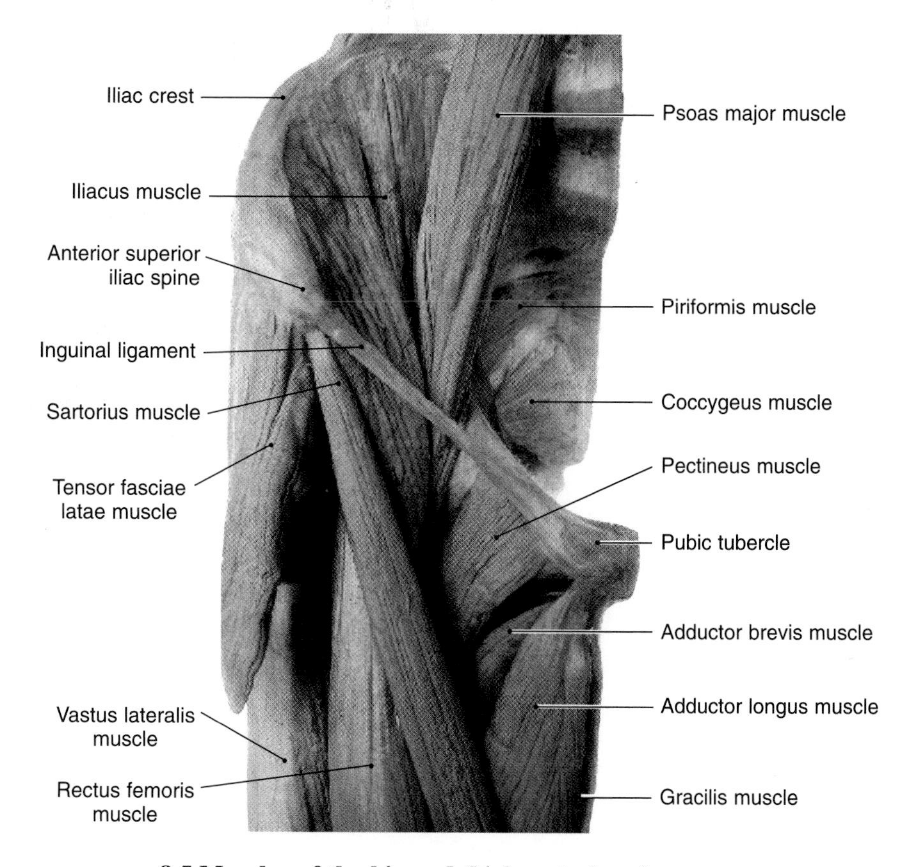

8.5 Muscles of the hip and thigh, anterior view

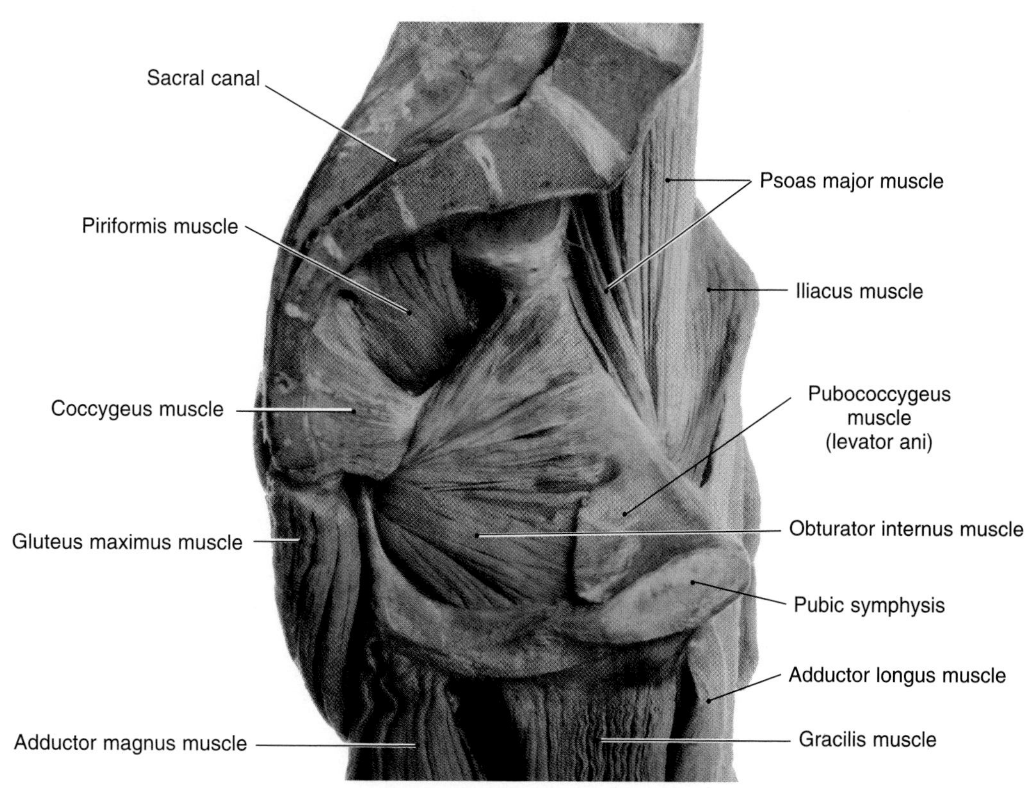

8.6 Muscles of the left half of the pelvis, medial view

8.7 The Right Thigh and Leg, Anterior View

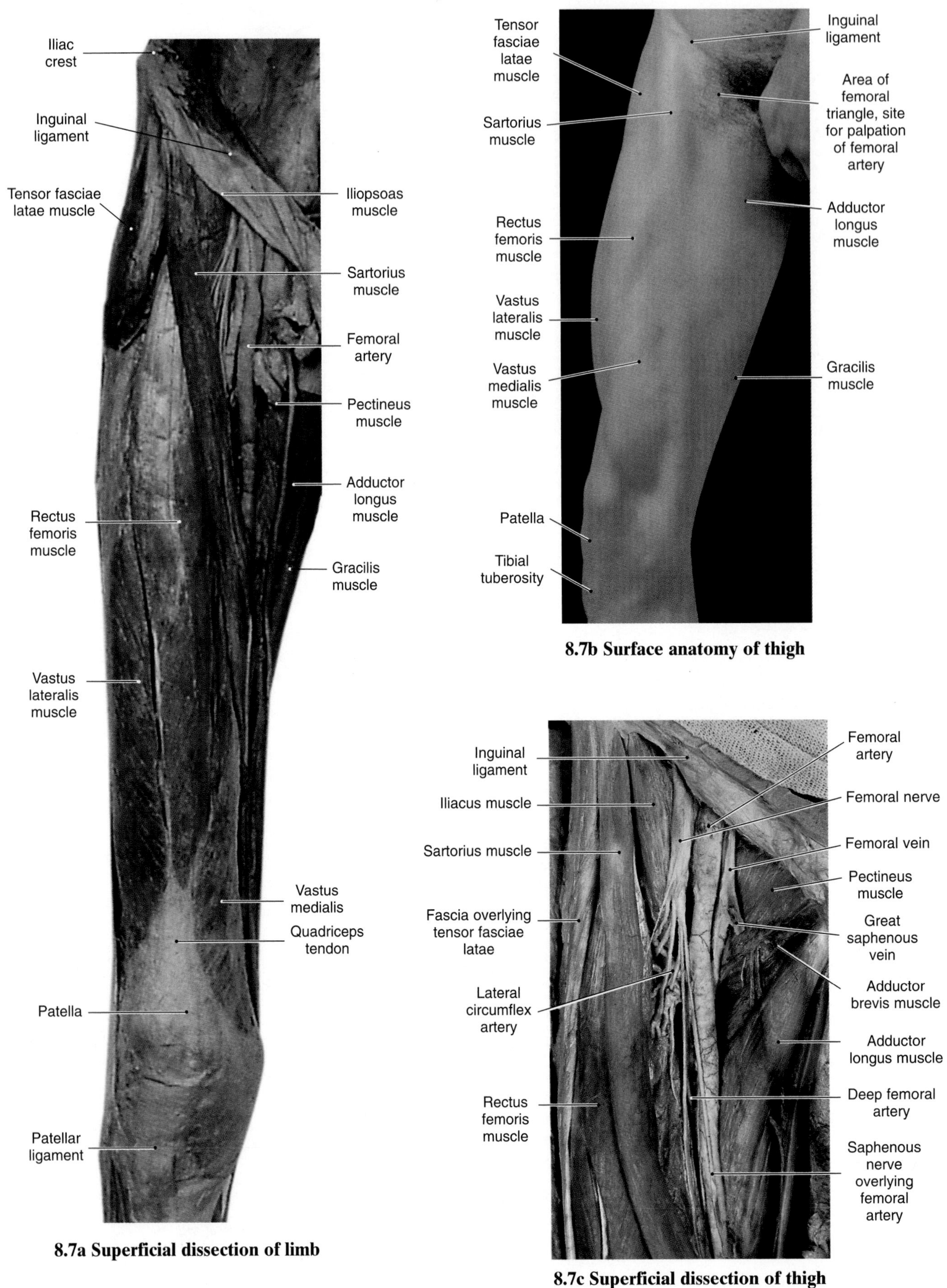

8.7a Superficial dissection of limb

8.7b Surface anatomy of thigh

8.7c Superficial dissection of thigh

8.8 The Right Thigh and Leg, Posterior View

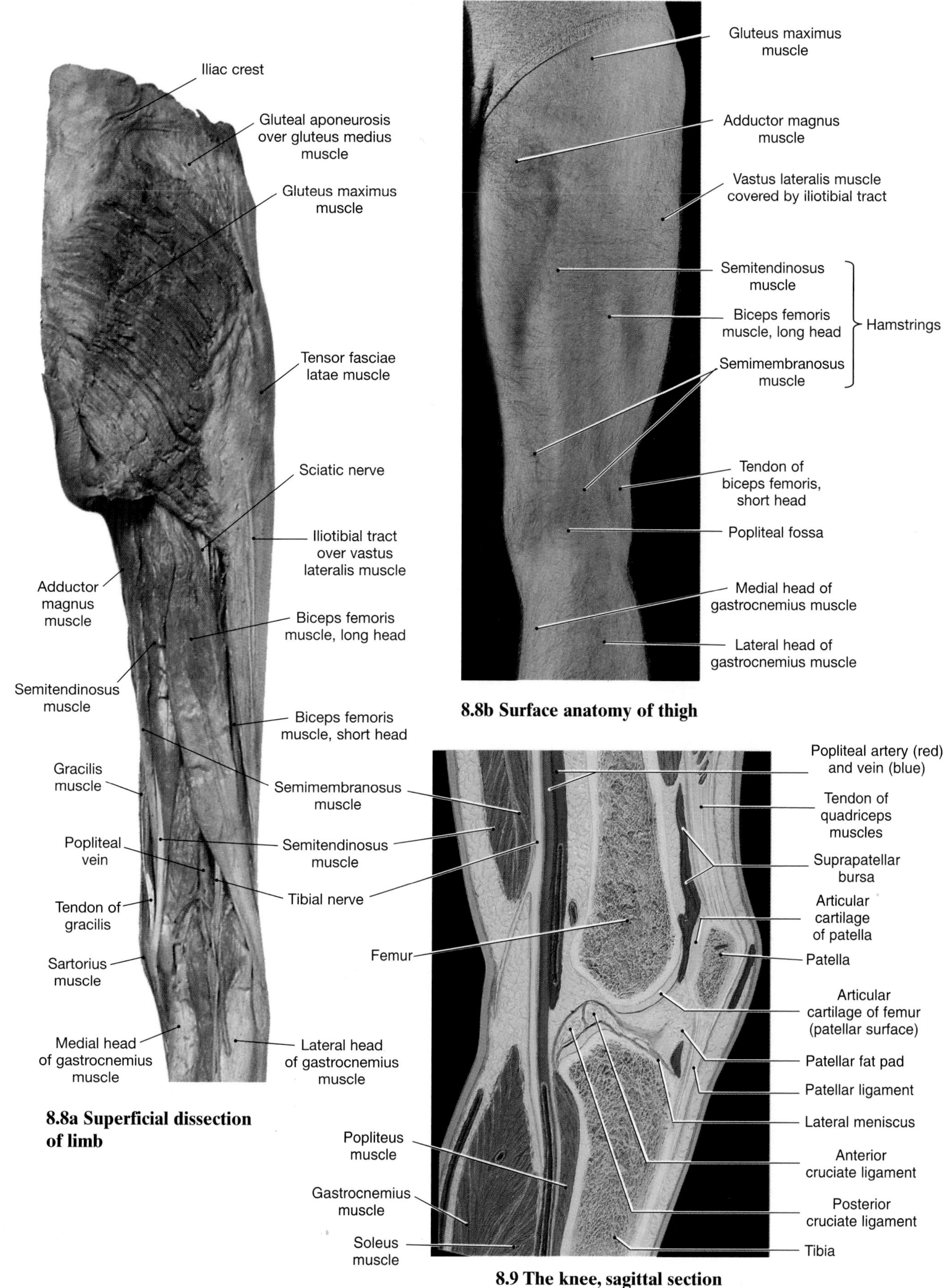

8.8a Superficial dissection of limb

8.8b Surface anatomy of thigh

8.9 The knee, sagittal section

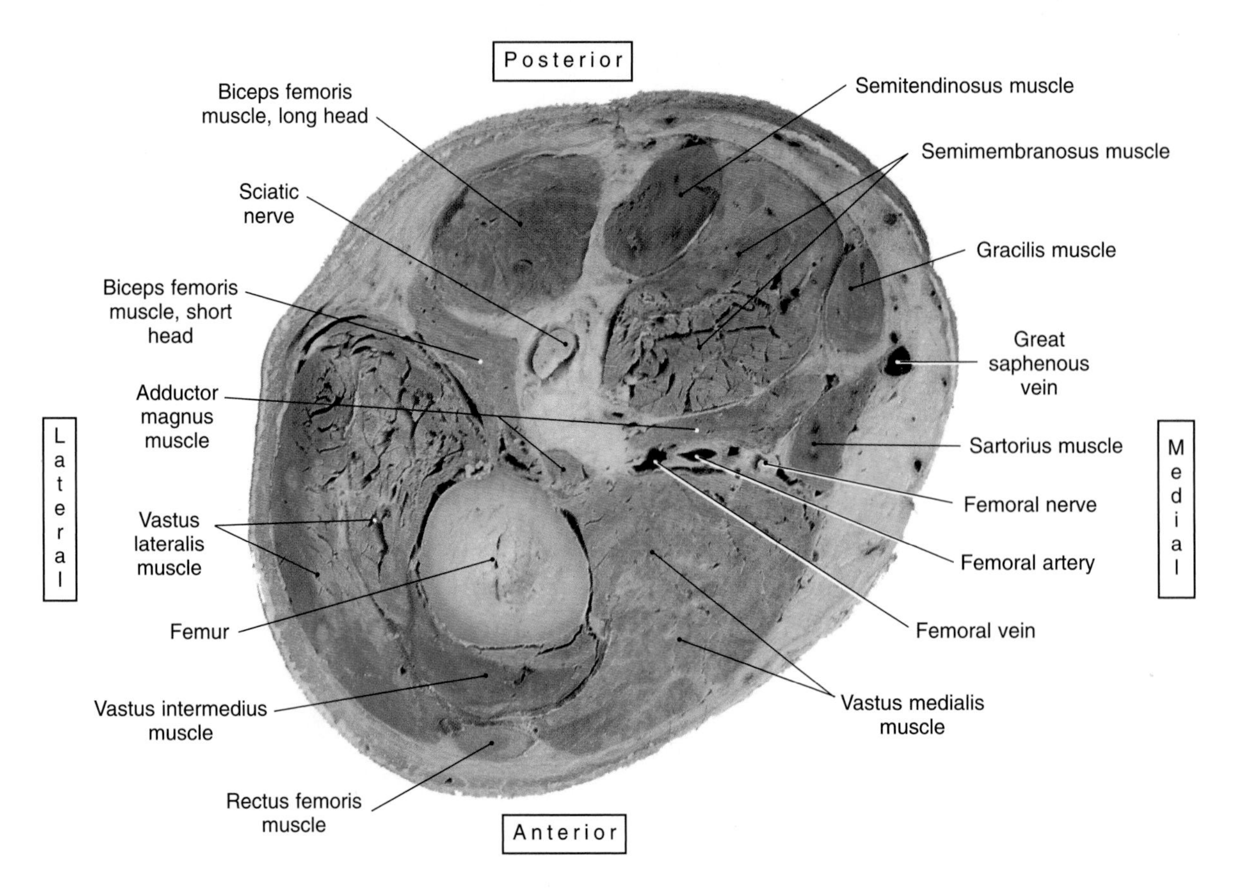

8.10 Horizontal section through the thigh

8.11 The Knee Joint

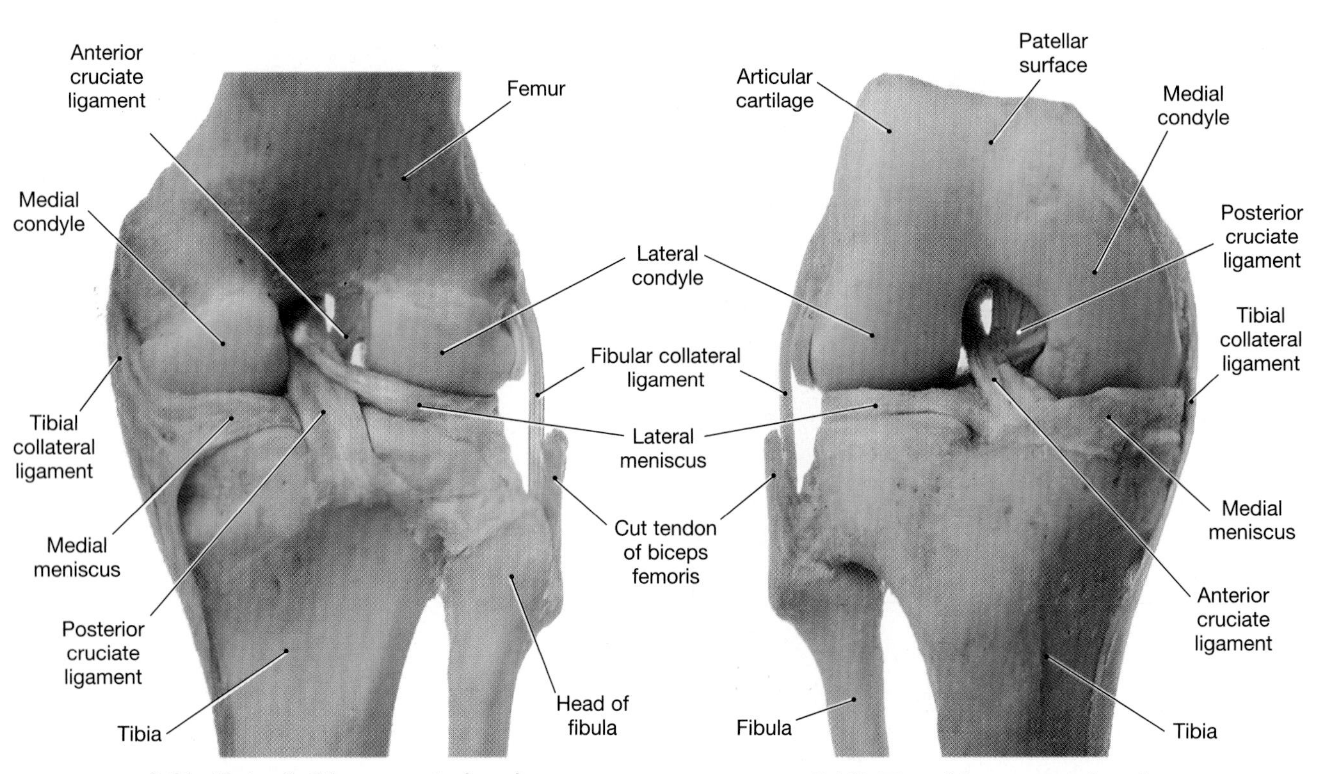

8.11a Extended knee, posterior view

8.11b Flexed knee, anterior view

8.12 The Right Leg and Foot, Anterior View

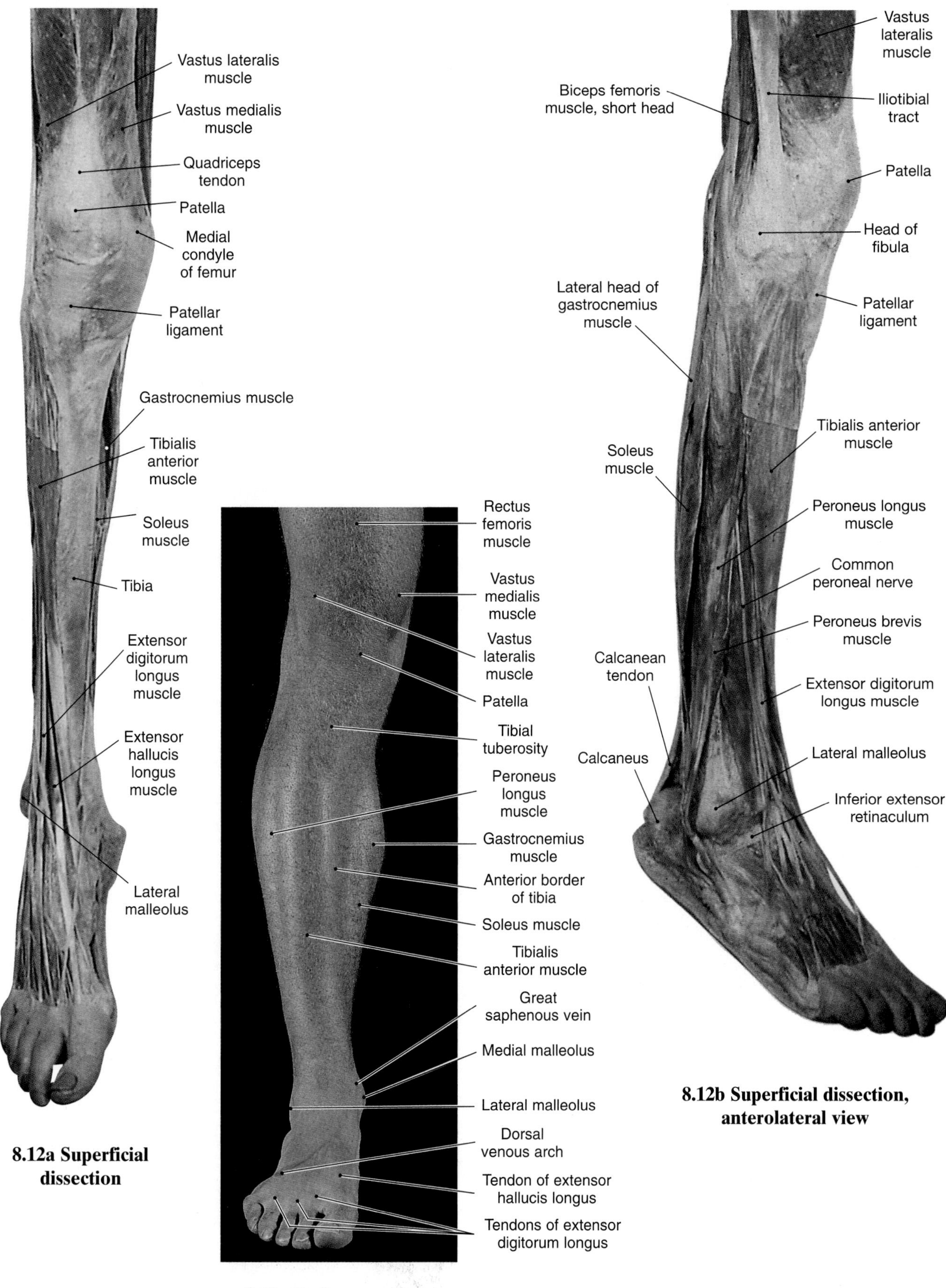

8.12a Superficial dissection

8.12c Surface anatomy

8.12b Superficial dissection, anterolateral view

8.13 The Right Leg and Foot, Posterior View

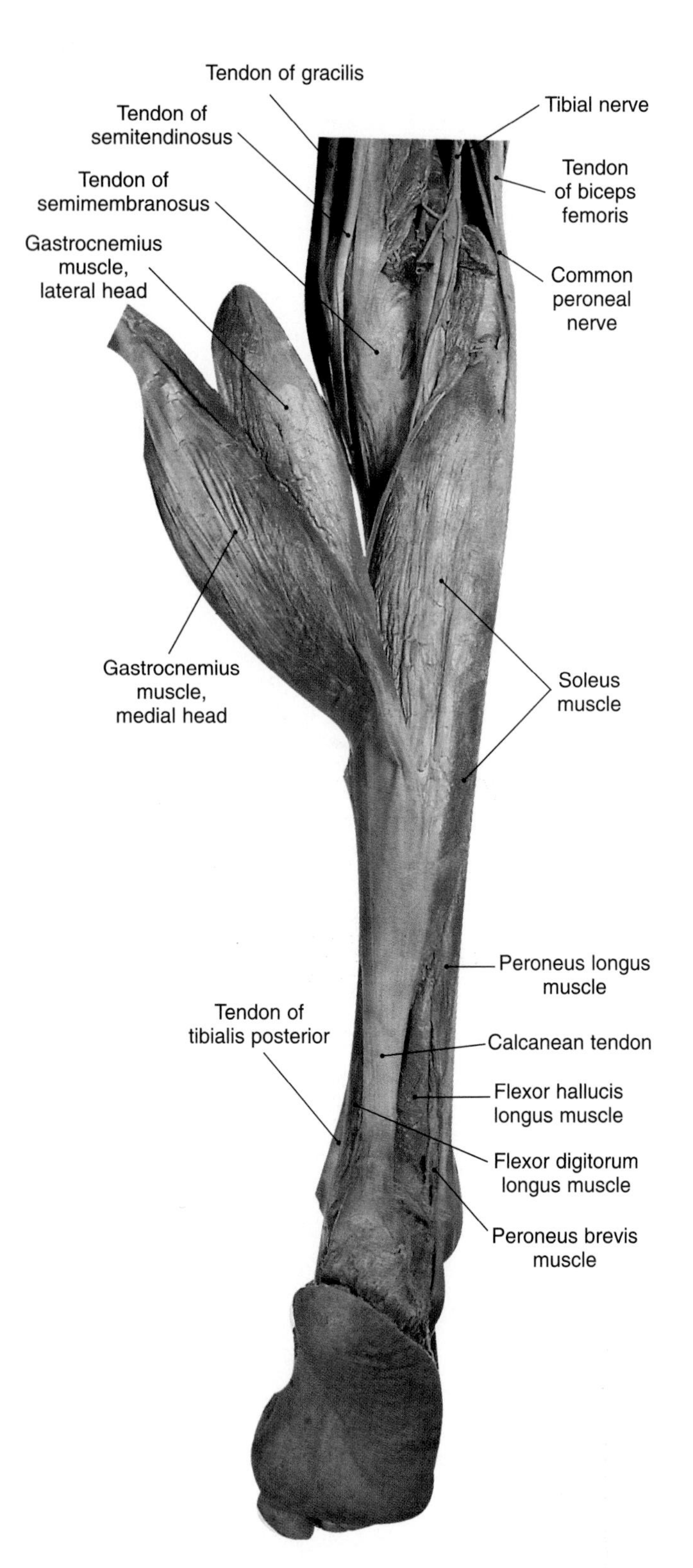

8.13a Superficial dissection

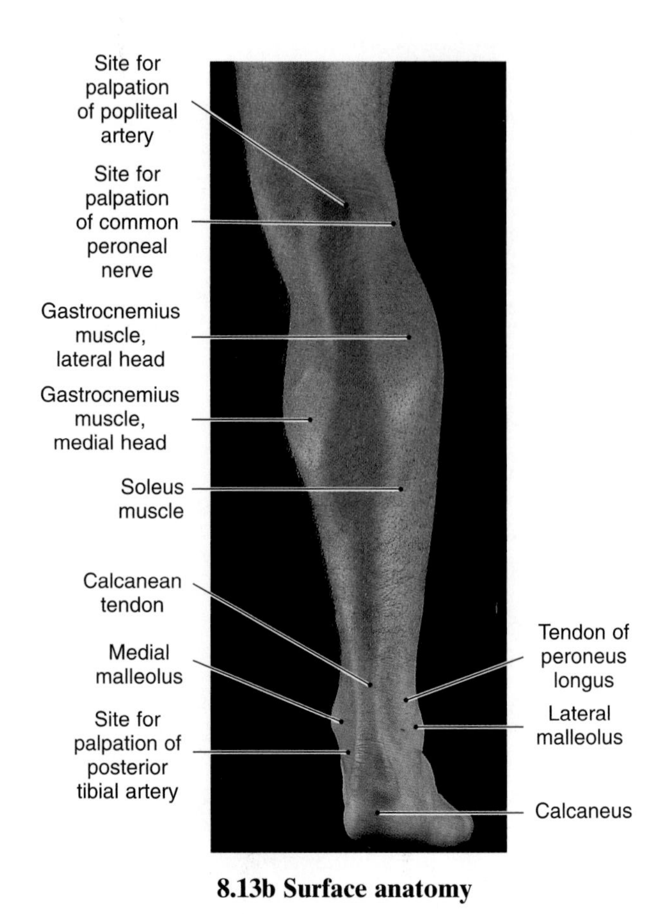

8.13b Surface anatomy

8.14 The Right Foot

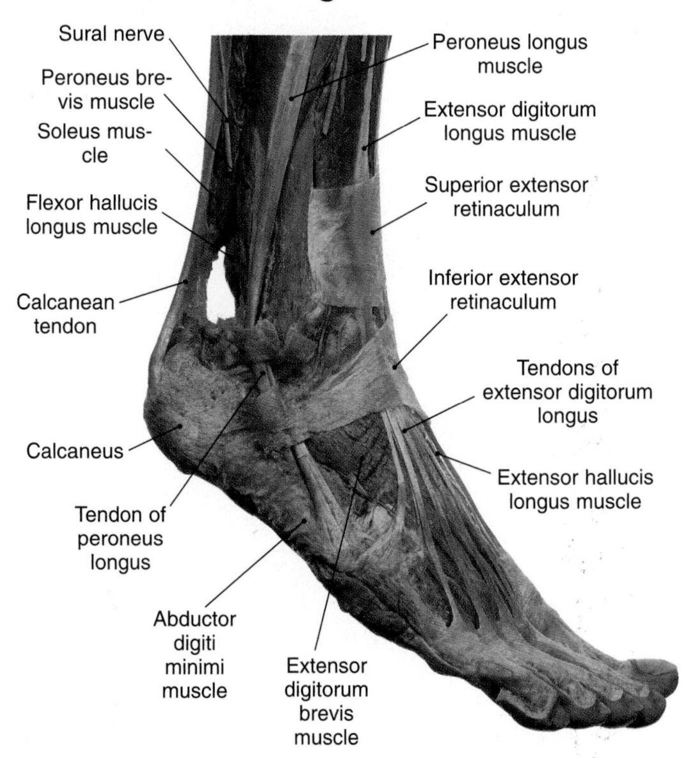

8.14a Superficial dissection, lateral view

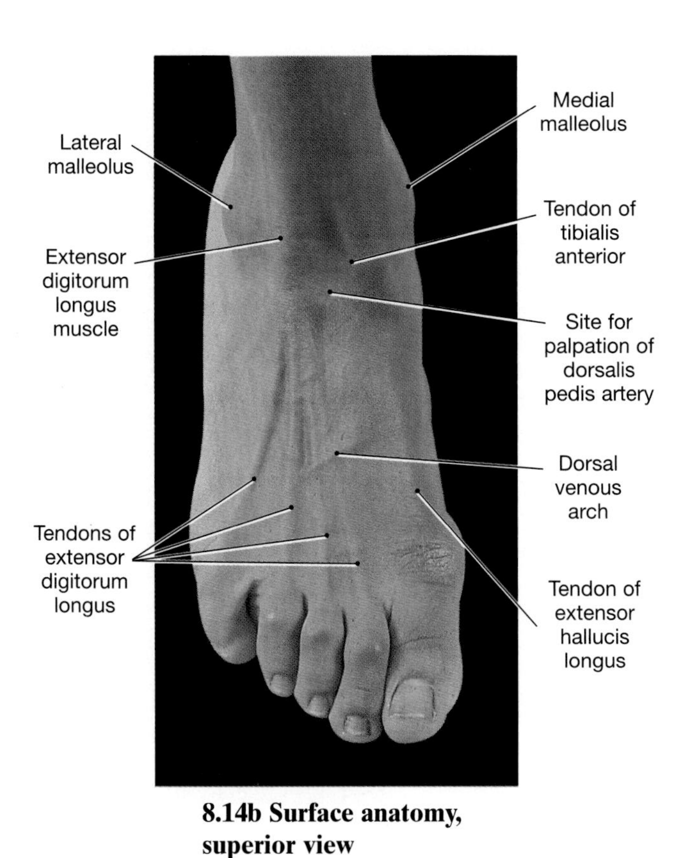

8.14b Surface anatomy, superior view

Peroneus brevis muscle
Superior extensor retinaculum
Lateral malleolus of fibula
Inferior extensor retinaculum
Tendons of extensor digitorum longus
Dorsal interosseus muscles
Tendons of extensor digitorum brevis
Medial malleolus of tibia
Tendon of tibialis anterior
Tendon of extensor hallucis longus
Abductor hallucis muscle
Tendon of extensor hallucis brevis

8.14c Superficial dissection, superior view

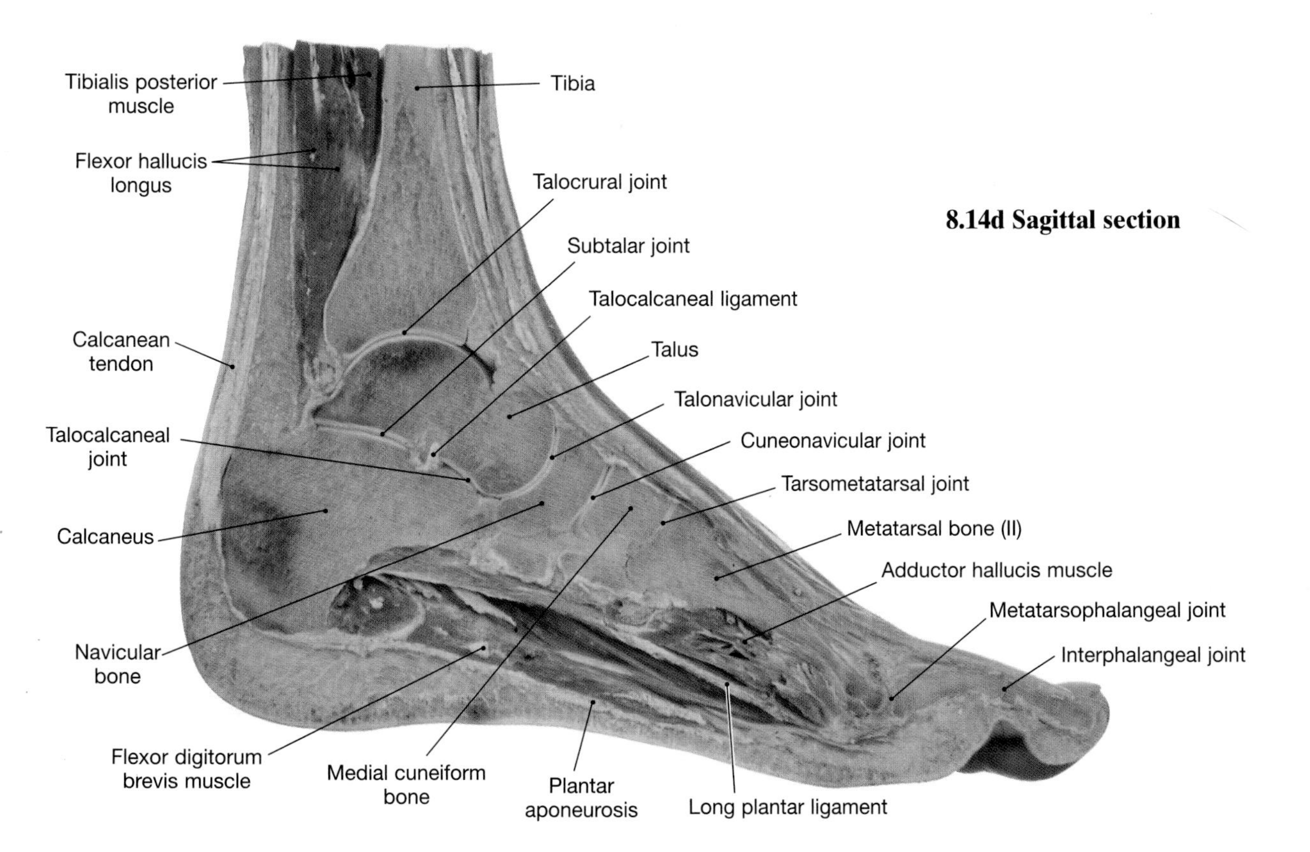

8.14d Sagittal section

Scanning Atlas

SCAN 1 MRI SCANS OF THE BRAIN

Horizontal Plane (Superior-to-Inferior Sequence)

1a

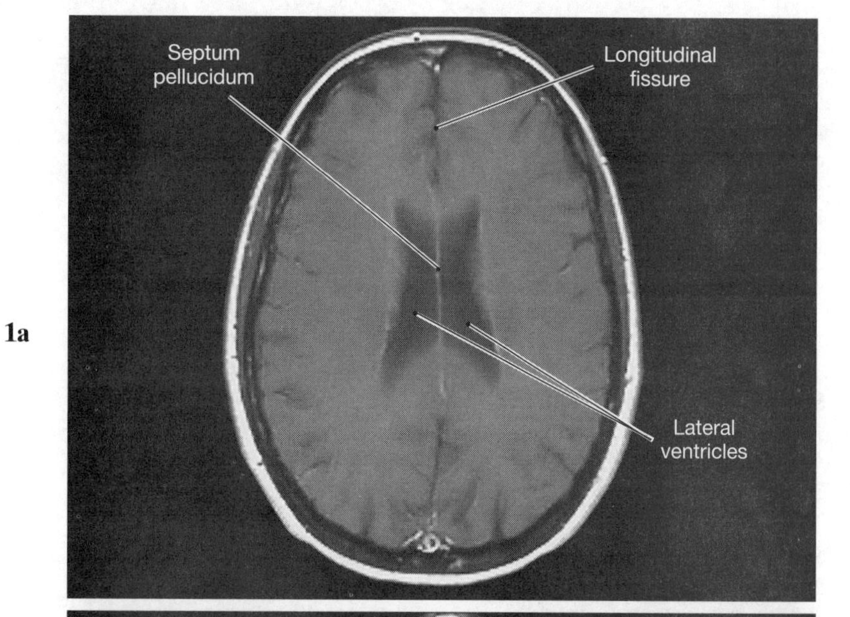

1b

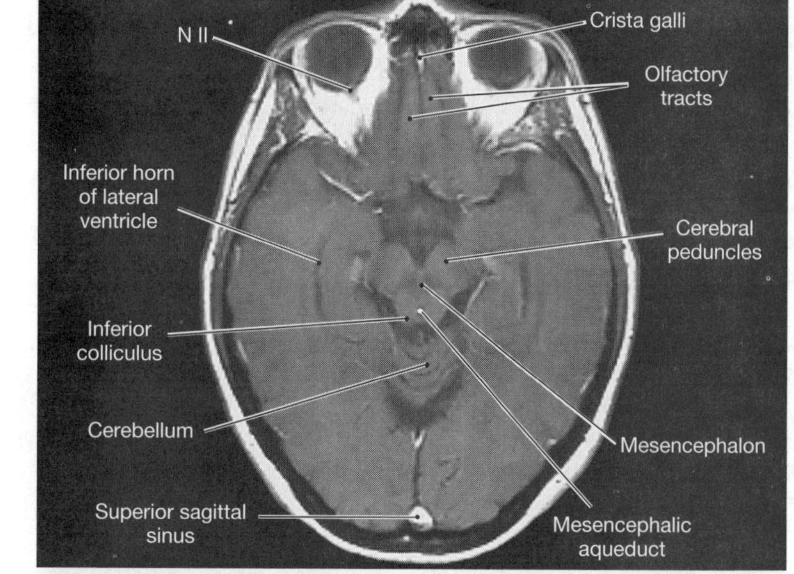

1c

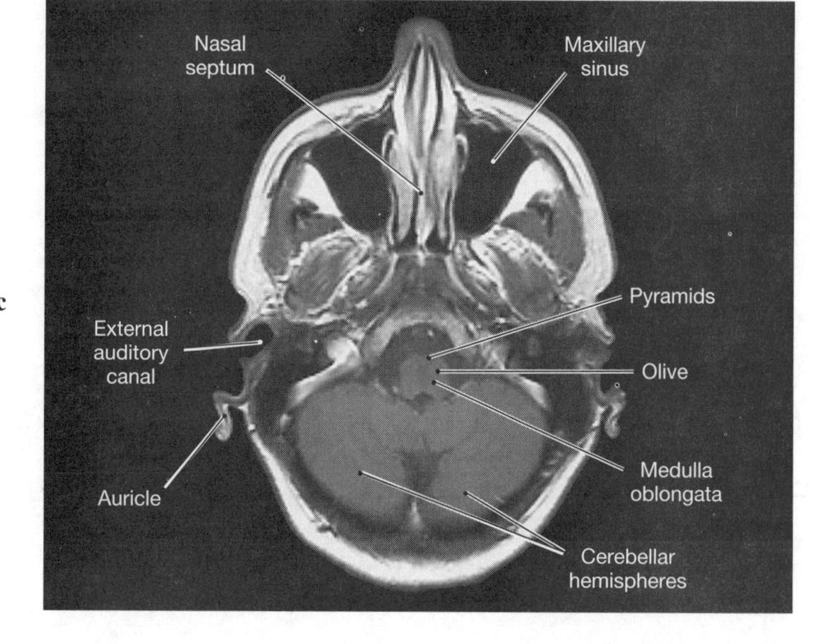

Sagittal Plane

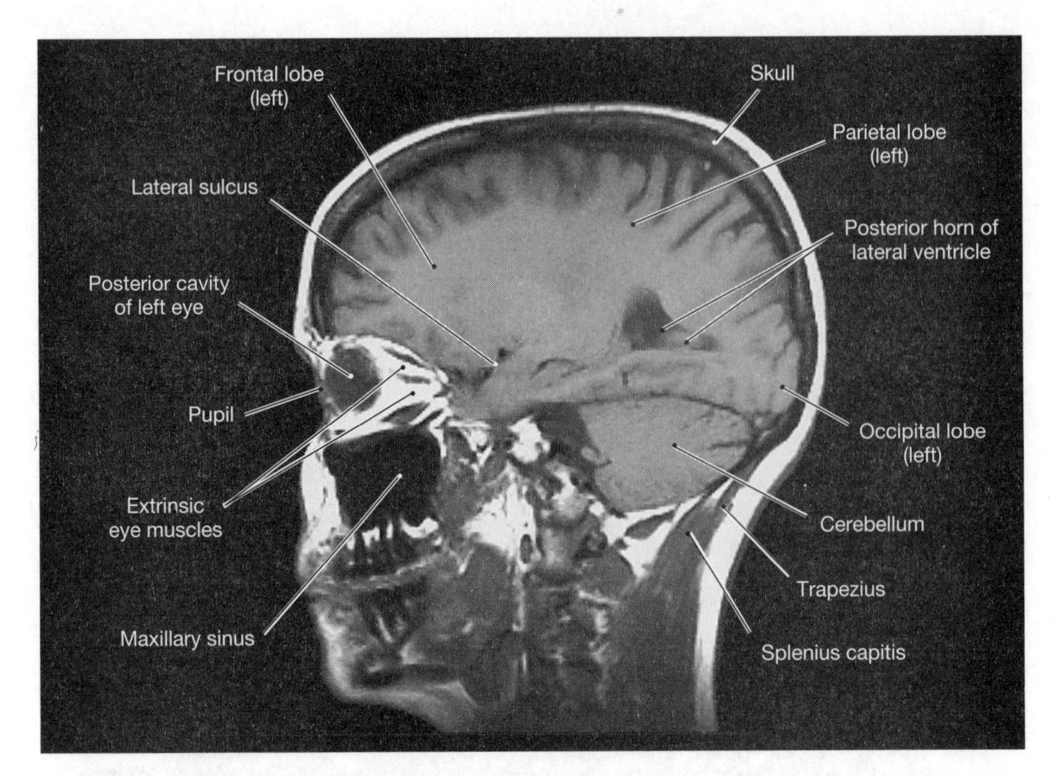

1d

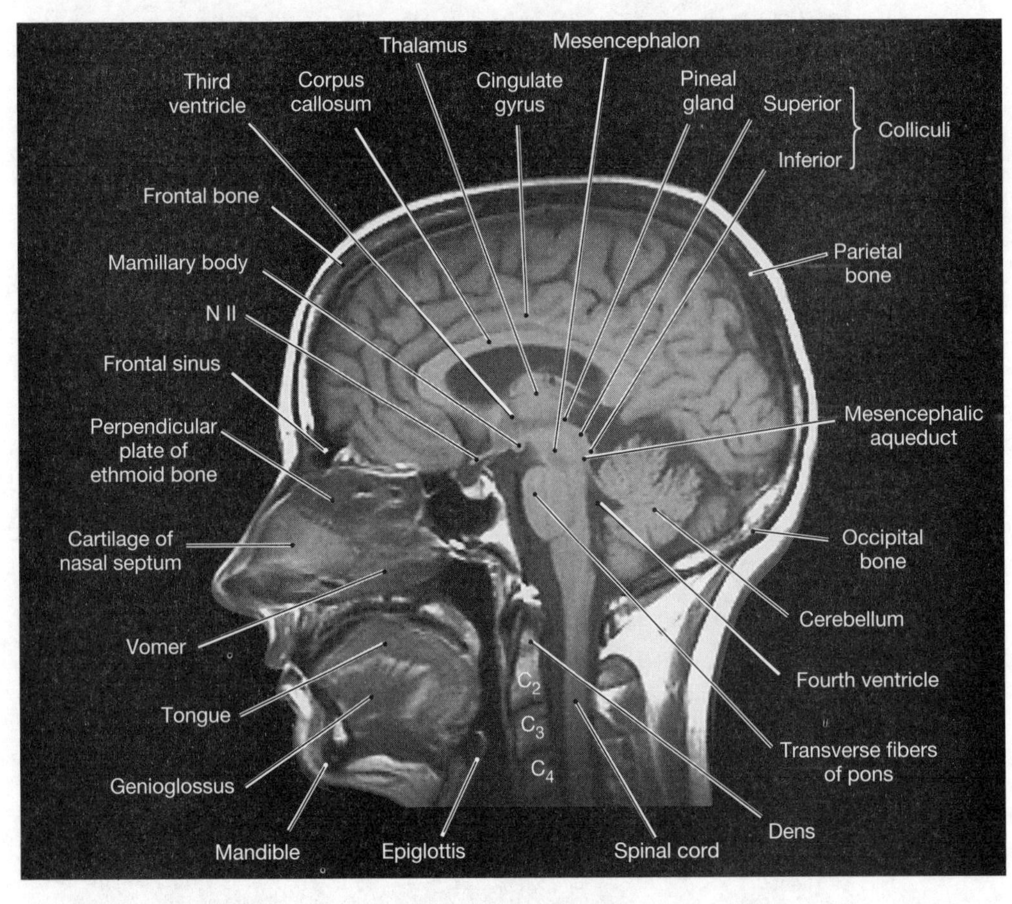

1e

SCAN 2 MRI SCANS OF THE BRAIN (continued)
Frontal Plane (Anterior-to-Posterior Sequence)

2a

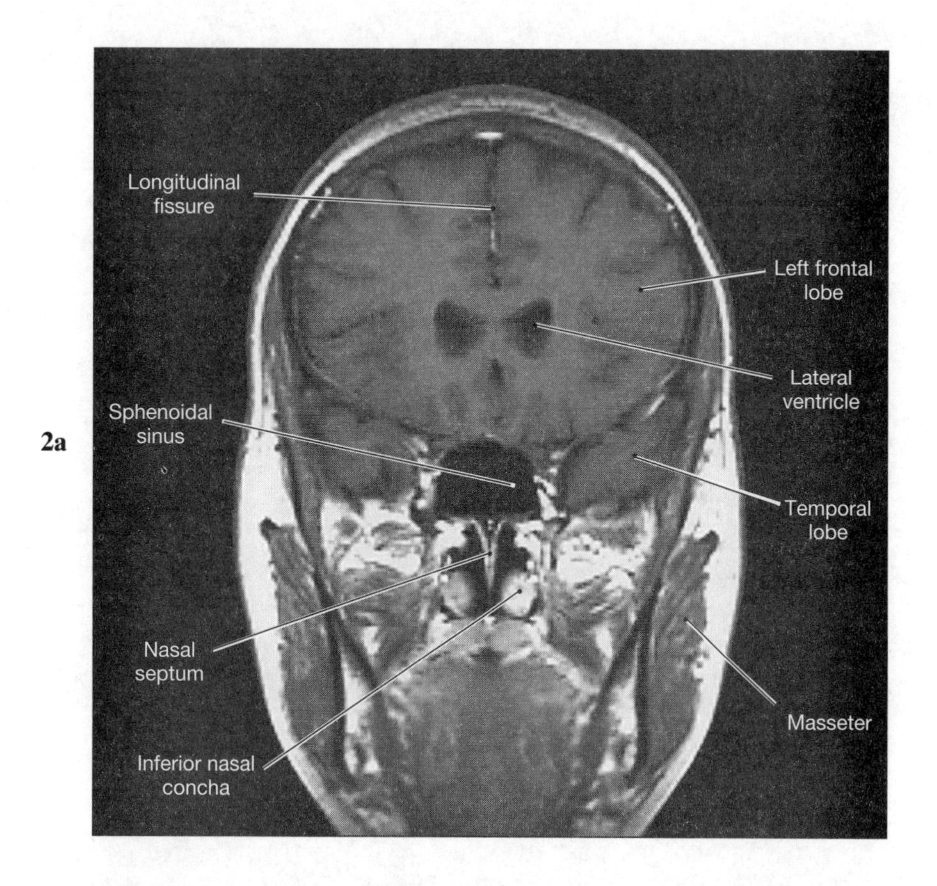

2b

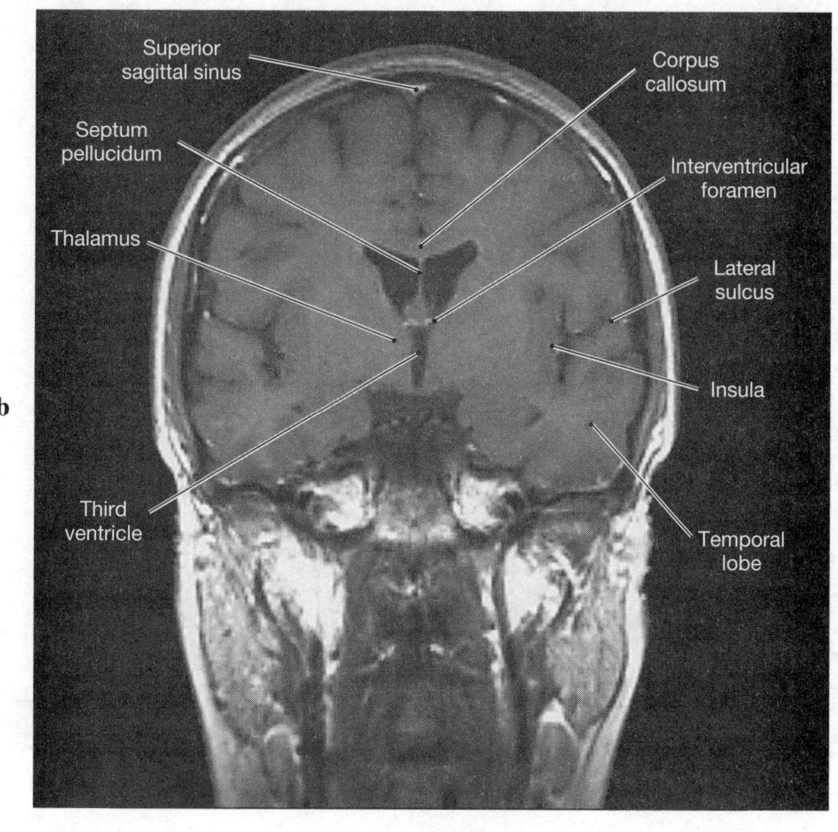

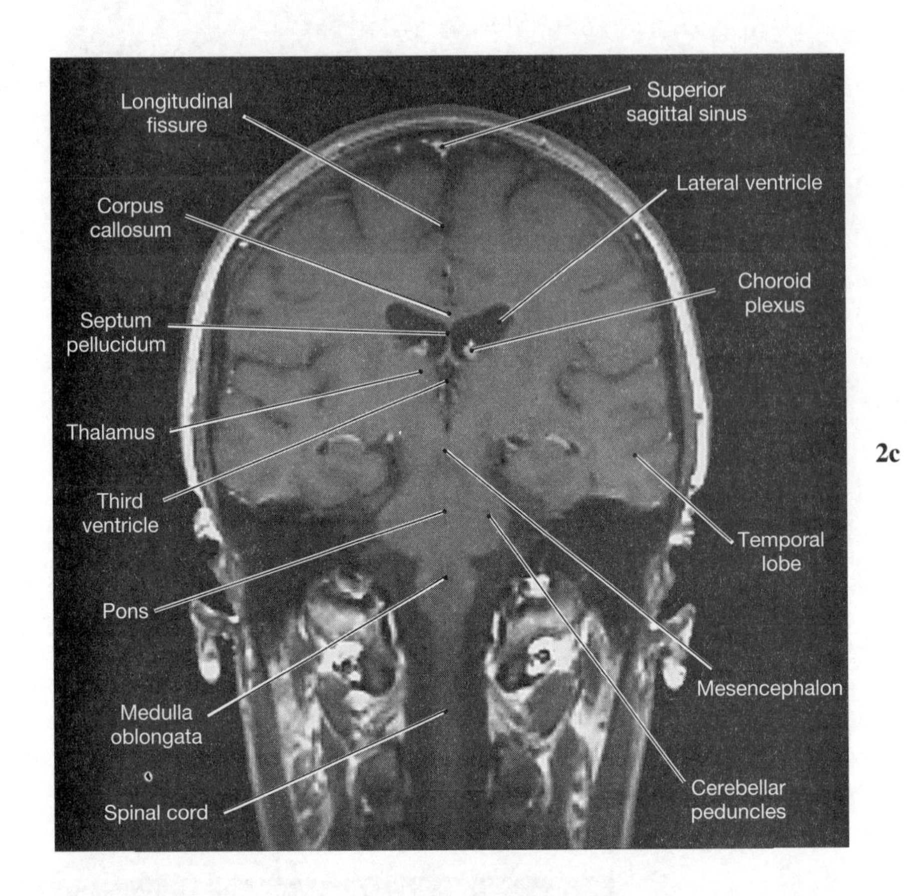
Longitudinal fissure
Superior sagittal sinus
Lateral ventricle
Corpus callosum
Choroid plexus
Septum pellucidum
Thalamus
Third ventricle
Temporal lobe
Pons
Mesencephalon
Medulla oblongata
Spinal cord
Cerebellar peduncles

2c

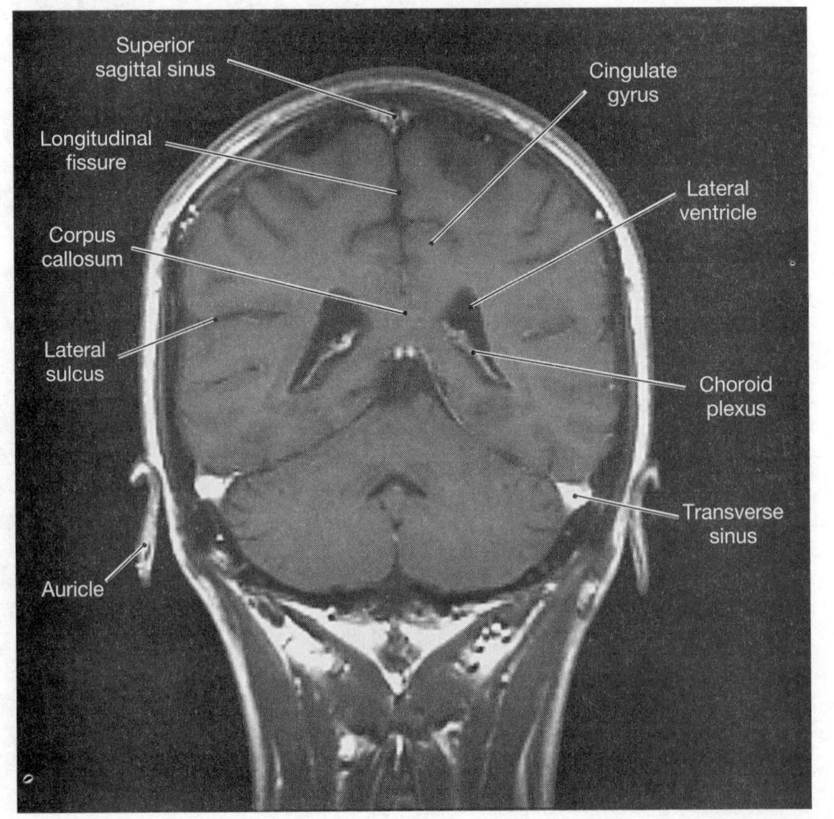
Superior sagittal sinus
Cingulate gyrus
Longitudinal fissure
Lateral ventricle
Corpus callosum
Lateral sulcus
Choroid plexus
Transverse sinus
Auricle

2d

SCAN 3 MRI SCANS OF THE VERTEBRAL COLUMN

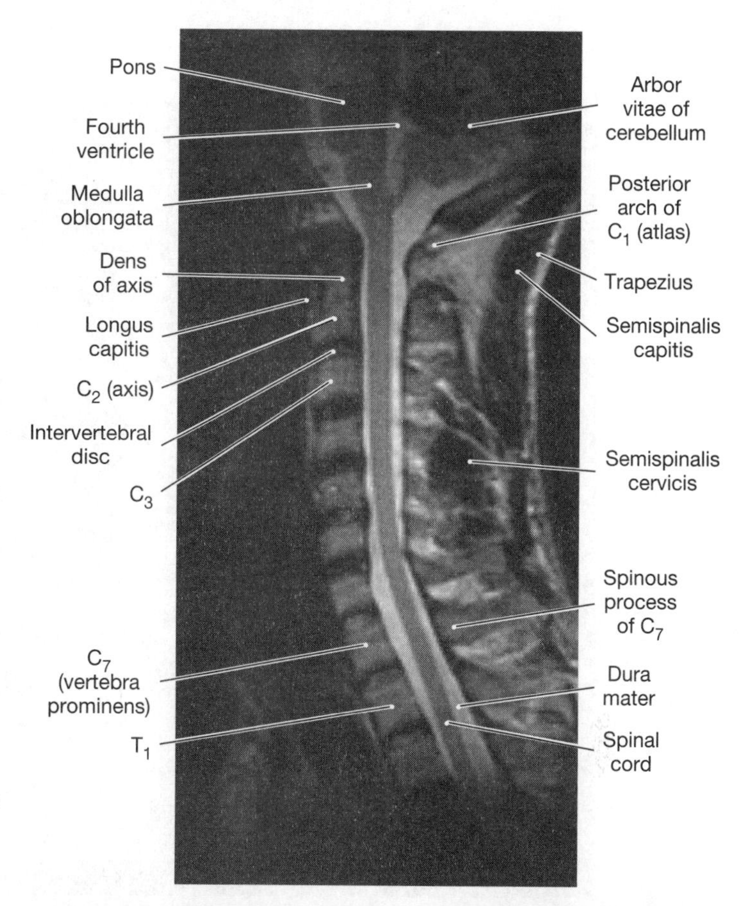

3a Cervical region, sagittal plane

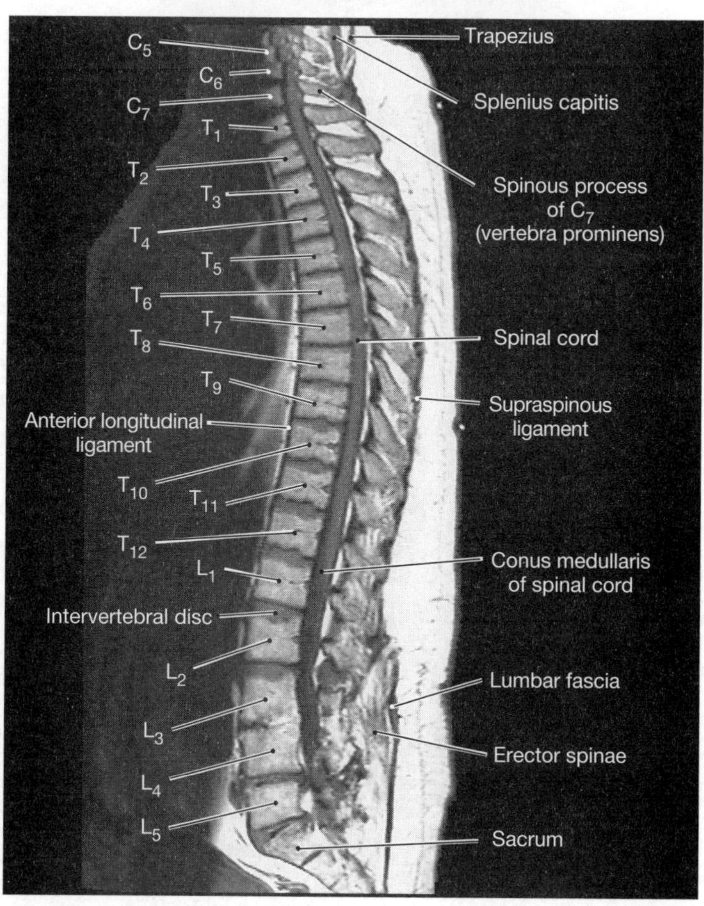

3b Vertebral column, sagittal plane

SCAN 4 MRI SCAN OF THE PELVIS AND HIP
Frontal Plane

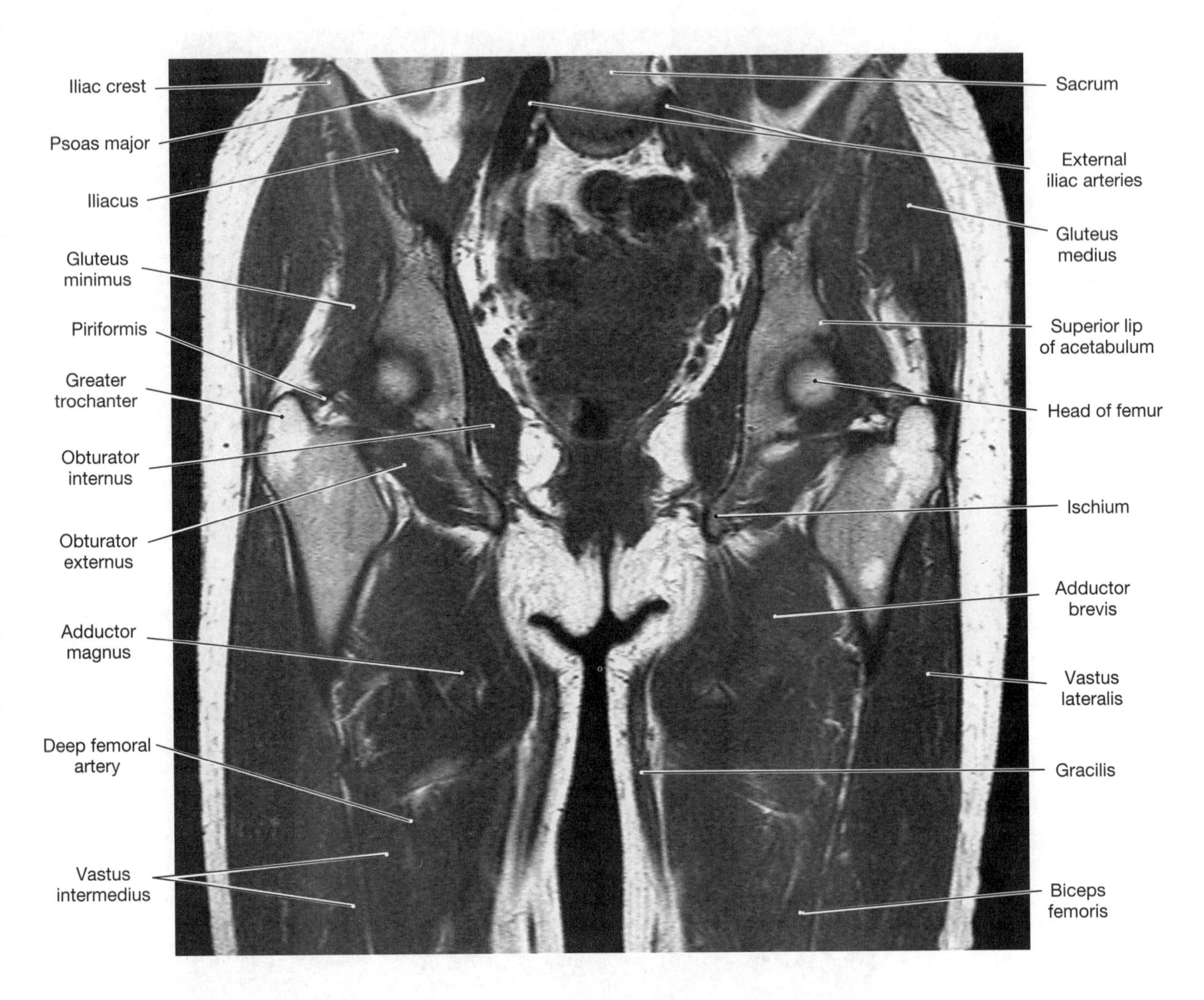

4

SCAN 5 MRI SCANS OF THE KNEE
Horizontal Plane (Superior-to-Inferior Sequence)

5a

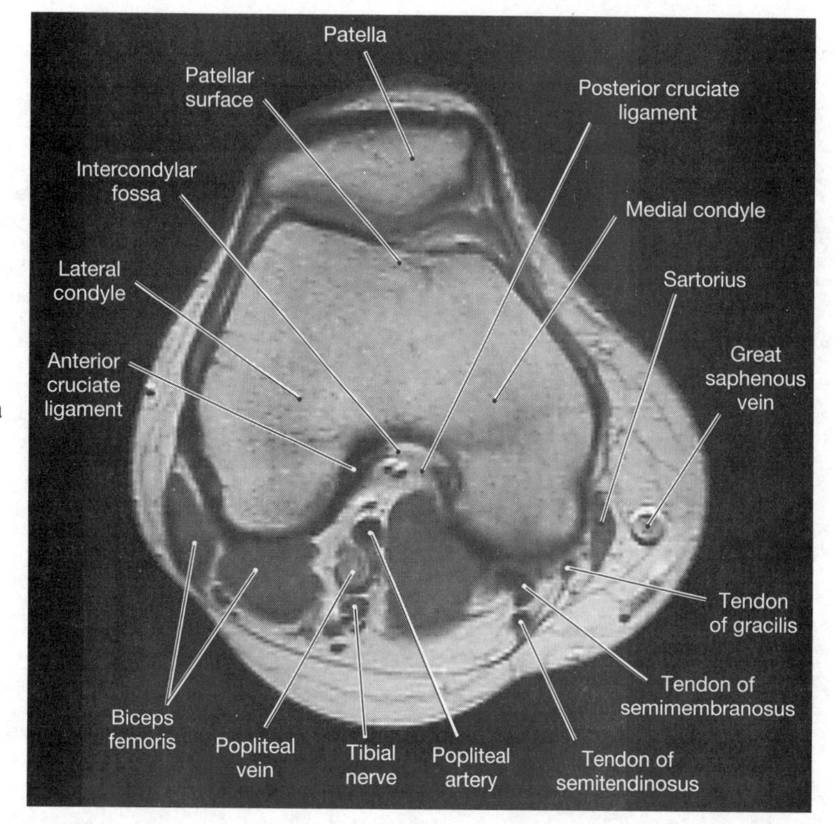

5b

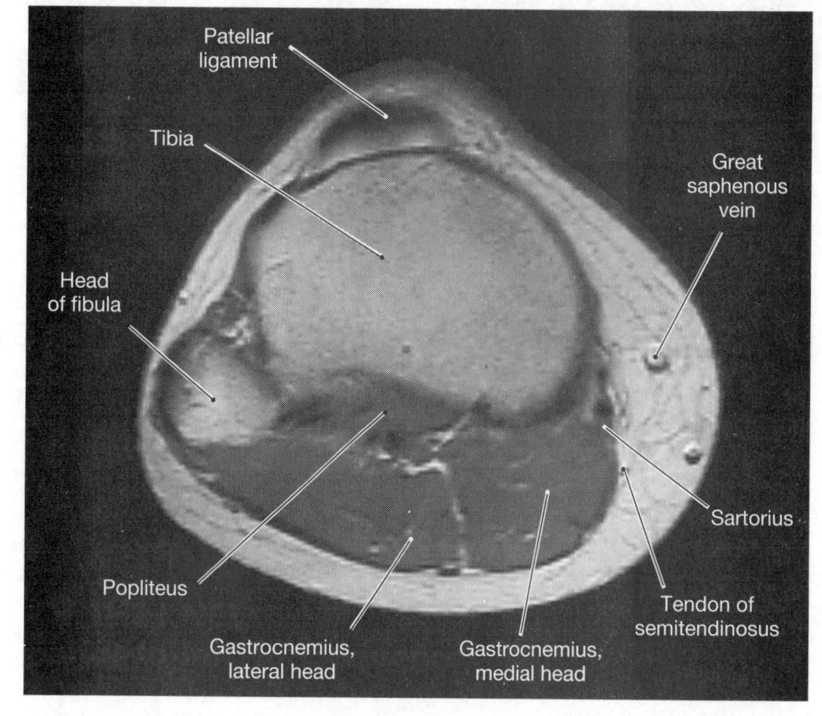

Sagittal Plane (Lateral-to-Medial Sequence)

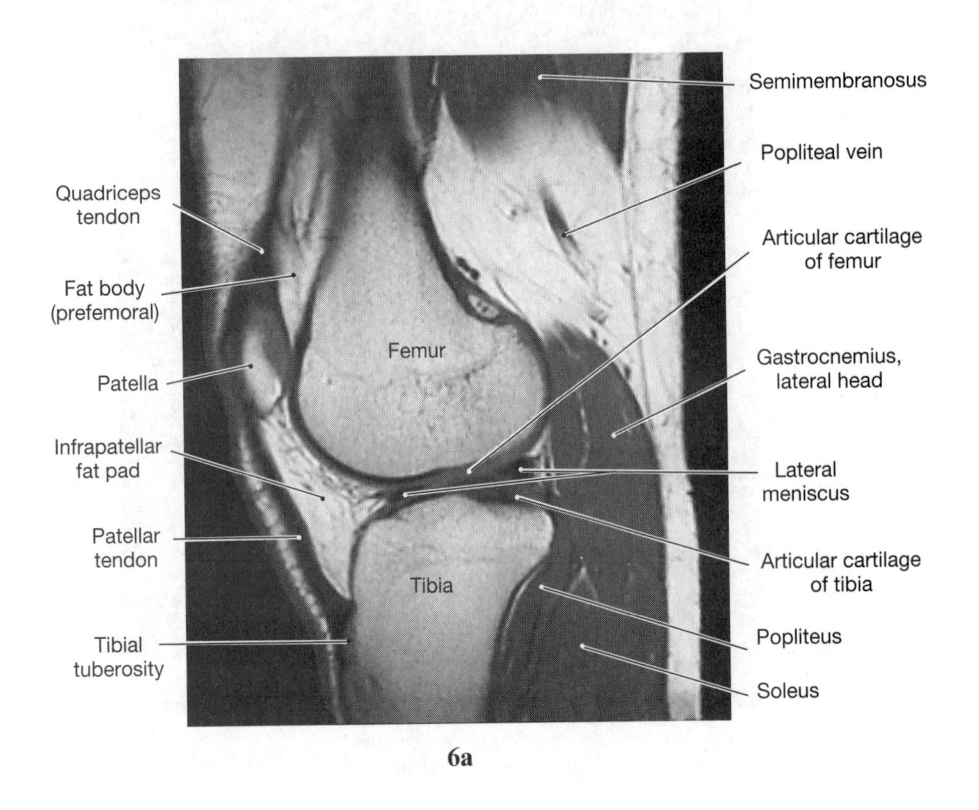

6a

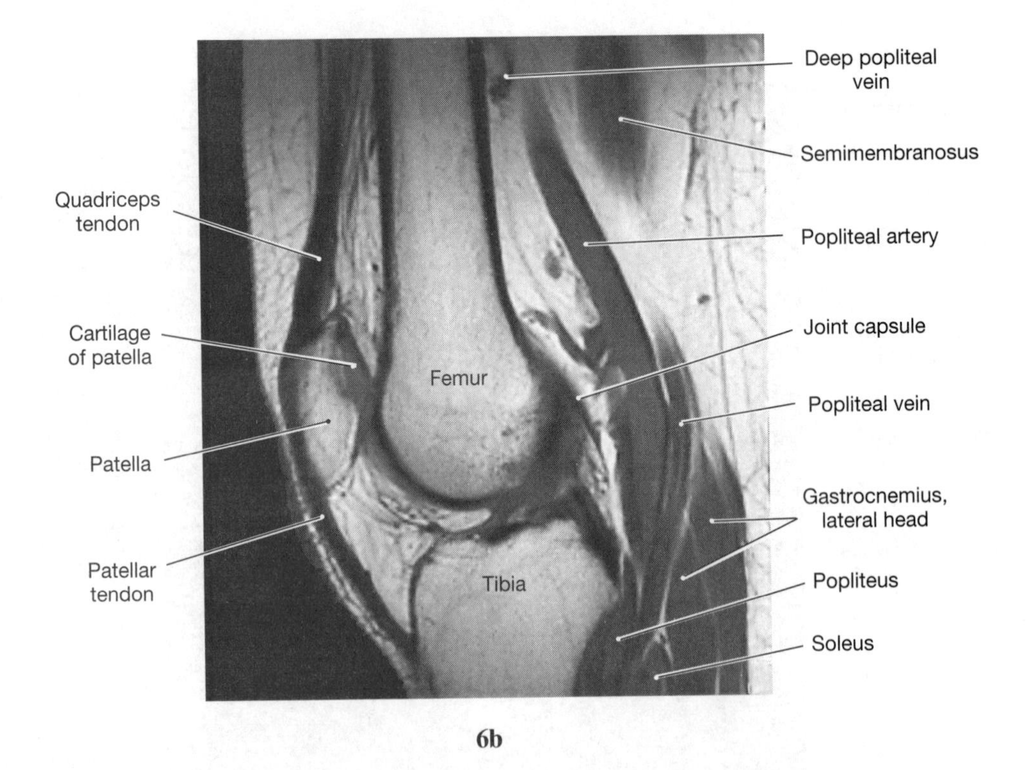

6b

SCAN 7 MRI SCANS OF THE KNEE (continued)
Frontal Plane (Posterior-to-Anterior Sequence)

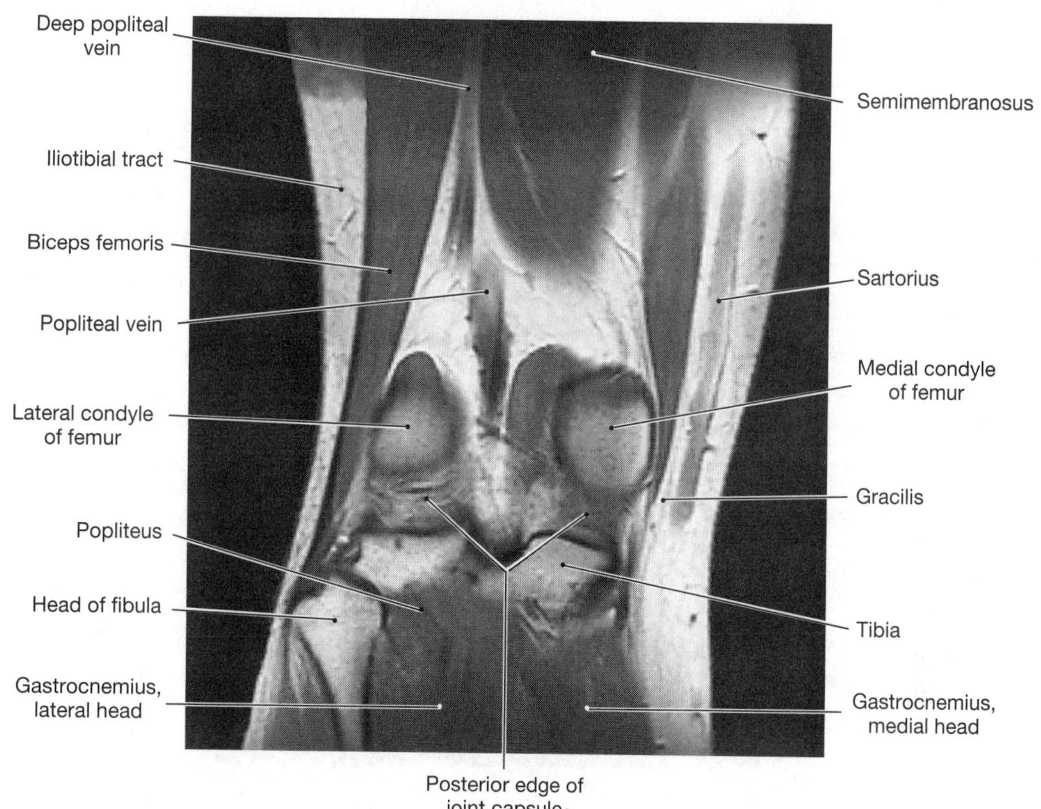

7a

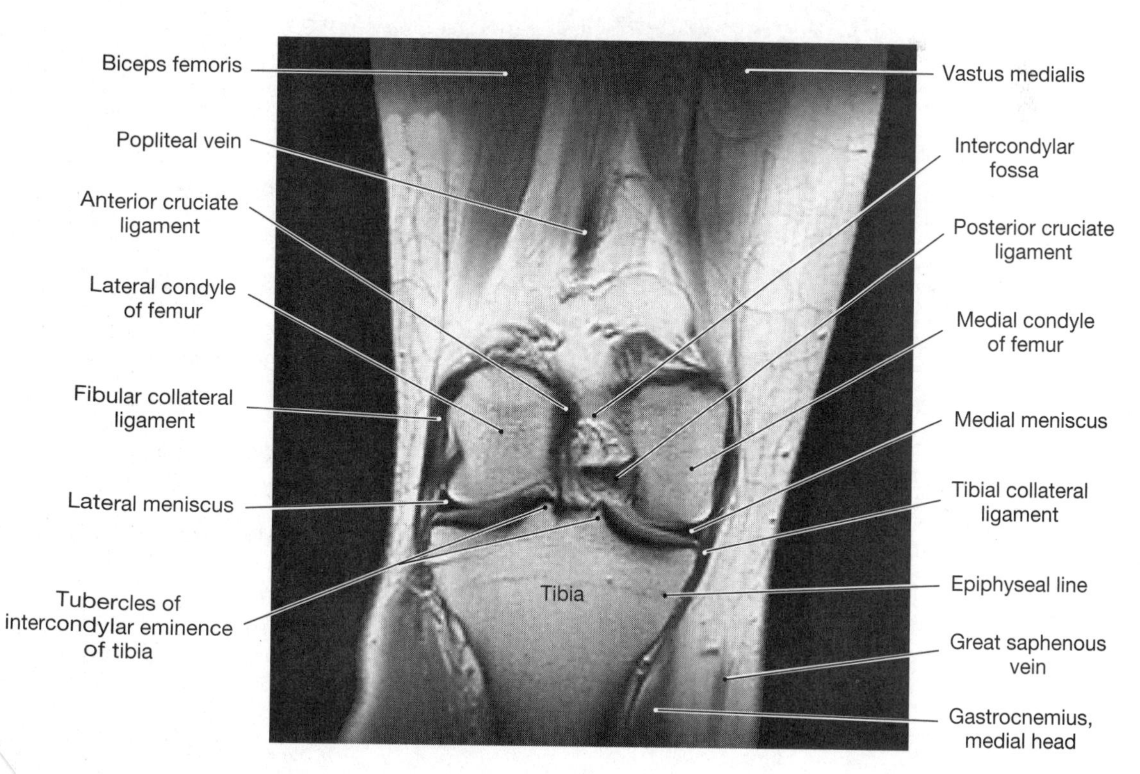

7b

SCAN 8 MRI SCANS OF THE ANKLE

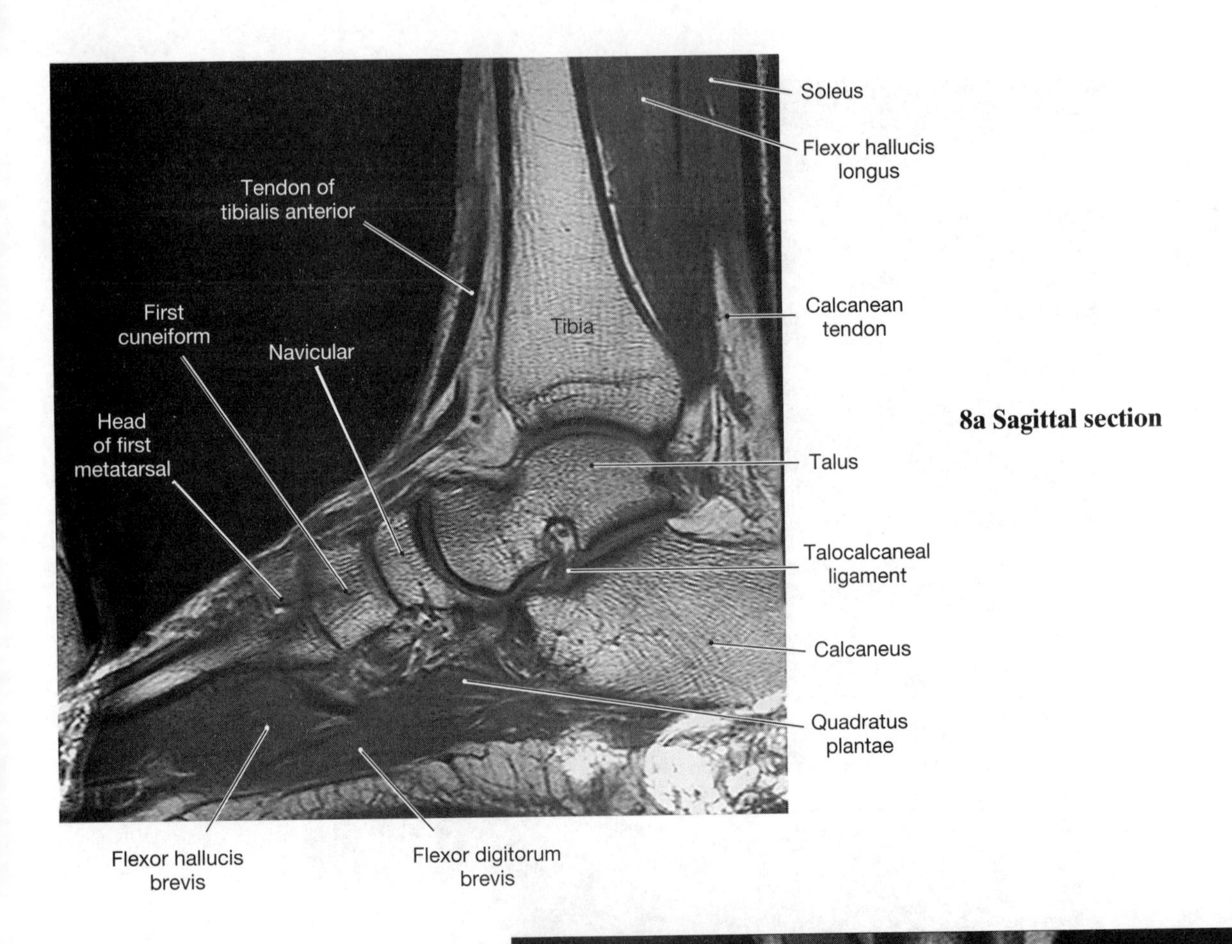

8a Sagittal section

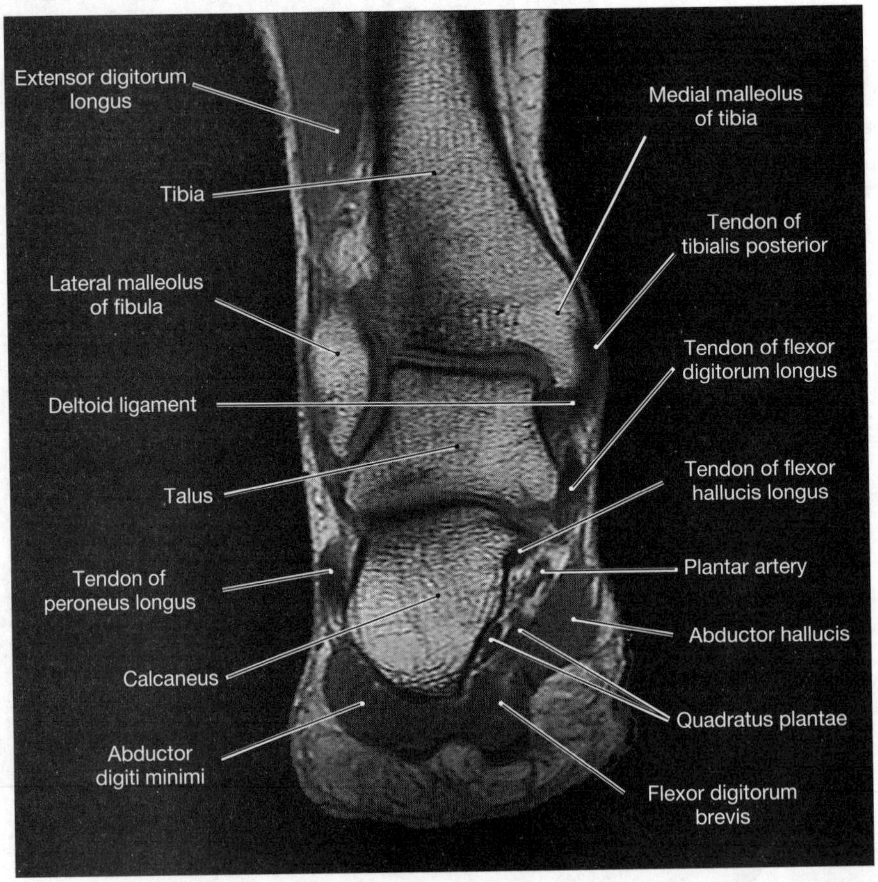

8b Coronal section

Extensor digitorum longus
Medial malleolus of tibia
Tibia
Tendon of tibialis posterior
Lateral malleolus of fibula
Tendon of flexor digitorum longus
Deltoid ligament
Talus
Tendon of flexor hallucis longus
Plantar artery
Tendon of peroneus longus
Abductor hallucis
Calcaneus
Quadratus plantae
Abductor digiti minimi
Flexor digitorum brevis

SCAN 9 REPRESENTATIVE SERIAL CT SCANS OF THE TRUNK
Horizontal Plane (Superior-to-Inferior Sequence)

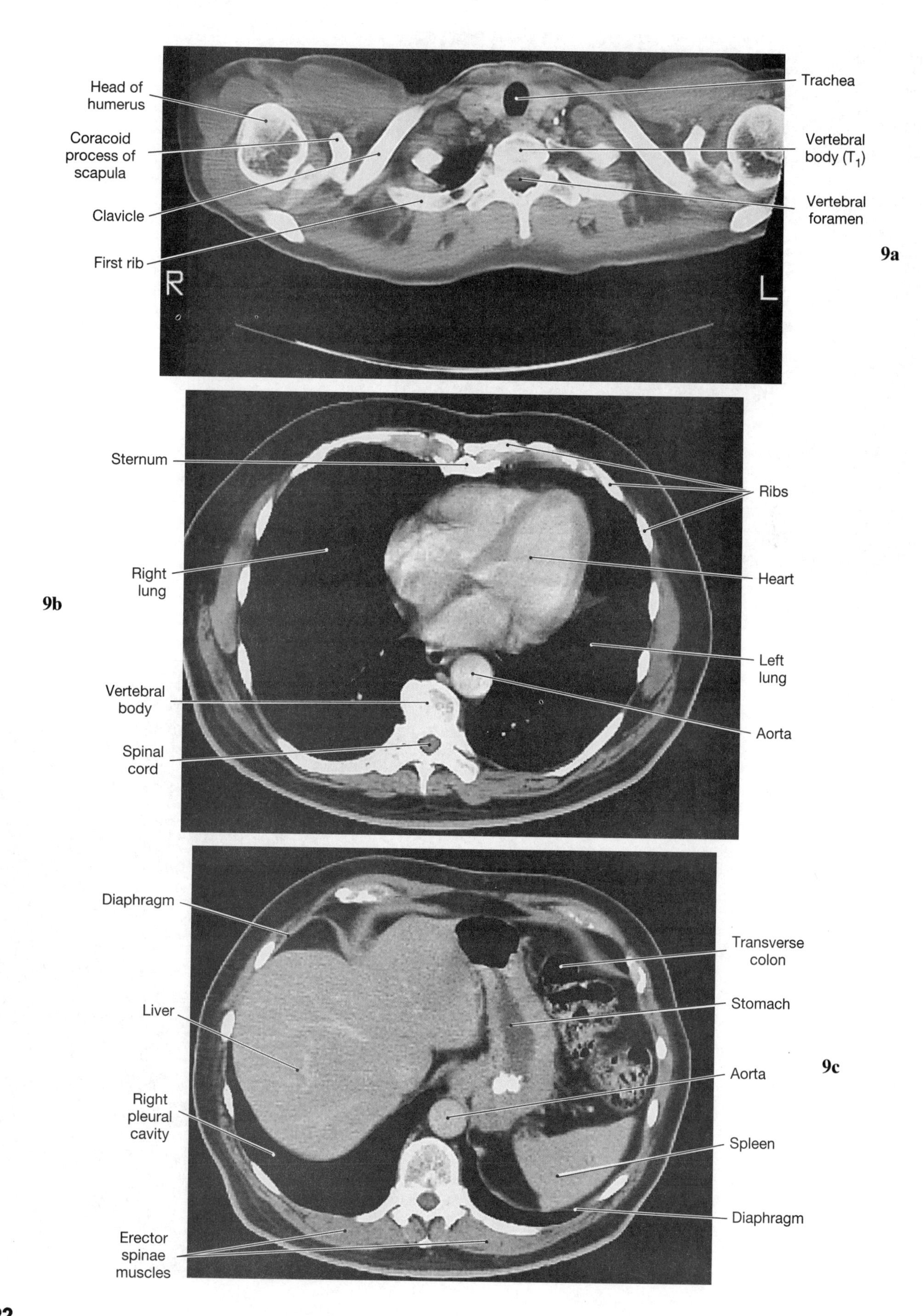

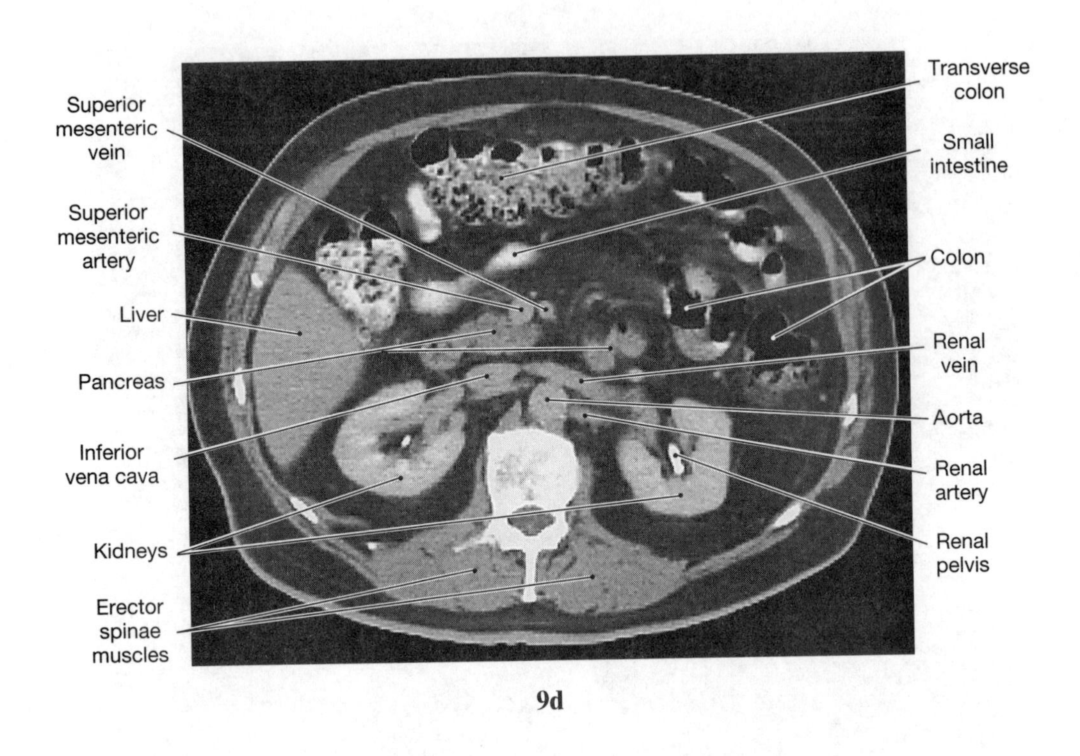
Superior
mesenteric
vein
Superior
mesenteric
artery
Liver
Pancreas
Inferior
vena cava
Kidneys
Erector
spinae
muscles
Transverse
colon
Small
intestine
Colon
Renal
vein
Aorta
Renal
artery
Renal
pelvis

9d

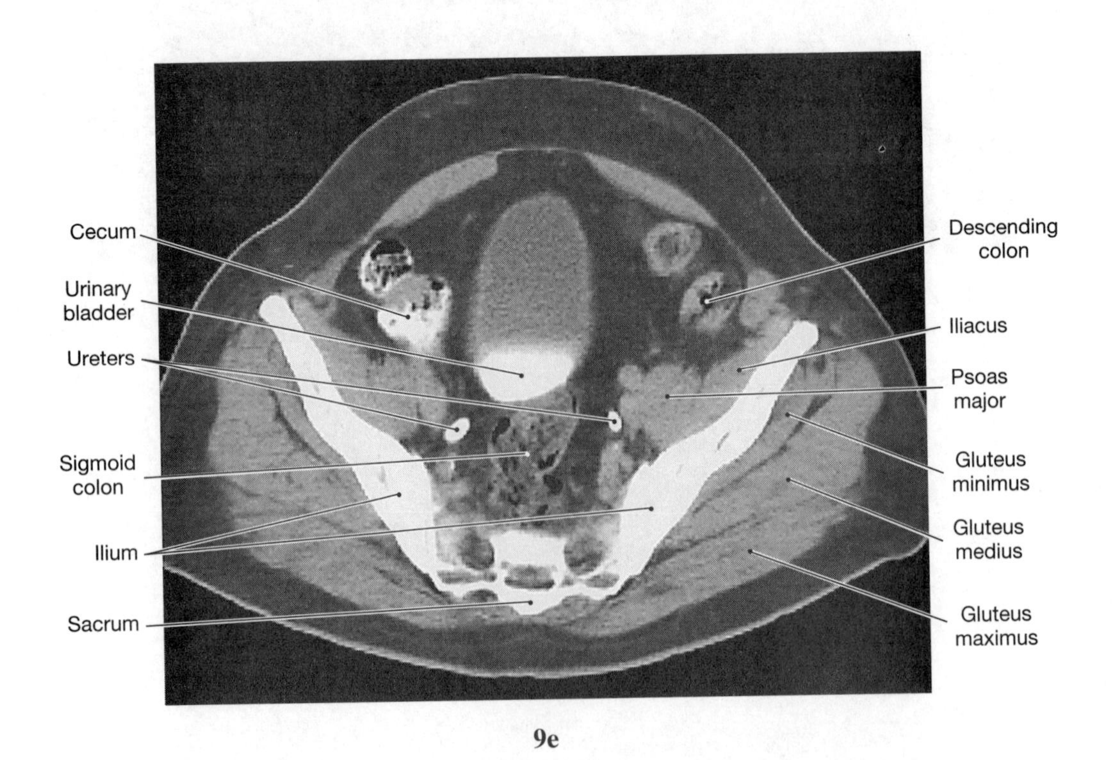
Cecum
Urinary
bladder
Ureters
Sigmoid
colon
Ilium
Sacrum
Descending
colon
Iliacus
Psoas
major
Gluteus
minimus
Gluteus
medius
Gluteus
maximus

9e

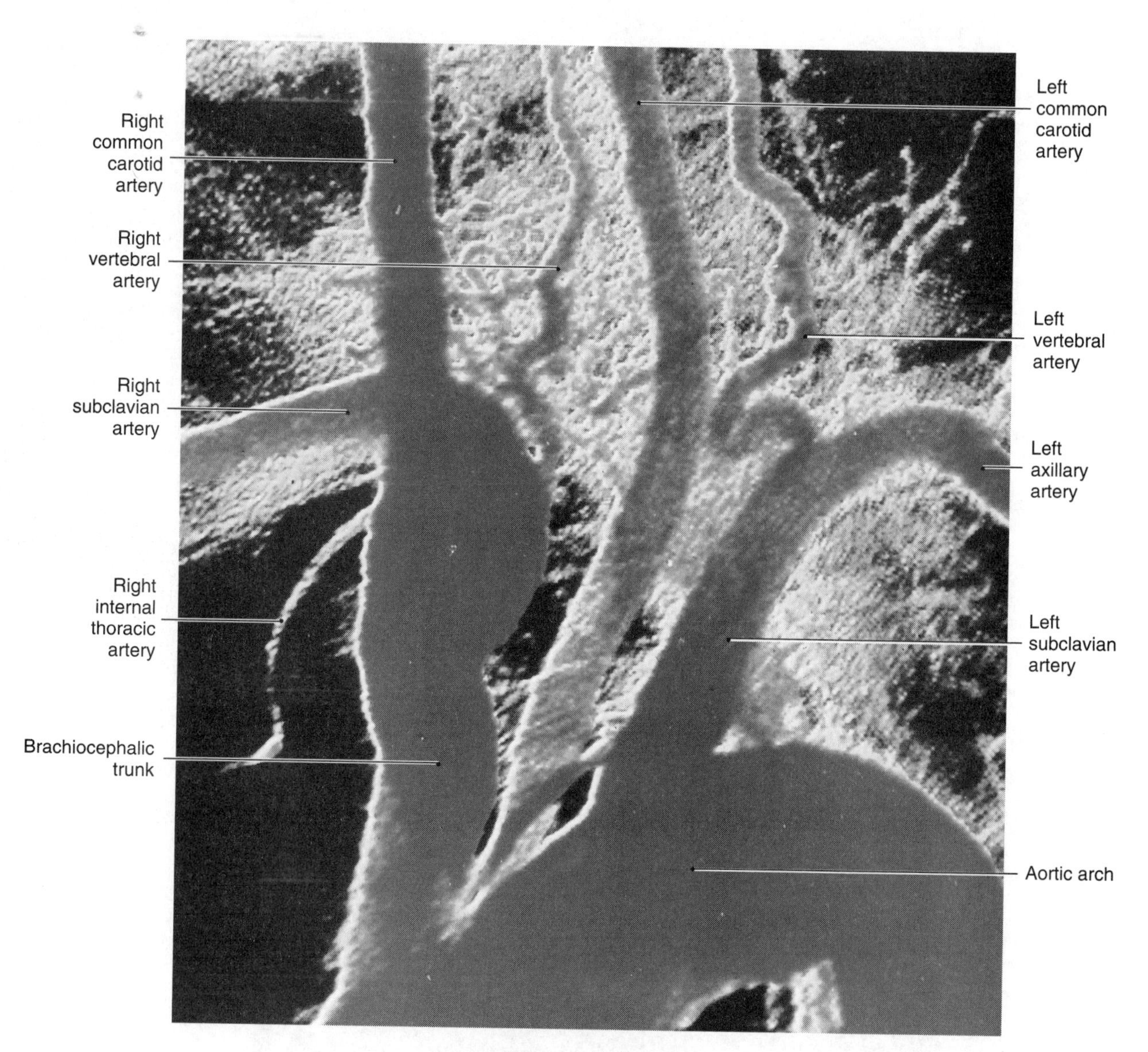
Right common carotid artery
Right vertebral artery
Right subclavian artery
Right internal thoracic artery
Brachiocephalic trunk
Left common carotid artery
Left vertebral artery
Left axillary artery
Left subclavian artery
Aortic arch

10 Aortic angiogram

11 Cerebral angiogram

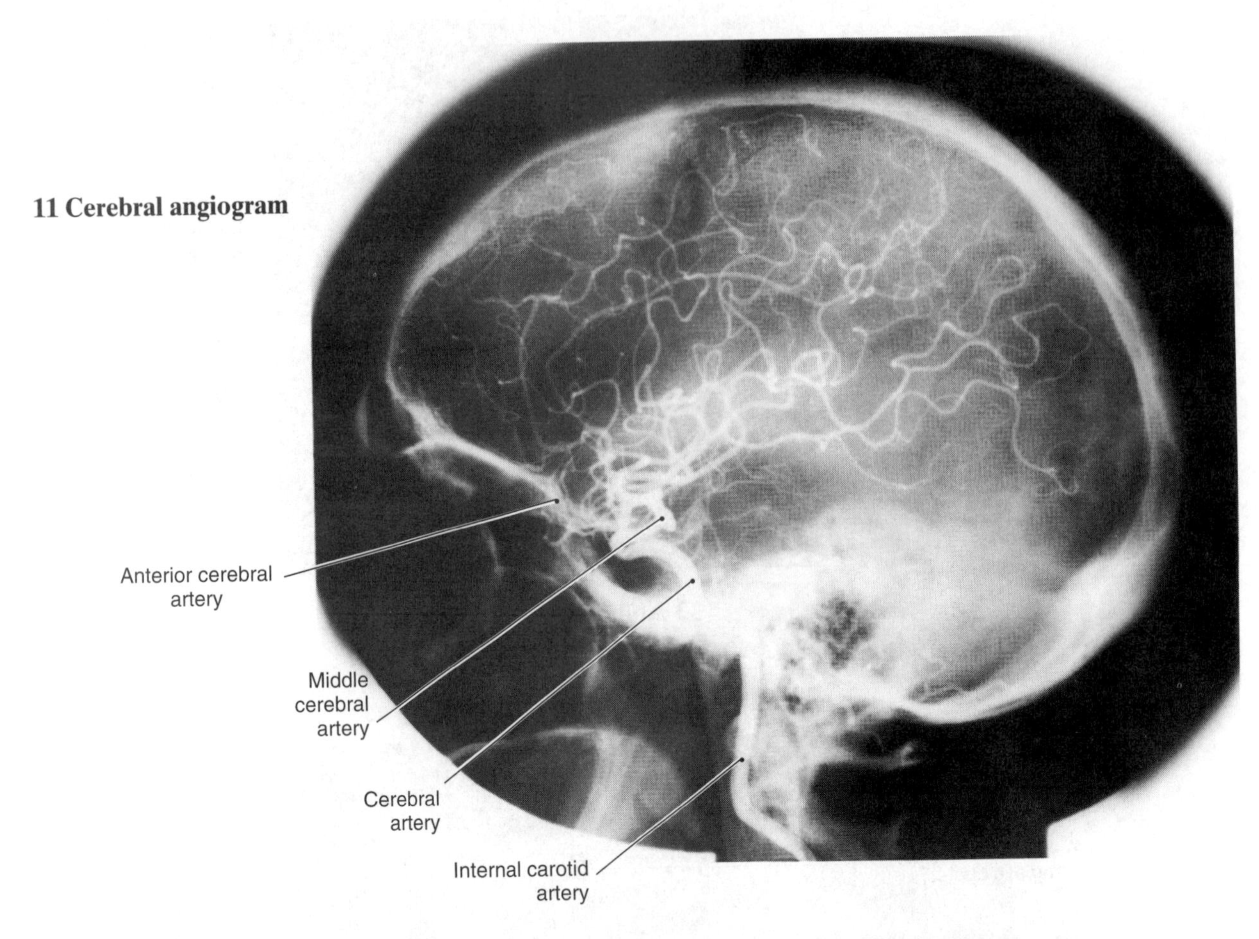

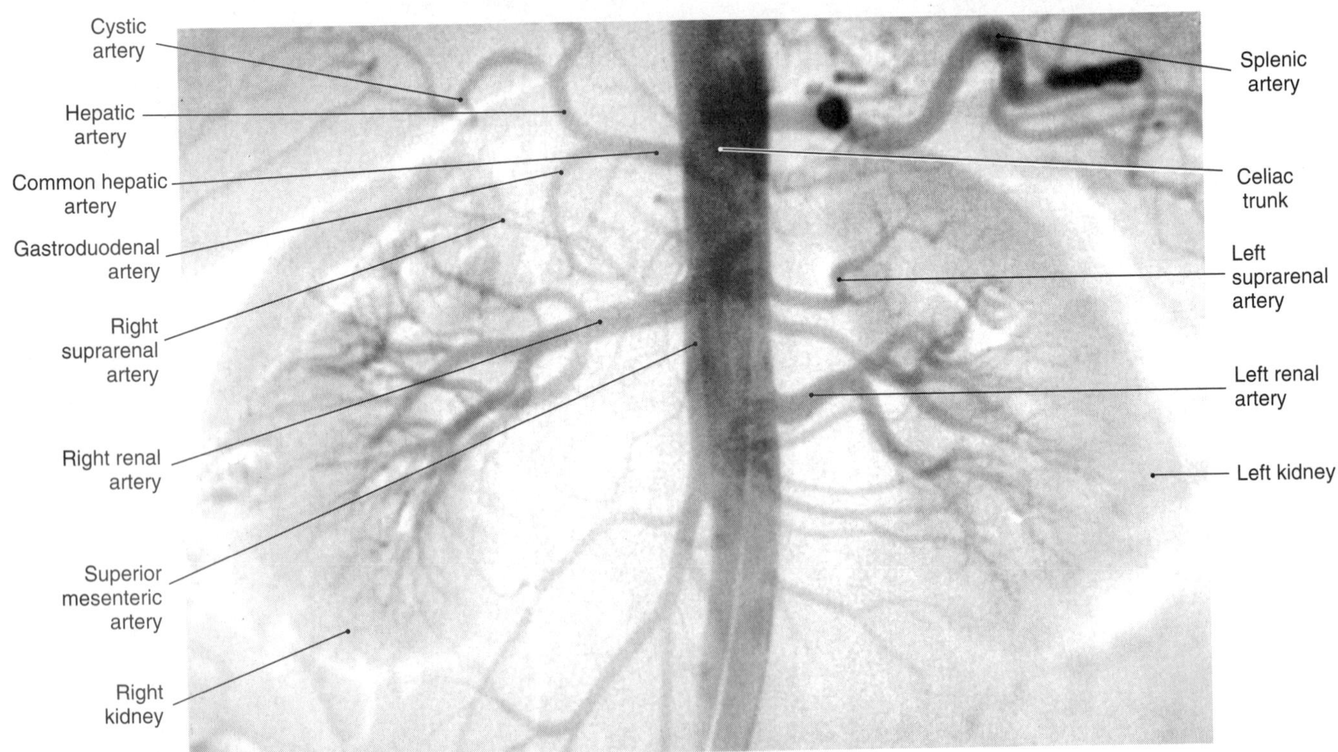

12 Abdominal angiogram

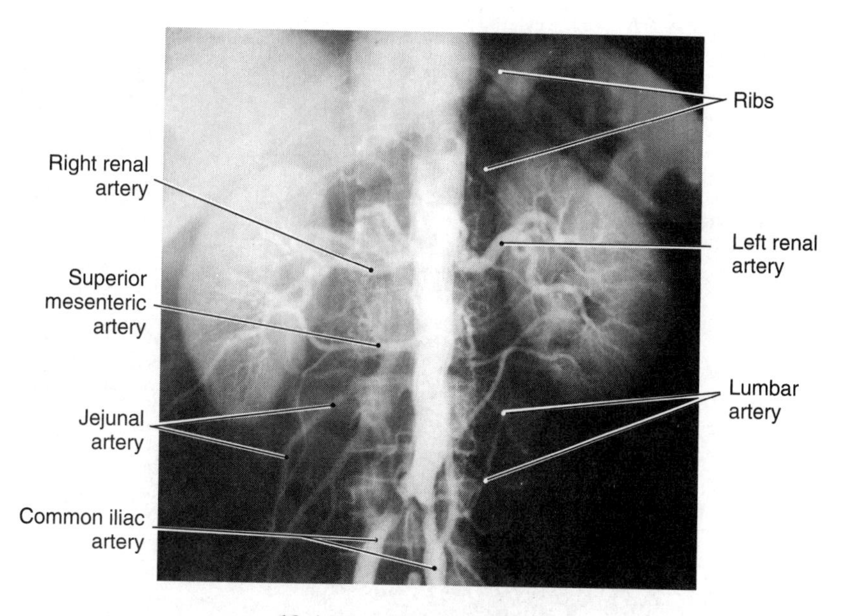

13 Abdominal arteriogram

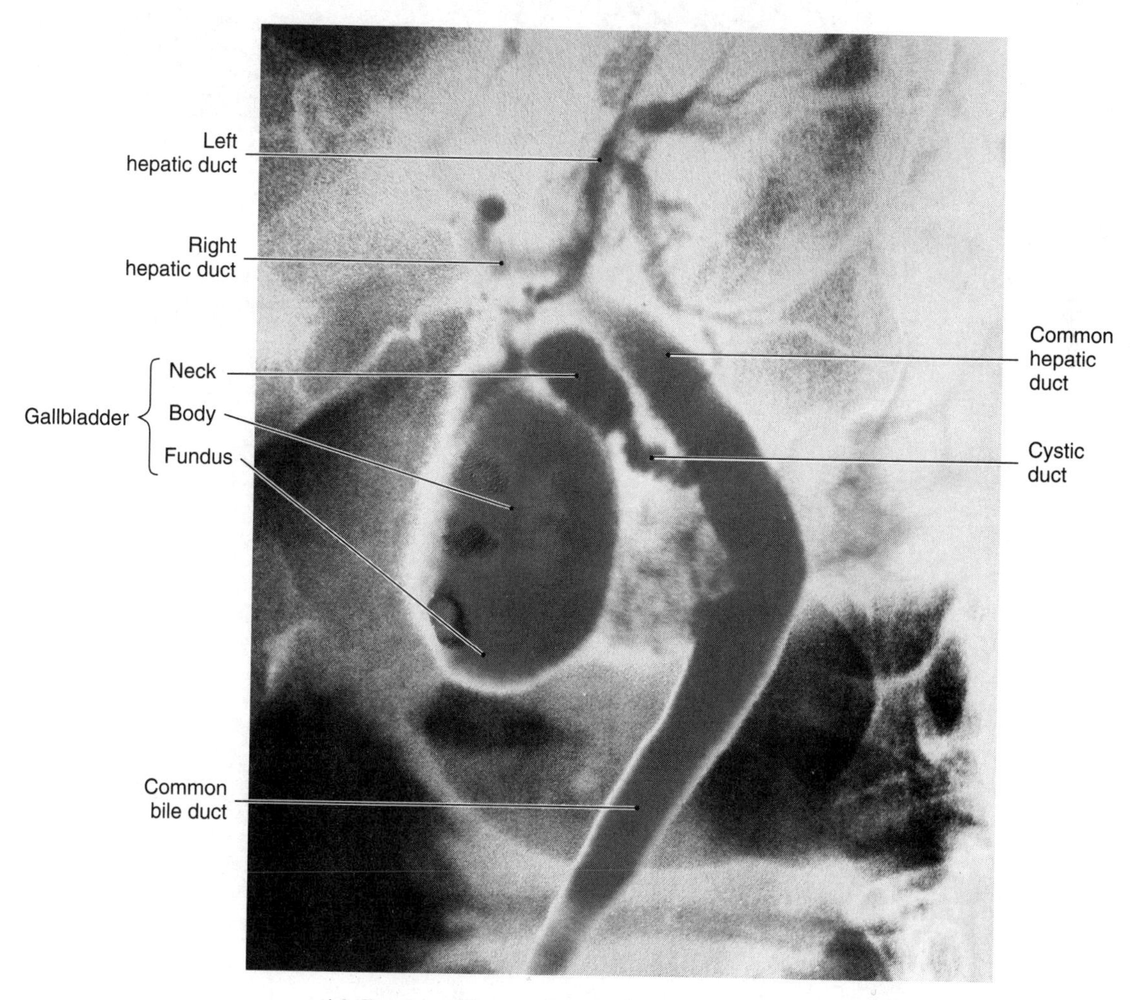

14 Contrast X-ray of gallbladder and bile duct system

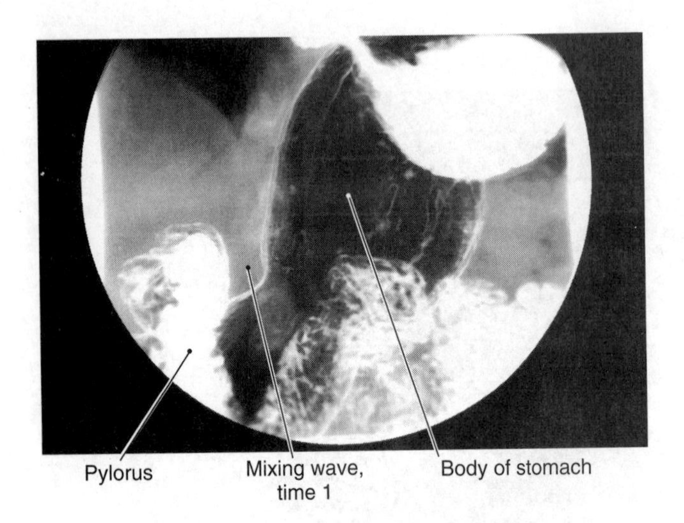

15a Gastric motility, anterior view, time 1

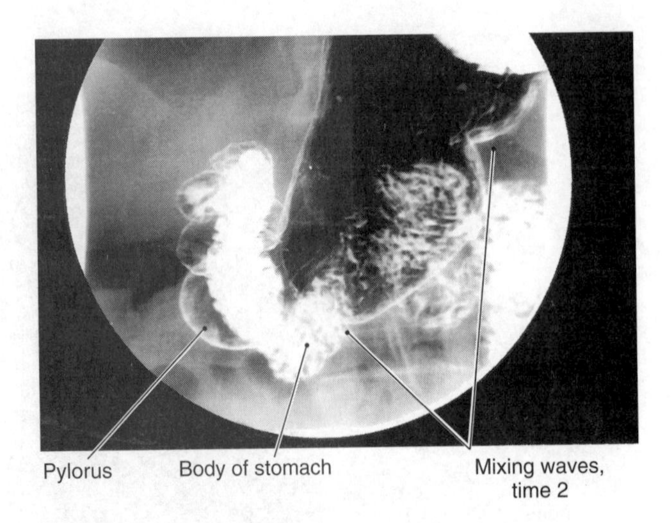

15b Gastric motility, anterior view, time 2

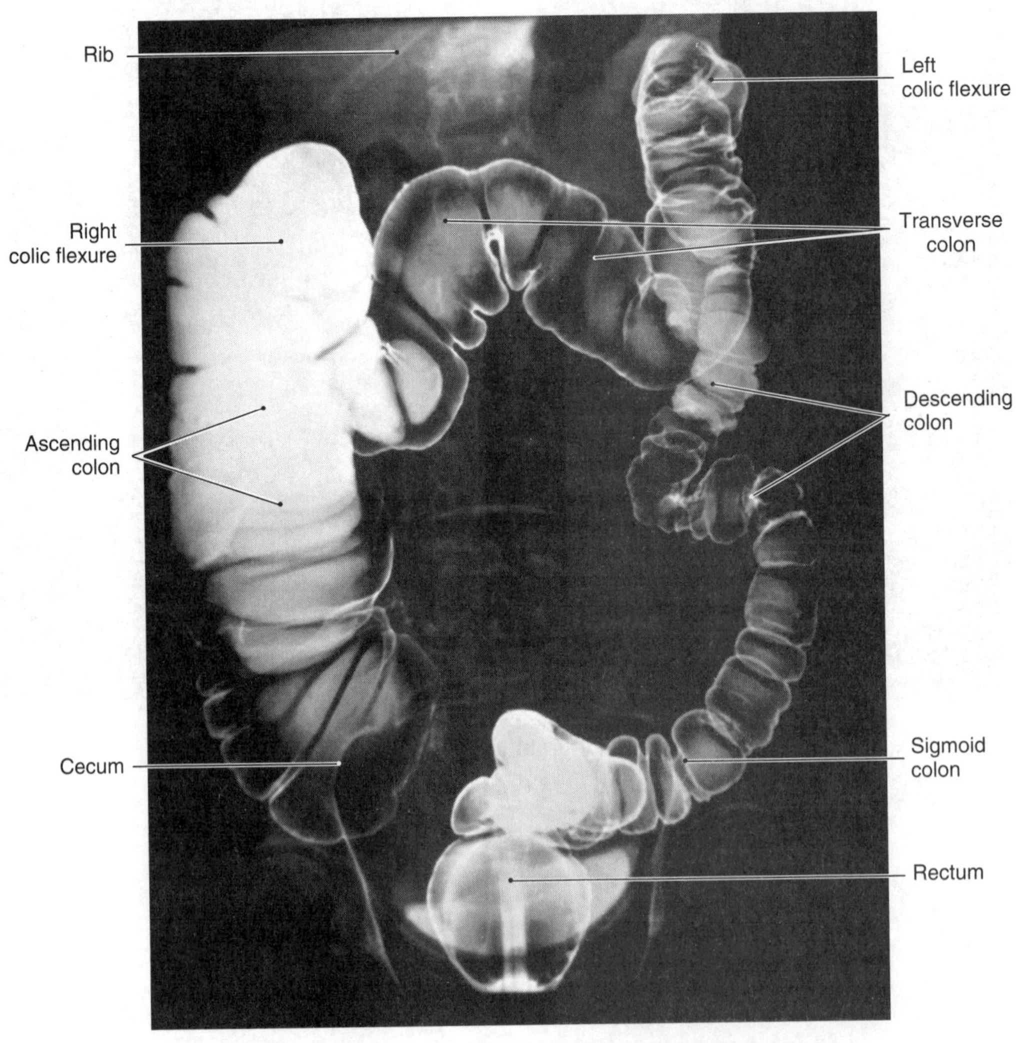

16 Contrast X-ray of colon and rectum, anterior view

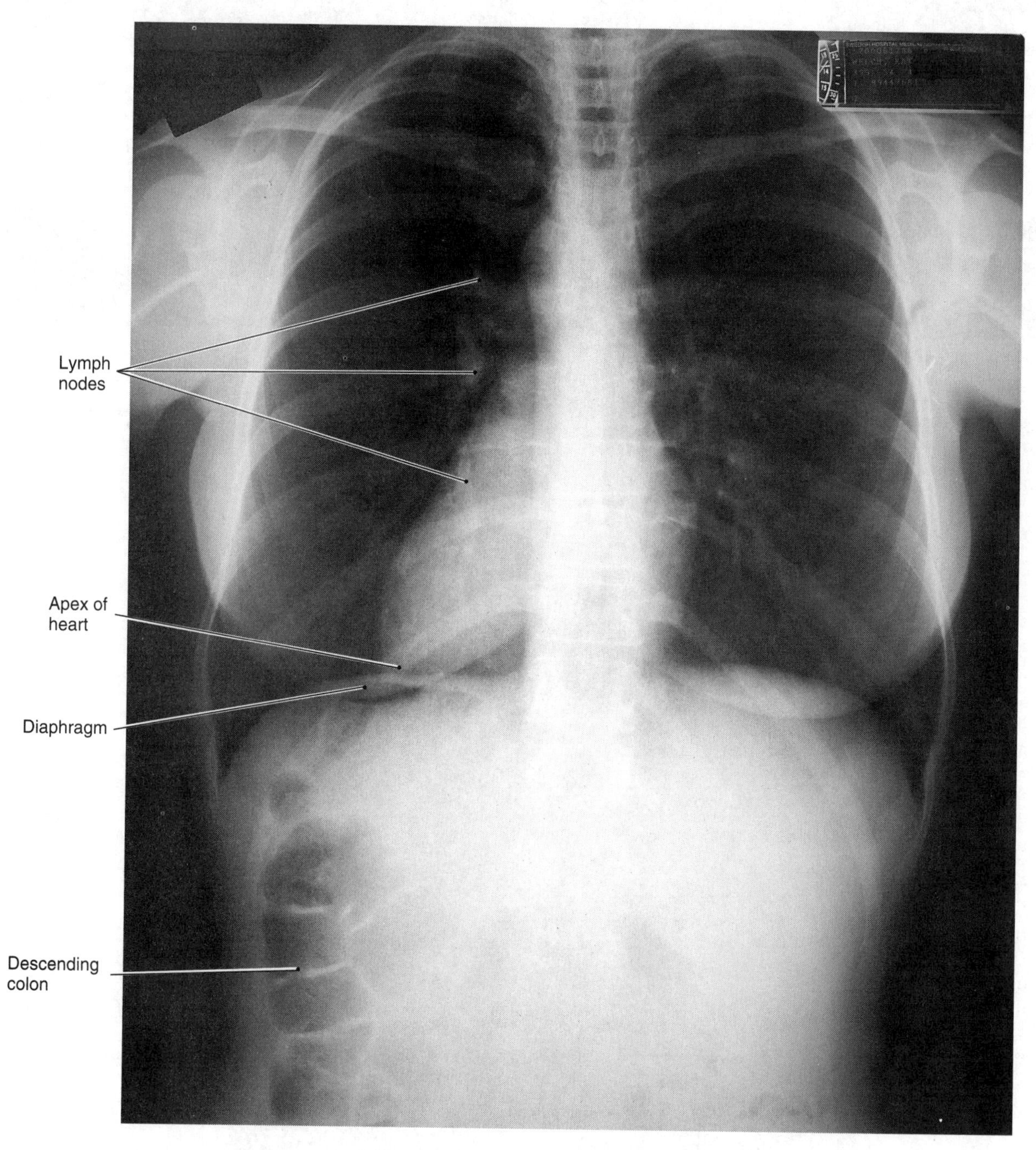

17 Lymphangiogram of thorax, posterior view

INDEX

D

H

I

J

K

L

O

P

Q

R

S

PHOTO CREDITS

A-3b Photo Researchers, Inc. **A-10a** Eric Grave/Science Source/Photo Researchers, Inc. **A-10b,c** Michael Abbey/Visuals Unlimited **A-10d** Arthur M. Siegelman/Visuals Unlimited **A-11a** Larry West/Photo Researchers, Inc. **A-11b** Cath Wadforth/Science Photo Library/Photo Researchers, Inc. **A-11c** Runk/Schoenberger/Grant Heilman Photography, Inc. **A-11d** John Shaw/Tom Stack & Associates **A-11e** A. M. Siegelman/Visuals Unlimited **A-12a** Professors P. Motta and A. Familiari/University *La Sapienza*, Rome/Science Photo Library/Photo Researchers, Inc. **A-12b** CNRI/Science Photo Library/Photo Researchers, Inc. **A-13b** Martin M. Rotker **A-17a,b** Courtesy of Elizabeth A. Abel, M.D., from the Leonard C. Winograd Memorial Slide Collection, Stanford University School of Medicine. **A-22aL,R** Harold Chen, M.D. **A-22b** Grantpix/Monkmeyer Press **A-22c** Science Photo Library/Photo Researchers, Inc. **A-23a** Science Photo Library/Custom Medical Stock Photo, Inc. **A-23b** National Medical Slide/Custom Medical Stock Photo, Inc. **A-23c** Princess Margaret Rose Orthopaedic Hospital/Science Photo Library/Photo Researchers, Inc. **A-24a** SIU Biomed Comm./Custom Medical Stock Photo, Inc. **A-24b,c** Courtesy of Dr. Eugene C. Wasson, III, and staff of Maui Radiology Consultants, Maui Memorial Hospital **A-25a,b,c** Smith & Nephew, Inc. **A-31** Barts Medical Library/Phototake NYC **A-32a** Kenneth E. Greer, M.D., University of Virginia Health Sciences Center **A-32b** Courtesy M. Hogeweg, Ophthalmologist. From *Trop. Doctor Supplement*, 1:51–21 (1992) **A-35** Jon Meyer/Custom Medical Stock Photo, Inc. **A-39a** Biophoto Associates/Photo Researchers, Inc. **A-39b** SIU/Custom Medical Stock Photo, Inc. **A-39c** Medtronic, Inc. **A-40** Jim Wehtje/PhotoDisc, Inc. **A-43a** Science Photo Library/Photo Researchers, Inc. **A-43bL** Hewlett-Packard Company **A-43c,d** Picker International, Inc. **A-48** From Cote, J., *Internat. J. Dermatol.* 30:500–501 (1991) **A-52** Ken Greer/Visuals Unlimited **A-54** AP/Wide World Photos **A-57** Dennis E. Feely, Stanley L. Erlandsen, and David G. Case **A-59** United Nations **A-65** Courtesy of U.S. Public Health Service

Surface Anatomy and Cadaver Atlas 1.1, 1.2, 2.1a,b,c, 2.2a,b,c,d,e, 2.3, 2.4b,c,d, 3.1, 3.2, 3.3, 3.4, 3.5 Ralph T. Hutchings **4.1** Patrick M. Timmons/Michael J. Timmons **4.2, 4.3** Ralph T. Hutchings **5.1a,b** Mentor Networks, Inc. **5.2a,b, 5.3a,b,c,d** Ralph T. Hutchings **6.1a,b,c** Mentor Networks, Inc. **6.2a,b** Ralph T. Hutchings **6.3a** Mentor Networks, Inc. **6.3b** Custom Medical Stock Photo, Inc. **6.3c** Ralph T. Hutchings **6.4a** Mentor Networks, Inc. **6.4b** Ralph T. Hutchings **6.5, 6.6a,b** Patrick M. Timmons/Michael J. Timmons **6.6c** Ralph T. Hutchings **7.1, 7.2, 7.3a** Mentor Networks, Inc. **7.3b** Custom Medical Stock Photo, Inc. **7.4a,b, 7.5a,b,c,d** Ralph T. Hutchings **7.6a,b,c,d,e** Courtesy of Michael J. Timmons and Ralph T. Hutchings/WARD'S Natural Science Establishment, Inc. **7.7a,b,c,d, 7.8a,b, 7.9a,b, 7.10a,b,c** Ralph T. Hutchings **7.10d,e** Courtesy of Michael J. Timmons and Ralph T. Hutchings/WARD'S Natural Science Establishment, Inc. **7.11a,b, 8.1, 8.2** Ralph T. Hutchings **8.3** Mentor Networks, Inc. **8.4, 8.5, 8.6, 8.7a** Ralph T. Hutchings **8.7b** Custom Medical Stock Photo, Inc. **8.7c, 8.8a** Ralph T. Hutchings **8.8b** Mentor Networks, Inc. **8.9** Patrick M. Timmons/Michael J. Timmons **8.10, 8.11a,b, 8.12a,b** Ralph T. Hutchings **8.12c** Mentor Networks, Inc. **8.13a** Ralph T. Hutchings **8.13b** Mentor Networks, Inc. **8.14a** Ralph T. Hutchings **8.14b** Mentor Networks, Inc. **8.14c,d** Ralph T. Hutchings

Scanning Atlas 1a,b,c,d,e, 2a,b,c,d, 3a,b, 4, 5a,b, 6a,b, 7a,b, 8a,b, 9a,b,c,d,e Courtesy of Dr. Eugene C. Wasson, III, and staff of Maui Radiology Consultants, Maui Memorial Hospital **10** CNRI/Science Photo Library/Photo Researchers, Inc. **11** Martin M. Rotker **12** Christopher J. Bodin, M.D., Tulane University Medical Center **13** Barry Slaven/P. Mode Photography/Visuals Unlimited **14** Dept. of Clinical Radiology, Salisbury District Hospital/Science Photo Library/Custom Medical Stock Photo, Inc. **15a,b** Christopher J. Bodin, M.D., Tulane University Medical Center **16** Scott Camazine/Brian Camazine/Photo Researchers, Inc. **17** Kathleen Welch, M.D.